Clifford algebras and Dirac operators in harmonic analysis

Already published

1   W.M.L. Holcombe *Algebraic automata theory*
2   K. Petersen *Ergodic theory*
3   P.T. Johnstone *Stone spaces*
4   W.H. Schikhof *Ultrametric calculus*
5   J-P. Kahane *Some random series of functions, second edition*
6   H. Cohn *Introduction to the construction of class fields*
7   J. Lambek & P.J. Scott *Introduction to higher-order categorical logic*
8   H. Matsumura *Commutative ring theory*
9   C.B. Thomas *Characteristic classed and the cohomology of finite groups*
10   M. Aschbacher *Finite group theory*
11   J.L. Alperin *Local representation theory*
12   P. Koosis *The logarithmic integral: I*
13   A. Pietsch *Eigenvalues and s-numbers*
14   S.J. Patterson *An introduction to the theory of the Riemann zeta-function*
15   H-J. Baues *Algebraic homotopy*
16   V.S. Varadarajan *Introduction to harmonic analysis on semisimple Lie groups*
17   W. Dicks & M.J. Dunwoody *Groups acting on graphs*
18   L.J. Corwin & F.P. Greenleaf *Representations of nilpotent Lie groups and their applications*
19   R. Frisch & R. Piccinini *Cellular structures and characters of finite groups*
20   H. Klingen *Introductory lectures on Siegel modular forms*
22   M.J. Collins *Representations and characters of finite groups*
24   H. Kunita *Stochastic flows and stochastic differential equations*
25   P. Wojtaszczyk *Banach spaces for analysts*
26   J.E. Gilbert & M.A.M. Murray *Clifford algebras and Dirac operators in harmonic analysis*
28   K. Goebel & W.A. Kirk *Topics in metric fixed point theory*
29   J.E. Humphreys *Reflection groups and Coxeter groups*
30   D.J. Benson *Representations and cohomology I*

Clifford algebras and Dirac operators in harmonic analysis

JOHN E. GILBERT

University of Texas

MARGARET A.M. MURRAY

Virginia Polytechnic Institute and State University

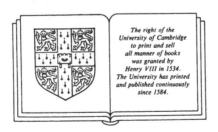

CAMBRIDGE UNIVERSITY PRESS

Cambridge

New York   Port Chester

Melbourne   Sydney

CAMBRIDGE UNIVERSITY PRESS
Cambridge, New York, Melbourne, Madrid, Cape Town, Singapore, São Paulo

Cambridge University Press
The Edinburgh Building, Cambridge CB2 8RU, UK

Published in the United States of America by Cambridge University Press, New York

www.cambridge.org
Information on this title: www.cambridge.org/9780521346542

First published 1991
This digitally printed version 2008

A catalogue record for this publication is available from the British Library

ISBN 978-0-521-34654-2 hardback
ISBN 978-0-521-07198-7 paperback

Contents

	Introduction	1
	Chapter 1 Clifford algebras	**5**
1	Quadratic spaces	6
2	Clifford algebras	8
3	Structure of Clifford algebras	22
4	Orthogonal transformations	32
5	Transformers, Clifford groups	38
6	Spin groups	46
7	The Euclidean case	49
8	$\text{Spin}(V, Q)$ as a Lie group	65
9	Spin groups as classical Lie groups	77
	Notes and remarks for chapter 1	85
	Chapter 2 Dirac operators and Clifford analyticity	**87**
1	Cauchy–Riemann operators	87
2	Dirac operators past and present	93
3	Clifford analyticity	97
4	Spaces of analytic functions	108
5	Spaces of Clifford analytic functions I: the upper half-space	119
6	Cauchy integrals and Hilbert transforms on Lipschitz domains	125
7	Spaces of Clifford analytic functions II: Lipschitz domains	135
	Notes and remarks for chapter 2	140

Chapter 3 Representations of Spin(V, Q) **143**
1   Elements of representation theory   144
2   Signature, fundamental representations   147
3   Class 1 representations   164
4   Polynomials of matrix argument   173
5   Harmonic polynomials of matrix argument   193
    Notes and remarks for chapter 3   201

Chapter 4 Constant coefficient operators of Dirac type   **203**
1   First-order systems: some general results   203
2   Operators of Dirac type   208
3   Rotation-invariant systems   213
4   The operators $\tilde{\delta}_\tau$ 220
5   Critical indices of subharmonicity   232
    Notes and remarks for chapter 4   244

Chapter 5 Dirac operators and manifolds   **247**
1   Local theory   248
2   Global theory   263
3   Dirac operators on hyperbolic and spherical space   272
4   Representation theory for Spin$_0(n, 1)$ 284
5   Asymptotics for heat kernels   296
6   The index theorem for Dirac operators   309
    Notes and remarks for chapter 5   317

References   **321**
Index   **328**

To

Vicki Gilbert,

for her love, patience, and understanding,

and to

Magdalene K. Murray,

for her courageous and independent spirit.

Introduction

In this book we present a comprehensive introduction to the use of Clifford algebras and Dirac operators in harmonic analysis and analysis more generally. In the past 30 years, Clifford algebras and Dirac operators have played a key role in three of the most important areas of mathematical research during that time: the boundedness of the Cauchy integral on Lipschitz surfaces, the realization of discrete series representations of semi-simple Lie groups, and the celebrated Atiyah–Singer index theorem. Much as an analyst would like to understand and appreciate these developments, however, there are formidable technical barriers to doing so, particularly for more classically trained analysts, as we have found to our cost over the years. Thus our aim from the outset has been to meld into a coherent and reasonably self-contained whole a body of ideas from classical singular integral theory, representation theory and analysis on manifolds, with a view to making this material accessible to more classically trained analysts.

Now the starting point for much of classical harmonic analysis is the study of the boundary regularity of harmonic functions in domains in Euclidean space. Classical Hardy space theory explores the consequences of the improved boundary regularity obtained when consideration is restricted to *analytic* functions in the plane. On the other hand, for $SL(2, \mathbb{R})$, the starting-point for representation theory of semi-simple Lie groups, some important unitary representations become irreducible only on restriction to *analytic* functions. It may be a dramatic overstatement to characterize analytic functions as those in the kernel of a first-order elliptic differential operator – the Cauchy–Riemann $\bar{\partial}$ operator – which factors the Laplacian and has rotation-invariant symbol; but it is precisely such properties that one looks for in differential operators on more general manifolds. For one can then develop a Hardy H^p theory on Euclidean space including an analysis of elliptic boundary value problems, as well as explicit realizations of semi-simple Lie groups on associated symmetric spaces. Index theorems arise in both cases, of course. Dirac operators and their generalization, the so-called *operators of Dirac type*, have such properties.

Much earlier, quite independently of all these analytic ideas, Clifford introduced his algebras as a common generalization of Grassmann's exterior algebra and Hamilton's quaternions, both of which sought to capture the geometric and algebraic properties of Euclidean space. Indeed, Clifford used the name 'geometric algebras' for his algebras quite appropriately, because the universal Clifford algebra for $\mathbb{R}^n$ is the minimal enlargement of $\mathbb{R}^n$ to an associative algebra capturing precisely the algebraic, geometric and metric properties of Euclidean space. It is not surprising, therefore, that the bundle formed by the Clifford algebra of the tangent space at each point of a manifold should be so important in the geometric analysis of that manifold.

In chapter 1 we present the general theory of Clifford algebras in an elementary and thoroughgoing fashion, which should be accessible to the algebra 'neophyte'; it is our aim to give a coherent account of material which is presently scattered throughout the literature with no one account being readily accessible. In chapter 2 we quickly review the classical Hardy space theory and its extension to minimally smooth domains, and then develop a higher-dimensional analogue for this theory based upon functions in the kernel of the Dirac operator. In chapter 3 we explore further the connections between Clifford algebras and representations of the spin group and of the rotation group. Then, in chapter 4, we define a more general notion of *operators of Dirac type*, and show that all of the important rotation-invariant geometric differential operators of Euclidean analysis are in fact of Dirac type. Finally, in chapter 5 we introduce and then study Clifford algebras and Dirac operators on more general manifolds, concluding with a recent simplified proof of the local Atiyah–Singer index theorem.

This book had its beginnings in the fall of 1985, when one of us (M.M.) was a visiting faculty member at the University of Texas at Austin. To whatever extent we have succeeded in our goal, we owe a debt of thanks to many of our friends and colleagues who have made this success possible. In particular, we wish to thank René Beerends, Klaus Bichteler, Chris Meaney and John Ryan for numerous helpful discussions, but most of all we wish to thank Kathy Davis, Gene Fabes and Ray Kunze for immeasurable help in the formulation of ideas going into the book, as well as in the writing of the book. One of us (M.M.) would like to acknowledge the particular help and support of her good friends and colleagues Daniel Farkas, and Carol and Frank Burch-Brown, without whom this work might never have come to fruition. Partial support from the National Science Foundation is acknowledged by both of us, too.

Finally, we wish to express our tremendous gratitude to Margaret Combs, whose patience, skill, and craftsmanship produced such a marvelous typescript.

1

Clifford algebras

Associated with any Euclidean space $\mathbb{R}^n$ or Minkowski space $\mathbb{R}^{p,q}$ is a universal Clifford algebra, denoted by $\mathfrak{A}_n$ and $\mathfrak{A}_{p,q}$, respectively. Roughly speaking, a Clifford algebra is an associative algebra with unit into which a given Euclidean or Minkowski space may be embedded, in which the corresponding quadratic form may be expressed as the negative of a square. The real numbers $\mathbb{R}$, the complex numbers $\mathbb{C}$, and the quaternions $\mathbb{H}$ are the simplest examples.

Our intent in this chapter is to give an elementary, coherent, and largely self-contained account of the theory of Clifford algebras. In sections 1 and 2 we present the definitions basic to all of our work. The balance of section 2 is devoted to three constructive proofs of the existence of universal Clifford algebras: two basis-free constructions using tensor algebras and exterior algebras, and a basis-dependent construction. The reader who is willing to accept the existence of Clifford algebras may wish to proceed directly to the statement of the major structural results in section 3. Sections 4, 5, and 6 explore the interconnections between Clifford algebras and orthogonal groups; the spin representation and spin groups will be studied in detail, with $\mathrm{Spin}(p,q)$ and $\mathrm{Spin}(p,q+1)$ both being realized in $\mathfrak{A}_{p,q}$ using the notion of transformers. The reader who is primarily interested in the analytic applications of Clifford algebras may wish to proceed directly to the discussion of the Euclidean case in section 7. Section 8 is a discussion of spin groups as Lie groups. In section 9 we construct various realizations of $\mathrm{Spin}(p,q)$, $p+q \leq 6$, whereby these groups are explicitly identified with classical Lie groups.

1 Quadratic spaces

Let V be a finite-dimensional vector space over the scalar field $\mathbb{F}$, where $\mathbb{F} = \mathbb{R}$ or $\mathbb{C}$. A *quadratic form* on V is a mapping $Q : V \to \mathbb{F}$ such that

(1.1)(i) $Q(\lambda v) = \lambda^2 Q(v)$, $\lambda \in \mathbb{F}$, $v \in V$,

(1.1)(ii) *the associated form*

$$B(v, w) = \tfrac{1}{2} \{ Q(v) + Q(w) - Q(v - w) \} , \qquad v, w \in V ,$$

is bilinear.

When such a Q exists, the pair (V, Q) is said to be a *quadratic space*; every vector space over $\mathbb{F}$ becomes a quadratic space with respect to the trivial quadratic form $Q \equiv 0$, for instance. The significance of condition (1.1)(ii) is that the form B defines an inner product on $V \times V$, and so all the usual geometric properties of inner product spaces can be exploited. Typically, a quadratic space arises from an inner product space, defining Q on V by, say, $Q(v) = (v \mid v)$ where $(\cdot \mid \cdot)$ is the inner product on $V \times V$. For example, if $(\cdot \mid \cdot)$ is the usual Euclidean inner product on $\mathbb{R}^n$ and $|v|^2 = (v \mid v)$, then both $(\mathbb{R}^n, |\cdot|^2)$ and $(\mathbb{R}^n, -|\cdot|^2)$ are real quadratic spaces with associated bilinear forms $(\cdot \mid \cdot)$ and $-(\cdot \mid \cdot)$ respectively. More generally, let p, q be non-negative integers with $p + q > 0$ and define a pseudo-Euclidean or Minkowski quadratic form on $\mathbb{R}^{p+q}$ by

(1.2)
$$Q_{p,q}(u) = -(u_1^2 + \cdots + u_p^2) + (u_{p+1}^2 + \cdots + u_{p+q}^2) , \quad u = (u_1, \ldots, u_{p+q}) ;$$

the corresponding real quadratic space we shall call *Minkowski* space and denote it by $(\mathbb{R}^{p,q}, Q_{p,q})$. Clearly $(\mathbb{R}^{n,0}, Q_{n,0})$ reduces to $(\mathbb{R}^n, -|\cdot|^2)$, while $(\mathbb{R}^{0,n}, Q_{0,n})$ is just $(\mathbb{R}^n, |\cdot|^2)$. By convention, $\mathbb{R}^{0,0} = \{0\}$. In the complex case, $(\mathbb{C}^n, Q_n)$ becomes a complex quadratic space on setting

(1.3) $\qquad Q_n(z) = z_1^2 + \cdots + z_n^2 , \qquad z = (z_1, \ldots, z_n) ;$

note that in (1.3) the associated form $B_n(z, w) = z_1 w_1 + \cdots + z_n w_n$ is complex linear in w, not conjugate-linear as in the usual inner product on $\mathbb{C}^n$.

Now let (V, Q) be an arbitrary quadratic space and $\{e_j\}$ a basis for V. Then

$$Q(v) = \sum_{j,k} B(e_j, e_k) v_j v_k , \qquad v = \sum_j v_j e_j ,$$

and if there is a basis which is B-*orthogonal* in the sense that $B(e_j, e_k) = 0$ when $j \neq k$, the expression for $Q(v)$ reduces to diagonal form

$$Q(v) = \sum_j Q(e_j) v_j^2 , \qquad v = \sum_j v_j e_j .$$

Such a basis is easily constructed. Let

$$\mathrm{Rad}(V, Q) = \{\, w \in V : B(v, w) = 0 \text{ , all } v \in V \,\} = V^{\perp}$$

be the *radical* of (V, Q). We say that (V, Q) is *non-degenerate* if $\mathrm{Rad}(V, Q) = \{0\}$; otherwise (V, Q) is *degenerate*, in which case V can be written as the B-orthogonal direct sum

$$V = \mathrm{Rad}(V, Q) \oplus \mathrm{Rad}(V, Q)^{\perp}$$

of $\mathrm{Rad}(V, Q)$ and its B-orthogonal complement. Clearly $(\mathrm{Rad}(V, Q)^{\perp}, Q)$ is non-degenerate, and B-orthogonal bases $\{e_j\}$, $Q(e_j) \neq 0$, can be constructed in the usual way for $\mathrm{Rad}(V, Q)^{\perp}$, or for V if (V, Q) is already non-degenerate. Using (1.1)(i) to normalize the e_j, we can also assume that $Q(e_j) = \pm 1$ when $\mathbb{F} = \mathbb{R}$, while $Q(e_j) = 1$ when $\mathbb{F} = \mathbb{C}$. Now augment this basis by any basis of $\mathrm{Rad}(V, Q)$ if (V, Q) is degenerate. Since Q is trivial on $\mathrm{Rad}(V, Q)$, we thus obtain a basis $\{e_j\}$ for V such that

(1.4)

(i) $B(e_j, e_k) = 0$, $j \neq k$,

(ii) $\{e_j : Q(e_j) = 0\}$ *is a basis for* $\mathrm{Rad}(V, Q)$,

(iii) $\{e_j : Q(e_j) \neq 0\}$ *is a basis for* $\mathrm{Rad}(V, Q)^{\perp}$ *such that* $Q(e_j) = \pm 1$
   when $\mathbb{F} = \mathbb{R}$, *while* $Q(e_j) = 1$ *when* $\mathbb{F} = \mathbb{C}$.

With some abuse of customary terminology, a basis for V satisfying (1.4)(i), (ii), (iii) will be said to be a *normalized basis*; many of the algebraic constructions to be discussed are conveniently given using such a basis. For instance, from such a basis it follows that every quadratic space (V, Q) is the sum of the particular examples given already. More precisely, we have the following.

(1.5) Theorem.
   Let (V, Q) be a quadratic space with B-orthogonal decomposition

$$V = \mathrm{Rad}(V, Q) \oplus \mathrm{Rad}(V, Q)^{\perp}.$$

Then

(a) $Q \equiv 0$ *on* $\mathrm{Rad}(V, Q)$,

(b) *when* $\mathbb{F} = \mathbb{R}$, $(\mathrm{Rad}(V, Q)^{\perp}, Q)$ *is isomorphic to* $\mathbb{R}^{p,q}$ *where* p, q *depend only on* Q,

(c) *when* $\mathbb{F} = \mathbb{C}$, $(\mathrm{Rad}(V, Q)^{\perp}, Q)$ *is isomorphic to* $(\mathbb{C}^n, Q_n)$ *where* n *depends only on* Q.

Proof. Part (a) is an immediate consequence of the definition of

$\mathrm{Rad}(V,Q)$. If $\mathrm{Rad}(V,Q) \neq V$, choose a basis $\{e_j\}$ satisfying (1.4)(i), (ii), (iii). In the case $\mathbb{F} = \mathbb{R}$ this basis can be indexed so that

$$Q(u) = Q\Big(\sum_j u_j e_j\Big) = -(u_1^2 + \cdots + u_p^2) + (u_{p+1}^2 + \cdots + u_{p+q}^2) ,$$

and $\dim(V,Q)^\perp = p + q > 0$; by Sylvester's theorem, the values of p, q do not vary with the choice of basis. Part (b) is now clear, and part (c) is proved in the same way. ∎

2 Clifford algebras

As Clifford's paper introducing 'geometric algebras' shows, Clifford based his ideas on the common features he saw in the construction of Grassmann's algebra and Hamilton's quaternions. In the framework of modern algebra we shall derive both constructions simultaneously beginning with an arbitrary quadratic space (V,Q), V a finite-dimensional vector space over $\mathbb{F}$. Let $\mathbf{A}$ be an associative algebra over $\mathbb{F}$ with identity 1 and $\nu : V \to \mathbf{A}$ an $\mathbb{F}$-linear embedding of V into $\mathbf{A}$.

(2.1) Definition.
 The pair $(\mathbf{A}, \nu)$ is said to be a Clifford algebra for (V,Q) when
 (i) $\mathbf{A}$ is generated as an algebra by $\{\nu(v) : v \in V\}$ and $\{\lambda 1 : \lambda \in \mathbb{F}\}$,
 (ii) $(\nu(v))^2 = -Q(v)1 ,$ all $v \in V$.

Roughly speaking, therefore, condition (ii) ensures that $\mathbf{A}$ is an algebra in which there exists a 'square root' of the quadratic form $-Q$; condition (i) is a minimality restriction on the 'size' of $\mathbf{A}$.

Some simple examples illustrate how this definition contains the algebras whose structure prompted Clifford to introduce 'geometric algebras'.

(2.2) Examples.
 (i) When $Q \equiv 0$ on V, let $\mathbf{A}$ be the exterior algebra $\Lambda^*(V) = \sum_{k=0}^n \Lambda^k(V)$ with $n = \dim V$, $\Lambda^0(V) \cong \mathbb{F}$, and $\Lambda^1(V) \cong V$, and let $\nu : V \to \Lambda^1(V)$. Since every element of $\Lambda^k(V)$, $k \geq 2$, is of the form $v_1 \wedge \cdots \wedge v_k$, clearly $\Lambda^*(V)$ is generated by $\{\nu(v) : v \in V\}$ and $\{\lambda 1 : \lambda \in \mathbb{F}\}$; in addition,

$$(\nu(v))^2 = v \wedge v = 0 = -Q(v)1 .$$

Hence the Grassmann algebra $\Lambda^*(V)$ is a Clifford algebra for (V,Q) when $Q \equiv 0$.

(ii) Define the *Pauli matrices* in $\mathbb{C}^{2 \times 2}$ by

(2.3)
$$\sigma_0 = \begin{bmatrix} 1 & 0 \\ 0 & 1 \end{bmatrix} , \quad \sigma_1 = \begin{bmatrix} 1 & 0 \\ 0 & -1 \end{bmatrix} , \quad \sigma_2 = \begin{bmatrix} 0 & -i \\ i & 0 \end{bmatrix} , \quad \sigma_3 = \begin{bmatrix} 0 & 1 \\ 1 & 0 \end{bmatrix}$$

and *associated Pauli matrices* by

(2.4)
$$e_0 = \begin{bmatrix} 1 & 0 \\ 0 & 1 \end{bmatrix} , \quad e_1 = \begin{bmatrix} i & 0 \\ 0 & -i \end{bmatrix} , \quad e_2 = \begin{bmatrix} 0 & 1 \\ -1 & 0 \end{bmatrix} , \quad e_3 = \begin{bmatrix} 0 & i \\ i & 0 \end{bmatrix} .$$

As elements of the associative algebra $\mathbb{C}^{2 \times 2}$,

$$\sigma_0^2 = \sigma_1^2 = \sigma_2^2 = \sigma_3^2 = I , \qquad e_0^2 = I , \qquad e_1^2 = e_2^2 = e_3^2 = -I ,$$

while

(2.5)
$$\sigma_j \sigma_k = -i\sigma_\ell , \qquad e_j e_k = e_\ell$$

when $\{j, k, \ell\}$ is a cyclic permutation of $\{1, 2, 3\}$. These matrices will occur throughout the theory of Clifford algebras. For instance, set

$$\mathfrak{A}_{0,0} = \{\lambda\sigma_0 : \lambda \in \mathbb{R}\} ,$$

$$\mathfrak{A}_{1,0} = \left\{ \begin{bmatrix} x & y \\ y & x \end{bmatrix} : x, y \in \mathbb{R} \right\} , \qquad \mathfrak{A}_{0,1} = \left\{ \begin{bmatrix} x & y \\ -y & x \end{bmatrix} : x, y \in \mathbb{R} \right\} ,$$

and

$$\mathfrak{A}_{0,2} = \left\{ \begin{bmatrix} x_0 + ix_1 & x_2 + ix_3 \\ -x_2 + ix_3 & x_0 - ix_1 \end{bmatrix} : x_j \in \mathbb{R} \right\} = \left\{ \begin{bmatrix} z_1 & z_2 \\ -\bar{z}_2 & \bar{z}_1 \end{bmatrix} : z_j \in \mathbb{C} \right\} .$$

Then each of these is an associative subalgebra (over $\mathbb{R}$) of $\mathbb{C}^{2 \times 2}$ having an identity, and

(2.6)
$$\mathfrak{A}_{0,0} \cong \mathbb{R} , \quad \mathfrak{A}_{1,0} \cong \mathbb{R} \oplus \mathbb{R} , \quad \mathfrak{A}_{0,1} \cong \mathbb{C} , \quad \mathfrak{A}_{0,2} \cong \mathbb{H}$$

where $\mathbb{H}$ is Hamilton's algebra of quaternions. As the notation suggests, $\mathfrak{A}_{p,q}$ also is a Clifford algebra for $\mathbb{R}^{p,q}$ with respective embeddings ν given by

$$0 \to 0 , \quad y \to y\sigma_3 , \quad y \to ye_2 , \quad (x_1, x_2) \to x_1 e_1 + x_2 e_2 .$$

In each case the proof amounts to a simple computation using properties of the σ_j and e_j. For instance, in $\mathfrak{A}_{0,2}$

$$\big(\nu(x_1, x_2)\big)^2 = (x_1 e_1 + x_2 e_2)^2 = -(x_1^2 + x_2^2)e_0$$

since the cyclic permutation property of $\{e_1, e_2, e_3\}$ ensures that $e_1 e_2 + e_2 e_1 = 0$; alternatively, by direct calculation we see that

$$\begin{bmatrix} ix_1 & x_2 \\ -x_2 & -ix_1 \end{bmatrix}^2 = -(x_1^2 + x_2^2) \begin{bmatrix} 1 & 0 \\ 0 & 1 \end{bmatrix} .$$

(iii) As an associative algebra over $\mathbb{C}$, the matrix algebra $\mathbb{C}^{2 \times 2}$ is a

Clifford algebra for $(\mathbb{C}^2, Q_2)$ with

$$\nu : \mathbb{C}^2 \longrightarrow \mathbb{C}^{2\times 2} , \qquad \nu : (z_1, z_2) \longrightarrow \begin{bmatrix} 0 & z_1 - iz_2 \\ z_1 + iz_2 & 0 \end{bmatrix} .$$

Again the proof is just a simple computation.

Some elementary properties of a Clifford algebra follow readily from its definition. Let (V, Q) be a quadratic space of dimension n and $(\mathbf{A}, \nu)$ a Clifford algebra for (V, Q). It is very convenient not to make distinction between $\lambda \in \mathbb{F}$ and $\lambda 1 \in \mathbf{A}$, or between $v \in V$ and $\nu(v) \in \mathbf{A}$. With this understanding, $\mathbf{A}$ is generated by $\mathbb{F}$ and V; in addition $v^2 = -Q(v)$. Similarly, if W is a subspace of V, then the subalgebra of $\mathbf{A}$ generated by $\mathbb{F}$ and W is a Clifford algebra for (W, Q). On the other hand, for arbitrary u, v in V

$$(u + v)^2 = -Q(u + v) = -2B(u, v) - Q(u) - Q(v)$$
$$= -2B(u, v) + u^2 + v^2 ;$$

but, by direct expansion,

$$(u + v)^2 = u^2 + uv + vu + v^2 .$$

Thus the algebra structure on $\mathbf{A}$ enables us to express the inner product B on $V \times V$ as

$$B(u, v) = -\tfrac{1}{2}(uv + vu) ;$$

in particular,

$$(2.7) \qquad e_j e_k + e_k e_j = -2Q(e_j)\delta_{jk} , \qquad 1 \le j, k \le n .$$

for any normalized basis $\{e_j\}_{j=1}^n$ of V. Often a Clifford algebra is defined as being the algebra generated by elements $e_1, \dots, e_n$ satisfying (2.7).

(2.8) Theorem.

Let $(\mathbf{A}, \nu)$ be a Clifford algebra for (V, Q) and $\{e_j\}_{j=1}^n$ a normalized orthogonal basis of V. Then

(i) $\mathbf{A}$ is spanned by all products

$$e_1^{m_1} e_2^{m_2} \cdots e_n^{m_n} , \qquad m_j = 0, 1,$$

where $e_1^0 \cdots e_n^0$ is interpreted as the identity in $\mathbf{A}$,

(ii) $\mathbf{A}$ has dimension at most 2^n.

Proof of 2.8(i) In view of (2.7)

$$(2.9) \qquad e_j e_k + e_k e_j = 0 \quad (j \ne k) , \qquad e_j^2 = -Q(e_j) .$$

Thus any product of powers of the e_j can be reduced to a scalar multiple of $e_1^{m_1} e_2^{m_2} \cdots e_n^{m_n}$ where $m_j = 0$ or 1 and $e_1^0 e_2^0 \cdots e_n^0$ is interpreted as the identity in **A**. The linear span of all such reduced products must then be **A**, since **A** is generated as an algebra by V and $\mathbb{F}$.

These reduced products are very convenient to use both in developing the general theory of Clifford algebras and in studying particular examples, so it is worth having a simple notation for them. Let $N_V = \{1, 2, \ldots, \dim V\}$ and for each non-empty subset α of N_V set

(2.10)
$$e_\alpha = e_{\alpha_1} \cdots e_{\alpha_k} \,, \qquad \alpha = \{\alpha_1, \ldots, \alpha_k\} \,, \quad 1 \le \alpha_1 < \cdots < \alpha_k \le \dim V \,;$$

by convention, $e_\emptyset$ is the identity 1 in **A** when $\emptyset$ is the empty subset of N_V.

Proof of 2.8(ii). Since there are 2^n subsets of N_V, **A** is spanned by the corresponding 2^n reduced products. Thus $\dim \mathbf{A} \le 2^n$. ∎

A Clifford algebra can have maximum dimension $2^{\dim V}$. For instance, all the examples in (2.2) have this property. Indeed, it is well-known that the exterior algebra $\Lambda^*(V)$ has dimension 2^n when $\dim V = n$, and that the reduced wedge products

(2.11) $\qquad e_\emptyset = 1 \,, \quad e_\alpha = e_{\alpha_1} \wedge \cdots \wedge e_{\alpha_k} \,, \quad \alpha = \{\alpha_1, \ldots, \alpha_k\} \subseteq N_V,$

form a basis for $\Lambda^*(V)$. Each of the $\mathfrak{A}_{p,q}$ in (2.6) also has maximum dimension as a (real) Clifford algebra for $\mathbb{R}^{p,q}$, while $\mathbb{C} \oplus \mathbb{C}$ and $\mathbb{C}^{2 \times 2}$ have maximum dimension as (complex) Clifford algebras for $(\mathbb{C}^1, Q_1)$ and $(\mathbb{C}^2, Q_2)$ respectively. On the other hand, $\mathfrak{A}_{0,2}$ is a Clifford algebra for $\mathbb{R}^{0,3}$ defining ν on $\mathbb{R}^{0,3}$ by

$$\nu : \mathbb{R}^{0,3} \to \mathfrak{A}_{0,2} \,, \qquad \nu(x_1, x_2, x_3) = x_1 e_1 + x_2 e_2 + x_3 e_3 \,;$$

as before, the anticommutation property $e_j e_k + e_k e_j = 0$, $j \ne k$, ensures that

$$\left(\nu(x_1, x_2, x_3) \right)^2 = -(x_1^2 + x_2^2 + x_3^2) e_0 \,.$$

But for this example $\dim \mathbf{A} = (1/2) 2^{\dim V}$. In fact, as we shall see later in section 3, the dimension of any Clifford algebra **A** for $\mathbb{R}^{p,q}$ will satisfy exactly one of

$\qquad$ (2.12)(a) $\dim \mathbf{A} = 2^{p+q}$,

$\qquad$ (2.12)(b) $\dim \mathbf{A} = (1/2) 2^{p+q}$, $p - q - 1 \equiv 0 \pmod 4$, $p + q$ *is odd,*

$\qquad\qquad$ *and* $e_1 e_2 \cdots e_{p+q} \in \mathbb{R}$ *for some normalized basis*

$\qquad\qquad$ $\{e_1, \ldots, e_{p+q}\}$ *of* $\mathbb{R}^{p,q}$

The Clifford algebra $\mathfrak{A}_{0,2}$ for $\mathbb{R}^{0,3}$ falls into category (2.12)(b).

Clifford algebras of maximum dimension have an important distinguishing property.

(2.13) Definition.

  A Clifford algebra $(\mathbf{A}, \nu)$ *for a quadratic space* (V, Q) *is said to be a universal Clifford algebra if for each Clifford algebra* $(\mathbf{B}, \mu)$ *for* (V, Q) *there is an algebra homomorphism* $\beta : \mathbf{A} \to \mathbf{B}$ *such that* $\mu = \beta \circ \nu$ *and* $\beta(1_{\mathbf{A}}) = 1_{\mathbf{B}}$.

  The fundamental result of this section is that for each quadratic space (V, Q) there is a universal Clifford algebra. Since V and $\mathbb{F}$ generate every Clifford algebra, it is virtually obvious that this universal Clifford algebra is unique up to isomorphism; hence we shall denote it by $\mathfrak{A}(V, Q)$, or, more briefly, $\mathfrak{A}$ when (V, Q) is understood, and speak of *the* universal Clifford algebra for (V, Q). In particular, we shall use the notation $\mathfrak{A}_{p,q}$, $\mathfrak{A}_n$, and $\mathcal{C}_n$ for the universal Clifford algebras $\mathfrak{A}(\mathbb{R}^{p,q}, Q_{p,q})$, $\mathfrak{A}(\mathbb{R}^{0,n}, Q_{0,n})$, and $\mathfrak{A}(\mathbb{C}^n, Q_n)$ respectively. The next result relates the dimension of a Clifford algebra to its universality.

(2.14) Theorem.

  A Clifford algebra $(\mathbf{A}, \nu)$ *for* (V, Q) *is universal when* $\dim \mathbf{A} = 2^{\dim V}$.

Proof. Let $\{x_j\}$ be a normalized basis for V and $(\mathbf{B}, \mu)$ an arbitrary Clifford algebra for (V, Q). Defining $e_j = \nu(x_j)$ in $\mathbf{A}$ and $f_j = \mu(x_j)$ in $\mathbf{B}$, we deduce from (2.7) that

(2.15) $e_j e_k + e_k e_j = 0 , \quad j \neq k , \quad e_j^2 = -Q(x_j)1_{\mathbf{A}} ,$

while

(2.16) $f_j f_k + f_k f_j = 0 , \quad j \neq k , \quad f_j^2 = -Q(x_j)1_{\mathbf{B}} .$

Now the reduced products $\{e_\alpha\}$ form a basis for $\mathbf{A}$ since $\mathbf{A}$ has maximum dimension, and the reduced products $\{f_\alpha\}$ span $\mathbf{B}$. Thus the linear extension of the map $\beta : e_\alpha \to f_\alpha$ is a well-defined linear mapping from $\mathbf{A}$ onto $\mathbf{B}$ such that $\mu = \beta \circ \nu$ on V and $\beta(1_{\mathbf{A}}) = 1_{\mathbf{B}}$. In addition, (2.15) and (2.16) ensure that

$$\beta(e_\alpha \cdot e_\delta) = \beta(e_1^{m_1} \cdots e_n^{m_n} \cdot e_1^{\ell_1} \cdots e_n^{\ell_n}) = \beta(e_\gamma) = f_\gamma = f_\alpha \cdot f_\delta ,$$

establishing the required algebra homomorphism property of β. Hence $(\mathbf{A}, \nu)$ is universal when $\dim \mathbf{A} = 2^{\dim V}$. ∎

(2.17) Theorem.

  For every quadratic space (V, Q) *there exists a universal Clifford algebra* $\mathfrak{A}(V, Q)$. *Furthermore, every universal Clifford algebra has maximum dimension.*

  To prove this theorem we give three explicit constructions of a universal Clifford algebra for (V, Q) having maximum dimension. Since universal Clifford algebras are unique up to isomorphism, it will then follow that a Clifford algebra is universal if and only if it has maximum dimension.

FIRST CONSTRUCTION.

In view of theorem (1.5), every (real) quadratic space (V, Q) can be written as an orthogonal finite direct sum

$$(V, Q) = (V_1, Q_1) \oplus \cdots \oplus (V_r, Q_r)$$

of quadratic spaces (V_j, Q_j) where $Q_j \equiv 0$ on V_j or (V_j, Q_j) is isomorphic to one of $\mathbb{R}^{1,0}$ or $\mathbb{R}^{0,1}$. In the complex case, $\mathbb{R}^{1,0}$ and $\mathbb{R}^{0,1}$ have to be replaced by $(\mathbb{C}^1, Q_1)$. As we have seen, each of these quadratic spaces has a corresponding two-dimensional Clifford algebra, so theorem (2.14) ensures that for each (V_j, Q_j) there is a universal Clifford algebra $\mathfrak{A}(V_j, Q_j)$ with $\dim \mathfrak{A}(V_j, Q_j) = 2^{\dim V_j}$. To complete the proof we shall combine these individual algebras into a universal Clifford algebra for (V, Q) using the following result which is of independent interest.

(2.18) Theorem.
Let $(\mathbf{A}_1, \nu_1)$, $(\mathbf{A}_2, \nu_2)$ be Clifford algebras for quadratic spaces (V_1, Q_1), (V_2, Q_2). Then there is a linear isomorphism ν from $V_1 \oplus V_2$ into $\mathbf{A}_1 \otimes \mathbf{A}_2$ and an associative multiplicative structure on $\mathbf{A}_1 \otimes \mathbf{A}_2$ so that $(\mathbf{A}_1 \otimes \mathbf{A}_2, \nu)$ is a Clifford algebra for $(V_1 \oplus V_2, \ Q_1 \oplus Q_2)$.

Applying theorem (2.18), we obtain a Clifford algebra $(\mathbf{A}, \nu)$ for (V, Q) with

$$\mathbf{A} = \mathfrak{A}(V_1, Q_1) \otimes \cdots \otimes \mathfrak{A}(V_r, Q_r), \quad \dim \mathbf{A} = 2^{\dim V_1} \cdots 2^{\dim V_r} = 2^{\dim V}.$$

Thus $\mathbf{A}$ has maximum dimension, and so $(\mathbf{A}, \nu)$ is a universal Clifford algebra for (V, Q).

To prove theorem (2.18) a $\mathbb{Z}(2)$-grading on a Clifford algebra is needed; this grading will be an integral part of its algebraic structure.

(2.19) Definition.
When $(\mathbf{A}, \nu)$ is a Clifford algebra for (V, Q) denote by $\mathbf{A}^+$ and $\mathbf{A}^-$ the subspaces of $\mathbf{A}$ generated by the products of even and odd numbers of elements of V, respectively.

Thus $\mathbf{A} = \mathbf{A}^+ \oplus \mathbf{A}^-$ and $\mathbb{F} \subseteq \mathbf{A}^+$, $V \subseteq \mathbf{A}^-$. Suppose now that $\{e_j\}$ is a normalized basis for V; and if $\alpha \subseteq N_V$, let $|\alpha|$ denote the cardinality of α. Then the set of reduced products e_α with $\alpha \subseteq N_V$ and $|\alpha|$ even spans $\mathbf{A}^+$, while the corresponding set of e_α with $\alpha \subseteq N_V$ and $|\alpha|$ odd spans

$\mathbf{A}^-$. Since the product of any two elements in $\{e_\alpha : \alpha \subseteq N_V, |\alpha| \text{ even}\}$ again belongs to this set, we see that $\mathbf{A}^+$ is a subalgebra of $\mathbf{A}$; similarly, $\mathbf{A}^-$ is an $\mathbf{A}^+$-module and products of elements from $\mathbf{A}^-$ belong to $\mathbf{A}^+$. In simple terms, therefore,

$$(2.20) \qquad \mathbf{A}^+\mathbf{A}^+ = \mathbf{A}^+, \quad \mathbf{A}^-\mathbf{A}^+ = \mathbf{A}^-, \quad \mathbf{A}^-\mathbf{A}^- = \mathbf{A}^+,$$

which expresses the $\mathbf{Z}(2)$-grading of $\mathbf{A}$ by $\mathbf{A}^+$, $\mathbf{A}^-$.

Proof of theorem (2.18). Define a linear mapping ν from $V_1 \oplus V_2$ into $\mathbf{A}_1 \otimes \mathbf{A}_2$ by

$$\nu : (v_1, v_2) \to v_1 \otimes 1_2 + 1_1 \otimes v_2, \qquad v_j \in V_j,$$

where 1_j is the identity in $\mathbf{A}_j$. This clearly is a linear isomorphism. To introduce a multiplicative structure on $\mathbf{A}_1 \otimes \mathbf{A}_2$, first define products of elements $a_1 \otimes a_2$, $a_1' \otimes a_2'$ where $a_j, a_j' \in \mathbf{A}_j^{\pm}$ by

$$(a_1 \otimes a_2)(a_1' \otimes a_2') = (-1)^{\varepsilon\varepsilon'}(a_1 a_1' \otimes a_2 a_2')$$

with

$$\varepsilon = \begin{cases} 0, & a_2 \in \mathbf{A}_2^+, \\ 1, & a_2 \in \mathbf{A}_2^-, \end{cases} \qquad \varepsilon' = \begin{cases} 0, & a_1' \in \mathbf{A}_1^+, \\ 1, & a_1' \in \mathbf{A}_1^-, \end{cases}$$

and then extend by linearity to all of $\mathbf{A}_1 \otimes \mathbf{A}_2$ using the fact that $\mathbf{A}_j = \mathbf{A}_j^+ \oplus \mathbf{A}_j^-$ and that every element of $\mathbf{A}_1 \otimes \mathbf{A}_2$ is a linear combination of elements of the form $a_1 \otimes a_2$. Straightforward calculations show that $\mathbf{A}_1 \otimes \mathbf{A}_2$ is an associative algebra with identity $1_1 \otimes 1_2$ under this multiplication; in addition, $\mathbf{A}_1 \otimes \mathbf{A}_2$ is generated by $\{\nu(v_1, v_2) : v_j \in V_j\}$ together with $\{\lambda(1_1 \otimes 1_2) : \lambda \in \mathbf{F}\}$.

To complete the proof we have to show that

$$(2.21) \qquad \big(\nu(v_1, v_2)\big)^2 = -\big(Q_1(v_1) + Q_2(v_2)\big)1_1 \otimes 1_2, \qquad v_j \in V_j.$$

It is precisely at this point that the form of the 'twisted' multiplicative structure on $\mathbf{A}_1 \otimes \mathbf{A}_2$ is needed. Indeed,

$$\big(\nu(v_1, v_2)\big)^2 = (v_1 \otimes 1_2 + 1_1 \otimes v_2)^2$$
$$= (v_1^2 \otimes 1_2 + 1_1 \otimes v_2^2) + (v_1 \otimes 1_2)(1_1 \otimes v_2) + (1_1 \otimes v_2)(v_1 \otimes 1_2)$$
$$= -\big(Q_1(v_1) + Q_2(v_2)\big)1_1 \otimes 1_2 + v_1 \otimes v_2 - v_1 \otimes v_2,$$

since $1_j \in \mathbf{A}_j^+$ and $v_j \in \mathbf{A}_j^-$. This establishes (2.21). ∎

SECOND CONSTRUCTION.

We shall obtain the universal Clifford algebra $\mathfrak{A}(V, Q)$ in very specific terms as a subalgebra of the algebra $\mathcal{L}(\Lambda^*(V))$ of linear transformations of $\Lambda^*(V)$. Recall that $\Lambda^*(V) = \sum_{\ell=0}^n \Lambda^\ell(V)$ where $n = \dim V$, $\Lambda^0(V) =$

$\mathbb{F}$, $V = \Lambda^1(V)$ and each element of $\Lambda^\ell(V)$ is a sum of wedge products $v_1 \wedge \cdots \wedge v_\ell$. For each v in V we define M_v and δ_v in $\mathcal{L}(\Lambda^*(V))$ as the linear extensions of

$$M_v(1) = v , \qquad M_v(v_1 \wedge \cdots \wedge v_\ell) = v \wedge v_1 \wedge \cdots \wedge v_\ell$$

and

$$\delta_v(1) = 0 , \qquad \delta_v(v_1 \wedge \cdots \wedge v_\ell) = \sum_{k=1}^{\ell} (-1)^{k-1} B(v, v_k) v_1 \wedge \cdots \wedge \hat{v}_k \wedge \cdots \wedge v_\ell ,$$

$\hat{v}_k$ meaning as usual that v_k is omitted from the product. The operator M_v is just exterior multiplication by v, while δ_v is interior multiplication with respect to the inner product induced on $V \times V$ by B; clearly δ_v is the adjoint of M_v. In the physics literature the operators M_v, δ_v are referred to as *creation* and *annihilation* operators, respectively. Now define $\nu : V \to \mathcal{L}(\Lambda^*(V))$ by $\nu(v) = M_v - \delta_v$; by definition $\nu(v)1 = v$, so ν clearly is a linear isomorphism. We shall show that the subalgebra $\mathbf{A}$ of $\mathcal{L}(\Lambda^*(V))$ generated by $\{\nu(v) : v \in V\}$ and $\{\lambda 1 : \lambda \in \mathbb{F}\}$ is a Clifford algebra for (V, Q) of maximum dimension. This will then be a universal Clifford algebra for (V, Q).

The proof rests on three simple properties of the operators M_v, δ_v:

$$(2.22) \qquad M_v^2 = \delta_v^2 = 0 , \qquad \delta_v(\omega \wedge z) = B(v, \omega)z - \omega \wedge \delta_v(z)$$

for any $\omega \in V$ and $z \in \Lambda^*(V)$. The first of these is an immediate consequence of the property $v \wedge v = 0$. To establish the second, observe that

$$\delta_v^2(v_1 \wedge \cdots \wedge v_\ell)$$

$$= \sum_{k=1}^{\ell} (-1)^{k-1} B(v, v_k) \delta_v(v_1 \wedge \cdots \wedge \hat{v}_k \wedge \cdots \wedge v_\ell)$$

$$= \sum_{1 \le j < k \le \ell} (-1)^{j+k} B(v, v_j) B(v, v_k) v_1 \wedge \cdots \wedge \hat{v}_j \wedge \cdots \wedge \hat{v}_k \wedge \cdots \wedge v_\ell$$

$$- \sum_{1 \le k < j \le \ell} (-1)^{j+k} B(v, v_j) B(v, v_k) v_1 \wedge \cdots \wedge \hat{v}_k \wedge \cdots \wedge \hat{v}_j \wedge \cdots \wedge v_\ell .$$

Symmetry in j, k ensures that the last expression vanishes, so $\delta_v^2 = 0$.

Finally, for any ω in V and $z = z_1 \wedge \cdots \wedge z_\ell$ in $\Lambda^*(V)$,

$$\delta_v(\omega \wedge z) = \delta_v(\omega \wedge z_1 \wedge \cdots \wedge z_\ell)$$

$$= B(v,\omega)z_1 \wedge \cdots \wedge z_\ell$$

$$- \sum_{k=1}^{\ell} (-1)^{k-1} B(v,z_k)\omega \wedge z_1 \wedge \cdots \wedge \hat{z}_k \wedge \cdots \wedge z_\ell$$

$$= B(v,\omega)z - \omega \wedge \delta_v(z) .$$

This proves (2.22), from which we deduce

$$(M_v - \delta_v)^2(z) = (M_v^2 + \delta_v^2 - \delta_v M_v - M_v \delta_v)z$$

$$= -\delta_v(v \wedge z) - v \wedge \delta_v(z) = -Q(v)z .$$

Thus $\nu(v)^2 = (M_v - \delta_v)^2 = -Q(v)I$, and so $(\mathbf{A}, \nu)$ is a Clifford algebra for (V, Q). To estimate the dimension of $\mathbf{A}$, define $\theta : \mathbf{A} \to \Lambda^*(V)$ as the linear extension of the mapping

(2.23) $\theta : \nu(v_1) \cdots \nu(v_\ell) \to (M_{v_1} + \delta_{v_1}) \cdots (M_{v_\ell} + \delta_{v_\ell})1 .$

By definition,

$$\theta : \nu(v_1) \cdots \nu(v_\ell) \to v_1 \wedge \cdots \wedge v_\ell + \sum_{k=0}^{\ell-1} \xi_k$$

for some $\xi_k \in \Lambda^k(V)$. Since θ maps $\{\lambda I : \lambda \in \mathbf{F}\}$ onto $\Lambda^0(V)$ and $\{\nu(v) : v \in V\}$ onto $\Lambda^1(V)$, an obvious induction argument shows that θ maps $\mathbf{A}$ onto $\Lambda^*(V)$. But then $\dim \mathbf{A} \geq \dim \Lambda^*(V) = 2^n$. Hence $\mathbf{A}$ must have maximum dimension, and so be a universal Clifford algebra for (V, Q).

When $Q \equiv 0$ on V, the realization of $\mathfrak{A}(V, Q)$ in the second construction reduces to the left regular representation of $\Lambda^*(V)$ on itself because $\delta_v \equiv 0$ in this case. Even when $Q \not\equiv 0$, however, the mapping θ defined by (2.23) identifies $\mathfrak{A}(V, Q)$ *linearly* with $\Lambda^*(V)$ in a completely explicit manner.

THIRD CONSTRUCTION.

This construction will establish directly the existence of a Clifford algebra for (V, Q) having the universal property rather than one having maximum dimension, but it does use the second construction in the process.

Let $T_k(V)$ be the k-fold tensor product $V \otimes \cdots \otimes V$ with $T_0(V) = \mathbf{F}$ and $T_1(V) = V$, and let $T(V) = \sum_{k=0}^{\infty} T_k(V)$ be the tensor algebra over V. Let I_Q be the two-sided ideal generated in $T(V)$ by $\{v \otimes v + Q(v)1 :$

$v \in V\}$, so that I_Q is spanned by

$$\left\{ u \otimes \left(v \otimes v + Q(v)1\right) \otimes \omega : u, \omega \in T(V), \ v \in V \right\},$$

let $\mathfrak{A}$ be the quotient algebra $T(V)/I_Q$, and let $\pi : T(V) \to \mathfrak{A}$ be the associated homomorphism. Define $\nu : V \to \mathfrak{A}$ by $\nu(v) = \pi(v)$. We shall prove that $(\mathfrak{A}, \nu)$ is the required algebra. Now certainly $\mathfrak{A}$ is an associative algebra with identity $1_{\mathfrak{A}} = \pi(1)$ and is generated by $\{\nu(v) : v \in V\}$ and $\{\lambda\pi(I) : \lambda \in \mathbb{F}\}$, since $T(V)$ is generated by V and $\mathbb{F}$. Also by construction, $\nu(v)^2 = -Q(v)1_{\mathfrak{A}}$. Thus $(\mathfrak{A}, \nu)$ will be a Clifford algebra for (V, Q) provided $\nu : V \to \mathfrak{A}$ is 1–1. To establish this and the universal property, first let $(\mathbf{A}, \alpha)$ be an arbitrary Clifford algebra for (V, Q). By the universal property of $T(V)$ with respect to associative algebras, there is an algebra homomorphism $\alpha' : T(V) \to \mathbf{A}$ such that $\alpha'(1) = 1_{\mathbf{A}}$ and $\alpha'(v) = \alpha(v)$, $v \in V$. But then

$$\alpha'\left(u\left(v \otimes v + Q(v)1\right)\omega\right) = \alpha'(u)\left(\alpha'(v)^2 + Q(v)1_{\mathbf{A}}\right)\alpha'(\omega)$$
$$= \alpha'(u)\left(\alpha(v)^2 + Q(v)1_{\mathbf{A}}\right)\alpha'(\omega) = 0$$

for all u, ω in $T(V)$ and v in V. Thus $I_Q \subseteq \ker(\alpha')$; hence $\alpha' : T(V) \to \mathbf{A}$ factors through $\mathfrak{A}$, so there is an algebra homomorphism $\beta : \mathfrak{A} \to \mathbf{A}$ for which $\alpha' = \beta \circ \pi$. Clearly then $\beta(1_{\mathfrak{A}}) = 1_{\mathbf{A}}$ while $\beta \circ \nu = \alpha$. This establishes the universal property, leaving only the 1–1 property of ν to be established. Choose for $(\mathbf{A}, \alpha)$ the subalgebra of $\mathcal{L}(\Lambda^*(V))$ generated by $\{M_v - \delta_v : v \in V\}$ and $\{\lambda I : \lambda \in \mathbb{F}\}$. By what we have just proved, there is an algebra homomorphism β from $\mathfrak{A}$ into $\mathcal{L}(\Lambda^*(V))$ such that $\beta(\nu(v)) = M_v - \delta_v$, $v \in V$. Since then $\beta(\nu(v))1 = (M_v - \delta_v)1 = v$, clearly $\beta \circ \pi$ is the identity on V. Hence ν must be 1–1, completing the construction. ∎

Three operations on $\mathfrak{A}(V, Q)$ will be basic to all that follows. To illustrate the various ideas developed in this section, we shall give both basis-dependent and basis-free definitions. The *principal automorphism* on $\mathfrak{A}(V, Q)$, denoted by $(')$, is the algebra automorphism defined on basis elements e_α by

$$(2.24) \qquad e_\alpha' = (-1)^{|\alpha|} e_\alpha, \qquad \alpha \subseteq N_V,$$

where $|\alpha|$ is the cardinality of α. The *principal anti-automorphism*, denoted by $(*)$, is defined on basis elements e_α by

$$(2.25) \qquad e_\alpha^* = (-1)^{\frac{1}{2}|\alpha|(|\alpha|-1)} e_\alpha,$$

while *conjugation*, denoted by $(^-)$, is the composition

$$(2.26) \qquad \bar{e}_\alpha = (e_\alpha^*)' = (e_\alpha')^* = (-1)^{\frac{1}{2}|\alpha|(|\alpha|+1)} e_\alpha$$

of the principal automorphism and anti-automorphism. The term *reversion* is sometimes used instead of 'principal anti-automorphism' because

(2.27) $\qquad e_\alpha^* = (e_{\alpha_1} e_{\alpha_2} \cdots e_{\alpha_k})^* = e_{\alpha_k} \cdots e_{\alpha_2} e_{\alpha_1}$.

To prove this observe first that $e_j e_k = -e_k e_j$ when $j \neq k$; thus

$$e_{\alpha_k} \cdots e_{\alpha_2} e_{\alpha_1} = -e_{\alpha_k} \cdots e_{\alpha_3} e_{\alpha_1} e_{\alpha_2}$$
$$= (-1)^{1+2} e_{\alpha_k} \cdots e_{\alpha_4} e_{\alpha_1} e_{\alpha_2} e_{\alpha_3} = \cdots$$
$$= (-1)^{1+2+\cdots+(k-1)} e_{\alpha_1} e_{\alpha_2} \cdots e_{\alpha_k} = (-1)^{\frac{1}{2}|\alpha|(|\alpha|-1)} e_\alpha ,$$

establishing (2.27). That

$$(e_\alpha e_\beta)' = e_\alpha' e_\beta' , \qquad (e_\alpha e_\beta)^* = e_\beta^* e_\alpha^*$$

hold as the names suggest can be checked by easy computation. The alternative basis-free approach to defining (') and (*) makes these properties clearer, however. Indeed, define a linear mapping $' : V \to \mathfrak{A}(V, Q)$ by $v' = -v$. Then $(v')^2 = -Q(v)$, so, by the universal property, this linear mapping extends uniquely to an algebra automorphism on $\mathfrak{A}(V, Q)$ which clearly is given by (2.24) on e_α. Now define the anti-automorphism (*) on the tensor algebra $T(V) = \sum_{k=0}^\infty T_k(V)$ by

$$(v_1 \otimes \cdots \otimes v_k)^* = v_k \otimes \cdots \otimes v_1 ;$$

this leaves invariant the two-sided ideal I_Q generated in $T(V)$ by $\{v \otimes v + Q(v)I : v \in V\}$. Passing to the quotient $T(V)/I_Q$, we obtain an anti-automorphism on $\mathfrak{A}(V, Q)$ given by (2.27) on elements e_α.

If $\mathfrak{A}(V, Q)$ has a matrix realization, these three operations are often familiar ones from linear algebra.

(2.28) Examples.

(i) When $\mathbb{C}$ is a realization of $\mathfrak{A}_1$ under the embedding $\nu : x \to ix$, then

$$z' = \bar{z} \ , \quad z^* = z , \qquad z \in \mathbb{C} ,$$

where $\bar{z}$ is the usual conjugation on $\mathbb{C}$; in particular, conjugation on $\mathbb{C}$ as a Clifford algebra coincides with the usual notion of conjugation.

(ii) When $\mathbb{C}^{2 \times 2}$ is a realization of $\mathfrak{A}_{3,0}$ under the embedding

$$\nu : (x_0, x_1, x_2) \longrightarrow \begin{bmatrix} x_0 & x_1 + ix_2 \\ x_1 - ix_2 & -x_0 \end{bmatrix}$$

of $\mathbb{R}^{3,0}$ as the traceless hermitian matrices in $\mathbb{C}^{2\times 2}$, then

$$\begin{bmatrix} a & b \\ c & d \end{bmatrix}' = \begin{bmatrix} \bar{d} & -\bar{c} \\ -\bar{b} & \bar{a} \end{bmatrix} \quad , \quad \begin{bmatrix} a & b \\ c & d \end{bmatrix}^* = \begin{bmatrix} \bar{a} & \bar{c} \\ \bar{b} & \bar{d} \end{bmatrix} ,$$

$$\begin{bmatrix} a & b \\ c & d \end{bmatrix}^- = \begin{bmatrix} d & -b \\ -c & a \end{bmatrix} .$$

Consequently, in this interpretation of $\mathbb{C}^{2\times 2}$ as a universal Clifford algebra, A^* is the usual adjoint (= conjugate-transpose) of A, while $\bar{A}$ is the adjugate of A; in particular,

$$A\bar{A} = \bar{A}A = (\det A)I .$$

(iii) When $\mathbb{C}^{2\times 2}$ is a realization of $\mathfrak{A}_{1,2}$ under the embedding

$$\nu : (x_0, x_1, x_2) \longrightarrow \begin{bmatrix} x_0 & x_1 + ix_2 \\ -(x_1 - ix_2) & -x_0 \end{bmatrix} ,$$

then

$$\begin{bmatrix} a & b \\ c & d \end{bmatrix}' = \begin{bmatrix} \bar{d} & \bar{c} \\ \bar{b} & \bar{a} \end{bmatrix} \quad , \quad \begin{bmatrix} a & b \\ c & d \end{bmatrix}^* = \begin{bmatrix} \bar{a} & -\bar{c} \\ -\bar{b} & \bar{d} \end{bmatrix} ,$$

$$\begin{bmatrix} a & b \\ c & d \end{bmatrix}^- = \begin{bmatrix} d & -b \\ -c & a \end{bmatrix} .$$

So again $\bar{A}$ is the adjugate matrix of A, and

$$A\bar{A} = \bar{A}A = (\det A)I .$$

These expressions are easily established using only routine calculations. They also generalize, replacing $\mathbb{C}$ by any universal Clifford algebra $\mathfrak{A}(V_0, Q_0)$ and $\mathbb{C}^{2\times 2}$ by

$$M\left(2, \mathfrak{A}(V_0, Q_0)\right) = \left\{ \begin{bmatrix} a & b \\ c & d \end{bmatrix} : a, b, c, d \in \mathfrak{A}(V_0, Q_0) \right\} ,$$

where (V_0, Q_0) is a real, non-degenerate quadratic space.

(2.29) Theorem.
  Under the quadratic form

$$Q(v, x, y) = -\left(x^2 + y^2 + Q_0(v)\right) \qquad (v \in V_0) ,$$

on $V = V_0 \oplus \mathbb{R}^{2,0}$, there is a realization of $\mathfrak{A}(V, Q)$ as $M(2, \mathfrak{A}(V_0, Q_0))$ such that

(2.30) $$\begin{cases} \begin{bmatrix} a & b \\ c & d \end{bmatrix}' = \begin{bmatrix} d' & -c' \\ -b' & a' \end{bmatrix} \quad , \quad \begin{bmatrix} a & b \\ c & d \end{bmatrix}^* = \begin{bmatrix} \bar{a} & \bar{c} \\ \bar{b} & \bar{d} \end{bmatrix} , \\ \\ \qquad \begin{bmatrix} a & b \\ c & d \end{bmatrix}^- = \begin{bmatrix} d^* & -b^* \\ -c^* & a^* \end{bmatrix} , \end{cases}$$

with respect to the corresponding operations on $\mathfrak{A}(V_0, Q_0)$.

Just as theorem (2.29) reduces to (2.28)(ii) when $\mathfrak{A}(V_0, Q_0) = \mathbb{C}$, so the next result reduces to (2.28)(iii) when $\mathfrak{A}(V_0, Q_0) = \mathbb{C}$.

(2.31) Theorem.
Under the quadratic form
$$Q(v, x, y) = Q_0(v) + y^2 - x^2 \qquad (v \in V_0),$$
*on $V = V_0 \oplus \mathbb{R}^{1,1}$, there is a realization of $\mathfrak{A}(V, Q)$ as $M(2, \mathfrak{A}(V_0, Q_0))$
such that*

$$(2.32) \begin{cases} \begin{bmatrix} a & b \\ c & d \end{bmatrix}' = \begin{bmatrix} d' & c' \\ b' & a' \end{bmatrix}, & \begin{bmatrix} a & b \\ c & d \end{bmatrix}^* = \begin{bmatrix} \bar{a} & -\bar{c} \\ -\bar{b} & \bar{d} \end{bmatrix}, \\[3mm] \begin{bmatrix} a & b \\ c & d \end{bmatrix}^- = \begin{bmatrix} d^* & -b^* \\ -c^* & a^* \end{bmatrix}, \end{cases}$$

with respect to the corresponding operations on $\mathfrak{A}(V_0, Q_0)$.

Proof of theorem (2.29). Let $\{e_1, \ldots, e_n\}$ be a normalized basis for (V_0, Q_0), $n = \dim V_0$, and define $\gamma_0, \ldots, \gamma_{n+1}$ in $M(2, \mathfrak{A}(V_0, Q_0))$ by

$$\gamma_0 = \begin{bmatrix} 1 & 0 \\ 0 & -1 \end{bmatrix}, \; \gamma_j = \begin{bmatrix} 0 & e_j \\ -e_j & 0 \end{bmatrix} \; (1 \le j \le n), \; \gamma_{n+1} = \begin{bmatrix} 0 & 1 \\ 1 & 0 \end{bmatrix}.$$

Then
$$\gamma_j \gamma_k + \gamma_k \gamma_j = 2\delta_{jk} Q_0(e_j) \qquad (1 \le j, k \le n),$$
while
$$\gamma_0 \gamma_j + \gamma_j \gamma_0 = 2\delta_{j0}, \quad \gamma_{n+1}\gamma_j + \gamma_j \gamma_{n+1} = 2\delta_{j,n+1} \qquad (0 \le j \le n+1).$$
Clearly
(2.33)
$$\nu : (v, x, y) \to \begin{bmatrix} x & y+v \\ y+v' & -x \end{bmatrix} = x\gamma_0 + \sum_{j=1}^n v_j \gamma_j + y\gamma_{n+1} \quad \left(v = \sum_{j=1}^n v_j e_j \right)$$

defines a linear embedding of $V = V_0 \oplus \mathbb{R}^{2,0}$, in $M(2, \mathfrak{A}(V_0, Q_0))$ satisfying the Clifford condition with respect to the quadratic form
(2.34)
$$Q(v, x, y) = -(x^2 + y^2 + Q_0(v))$$
on V. Now the universal Clifford algebra $\mathfrak{A}(V_0, -Q_0)$ is realized in $M(2, \mathfrak{A}(V_0, Q_0))$ by
$$\mathfrak{A}^+(V_0, -Q_0) = \left\{ \begin{bmatrix} z & 0 \\ 0 & z' \end{bmatrix} : z \in \mathfrak{A}^+(V_0, Q_0) \right\},$$
and
$$\mathfrak{A}^-(V_0, -Q_0) = \left\{ \begin{bmatrix} 0 & \varsigma \\ \varsigma' & 0 \end{bmatrix} : \varsigma \in \mathfrak{A}^-(V_0, Q_0) \right\};$$

these are generated by $\{\gamma_j\}_{j=1}^n$. On the other hand, the universal Clifford algebra $\mathfrak{A}_{2,0}$ of $\mathbb{R}^{2,0}$ is realized in $M(2, \mathfrak{A}(V_0, Q_0))$ by

$$M(2, \mathbb{R}) = \left\{ \begin{bmatrix} a & b \\ c & d \end{bmatrix} : a, b, c, d \in \mathbb{R} \right\} \; ;$$

and this algebra is generated by $\{\gamma_0, \gamma_{n+1}\}$. Jointly, therefore, these two generating sets generate all of $M(2, \mathfrak{A}(V_0, Q_0))$. Since this last algebra has maximum dimension $4 \cdot 2^{\dim V_0} = 2^{\dim V}$, $M(2, \mathfrak{A}(V_0, Q_0))$ is a realization of $\mathfrak{A}(V, Q)$. Straightforward but lengthy calculations establish (2.30). ∎

Proof of theorem (2.31). Define the basic elements $\gamma_0, \ldots, \gamma_{n+1}$ now by

$$\gamma_0 = \begin{bmatrix} 1 & 0 \\ 0 & -1 \end{bmatrix}, \quad \gamma_j = \begin{bmatrix} 0 & e_j \\ e_j & 0 \end{bmatrix} \quad (1 \le j \le n), \quad \gamma_{n+1} = \begin{bmatrix} 0 & 1 \\ -1 & 0 \end{bmatrix},$$

and let

(2.35) $$\nu : (v, x, y) \longrightarrow \begin{bmatrix} x & y + v \\ -y - v' & -x \end{bmatrix}$$

be the corresponding embedding of $V = V_0 \oplus \mathbb{R}^{1,1}$ in $M(2, \mathfrak{A}(V_0, Q_0))$. The Clifford condition is satisfied with respect to the quadratic form

(2.36) $$Q(v, x, y) = y^2 - x^2 + Q_0(v)$$

on V. Only trivial modifications to the proof of the previous theorem are required to complete the present proof. ∎

Restricting theorems (2.29) and (2.31) to subspaces of $V_0 \oplus \mathbb{R}^{2,0}$ and $V_0 \oplus \mathbb{R}^{1,1}$ respectively, we deduce the following.

(2.37) Theorem.
Under the quadratic form

$$Q(v, x) = -\left(x^2 + Q_0(v)\right) \qquad (v \in V_0)$$

on $V = V_0 \oplus \mathbb{R}^{1,0}$, there is a realization of $\mathfrak{A}(V, Q)$ as the subalgebra

$$\left\{ \begin{bmatrix} z & \zeta \\ \zeta' & z' \end{bmatrix} : z, \zeta \in \mathfrak{A}(V_0, Q_0) \right\}$$

of $M(2, \mathfrak{A}(V_0, Q_0))$.

(2.38) Theorem.
Under the quadratic form

$$Q(v, x) = Q_0(v) + x^2 \qquad (v \in V_0)$$

on $V = V_0 \oplus \mathbb{R}^{0,1}$, *there is a realization of* $\mathfrak{A}(V,Q)$ *as the subalgebra*

$$\left\{ \begin{bmatrix} z & \zeta \\ -\zeta' & z' \end{bmatrix} : z, \zeta \in \mathfrak{A}(V_0, Q_0) \right\}$$

of $M(2, \mathfrak{A}(V_0, Q_0))$.

Repeated use of these last two theorems actually provides yet another proof of theorem (2.17), but the main use of all four theorems will be in recognizing various matrix algebras as Clifford algebras and in constructing the spin groups which will play such an important role later.

3 Structure of Clifford algebras

The fundamental use of Clifford algebras in studying Dirac operators and Lie groups arises from basic structural identifications of $\mathfrak{A}(V,Q)$ to be made in this section. It is not surprising that these identifications depend on whether $\mathbb{F}$ is $\mathbb{R}$ or $\mathbb{C}$, but it is surprising that they depend on the parity of the dimension of V. For instance, example (2.2)(iii) shows that

$$(3.1) \qquad \mathfrak{A}(\mathbb{C}^1, Q_1) \equiv \mathbb{C} \oplus \mathbb{C} , \qquad \mathfrak{A}(\mathbb{C}^2, Q_2) \cong \mathbb{C}^{2\times2} ,$$

so that the universal Clifford algebra $\mathbb{C}_2$ is isomorphic to the full matrix algebra $\mathbb{C}^{2\times2}$, whereas $\mathbb{C}_1$ is isomorphic to the direct sum of *two* full matrix algebras, regarding $\mathbb{C}$ as $\mathbb{C}^{1\times1}$. These results are typical of the structure of $\mathfrak{A}(V,Q)$ for any non-degenerate complex quadratic space (V,Q).

(3.2) Theorem.
Let $\mathfrak{A}(V,Q)$ be the universal Clifford algebra for a non-degenerate complex quadratic space (V,Q). Then

(i) $\mathfrak{A}(V,Q)$ is isomorphic to the full matrix algebra $\mathbb{C}^{2^m \times 2^m}$ when $\dim V = 2m$,

(ii) $\mathfrak{A}(V,Q)$ is isomorphic to the direct sum $\mathbb{C}^{2^m \times 2^m} \oplus \mathbb{C}^{2^m \times 2^m}$ when $\dim V = 2m + 1$.

From theorem (3.2) the structural identification of $\mathfrak{A}(V,Q)$ for any complex quadratic space (V,Q) follows easily. For (V,Q) can be written as a B-orthogonal direct sum

$$(V,Q) = \mathrm{Rad}(V,Q) \oplus \mathrm{Rad}(V,Q)^\perp$$

of quadratic spaces where in one case $Q \equiv 0$, while in the other Q is non-degenerate. But as we have seen

$$\mathfrak{A}\big(\mathrm{Rad}(V,Q)\big) \cong \Lambda^*\big(\mathrm{Rad}(V,Q)\big) .$$

Together with theorems (2.18) and (3.2), this identifies $\mathfrak{A}(V,Q)$ with one of the tensor products

$$\Lambda^*(\mathbb{C}^k) \otimes \mathbb{C}^{2^m \times 2^m} \, , \qquad \Lambda^*(\mathbb{C}^k) \otimes (\mathbb{C}^{2^m \times 2^m} \oplus \mathbb{C}^{2^m \times 2^m})$$

according as $\dim V = k + 2m$ or $\dim V = k + 2m + 1$. It must be remembered, of course, that the multiplicative structure on these tensor products is the twisted multiplicative structure defined in the proof of theorem (2.18), thus eliminating any simple-minded proof based on that theorem. Nonetheless, this failure brings out the essential element of a correct proof. Indeed, suppose $\dim V = 2m$ and identify (V,Q) with the m-fold B-orthogonal direct sum

$$(V,Q) = (\mathbb{C}^2, Q_2) \oplus \cdots \oplus (\mathbb{C}^2, Q_2) \, .$$

Then, by (3.1) and theorem (2.18), $\mathfrak{A}(V,Q)$ is isomorphic to the m-fold tensor product

$$\mathfrak{A}(V,Q) \cong \mathbb{C}^{2 \times 2} \otimes \cdots \otimes \mathbb{C}^{2 \times 2}$$

which in turn is isomorphic to $\mathbb{C}^{2^m \times 2^m}$. Because of the twisted multiplicative structure on the tensor product, however, this last isomorphism is only a linear isomorphism when $\mathbb{C}^{2^m \times 2^m}$ has its usual multiplicative structure. Thus, the linear isomorphism from $V \cong \mathbb{C}^2 \oplus \cdots \oplus \mathbb{C}^2$ into $\mathbb{C}^{2^m \times 2^m}$ $(\cong \mathbb{C}^{2 \times 2} \otimes \cdots \otimes \mathbb{C}^{2 \times 2})$ defined by

$$(3.3) \quad \mu : v = (v_1, \ldots, v_m) \to v_1 \otimes 1_2 \otimes \cdots \otimes 1_m + \cdots + 1_1 \otimes \cdots \otimes 1_{m-1} \otimes v_m$$

as in the proof of theorem (2.18) does *not* have the property $\mu(v)^2 = -Q(v)$ with respect to the usual multiplication on $\mathbb{C}^{2^m \times 2^m}$. The two proofs of theorem (3.2) that we shall give both amount to defining a linear isomorphism μ which does have the property $\mu(v)^2 = -Q(v)$ in $\mathbb{C}^{2^m \times 2^m}$. Hence, by the universal property, μ extends to an algebra homomorphism β from $\mathfrak{A}(V,Q)$ into $\mathbb{C}^{2^m \times 2^m}$. Had we been able to apply theorem (2.18), we would have been able to deduce *at once* that the range of this homomorphism is all of $\mathbb{C}^{2^m \times 2^m}$, establishing the required isomorphism because $\dim \mathfrak{A}(V,Q) = 2^{2m} = \dim \mathbb{C}^{2^m \times 2^m}$. Thus the second step in both proofs is to show that the dimension of $\beta(\mathfrak{A}(V,Q)) = 2^{2m}$. In the first proof this is done directly by dimension and structural arguments which will be useful in both the real and the complex case, while the second uses a Wedderburn-type argument which brings out interesting algebraic properties of $\mathfrak{A}(V,Q)$. Similar comments apply also when V has odd dimension.

As preparation for the proofs of theorem (3.2) we prove several useful results. For the moment no restriction is placed on the scalar field, i.e., $\mathbb{F}$ can be $\mathbb{R}$ or $\mathbb{C}$, and $(\mathbf{A}, \nu)$ will be an arbitrary Clifford algebra for a quadratic space (V, Q) over $\mathbf{F}$. Fix a normalized basis $\{e_j\}$ in V so that the set $\{e_\alpha : \alpha \subseteq N_V\}$ of reduced products spans $\mathbf{A}$.

(3.4) Lemma.

   For each $\alpha \subseteq N_V$ and basis element e_j

   (i) $e_j e_\alpha = (-1)^{|\alpha|} e_\alpha e_j, \quad j \notin \alpha,$

   (ii) $e_j e_\alpha = (-1)^{|\alpha|-1} e_\alpha e_j, \quad j \in \alpha,$

where $|\alpha|$ is the cardinality of α. In particular, $e_j e_\alpha = e_\alpha e_j$ for all $j = 1, \ldots, \dim V$, if and only if $\alpha = \emptyset$ or N_V; the latter case can occur only when V is odd-dimensional.

Proof. All the results follow from successive use of the anti-commutation property $e_j e_k + e_k e_j = 0, \ j \neq k$. For, if $\alpha = \{\alpha_1, \ldots, \alpha_k\}$ and $j \notin \alpha$,

$$e_j e_\alpha = e_j e_{\alpha_1} \cdots e_{\alpha_k} = (-1) e_{\alpha_1} e_j e_{\alpha_2} \cdots e_{\alpha_k}$$
$$= (-1)^2 e_{\alpha_1} e_{\alpha_2} e_j \cdots e_{\alpha_k} = \cdots = (-1)^{|\alpha|} e_\alpha e_j \ ;$$

whereas, if $j \in \alpha$, one fewer (-1) factor will arise from the need to permute e_j with all but one of the basis elements in e_α. Thus $e_j e_\alpha = e_\alpha e_j$ if and only if $|\alpha|$ is even and $j \notin \alpha$, or $|\alpha|$ is odd and $j \in \alpha$. Hence $e_j e_\alpha = e_\alpha e_j$ for all $j = 1, \ldots, \dim V$, if and only if $\alpha = \emptyset$ or $\alpha = N_V$ and V is odd-dimensional. ∎

   The particular reduced product $e_1 e_2 \cdots e_{2m+1}, \ \dim V = 2m + 1$, will have a special role to play in the structural theory, so we shall denote it by e_V. The name *pseudo-scalar* is often used for e_V.

(3.5) Lemma.

   Suppose (V, Q) is non-degenerate and $\sum_{\beta \subseteq N_V} \lambda_\beta e_\beta = 0$ for a choice of $\lambda_\beta \in \mathbf{F}$. Then either $\lambda_\emptyset = 0$ or $\lambda_\emptyset + \lambda_V e_V = 0$; the latter case can only occur when V is odd-dimensional.

Proof. The non-degeneracy ensures that each e_j is invertible in $\mathbf{A}$ since $e_j^2 = \pm 1$. But then, if $\sum_{\beta \subseteq N_V} \lambda_\beta e_\beta = 0$, certainly

$$0 = \frac{1}{2} \left\{ e_1 \left(\sum_{\beta \subseteq N_V} \lambda_\beta e_\beta \right) e_1^{-1} + \left(\sum_{\beta \subseteq N_V} \lambda_\beta e_\beta \right) \right\}$$

$$= \sum_{\beta \subseteq N} \lambda_\beta \frac{1}{2} (e_1 e_\beta e_1^{-1} + e_\beta) = \sum_{\beta \in C} \lambda_\beta e_\beta$$

where

$$C = \{ \beta \subseteq N_V : e_1 e_\beta = e_\beta e_1 \} \ .$$

Repeating this successively with $e_2, \ldots, e_{\dim V}$ in place of e_1, we deduce that $\sum_{\beta \in D} \lambda_\beta e_\beta = 0$, the sum now being taken just over the set

$$D = \{ \beta \subseteq N_V : e_j e_\beta = e_\beta e_j \text{ for all } j \in N_V \} .$$

By lemma (3.4) the only possible choices for β are $\beta = \emptyset$ or, when V is odd-dimensional, $\beta = N_V$. Hence $\lambda_\emptyset = 0$ or V is odd-dimensional and $\lambda_\emptyset + \lambda_V e_V = 0$. ∎

(3.6) Corollary.

Let (V, Q) be a non-degenerate quadratic space and suppose that either $\dim V$ is even, or $\dim V$ is odd and $e_V \notin \mathbb{F}$. Then every Clifford algebra for (V, Q) has maximum dimension.

Proof. It is enough to show that the spanning set $\{e_\alpha : \alpha \subseteq N_V\}$ is linearly independent. Suppose then that $\sum_\alpha \lambda_\alpha e_\alpha = 0$ for some choice of $\{\lambda_\alpha\} \subseteq \mathbb{F}$. Now each reduced product is invertible because (V, Q) is non-degenerate. Thus, for fixed $\delta \subseteq N_V$,

$$\lambda_\delta + \sum_{\alpha \neq \delta} \lambda_\alpha e_\alpha e_\delta^{-1} = \left(\sum_\alpha \lambda_\alpha e_\alpha \right) e_\delta^{-1} = 0 .$$

Hence $\lambda_\delta = 0$ or V is odd-dimensional and $\lambda_\delta + \lambda_\gamma e_V = 0$ for some $\gamma \subseteq N_V$. In the latter case, however, e_V would have to belong to $\mathbb{F}$ contrary to hypothesis. This establishes linear independence. ∎

We can now give a direct constructive proof of theorem (3.2).

First proof of (3.2)(i). When $\dim V = 2m$, define $E_1, \ldots, E_{2m}$ in the m-fold tensor product $\mathbb{C}^{2 \times 2} \otimes \cdots \otimes \mathbb{C}^{2 \times 2} \cong \mathbb{C}^{2^m \times 2^m}$ by

$$(3.7) \quad \left\{ \begin{array}{l} E_s = \sigma_1 \otimes \cdots \otimes \sigma_1 \otimes \sigma_2 \otimes I \otimes \cdots \otimes I , \\ E_{m+s} = \sigma_1 \otimes \cdots \otimes \sigma_1 \otimes \sigma_3 \otimes I \otimes \cdots \otimes I , \end{array} \right\} \quad 1 \leq s \leq m ,$$

there being $s - 1$ factors σ_1 in each product. The linear isomorphism

$$(3.8) \quad \mu : \lambda_1 e_1 + \cdots + \lambda_{2m} e_{2m} \longrightarrow i\lambda_1 E_1 + \cdots + i\lambda_{2m} E_{2m}$$

will be the required substitute for (3.3) since routine calculations using the properties

$$(3.9) \quad \sigma_1^2 = \sigma_2^2 = \sigma_3^2 = I , \qquad \sigma_j \sigma_k + \sigma_k \sigma_j = 0 \quad (j \neq k) ,$$

of the Pauli matrices show that

$$(3.10) \quad E_j^2 = I \otimes \cdots \otimes I , \qquad E_j E_k + E_k E_j = 0 \quad (j \neq k) .$$

Indeed, the first equality in (3.10) follows immediately from the first chain of equalities in (3.9). To establish the second when, say, $1 \leq j \leq m$

and $m + j < k \leq 2m$, note that

$$E_j E_k = \underbrace{\sigma_1^2 \otimes \cdots \otimes \sigma_1^2}_{j-1} \otimes \sigma_2\sigma_1 \otimes \underbrace{\sigma_1 \otimes \cdots \otimes \sigma_1}_{k-m-j-1} \otimes \sigma_3 \otimes \underbrace{I \otimes \cdots \otimes I}_{2m-k}$$

$$= \underbrace{I \otimes \cdots \otimes I}_{j-1} \otimes \sigma_2\sigma_1 \otimes \underbrace{\sigma_1 \otimes \cdots \otimes \sigma_1}_{k-m-j-1} \otimes \sigma_3 \otimes \underbrace{I \otimes \cdots \otimes I}_{2m-k}$$

while

$$E_k E_j = \underbrace{\sigma_1^2 \otimes \cdots \otimes \sigma_1^2}_{j-1} \otimes \sigma_1\sigma_2 \otimes \underbrace{\sigma_1 \otimes \cdots \otimes \sigma_1}_{k-m-j-1} \otimes \sigma_3 \otimes \underbrace{I \otimes \cdots \otimes I}_{2m-k}$$

$$= \underbrace{I \otimes \cdots \otimes I}_{j-1} \otimes \sigma_1\sigma_2 \otimes \underbrace{\sigma_1 \otimes \cdots \otimes \sigma_1}_{k-m-j-1} \otimes \sigma_3 \otimes \underbrace{I \otimes \cdots \otimes I}_{2m-k}$$

$$= -E_j E_k$$

using the second equality in (3.9); proofs for the other cases are correspondingly straightforward. But then

$$\mu\left(\sum_j \lambda_j e_j\right)^2 = \left(\sum_j i\lambda_j E_j\right)^2 = -\sum_j \lambda_j^2 = -Q\left(\sum_j \lambda_j e_j\right),$$

so that when $\mathbf{A}$ is the algebra generated by $\{\sum_j i\lambda_j E_j : \lambda_j \in \mathbb{C}\}$ and $\{\lambda(I \otimes \cdots \otimes I) : \lambda \in \mathbb{F}\}$ the pair $(\mathbf{A}, \nu)$ is a Clifford algebra for (V, Q). By corollary (3.6), however, this must be of maximum dimension because V is even-dimensional. Hence $\mathbf{A} = \mathbb{C}^{2\times2} \otimes \cdots \otimes \mathbb{C}^{2\times2}$, and μ extends to an algebra *isomorphism* from $\mathfrak{A}(V, Q)$ onto $\mathbb{C}^{2\times2} \otimes \cdots \otimes \mathbb{C}^{2\times2} \cong \mathbb{C}^{2^m \times 2^m}$.

First proof of (3.2) (ii). When $\dim V = 2m + 1$ define $F_1, \ldots, F_{2m}$ in the direct sum of two copies of $\mathbb{C}^{2\times2} \otimes \cdots \otimes \mathbb{C}^{2\times2}$ by

$$F_j = E_j \oplus E_j \qquad (1 \leq j \leq 2m),$$

with E_j as in (3.7); the remaining basis element F_{2m+1} is defined by

$$F_{2m+1} = (\sigma_1 \otimes \cdots \otimes \sigma_1) \oplus \left(-(\sigma_1 \otimes \cdots \otimes \sigma_1)\right).$$

The proof now follows the same lines as that for part (i), with the substitute for (3.3) this time being

$$\mu : \lambda_1 e_1 + \cdots + \lambda_{2m+1} e_{2m+1} \longrightarrow i\lambda_1 F_1 + \cdots + i\lambda_{2m+1} F_{2m+1}.$$

Indeed, the first of the equalities in

$$F_j^2 = (I \otimes \cdots \otimes I) \oplus (I \otimes \cdots \otimes I), \qquad F_j F_k + F_k F_j = 0 \quad (j \neq k),$$

is clear, while the second follows from (3.10) since $\sigma_1 \otimes \cdots \otimes \sigma_1$ anticommutes with each E_j. Consequently, when $\mathbf{A}$ is the algebra generated by $\{i\lambda_j F_j : \lambda_j \in \mathbb{F}\}$ and $\{\lambda(I \otimes \cdots \otimes I) \oplus (I \otimes \cdots \otimes I) : \lambda \in \mathbb{C}\}$, the pair $(\mathbf{A}, \nu)$ is a Clifford algebra for (V, Q) which by corollary (3.6) will be of maximum dimension provided that $(-1)^m i\, F_1 F_2 \cdots F_{2m+1}$ is not a complex scalar.

To evaluate $F_1 F_2 \cdots F_{2m+1}$ first note that

$$E_1 E_2 \cdots E_m = \sigma_2(\sigma_1)^{m-1} \otimes \sigma_2(\sigma_1)^{m-2} \otimes \cdots \otimes \sigma_2\sigma_1 \otimes \sigma_2 \;,$$

while

$$E_{m+1}E_{m+2} \cdots E_{2m} = \sigma_3(\sigma_1)^{m-1} \otimes \sigma_3(\sigma_1)^{m-2} \otimes \cdots \otimes \sigma_3\sigma_1 \otimes \sigma_3 \;.$$

Thus, using (2.5) together with the fact that $\sigma_1^2 = I$, we obtain

$$\begin{aligned}
E_1 E_2 \cdots E_{2m} &= \sigma_2(\sigma_1)^{m-1}\sigma_3(\sigma_1)^{m-1} \otimes \sigma_2(\sigma_1)^{m-2} \\
&\qquad \sigma_3(\sigma_1)^{m-2} \otimes \cdots \otimes \sigma_2\sigma_3 \\
&= (-1)^{m-1}\sigma_2\sigma_3 \otimes (-1)^{m-2}\sigma_2\sigma_3 \otimes \cdots \otimes \sigma_2\sigma_3 \\
&= \alpha_m\sigma_1 \otimes \sigma_1 \otimes \cdots \otimes \sigma_1 \;,
\end{aligned}$$

where $\alpha_m = (-1)^{\frac{1}{2}m(m+1)}i^m$. In this case

$$F_1 F_2 \cdots F_{2m+1} = \alpha_m(I \otimes \cdots \otimes I) \oplus \bigl(-(I \otimes \cdots \otimes I)\bigr)$$

does not belong to $\mathbb{F}$. Corollary (3.6) now ensures that the Clifford algebra $(\mathbf{A}, \nu)$ has maximum dimension 2^{2m+1}, so $\mathbf{A}$ is the direct sum of two copies of $\mathbb{C}^{2\times 2} \otimes \cdots \otimes \mathbb{C}^{2\times 2}$. This completes the first proof of theorem (3.2). ∎

The second proof rests on results describing the simplicity or semi-simplicity of a Clifford algebra. It will be instructive to present this second proof as much as possible parallel to the first one, so once again for the moment (V, Q) is an arbitrary quadratic space over $\mathbb{F} = \mathbb{R}$ or $\mathbb{C}$, and $(\mathbf{A}, \nu)$ a Clifford algebra for (V, Q). Let $\{e_j\}$ be a normalized basis of V.

(3.11) Lemma.

The center $\mathfrak{Z}_\mathbf{A}$ of $(\mathbf{A}, \nu)$ consists of all x in $\mathbf{A}$ such that $e_j x = x e_j$ for all $j \in N_V$.

Proof. Clearly $e_j x = x e_j$ for all e_j when x belongs to the center $\mathfrak{Z}_\mathbf{A}$. Conversely, if $e_j x = x e_j$ for all e_j, x will commute with every reduced product $e_\alpha = e_{\alpha_1} \cdots e_{\alpha_k}$; since these reduced products span $\mathbf{A}$, every such x must lie in $\mathfrak{Z}_\mathbf{A}$. ∎

Lemma (3.4) now gives $\mathfrak{Z}_\mathbf{A}$ explicitly at least for non-degenerate (V, Q).

(3.12) Corollary.

Suppose (V, Q) is a non-degenerate quadratic space and $(\mathbf{A}, \nu)$ a Clif-

ford algebra for (V,Q). Then $3_\mathbf{A} = \mathbb{F}$ when V is even-dimensional, whereas $3_\mathbf{A} = \mathbb{F} \oplus \mathbb{F}e_V$ when V is odd-dimensional.

The non-degeneracy assumption in the corollary above is essential, since lemma (3.11) ensures that every wedge product $e_j \wedge e_k$ lies in the center of $\Lambda^*(V)$.

Proof of (3.12). Non-degeneracy ensures that each e_j is invertible in $\mathbf{A}$. Let $x = \sum_{\beta \subseteq N_V} \lambda_\beta e_\beta$ be an element in $3_\mathbf{A}$. Then $e_j x = x e_j$ for all e_j, and so in particular

$$x = \tfrac{1}{2}\{e_1 x e_1^{-1} + x\} = \sum_{\beta \subseteq N_V} \tfrac{1}{2}\lambda_\beta(e_1 e_\beta e_1^{-1} + e_\beta) = \sum_{\beta \in C} \lambda_\beta e_\beta$$

where $C = \{\beta \subseteq N_V : e_1 e_\beta = e_\beta e_1\}$. Repeating this successively with $e_2, \ldots, e_{\dim V}$ in place of e_1, we deduce that $x = \sum_{\beta \in D} \lambda_\beta e_\beta$ where

$$D = \{\ \beta \subseteq N_V : e_j e_\beta = e_\beta e_j\ ,\quad \text{for all } j \in N_V\ \}.$$

By lemma (3.4) β must be empty, or else $\beta = N_V$ and V is odd-dimensional. Hence $x = \lambda I$ or, if V is odd-dimensional, $x = \lambda I + \lambda_V e_V$. This proves the corollary. ∎

Next we consider the ideal structure of $(\mathbf{A}, \nu)$, beginning with the following result.

(3.13) Lemma.
Suppose (V,Q) is non-degenerate and I is a non-trivial ideal in $\mathbf{A}$. Then I contains the identity 1, or V is odd-dimensional and I contains an element of the form $1 + \lambda_V e_V$ for some λ_V in $\mathbb{F}$.

Proof. Let $x = \sum_{\alpha \subseteq N_V} \mu_\alpha e_\alpha$ be a non-zero element in I with, say, $\mu_\delta \neq 0$, for some $\delta \subseteq N_V$. But then, since $\mu_\delta e_\delta$ is invertible, the element

$$x_0 = (\mu_\delta e_\delta)^{-1} x = \sum_{\alpha \subseteq N_V} (\mu_\delta^{-1}\mu_\alpha)e_\delta^{-1}e_\alpha$$
$$= 1 + \sum_{\beta \in B} \lambda_\beta e_\beta$$

also belongs to I where, crucially, B does not contain the empty set $\emptyset$. The ideal I thus also contains the element

$$x_1 = \tfrac{1}{2}\{e_1 x_0 e_1^{-1} + x_0\} = 1 + \sum_{\beta \in B} \tfrac{1}{2}\lambda_\beta(e_1 e_\beta e_1^{-1} + e_\beta)$$
$$= 1 + \sum_{\beta \in B_1} \lambda_\beta e_\beta$$

where now

$$B_1 = \{\beta \in B : e_1 e_\beta = e_\beta e_1\}.$$

Repeating this successively with $e_2, \ldots, e_{\dim V}$ in place of e_1, we finally obtain an element y in I such that $y = 1 + \sum_{\beta \in C} \lambda_\beta e_\beta$ where

$$C = \{\beta \in B : e_j e_\beta = e_\beta e_j , \quad \text{for all } j \in N_V\} .$$

Since $\emptyset \notin B$, lemma (3.4) ensures that either C is empty, or V is odd-dimensional and C consists solely of the set N_V. The lemma follows immediately. ∎

Thus if (V, Q) is an even-dimensional non-degenerate quadratic space, any Clifford algebra $(\mathbf{A}, \nu)$ for (V, Q) is central and simple, in the sense that $\mathfrak{Z}_\mathbf{A} = \mathbb{F}$ and $I = \mathbf{A}$ when I is a non-trivial ideal in $\mathbf{A}$. In view of Wedderburn's theorem for central simple algebras therefore, it is now not so surprising that $\mathbf{A}$ must be isomorphic to $\mathbb{C}^{2^m \times 2^m}$ when V is a complex space with dimension $2m$. A proof using just Clifford algebra ideas can be given as follows.

Second proof of (3.2)(i). By corollary (3.6), $\mathbf{A}$ has maximum dimension; consequently, we can assume $\mathbf{A} = \mathfrak{A}(V, Q)$.

Let $\{e_j : j = 1, \ldots, 2m\}$ be a normalized basis for V as usual and let (V', Q') be the complex quadratic space obtained by restricting Q to the subspace V' of V having basis $\{e_j : j = 1, \ldots, m\}$. Since the universal Clifford algebra $\mathfrak{A}' = \mathfrak{A}(V', Q')$ has dimension 2^m, we shall exhibit the isomorphism $\mathfrak{A}(V, Q) \cong \mathbb{C}^{2^m \times 2^m}$ by constructing an algebra isomorphism γ from $\mathfrak{A}(V, Q)$ onto the algebra $\mathcal{L}(\mathfrak{A}')$ of all linear transformations on $\mathfrak{A}'$. When $\{e_\alpha : \alpha \subseteq N_{V'}\}$ is a basis for $\mathfrak{A}'$, define $\delta : V \to \mathcal{L}(\mathfrak{A}')$ as the linear extension of

(i) $\delta(e_j)e_\alpha = e_\alpha e_j$, (ii) $\delta(e_{m+j})e_\alpha = (-1)^{|\alpha|} i e_j e_\alpha$ $(1 \leq j \leq m)$;

it is at this point in the proof that the hypothesis $\mathbb{F} = \mathbb{C}$ is used. Straightforward calculations now show that $\delta(v)^2 = -Q(v)I_{\mathfrak{A}'}$, $v \in V$. Hence δ lifts to an algebra homomorphism γ from $\mathfrak{A}(V, Q)$ into $\mathcal{L}(\mathfrak{A}')$ such that $\gamma(1_{\mathfrak{A}}) = I_{\mathfrak{A}'}$. Since the image of γ is a Clifford algebra for (V, Q), it must have maximum dimension by corollary (3.6). Consequently, γ is an isomorphism from $\mathfrak{A}(V, Q)$ onto $\mathcal{L}(\mathfrak{A}(V', Q'))$. This completes the second proof of (3.2)(i). ∎

Second proof of (3.2)(ii). When $\dim V = 2m + 1$, lemma (3.4) ensures that

$$e_V e_V = e_1 \ldots e_{2m+1} e_1 \ldots e_{2m+1} = (-1)^{2m} e_1^2 e_2 \ldots e_{2m+1} e_2 \ldots e_{2m+1}$$

$$= \cdots = (-1)^{2m}(-1)^{2m-1} \cdots (-1) e_1^2 e_2^2 \cdots e_{2m+1}^2$$

$$= (-1)^{(m+1)(2m+1)} .$$

Now set $P = e_V$ when m is odd and $P = i e_V$ when m is even (it is here that the hypothesis $\mathbb{F} = \mathbb{C}$ is used). Then P is a central element of $\mathbf{A}$ such that $P^2 = 1$, and so

(3.14) $$ e_+ = \tfrac{1}{2}(1 + P) , \qquad e_- = \tfrac{1}{2}(1 - P) $$

are central idempotents in $\mathbf{A}$; in addition,

(3.15) $$ e_+ e_- = e_- e_+ = \tfrac{1}{4}(1 - P^2) = 0 . $$

Let I_+ (resp. I_-) be the ideal generated in $\mathbf{A}$ by e_+ (resp. e_-). Because of (3.14), I_+ and I_- are subalgebras of $\mathbf{A}$ having identity e_+ and e_- respectively, while (3.15) ensures that $\mathbf{A} = I_+ \oplus I_-$ is a direct sum decomposition of $\mathbf{A}$. The proof will be completed by showing that I_+ and I_- are both Clifford algebras isomorphic to $\mathbb{C}^{2^m \times 2^m}$.

Let (V', Q') be the complex quadratic space obtained by restricting Q to the subspace V' of V having basis $\{e_j : j = 1, \dots, 2m\}$, and let $\mathfrak{A}(V', Q')$ be its universal Clifford algebra. Define a linear isomorphism $\delta_+ : V' \to I_+$ from V' into I_+ by the linear extension of $\delta(e_j) = e_j e_+$. Then

$$ \delta_+(e_j)^2 = (e_j e_+)(e_j e_+) = e_j^2 e_+^2 = -Q(e_j) e_+ , $$

and δ_+ lifts to an algebra homomorphism from $\mathfrak{A}(V', Q')$ into I_+. Since V' is even-dimensional, this homomorphism will be an isomorphism, as the second proof of (3.2)(i) shows; in particular, $\dim I_+ \geq 2^{2m}$. Similarly, defining $\delta_- : V' \to I_-$ in an analogous way, we obtain $\dim I_- \geq 2^{2m}$. It follows that

$$ \dim I_+ = \dim I_- = 2^{2m} , $$

and that each is isomorphic to $C(V', Q') \cong \mathbb{C}^{2^m \times 2^m}$. Hence

$$ \mathbf{A} = I_+ \oplus I_- \cong \mathbb{C}^{2^m \times 2^m} \oplus \mathbb{C}^{2^m \times 2^m} , $$

completing the proof of theorem (3.2)(ii) once again. ∎

We conclude this section with a characterization of the structure of real Clifford algebras. Suppose that (V, Q) is a non-degenerate quadratic space over $\mathbb{R}$, which is isomorphic to $\mathbb{R}^{p,q}$ with $p + q = n$. Now $\mathfrak{A}(V, Q)$ is isomorphic to a subalgebra of $\mathfrak{C}_n$ (considered as an algebra over $\mathbb{R}$). To see this, let $\{e_1, e_2, \dots, e_n\}$ be a normalized basis for $(\mathbb{C}^n, Q_n)$, so that $\{e_\alpha : \alpha \subseteq \{1, 2, \dots, n\}\}$ spans $\mathfrak{C}_n$ as a complex vector space. Let $\{f_1, \dots, f_n\}$ be a normalized basis for (V, Q), ordered so that

$$ Q(f_1) = \cdots = Q(f_p) = -1 , \qquad Q(f_{p+1}) = \cdots = Q(f_{p+q}) = 1 . $$

Then the map $\mu : V \to \mathfrak{C}_n$, given by

$$ \mu(f_j) = i e_j , \quad 1 \leq j \leq p , \qquad \mu(f_{p+j}) = e_{p+j} , \quad 1 \leq j \leq q $$

and extended by linearity, satisfies $\mu(v)^2 = -Q(v)$ for all $v \in V$. By universality, μ extends to an injection β of $\mathfrak{A}(V,Q)$ into $\mathfrak{C}_n$, which is a homomorphism of *real* algebras. Thus $\mathfrak{A}(V,Q)$ is isomorphic to a 2^n-dimensional *real* subalgebra of $\mathfrak{C}_n$. The following result is now immediate from theorem (3.2).

(3.16) Corollary.
Let $\mathfrak{A}(V,Q)$ be the universal Clifford algebra for a non-degenerate real quadratic space (V,Q). Then

(i) $\mathfrak{A}(V,Q)$ is isomorphic to a real subalgebra of the matrix algebra $\mathbb{C}^{2^m \times 2^m}$ when $\dim V = 2m$,

(ii) $\mathfrak{A}(V,Q)$ is isormorphic to a real subalgebra of the direct sum $\mathbb{C}^{2^m \times 2^m} \oplus \mathbb{C}^{2^m \times 2^m}$ when $\dim V = 2m+1$.

As an easy consequence of our work in this section, we can characterize all possible Clifford algebras for (V,Q) in terms of dimension:

(3.17) Theorem.
Let A be a Clifford algebra for a non-degenerate real quadratic space (V,Q). Suppose further that (V,Q) is isomorphic to $\mathbb{R}^{p,q}$, and let $\{e_1, e_2, \ldots, e_{p+q}\}$ be a normalized basis for (V,Q) such that $e_j^2 = +1$ for $1 \le j \le p$, $e_{j+p}^2 = -1$ for $1 \le j \le q$. Then exactly one of the following is true:

(i) A has maximum dimension;

(ii) $\dim A = (1/2)2^{p+q}$, $p+q$ is odd, $p - q - 1 \equiv 0 \pmod 4$, and $e_1 e_2 \cdots e_{p+q} \in \mathbb{R}$.

Proof. Assume that $\dim A < 2^{p+q}$, i.e., (i) is false. By corollary (3.6), it follows that $p + q$ is odd and $e_V = e_1 e_2 \cdots e_{p+q} \in \mathbb{R}$. In particular, the set $\{e_\alpha : \alpha \subseteq N_V\}$ is linearly dependent. In fact, whenever $\alpha \subseteq N_V$ with $|\alpha|$ odd, e_α is a real multiple of $e_{N_V \setminus \gamma} \in A^+$. Thus $\{e_\alpha : \alpha \subseteq N_V, |\alpha| \text{ even}\}$ spans A, and is, in fact a basis of A. To see this, note that $\sum_{|\beta| \text{ even}} \lambda_\beta e_\beta = 0$ implies that, for each $\alpha \subseteq N_V$ with $|\alpha|$ even,

$$0 = e_\alpha^{-1} \left(\sum_{|\beta| \text{ even}} \lambda_\beta e_\beta \right) = \lambda_\alpha e_\phi + \sum_{\substack{|\beta| \text{ even} \\ \alpha \ne \beta}} \lambda_\beta e_\alpha^{-1} e_\beta$$

from which it follows that $\lambda_\alpha = 0$ by (3.5). Thus $\{e_\alpha : \alpha \subseteq N_V, |\alpha| \text{ even}\}$ is linearly independent, so $\dim A = (1/2)2^{p+q}$. Setting $p + q = 2m + 1$ and using (2.25), we obtain

$$(3.18) \qquad e_V^2 = (-1)^{m(2m+1)} e_V e_V^* = (-1)^{2m^2 + m + q} .$$

Since $e_V \in \mathbb{R}$, it follows that $e_V^2 = 1$, so $m + q$ must be even. Thus $2m - 2q \equiv 0 \pmod{4}$. Since $2m - 2q = p - q - 1$, we obtain (ii). ∎

Finally, we can characterize the universal Clifford algebra $\mathfrak{A}(V, Q)$ in terms of its ideal structure.

(3.19) Theorem.

Let (V, Q) be a non-degenerate real quadratic space as in theorem (3.17). Then the universal Clifford algebra $\mathfrak{A}(V, Q)$ is simple, unless $p + q$ is odd and $p - q - 1 \equiv 0 \pmod{4}$. In this case, $\mathfrak{A}(V, Q)$ has exactly two non-trivial proper ideals, I_+ and I_-, and $\mathfrak{A}(V, Q) = I_+ \oplus I_-$.

Proof. By lemma (3.13), $\mathfrak{A}(V, Q)$ is simple when $p + q$ is even; if $p + q = 2m + 1$ is odd and I is a non-trivial ideal of A, then I contains an element of the form $1 + \lambda_V e_V$ for some $\lambda_V \in \mathbb{R}$. Using (3.18), we deduce that

$$(1 + \lambda_V e_V)(1 - \lambda_V e_V) = 1 - \lambda_V^2 e_V^2 = 1 - \lambda_V^2 (-1)^{m+q} .$$

When $m + q$ is odd, every element of the form $1 + \lambda_V e_V$ is invertible, so $\mathfrak{A}(V, Q)$ must be simple. When $m + q$ is even, i.e., $p - q - 1 \equiv 0 \pmod{4}$, $1 + \lambda_V e_V$ is invertible unless $\lambda_V = \pm 1$. As in the second proof of (3.2)(ii), we see that

$$e_+ = \frac{1}{2}(1 + e_V) , \qquad e_- = \frac{1}{2}(1 - e_V)$$

are central idempotents in $\mathfrak{A}(V, Q)$ such that $e_+ e_- = 0$; letting I_+ (resp. I_-) be the ideal generated by e_+ (resp. e_-), we see as before that $\mathfrak{A}(V, Q) = I_+ \oplus I_-$. Clearly, any *other* non-trivial ideal of $\mathfrak{A}(V, Q)$ would necessarily contain an *invertible* element of the form $1 + \lambda_V e_V$, so I_+ and I_- are the only proper non-trivial ideals. ∎

4 Orthogonal transformations

An $\mathbb{F}$-linear transformation, $\mathbb{F} = \mathbb{R}$ or $\mathbb{C}$, between quadratic spaces (V, Q) and (V', Q') is said to be an *orthogonal transformation* when it preserves the quadratic forms, i.e., when $S : V \to V'$ is a linear transformation such that

(4.1) $Q'(Sv) = Q(v) , \qquad v \in V ;$

in practice it is often more convenient to use (4.1) in its equivalent form

(4.2) $B'(Su, Sv) = B(u, v) , \qquad u, v \in V ,$

where B, B' are the bilinear forms associated with Q and Q' respectively. The set of all orthogonal transformations $S : V \to V'$ will be denoted by $O(V, V')$; when $(V, Q) = (V', Q')$ the notation $O(V, V')$ will be replaced by $O(V)$ or by $O(V, Q)$ if there is a need to emphasize the quadratic form Q on V. Clearly the composition of T in $O(V, V')$ and S in $O(V', V'')$ is an orthogonal transformation $ST : V \to V''$. Throughout the remainder of this section *all quadratic spaces are assumed to be non-degenerate.* One effect of this assumption is that each S in $O(V, V')$ is 1–1 since equality (4.2) ensures that $Sv = 0$ only when $v \in \mathrm{Rad}(V, Q)$; in particular, if S maps V *onto* V', then S is an *orthogonal isomorphism* and S^{-1} belongs to $O(V', V)$.

(4.3) Examples.
    (i) Let (V, Q) be a real quadratic space and $\{e_j : j = 1, \ldots, p + q\}$ a normalized basis for V indexed so that

$$Q(v) = Q\left(\sum_j \lambda_j e_j\right) = -(\lambda_1^2 + \cdots + \lambda_p^2) + (\lambda_{p+1}^2 + \cdots + \lambda_{p+q}^2) \ .$$

Then $S : \sum_j \lambda_j e_j \to (\lambda_1, \ldots, \lambda_{p+q})$ is an orthogonal isomorphism from (V, Q) onto $(\mathbb{R}^{p,q}, Q_{p,q})$. Similarly, when (V, Q) is a complex quadratic space of dimension n, there is a corresponding orthogonal isomorphism from (V, Q) onto $(\mathbb{C}^n, Q_n)$.
    (ii) Fix v in (V, Q) with $Q(v) \neq 0$ and define $S_v : V \to V$ by

$$S_v(u) = u - \frac{2B(u, v)}{Q(v)}\, v \ , \qquad u \in V \ .$$

Since

$$B\big(S_v(u),\ S_v(w)\big) = B(u, w) \ , \qquad u, w \in V \ ,$$

each such S_v belongs to $O(V, Q)$. Geometrically, $S_v(u)$ is a *reflection* of u in the subspace $(\mathbb{F}v)^\perp$ of V which is B-orthogonal to $\mathbb{F}v$. Indeed,

$$u = \frac{B(u, v)}{Q(v)}\, v + \left(u - \frac{B(u, v)}{Q(v)}\, v\right) \ , \qquad u \in V \ ,$$

is the decomposition of u with respect to the B-orthogonal direct sum $V = (\mathbb{F}v) \oplus (\mathbb{F}v)^\perp$, and

$$S_v(u) = -\frac{B(u, v)}{Q(v)}\, v + \left(u - \frac{B(u, v)}{Q(v)}\, v\right) \ .$$

Clearly, $(S_v \circ S_v)(u) = u$, i.e., $S_v^2 = I$ on V.
    Under composition $O(V, Q)$ is a group which we shall call the *orthogonal group* associated with (V, Q). Example (4.3)(ii) shows that $O(V, Q)$ contains all reflections $\{S_v : Q(v) \neq 0\}$, hence all finite products of such reflections. In fact, more is true.

(4.4) Theorem.

Each S in $O(V, Q)$, $S \neq I$, can be written as the product of at most $2n$ reflections, $n = \dim V$.

The proof of theorem (4.4) is by induction on the dimension of V. When $n = 1$ the group $O(V, Q)$ reduces to $\{\pm I\}$ and $-I$ can be written as the reflection S_v for any non-zero v in V. So suppose that theorem (4.4) has been established for all quadratic spaces of dimension less than n. To establish the result for an n-dimensional space the following lemma is needed.

(4.5) Lemma.

Let u, v be elements of (V, Q) with $Q(u) = Q(v) \neq 0$. Then u can be mapped to v by the product of at most two reflections.

Proof. Since

$$B(u - v, \ u + v) = Q(u) - Q(v) = 0 \ ,$$

the element $u - v$ is B-orthogonal to $u + v$. In this case

$$Q(u + v) + Q(u - v) = 2\{Q(u) + Q(v)\} \ ,$$

and so at least one of $Q(u+v)$, $Q(u-v)$ is non-zero. Now, if $Q(u-v) \neq 0$,

$$S_{u-v}(u) = S_{u-v}\left(\tfrac{1}{2}(u - v) + \tfrac{1}{2}(u + v)\right) = -\tfrac{1}{2}(u - v) + \tfrac{1}{2}(u + v) = v \ .$$

On the other hand, if $Q(u + v) \neq 0$,

$$S_{u+v}(u) = S_{u+v}\left(\tfrac{1}{2}(u - v) + \tfrac{1}{2}(u + v)\right) = \tfrac{1}{2}(u - v) - \tfrac{1}{2}(u + v) = -v \ ,$$

so that $S_v \circ S_{u+v} : u \to v$. This proves the lemma. ∎

To complete the proof of theorem (4.4) let $\{e_j\}_{j=1}^n$ be a B-orthogonal basis for V and (V', Q') the quadratic space obtained by restricting Q to the subspace of V spanned by $\{e_j : 1 \leq j < n\}$. Now, by the preceding lemma, for each T in $O(V, Q)$ there exists an S in $O(V, Q)$ which is the product of at most two reflections and such that $(S \circ T)e_n = e_n$, because $Q(Te_n) = Q(e_n) \neq 0$. Consequently, $S \circ T$ belongs to $O(V', Q')$. If $S \circ T \neq I$, the induction hypothesis ensures that $T = S^{-1} \circ T_1$ where T_1 is the product of at most $2(n - 1)$ reflections, while if $S \circ T = I$ already $T = S^{-1}$. Since any reflection is self-inverse, i.e., $S_v = S_v^{-1}$, it follows that T must be the product of at most $2n$ reflections, completing the proof. ∎

The group $O(V, Q)$ and various groups derived from it are fundamental to all that follows because they will arise both geometrically as transformation groups on the particular differentiable manifolds to be considered and analytically as 'symmetry groups' for important differential operators defined on such manifolds.

For the remainder of the section (V, Q) will be assumed to be a *real* quadratic space. The group $O(V, Q)$ often arises then as a subgroup of two important groups of transformations, the rigid motions or isometries and the conformal or Möbius transformations.

(4.6) Definition.
A mapping from V onto V is said to be a rigid motion (or isometry) when it preserves the distance between pairs of points, i.e., when

$$Q(Tu - Tv) = Q(u - v) , \qquad u, v \in V .$$

The set of all rigid motions will be denoted by $E(V, Q)$; clearly it is a semigroup under composition, containing the composition

(4.7) $$T_{(A,w)} : x \longrightarrow Ax + w , \qquad x \in V ,$$

of *rotation* by A in $O(V, Q)$ and *translation* by $w \in V$. Consequently, $E(V, Q)$ contains the reflection $T : x \to S_v x + w$ of V in the hyper-plane $(\mathbb{R}v)^{\perp} + w$, $Q(v) \neq 0$, hence all finite products of such reflections. Just as each S in $O(V, Q)$ is the finite product of the reflections S_v, $Q(v) \neq 0$, so $E(V, Q)$ can be characterized by the following.

(4.8) Theorem.
Each T in $E(V, Q)$ is a composition of finitely many reflections $x \to S_v x + w$ in hyper-planes.

Since the composition of finitely many such reflections is of the form $T_{(A,w)}$ for some A in $O(V, Q)$ and w in V, we deduce the following.

(4.9) Corollary.
A mapping T from V onto V is a rigid motion if and only if it is a composition $T_{(A,w)}$ of rotation by A in $O(V, Q)$ and translation by w in V. In particular, $E(V, Q)$ is a group isomorphic to the semi-direct product

$$O(V, Q) \circledS V = \{(A, w) : A \in O(V, Q) , \ w \in V\}$$

in which group multiplication is

$$(A_1, w_1)(A_2, w_2) = (A_1 A_2, A_1 w_2 + w_1) .$$

Proof of theorem (4.8). Each T in $E(V, Q)$ can be written as $Tv = Sv + T(0)$, $v \in V$, where S is an isometry such that $S(0) = 0$, $Q(Sv) = Q(v)$. Consequently, if S is known in advance to be linear, then S will be orthogonal and (4.4) can be applied, completing the proof; since linearity is not a part of (4.7), however, a separate proof is needed. Now, for any u, v in V,

$$B(Su, Sv) = \tfrac{1}{2}\{Q(Su) + Q(Sv) - Q(Su - Sv)\}$$
$$= \tfrac{1}{2}\{Q(u) + Q(v) - Q(u - v)\} = B(u, v) ,$$

so S also preserves inner products. This ensures that the image $\{Se_j\}$ of any normalized basis $\{e_j\}$ of V is again a normalized basis. Consequently, for every $v = \sum_j v_j e_j$ in V, there exist $x_j \in \mathbf{R}$ such that

$$Sv = \sum_j x_j (Se_j) \quad , \quad x_j = B(Sv, Se_j) = B(v, e_j) = v_j .$$

Hence

$$Sv = S\left(\sum_j v_j e_j\right) = \sum_j v_j (Se_j)$$

establishing linearity.

Proof of corollary (4.9). The identification of each T in $E(V, Q)$ with an isometry $T_{(A,w)}$ was part of the previous proof. But $T_{(A,w)}$ is invertible within $E(V, Q)$; in fact,

$$(T_{(A,w)})^{-1} = T_{(B,z)} , \quad B = A^{-1} , \quad z = -A^{-1}w .$$

Thus $E(V, Q)$ is a group. On the other hand,

$$T_{(A_1,w_1)} \circ T_{(A_2,w_2)} : v \longrightarrow (A_1 A_2)v + (A_1 w_2 + w_1) ,$$

so $T_{(A,w)} \to (A, w)$ defines an isomorphism from $E(V, Q)$ onto $O(V, Q) \circledS V$, completing the proof. ∎

The group of rigid motions $E(n)$ associated with Euclidean space $\mathbf{R}^n$ is known as the *Euclidean motion group*, not surprisingly; its counterpart $E(3, 1)$ associated with Minkowski space $\mathbf{R}^{3,1}$ is often called the *Poincaré group*, although this term is sometimes reserved for the two-fold covering group of $E(3, 1)$. More generally, $E(p, q)$ will denote the group of rigid motions on $\mathbf{R}^{p,q}$.

Finally, we come to the conformal group $C(V, Q)$ for (V, Q), which contains the motion group $E(V, Q)$ and the orthogonal group $O(V, Q)$ as subgroups. For simplicity here, we shall assume that (V, Q) is isomorphic to Euclidean space $\mathbf{R}^n$. Now a smooth mapping $\phi : U_0 \to U_1$ between open sets in V is said to be *conformal* on U_0 when for each x in U_0 there exist $\lambda > 0$ and $A \in O(V, Q)$ so that $\phi'(x) = \lambda A$; thus $\phi : U_0 \to U_1$ is conformal on U_0 when there is a real-valued function ψ on U_0 such that

(4.10) $|\phi'(x)v|^2 = e^{\psi(x)}|v|^2 , \qquad v \in V , \ x \in U_0 .$

By the chain rule, the composition of conformal mappings is again conformal whenever the composition is defined. For example, each rigid motion $T_{(A,w)} : x \to Ax + w$ obviously has derivative $T'_{(A,w)} = A$, so $E(V, Q)$ is a group of conformal transformations on V. Every dilation $a_\lambda : x \to \lambda x$, $\lambda > 0$, of V also will be conformal on V, since $a'_\lambda = \lambda I$. Thus along with the group of rigid motions, the group $\{a_\lambda : \lambda > 0\}$ generates the important group $P(V, Q)$ of similarities

$$(4.11) \qquad T_{(A,w,\lambda)} : x \to \lambda Ax + w , \qquad x \in V ,$$

of V, each of which is conformal on all of V. On the other hand, inversion

$$(4.12) \qquad J : x \longrightarrow \frac{x}{|x|^2}$$

is well-defined for non-zero x in V, and has derivative

$$J'(x) = \frac{1}{|x|^2} \left(I - \frac{2x \otimes x}{|x|^2} \right) = \frac{1}{|x|^2} S_x .$$

Consequently, J is conformal on $V \setminus \{0\}$. With all this in mind, we make the following definition.

(4.13) Definition.
   The conformal group $C(V, Q)$ of (V, Q) is the group of transformations of V generated by the group $P(V, Q)$ together with the inversion J, i.e., by translations, rotations, dilations and inversion.

   A *conformal sphere* in V is any set

$$(4.14) \qquad \Sigma_w = \left\{ x \in V : \alpha |x|^2 - 2x \cdot v + \beta = 0 \right\}$$

with

$$(4.15) \qquad w = (\alpha, v, \beta) \in \mathbb{R} \oplus V \oplus \mathbb{R} , \qquad w \neq 0 .$$

The notion of conformal sphere includes spheres in the usual sense (the case $\alpha \neq 0$) as well as hyper-planes (the case $\alpha = 0$). By *reflection* (or *inversion*) in a conformal sphere Σ_w, we mean the function ϕ defined by

$$\phi(x) = v + \frac{|v|^2 - \alpha\beta}{|v - x|^2} (v - x)$$

when Σ_w is a sphere, and

$$\phi(x) = S_v(x) + \left(\frac{\beta}{2|v|^2} \right) v$$

when Σ_w is a hyper-plane (i.e., $w = (0, v, \beta)$). These are not defined on all of V, but they are conformal where they are defined. We then define a *Möbius transformation* on (V, Q) to be any finite composition of reflections in conformal spheres. Since any such reflection is self-inverse, the set of all Möbius transformations is a subgroup, called the *Möbius group*, of $C(V,Q)$. In view of theorems (4.4) and (4.8), the Möbius group contains $O(V,Q)$ and $E(V,Q)$. In fact, there is a characterization of the connected component of the Möbius group which provides the appropriate analogue to these results.

(4.16) Theorem.
The connected component of the Möbius group coincides with the set of sense-preserving conformal transformations, i.e., with the subgroup of $C(V,Q)$ generated by translations, proper rotations, dilations, and inversion.

The proof of this result is somewhat beyond the scope of the present work. We do remark, however, that there are analogous results in the case where (V,Q) is an arbitrary non-degenerate quadratic space.

5 Transformers, Clifford groups

Using the multiplicative structure on a Clifford algebra we shall construct covering groups for various of the orthogonal groups introduced earlier. Throughout this section (V,Q) will be an arbitrary non-degenerate quadratic space and $\mathfrak{A} = \mathfrak{A}(V,Q)$ its universal Clifford algebra, no restrictions being placed on $\mathbb{F}$ unless specifically indicated.

Now, because of the Clifford condition $v^2 = -Q(v)$ on the elements of V, v will be invertible in $\mathfrak{A}$ if and only if $Q(v) \neq 0$. Consequently, the set

(5.1) $\{ v_1 \cdots v_k : v_j \in V , \; Q(v_j) \neq 0 \}$

of all finite products of such elements is a subgroup of $GL(\mathfrak{A})$. It is a subgroup of the *Clifford group* $\Gamma(V,Q)$ in $\mathfrak{A}$ to be associated with (V,Q). On the other hand, the orthogonal group $O(V,Q)$ was identified in theorem (4.4) as the subgroup

(5.2) $O(V,Q) = \{ S_{v_1} \cdots S_{v_k} : v_j \in V , \; Q(v_j) \neq 0 \}$

of $GL(V)$ of all finite products of reflections. One of the principal goals of this section will be to extend the mapping $v \to S_v$ to a covering homomorphism from (5.1) onto $O(V,Q)$, but the route will be far from direct.

The simple case of $\mathbb{R}^{2,0}$ illustrates the basic ideas involved. Its universal Clifford algebra $\mathfrak{A}_{2,0}$ can be identified with $\mathbb{R}^{2\times 2}$ under the embedding

(5.3) $\nu : x = (x_1, x_2) \longrightarrow x_1\sigma_1 + x_2\sigma_3 = X ,$

where X is the traceless symmetric matrix

(5.4)
$$X = \begin{bmatrix} x_1 & x_2 \\ x_2 & -x_1 \end{bmatrix}$$

in $\mathbb{R}^{2\times 2}$. Thus

(5.5) $\qquad Q_{2,0}(x) = \det(\nu(x)) = \det(X), \qquad x \in \mathbb{R}^{2,0}$.

This embedding extends to an embedding

(5.6) $\qquad \nu : z = (x_1, x_2, y) \longrightarrow Z = yI + X = \begin{bmatrix} y + x_1 & x_2 \\ x_2 & y - x_1 \end{bmatrix}$

of $\mathbb{R}^{2,1}$ as the set of all symmetric matrices in $\mathbb{R}^{2\times 2}$. Again

(5.7) $\qquad Q_{2,1}(z) = \det(\nu(x)) = \det Z, \qquad z \in \mathbb{R}^{2,1}$,

although the Clifford condition is not now satisfied. In both cases, however, the condition $Q(v) \neq 0$ is equivalent to $\nu(v) \in GL(2, \mathbb{R})$.

(5.8) Theorem.
 When $\mathfrak{A}_{2,0}$ is identified with $\mathbb{R}^{2\times 2}$ by (5.3), $\mathfrak{A}_{2,0}$ coincides with the set

$$\Lambda = \{Z_1 \cdots Z_k : Z_j = \nu(z_j), \ z_j \in \mathbb{R}^{2,1}\}$$

of all finite products of elements from the image of $\mathbb{R}^{2,1}$ in $\mathfrak{A}_{2,0}$, while $GL(2, \mathbb{R})$ coincides with the set

$$\Gamma = \{Z_1 \cdots Z_k : Z_j = \nu(z_j), \ Q_{2,1}(z_j) \neq 0\}$$

of invertible such products.

In the terminology to be introduced shortly, $GL(2, \mathbb{R})$ is the *Clifford group* of $\mathfrak{A}_{2,0}$, and $\mathfrak{A}_{2,0}$ itself is the *Clifford semigroup* of $\mathfrak{A}_{2,0}$.

Proof of theorem (5.8). It is more convenient to deal with the invertible case first. Obviously, $GL(2, \mathbb{R}) \supseteq \Gamma$. Conversely, by the polar decomposition of matrices, each A in $GL(2, \mathbb{R})$ can be written uniquely as $A = BZ$ with B in $O(2, \mathbb{R})$ and Z a positive-definite symmetric matrix in $GL(2, \mathbb{R})$. But all such Z are already in Γ, so it is enough to show that $\Gamma \supseteq O(2, \mathbb{R})$. It is at this point that the subgroup (5.1) of Γ determined by $\mathbb{R}^{2,0}$ enters. For if

$$k_\theta = \begin{bmatrix} \cos\theta & -\sin\theta \\ \sin\theta & \cos\theta \end{bmatrix}, \qquad 0 \leq \theta < 2\pi,$$

then

(5.9) $\qquad SO(2, \mathbb{R}) = \{k_\theta : 0 \leq \theta < 2\pi\}$

and

(5.10) $O(2, \mathbb{R}) = SO(2, \mathbb{R}) \; \cup \; \begin{bmatrix} 0 & 1 \\ 1 & 0 \end{bmatrix} SO(2, \mathbb{R}) \; .$

On the other hand,

(5.11) $k_\theta = \begin{bmatrix} 0 & 1 \\ 1 & 0 \end{bmatrix} \begin{bmatrix} \sin\theta & \cos\theta \\ \cos\theta & -\sin\theta \end{bmatrix} .$

Consequently, $SO(2, \mathbb{R})$ consists of *all even* finite products in $\mathfrak{A}_{2,0}$ of elements X, $\det X = -1$, from the image of $\mathbb{R}^{2,0}$, while $O(2, \mathbb{R})$ consists of *all* finite products of such elements. Hence $\Gamma = GL(2, \mathbb{R})$.

To complete the proof we have to show that every A in $\mathbb{R}^{2\times 2}$ is a finite product of symmetric matrices even when $\det A = 0$. So suppose

$$A = \begin{bmatrix} \alpha & \beta \\ \gamma & \delta \end{bmatrix} , \qquad \alpha\delta = \gamma\beta \; .$$

Then, if $\alpha \neq 0$,

$$A = \begin{bmatrix} \alpha & 0 \\ 0 & \gamma \end{bmatrix} \begin{bmatrix} 1 & 1 \\ 1 & 1 \end{bmatrix} \begin{bmatrix} 1 & 0 \\ 0 & 0 \end{bmatrix} \begin{bmatrix} 1 & \beta/\alpha \\ \beta/\alpha & 1 \end{bmatrix} .$$

On the other hand, if $\alpha = 0$, at least one of γ, β must also be zero. But then

$$\begin{bmatrix} 0 & 0 \\ \gamma & \delta \end{bmatrix} = \begin{bmatrix} 0 & 0 \\ 0 & 1 \end{bmatrix} \begin{bmatrix} \delta & \gamma \\ \gamma & \delta \end{bmatrix} , \qquad \begin{bmatrix} 0 & \beta \\ 0 & \delta \end{bmatrix} = \begin{bmatrix} \delta & \beta \\ \beta & \delta \end{bmatrix} \begin{bmatrix} 0 & 0 \\ 0 & 1 \end{bmatrix} .$$

This completes the proof.   ∎

Now let $\sigma(A)$ be the modified similarity transformation

(5.12) $\sigma(A) : Z \longrightarrow (1/\det A) A Z A^t \; , \qquad A \in GL(2, \mathbb{R}),$

on $\mathbb{R}^{2\times 2}$, where A^t is the transpose of A. The restriction of $\sigma(A)$ to the space of symmetric matrices in $\mathbb{R}^{2\times 2}$ will determine a linear transformation of $\mathbb{R}^{2,1}$; in fact, this will be an orthogonal transformation of $\mathbb{R}^{2,1}$ because $\det(\sigma(A)Z) = \det Z$. Thus $A \to \sigma(A)$ is a homomorphism from the Clifford group Γ into the orthogonal group $O(2,1)$ associated with $\mathbb{R}^{2,1}$. What is more, by restricting $\sigma(A)$ to traceless, symmetric matrices, we see that $\sigma(A)$ is an orthogonal transformation of $\mathbb{R}^{2,0}$ when A is in the subgroup

(5.13) $\{ X_1 \cdots X_k : X_j = \nu(x_j) \; , \; Q_{2,0}(x_j) \neq 0 \}$

of $GL(2, \mathbb{R})$ of all finite products in $\mathfrak{A}_{2,0}$ of elements from the image of $\mathbb{R}^{2,0}$ (see (5.10), (5.11)). Thus $A \to \sigma(A)$ is a homomorphism from (5.13) into $O(2, \mathbb{R})$. It can (and will later) be shown that σ actually is a covering homomorphism from Γ onto $SO(2,1)$ and from the subgroup (5.13) of Γ onto $O(2, \mathbb{R})$.

In order to proceed with the general construction, we introduce a general substitute for the determinant function on $\mathbb{R}^{2\times 2}$. This is the *norm* function

(5.14) $$\Delta : \mathfrak{A} \longrightarrow \mathfrak{A} , \qquad \Delta(x) = \bar{x}x .$$

Since $(\lambda x)^- = \lambda\bar{x}$ for all $\lambda \in \mathbb{F}$ and $x \in \mathfrak{A}$, clearly $\Delta(\lambda x) = \lambda^2\Delta(x)$; but, in general, Δ is not a quadratic form on $\mathfrak{A}$ since Δ is not $\mathbb{F}$-valued on all of $\mathfrak{A}$ when $\dim V > 2$. For, if $\dim V \geq 3$ and $x = 1 + e_1e_2e_3$, then $x = \bar{x}$ and

$$\Delta(x) = x^2 = 1 + 2e_1e_2e_3 + (e_1e_2e_3)^2 = \lambda + 2e_1e_2e_3$$

for some $\lambda \in \mathbb{F}$; however, $e_1e_2e_3$ cannot be in $\mathbb{F}$ because both 1 and $e_1e_2e_3$ belong to the basis of all reduced products for $\mathfrak{A}(V,Q)$. Hence $\Delta(x) \notin \mathbb{F}$. Nonetheless, on the subset

(5.15) $$\mathfrak{N} = \mathfrak{N}(V,Q) = \{ x \in \mathfrak{A}(V,Q) : \Delta(x) \in \mathbb{F} \}$$

on which Δ is $\mathbb{F}$-valued the basic properties of the norm function mirror those of the determinant.

(5.16) Theorem.
(i) The set $\mathfrak{N}$ is a multiplicative semigroup in $\mathfrak{A}$ which is closed under scalar multiplication as well as under the principal automorphism and conjugation; furthermore,

$$\Delta(xy) = \Delta(x)\Delta(y) , \quad \Delta(x') = \Delta(\bar{x}) = \Delta(x^*) = \Delta(x) , \qquad x, y \in \mathfrak{N} .$$

(ii) An x in $\mathfrak{N}$ is invertible if and only if $\Delta(x) \neq 0$, and then

(5.17) $$x^{-1} = (1/\Delta(x))\bar{x} , \quad \Delta(x^{-1}) = 1/\Delta(x) .$$

Clearly, $\mathfrak{N}(V,Q)$ always contains $\mathbb{F} \oplus V$ because

(5.18) $$\Delta(\lambda + v) = (\lambda - v)(\lambda + v) = \lambda^2 - v^2 = \lambda^2 + Q(v) .$$

Consequently, the Clifford semigroup

(5.19) $$\Lambda(V,Q) = \{w_1 \cdots w_k : w_j \in \mathbb{F} \oplus V\}$$

of all finite products in $\mathfrak{A}$ of elements from $\mathbb{F} \oplus V$ is a sub-semigroup of $\mathfrak{N}(V,Q)$. This Clifford semigroup contains every reduced product $e_{\alpha_1} \cdots e_{\alpha_k}$ in $\mathfrak{A}(V,Q)$, and in the case of $\mathbb{R}^{2,0}$ coincides with $\mathfrak{A}(V,Q)$; but in general, $\Lambda(V,Q) \neq \mathfrak{A}(V,Q)$ since Δ is not always $\mathbb{F}$-valued.

Proof of theorem (5.16). If $x, y \in \mathfrak{N}$, then

$$\Delta(xy) = \overline{xy}\, xy = \bar{y}(\bar{x}x)y = \Delta(x)\Delta(y)$$

since $\Delta(x), \Delta(y) \in \mathbb{F}$. If $\Delta(x) \neq 0$, then x is easily seen to be invertible and (5.17) is immediate; moreover, in this case, $\bar{x} = \Delta(x)x^{-1}$ so that

$$\Delta(\bar{x}) = \Delta(x)^2\Delta(x^{-1}) = \Delta(x) .$$

If, on the other hand, $\Delta(x) = 0$, then x is a zero divisor in $\mathfrak{A}$, hence not invertible; consequently $\bar{x}$ is not invertible, which forces $\Delta(\bar{x}) = \Delta(x) = 0$. Finally,

$$\Delta(x') = \overline{x'}\,x' = (\bar{x}x)' = \Delta(x)' = \Delta(x) \quad , \quad \Delta(x^*) = \Delta(\bar{x}') = \Delta(x) \; ,$$

whenever $x \in \mathfrak{N}$. ∎

From (5.16) and (5.19) we obtain immediately a multiplicative group, the so-called *Clifford group* $\Gamma(V, Q)$ of $\mathfrak{A}(V, Q)$, which is the general analogue of the group $GL(2, \mathbb{R})$ in $\mathfrak{A}_{2,0}$.

(5.20) Corollary.
 The Clifford group

$$\Gamma(V, Q) = \big\{ w_1 \cdots w_k : w_j \in \mathbb{F} \oplus V \; , \; \Delta(w_j) \neq 0 \big\}$$

of all invertible finite products in $\mathfrak{A}(V, Q)$ of elements from $\mathbb{F} \oplus V$ is a multiplicative group in $\mathfrak{A}$ which is closed under the principal automorphism and anti-automorphism as well as under conjugation.

To establish the connection with various orthogonal groups associated with (V, Q), we use a modified similarity transformation analogous to (5.12), regarding the quadratic space $(\mathbb{F} \oplus V, \Delta)$ as an extension of (V, Q), just as $\mathbb{R}^{2,1}$ was earlier regarded as an extension of $\mathbb{R}^{2,0}$ (see (5.18)).

(5.21) Definition.
 An element A in $\mathfrak{A}(V, Q)$ is said to be a transformer if to each w in $\mathbb{F} \oplus V$ there corresponds a z in $\mathbb{F} \oplus V$ so that $Aw = zA'$.

Obviously every element of $\mathbb{F}$ is a transformer since

$$(5.22) \qquad \lambda w = w\lambda = w\lambda' \; , \qquad w \in \mathbb{F} \oplus V \; , \; \lambda \in \mathbb{F} \; .$$

In studying further this notion it is worth singling out the properties of $(\mathbb{F} \oplus V, \Delta)$ that are needed:

   (5.23)(i) $V \subseteq \mathbb{F} \oplus V \subseteq \mathfrak{N}(V, Q)$,

   (5.23)(ii) $(\mathbb{F} \oplus V, \Delta)$ *is a non-degenerate quadratic space, which is closed under the principal automorphism,*

   (5.23)(iii) *on $(\mathbb{F} \oplus V, \Delta)$, the principal anti-automorphism is the identity, i.e., $w^* = w$ for $w \in \mathbb{F} \oplus V$.*

The first result is the principal step along the path to the characterization of the Clifford group.

(5.24) Theorem.
The set of all transformers is a sub-semigroup of $\mathfrak{N}(V, Q)$ which is closed under the principal automorphism.

Proof. By definition, to each transformer A and w in $\mathbb{F} \oplus V$ there corresponds z in $\mathbb{F} \oplus V$ so that $Aw' = zA'$. But then
$$A'w = (Aw')' = (zA')' = z'A .$$
Since $z' \in \mathbb{F} \oplus V$ (see (5.23)(iii)), it follows that A' is a transformer. Moreover, if A, B are transformers, then for each w in $\mathbb{F} \oplus V$ there exist η, ζ in $\mathbb{F} \oplus V$ such that
$$A(Bw) = A(\eta B') = (\zeta A')B' = \zeta (AB)' .$$
Hence, the set of all transformers is a multiplicative semigroup which is closed under the principal automorphism. Finally, to show that Δ is $\mathbb{F}$-valued on this set, we shall use the following lemma, which is of some independent interest.

(5.25) Lemma.
An x in $\mathfrak{A}$ satisfies $xv = vx'$ for all v in V if and only if $x \in \mathbb{F}$.

Proof of lemma (5.25). As noted already in (5.22), the equality $xv = vx'$, $v \in V$, holds for each x in $\mathbb{F}$. Conversely, let $\{e_j\}$ be a B-orthogonal basis for V and $\{e_\alpha\}$ the corresponding basis of reduced products for $\mathfrak{A}$. Then, by (3.4),
$$e_j e'_\alpha = (-1)^{|\alpha|} e_j e_\alpha = \begin{cases} e_\alpha e_j , & j \notin \alpha , \\ -e_\alpha e_j , & j \in \alpha ; \end{cases}$$
consequently, for any $x = \sum_\alpha \lambda_\alpha e_\alpha$ in $\mathfrak{A}$,
$$e_j x' = \left(\sum_{j \notin \alpha} \lambda_\alpha e_\alpha \right) e_j - \left(\sum_{j \in \alpha} \lambda_\alpha e_\alpha \right) e_j ,$$
while
$$x e_j = \left(\sum_{j \notin \alpha} \lambda_\alpha e_\alpha \right) e_j + \left(\sum_{j \in \alpha} \lambda_\alpha e_\alpha \right) e_j .$$
Thus $e_j x' = x e_j$ can hold only if $\lambda_\alpha = 0$ for those α containing j; hence $vx' = xv$, $v \in V$, can hold only when $x = \lambda_\emptyset$, i.e., $x \in \mathbb{F}$. ∎

To complete the proof of theorem (5.24), fix a transformer A and w in $\mathbb{F} \oplus V$. Then there is a z in $\mathbb{F} \oplus V$ so that
$$\Delta(A)w = \bar{A}(zA') = (\bar{A}z)A' = (zA')^* A'$$
$$= (Aw)^* A' = (wA^*)A' = w(\Delta(A))' .$$

using (5.23)(iii). Now, since $V \subseteq \mathbb{F} \oplus V$ (see (5.23)(i)), we have

$$\Delta(A)v = v\big(\Delta(A)\big)' , \qquad v \in V ;$$

thus lemma (5.25) ensures that $\Delta(A) \in \mathbb{F}$, completing the proof. ∎

The significance of the idea of transformer can now be seen. For if A is an invertible transformer, then by (5.24) A' also will be invertible, and the modified similarity transformation

(5.26) $\sigma(A) : w \to Aw(A')^{-1} , \qquad w \in \mathbb{F} \oplus V ,$

will define a linear transformation on $\mathbb{F} \oplus V$. This defines a homomorphism which by (5.25) has kernel

(5.27) $\ker \sigma = \{\lambda \in \mathbb{F} : \lambda \neq 0\} .$

On the other hand, since $A^* A' = (\bar{A} A)' = \Delta(A)$, σ can also be written as

(5.28) $\sigma(A) : w \longrightarrow \dfrac{1}{\Delta(A)} AwA^* , \qquad w \in \mathbb{F} \oplus V ,$

which is exactly how (5.12) appears because of the form the principal anti-automorphism takes when $\mathfrak{A}_{2,0}$ is realized as $\mathbb{R}^{2 \times 2}$ (see (2.30)). Furthermore, each such $\sigma(A)$ is orthogonal on $\mathbb{F} \oplus V$ since (5.23)(iii) ensures that $\bar{w} = (w^*)' = w'$ and hence that

$$\Delta\big(\sigma(A)w\big) = \big(\sigma(A)w\big)' \sigma(A)w = \big(A'w'A^{-1}\big)\big(Aw(A')^{-1}\big)$$
$$= A'w'w(A')^{-1} = A'\Delta(w)(A')^{-1} = \Delta(w)$$

(we have used the fact that $\Delta(w) \in \mathbb{F}$, i.e., (5.23)(i), in the last step). Consequently, $\sigma : A \to \sigma(A)$ actually is a homomorphism from the group of invertible transformers into $O(\mathbb{F} \oplus V, \Delta)$. This homomorphism leads directly to the most important result of this section.

(5.29) Theorem.

The Clifford group $\Gamma(V, Q)$ coincides with the group of invertible transformers in $\mathfrak{A}(V, Q)$; in addition, $\sigma : A \to \sigma(A)$ defines a covering homomorphism

$$\sigma : \Gamma(V, Q) \longrightarrow SO(\mathbb{F} \oplus V, \Delta)$$

whose kernel is $\{\lambda \in \mathbb{F} : \lambda \neq 0\}$.

Proof. The principal step in the proof is to relate each $w \in \mathbb{F} \oplus V$, $\Delta(w) \neq 0$, as a transformer $\sigma(w)$ on $\mathbb{F} \oplus V$ with the reflection S_w of $\mathbb{F} \oplus V$ determined by w. Now the bilinear form on $\mathbb{F} \oplus V$ associated to Δ is given by

(5.30) $B(w, z) = \frac{1}{2}(w\bar{z} + z\bar{w}) , \qquad w, z \in \mathbb{F} \oplus V .$

Consequently,

$$S_w : z \longrightarrow z - \left(\frac{2B(w,z)}{\Delta(w)}\right)w \ , \qquad z \in \mathbb{F} \oplus V \ ,$$

whenever $\Delta(w) \neq 0$. But, by (5.23)(iii),

$$\bar{w} = w' \ , \qquad ww' = w'w = \Delta(w) \ ,$$

and so

$$wz = -z'w' + 2B(w',z) = -z'w' + 2B(w,z')$$

$$= \left(-z' + \frac{2B(w,z')}{\Delta(w)}w\right)w' = S_w(-z')w' \ ,$$

i.e., $\sigma(w)z = S_w(-z')$, whenever $\Delta(w) \neq 0$. Hence the Clifford group is a subgroup of the group of all invertible transformers on $\mathbb{F} \oplus V$. We shall prove that these two groups coincide by showing that they have the same range $SO(\mathbb{F} \oplus V, \Delta)$ with respect to the homomorphism σ, since in both cases the kernel of σ is just $\{\lambda \in \mathbb{F} : \lambda \neq 0\}$ (see (5.27)).

First we must recognize the linear transformation $z \rightarrow -z'$, $z \in W$. But, by (5.30),

$$B(z,1) = \tfrac{1}{2}(\bar{z} + z) = \tfrac{1}{2}(z' + z) \ , \qquad z \in W \ ;$$

consequently,

$$-z' = z - 2B(z,1)1 = S_1(z) \ , \qquad z \in W \ .$$

Hence, as the composition $S_w \circ S_1$ of two reflections, $\sigma(w)$ is in $SO(\mathbb{F} \oplus V, \Delta)$, not just in $O(\mathbb{F} \oplus V, \Delta)$. Clearly σ now extends to a homomorphism from $\Gamma(V,Q)$ into $SO(\mathbb{F} \oplus V, \Delta)$ such that

$$\sigma(w') = \sigma(w)^{-1} = (S_w \circ S_1)^{-1} = S_1 \circ S_w$$

since $w'w \ (= \Delta(w))$ is in the kernel of σ; in particular, $\sigma(wz') = S_w \circ S_z$. On $\Gamma(V,Q)$, therefore, σ is a homomorphism having range all of $SO(\mathbb{F} \oplus V, \Delta)$ and kernel $\{\lambda \in \mathbb{F} : \lambda \neq 0\}$, since $SO(\mathbb{F} \oplus V, \Delta)$ consists of all *even* finite products of reflections. In particular, every element of $SO(\mathbb{F} \oplus V, \Delta)$ arises as $\sigma(A)$ for some invertible transformer A. Consequently, if there exists an invertible transformer B such that $\sigma(B)$ is in $O(\mathbb{F} \oplus V, \Delta)$ but not in $SO(\mathbb{F} \oplus V, \Delta)$, then $\sigma(B) = S_1 \circ \sigma(A)$ for some such A. Hence we can assume from the outset that $S_1 = \sigma(B)$, i.e.,

$$S_1(z) = -z' = Bz(B')^{-1} \ , \qquad z \in W \ .$$

For z in V, this reduces to $Bz = zB'$; and so B is in the kernel of σ and $\sigma(B) = I \neq S_1$, contradicting the original choice of B. Thus σ also maps the group of all invertible transformers onto $SO(\mathbb{F} \oplus V, \Delta)$, completing the proof. ∎

6 Spin groups

One of the most fundamental aspects of the theory of Clifford algebras
is the construction of various 'spin subgroups' of the Clifford group of $\mathfrak{A}$.
By restricting the homomorphism σ to such a subgroup, we thus obtain
a covering of a subgroup of $SO(\mathbb{F} \oplus V, \Delta)$, and the idea is to achieve this
as economically as possible by ensuring that the kernel of the mapping is
$\{\pm 1\}$ rather than $\{\lambda \in \mathbb{F} : \lambda \neq 0\}$. Covering homomorphisms with this
property will be said to be *two-fold coverings*. These constructions are
made possible because all non-zero multiples λv of any v in V, $Q(v) \neq 0$,
determine the same reflection

$$S_{\lambda v} : x \longrightarrow x - \frac{2B(x,v)}{Q(v)} \, v$$

of V as v does. Consequently, every A in $O(V,Q)$ is the finite product
of reflections S_v with v in $\Sigma_{\mathbb{F}}(V)$ where

(6.1)(i) $\Sigma_{\mathbb{R}}(V) = \{v \in V : Q(v) = \pm 1\}$

when $\mathbb{F} = \mathbb{R}$, and

(6.1)(ii) $\Sigma_{\mathbb{C}}(V) = \{v \in V : Q(v) = 1\}$

when $\mathbb{F} = \mathbb{C}$; every A in $SO(V,Q)$ is an even product of such reflections.
Similarly, every A in $SO(\mathbb{F} \oplus V, \Delta)$ is an even finite product of reflections
S_w with w in the corresponding $\Sigma_{\mathbb{F}}(\mathbb{F} \oplus V)$. Throughout this section
(V,Q) will be any real or complex, non-degenerate quadratic space unless
specifically restricted.

The prototypical spin group is the set

(6.2) $\mathrm{Spin}(V,Q) = \{v_1 \cdots v_{2k} : v_j \in \Sigma_{\mathbb{F}}(V)\}$

of all even finite products in $\mathfrak{A}$ of elements from $\Sigma_{\mathbb{F}}(V)$. Since the
inverse $-v/Q(v)$ of each v in $\Sigma_{\mathbb{F}}(V)$ is again in $\Sigma_{\mathbb{F}}(V)$, $\mathrm{Spin}(V,Q)$ is a
subgroup of $\Gamma(V,Q)$; it is contained also in the *even* subalgebra $\mathfrak{A}_+$ of
$\mathfrak{A}$. Thus on V each $\sigma(g)$, $g \in \mathrm{Spin}(V,Q)$, actually is a genuine similarity
transformation $\sigma(g) : v \to gvg^{-1}$ because the principal automorphism
reduces to the identity on $\mathfrak{A}_+$.

(6.3) Theorem.

The restriction of σ to $\mathrm{Spin}(V,Q)$ defines a two-fold covering homo-
morphism

(6.4) $\sigma : \mathrm{Spin}(V,Q) \longrightarrow SO(V,Q)$, $\sigma(g) : v \longrightarrow gvg^{-1}$.

Proof. As in the proof of the corresponding result for the Clifford group, the basic idea is to relate the transformer $\sigma(v)$ on V determined by each v in $\Sigma_{\mathbf{F}}(V)$ with the reflection S_v of V that v determines. Now

$$B(x,v) = -\tfrac{1}{2}(xv + vx) , \qquad x \in V ,$$

is the bilinear form on V associated with Q. Thus

$$vx = -xv - 2B(x,v) = -\left(x - \frac{2B(x,v)v}{Q(x)}\right)v .$$

Consequently, $\sigma(v) = -S_v$, and so (6.4) is a covering homomorphism. Its kernel consists of all λ in $\mathbf{F}$ which can be written as $\lambda = v_1 \cdots v_{2k}$ with $v_j \in \Sigma_{\mathbf{F}}(V)$. But then

$$\lambda^2 = \Delta(\lambda) = \Delta(v_1 \cdots v_{2k}) = \prod_j Q(v_j) = 1$$

whether $\mathbf{F} = \mathbb{R}$ or $\mathbb{C}$. Hence σ has kernel $\{\pm 1\}$ in $\mathrm{Spin}(V,Q)$. ∎

Had we taken all finite products of elements from $\Sigma_{\mathbf{F}}(V)$ instead of even finite products, the range of σ would have been $O(V,Q)$, not $SO(V,Q)$. Thus in a jocular vein we are led to the group

(6.5) $$\mathrm{Pin}(V,Q) = \{v_1 \cdots v_k : v_j \in \Sigma_{\mathbf{F}}(V)\}$$

of all finite products in $\mathfrak{A}$ of elements from $\Sigma_{\mathbf{F}}(V)$.

The same proof as for (6.3) gives the following.

(6.6) Theorem.
The restriction of σ to $\mathrm{Pin}(V,Q)$ defines a two-fold covering homomorphism

$$\sigma : \mathrm{Pin}(V,Q) \longrightarrow O(V,Q) \quad , \quad \sigma(g) : v \longrightarrow gv(g')^{-1} .$$

Replacing $\Sigma_{\mathbf{F}}(V)$ by $\Sigma_{\mathbf{F}}(\mathbf{F} \oplus V)$ we obtain

(6.7) $$\mathrm{Spoin}(V,Q) = \{w_1 \cdots w_k : w_j \in \Sigma_{\mathbf{F}}(\mathbf{F} \oplus V)\} ;$$

it too is a subgroup of $\Gamma(V,Q)$.

(6.8) Theorem.
When $\mathbf{F} = \mathbb{R}$

$$\mathrm{Spoin}(V,Q) = \{g \in \Gamma(V,Q) : \Delta(g) = \pm 1\} ,$$

whereas when $\mathbf{F} = \mathbb{C}$

(6.9) $$\mathrm{Spoin}(V,Q) = \{g \in \Gamma(V,Q) : \Delta(g) = 1\} .$$

Proof. Suppose $\mathbf{F} = \mathbb{R}$. Then, since $\Delta : \Gamma(V,Q) \to \mathbb{R}_*$ is a multiplicative real character,

(6.10) $$\{g \in \Gamma(V,Q) : \Delta(g) = \pm 1\}$$

clearly is a subgroup of $\Gamma(V, Q)$ containing $\text{Spoin}(V, Q)$. On the other hand, by (5.29), every g in $\Gamma(V, Q)$ can be written as

$$(6.11) \qquad g = w_1 \cdots w_k = |\Delta(g)|^{1/2} \prod_j \frac{w_j}{|\Delta(w_j)|^{1/2}} = |\Delta(g)|^{1/2} g_0$$

with g_0 in $\text{Spoin}(V, Q)$. The subgroup (6.10) thus coincides with $\text{Spoin}(V, Q)$. The proof for the case $\mathbb{F} = \mathbb{C}$ is the same, replacing (6.10) by the right-hand group in (6.9). ∎

(6.12) Theorem.

The restriction of σ to $\text{Spoin}(V, Q)$ defines a two-fold covering homomorphism

$$\sigma : \text{Spoin}(V, Q) \longrightarrow SO(\mathbb{F} \oplus V, \Delta) \quad , \quad \sigma(g) : w \longrightarrow gw(g')^{-1} \ .$$

Proof. The proof of theorem (5.29) shows that, whenever $w \in \Sigma_\mathbb{F}(\mathbb{F} \oplus V)$, $\sigma(w)$ is equal to $S_1 \circ S_{w'} = S_w \circ S_1$. Using the fact that $S_1^2 = I$, it is easy to see that σ maps $\text{Spoin}(V, Q)$ onto $SO(\mathbb{F} \oplus V, \Delta)$. The kernel of σ consists of all $\lambda \in \mathbb{F}$ which may be written as $\lambda = w_1 w_2 \cdots w_k$, with $w_j \in \Sigma_\mathbb{F}(\mathbb{F} \oplus V)$. A simple application of theorem (6.8) shows that the kernel of σ in $\text{Spoin}(V, Q)$ is just $\{\pm 1\}$. ∎

The motivation for introducing $\text{Pin}(V, Q)$ is clear from (6.6). What is not so clear perhaps is the need for the group $\text{Spoin}(V, Q)$, since it is isomorphic to $\text{Spin}(\mathbb{F} \oplus V, \Delta)$ under the algebra isomorphism

$$(6.13) \qquad\qquad \mathfrak{A}(V, Q) \cong \mathfrak{A}_+(\mathbb{F} \oplus V, \Delta) \ .$$

In fact, in any *theoretical discussion* $\text{Spin}(V, Q)$ is more convenient to study than $\text{Spoin}(V, Q)$ because of the awkward dual nature of the role that $\mathbb{F}$ is being forced to play in the latter case. But in practice, $\text{Spoin}(\cdot)$ realizes a *particular spin group* such as $\text{Spin}(p, q)$ inside the Clifford algebra $\mathfrak{A}_{p,q-1}$, whereas $\text{Spin}(\cdot)$ realizes it inside the larger algebra $\mathfrak{A}_{p,q}$. On the grounds of economy of size, therefore, $\text{Spoin}(\cdot)$ is often preferable to $\text{Spin}(\cdot)$ in specific constructions, as we shall see.

The connectedness of $SO(V, Q)$ depends both on the scalar field $\mathbb{F}$ and on Q. In turn this affects its two-fold covering $\text{Spin}(V, Q)$, forcing the introduction of yet another Spin group, $\text{Spin}_0(V, Q)$, in the disconnected case. When $\mathbb{F} = \mathbb{C}$, for instance, $SO(V, Q)$ has a matrix realization as $SO(n, \mathbb{C})$, $n = \dim_\mathbb{C} V$, with respect to any choice of normalized basis, and so will always be connected. By contrast, when $\mathbb{F} = \mathbb{R}$, $SO(V, Q)$ has a matrix realization as $SO(p, q)$, $p + q = \dim_\mathbb{R} V$. Such a group is connected if and only if p or q is zero, i.e., if and only if Q is negative-

or positive-definite; otherwise, it has two components. When $\mathbb{F} = \mathbb{R}$ and Q is not definite, we shall denote by $SO_0(V, Q)$ (resp. $\mathrm{Spin}_0(p, q)$) the connected component of $SO(V, Q)$ (resp. $\mathrm{Spin}(p, q)$) containing the identity.

(6.14) Definition.
Let (V, Q) be an indefinite real quadratic space. Then the subgroup
$$\{g \in \mathrm{Spin}(V, Q) : \Delta(g) = 1\}$$
of $\mathrm{Spin}(V, Q)$ *will be denoted by* $\mathrm{Spin}_0(V, Q)$.

Since $\Delta : \mathrm{Spin}(V, Q) \to \{\pm 1\}$ is always a real multiplicative character, $\mathrm{Spin}_0(V, Q)$ is just the kernel of Δ, hence a subgroup of $\mathrm{Spin}(V, Q)$ whether $\mathbb{F} = \mathbb{R}$ or $\mathbb{C}$ and Q is definite or indefinite. But it is only when $\mathbb{F} = \mathbb{R}$ and Q is indefinite, i.e., Q does not have constant sign on V, that $\mathrm{Spin}_0(V, Q)$ is a *proper* subgroup. When $\dim V > 2$, we obtain the following result, extending (6.3).

(6.15) Theorem.
When (V, Q) is a real indefinite quadratic space, the restriction of σ to $\mathrm{Spin}_0(V, Q)$ *defines a two-fold covering homomorphism*
$$\sigma : \mathrm{Spin}_0(V, Q) \longrightarrow SO_0(V, Q) \quad , \quad \sigma(g) : v \longrightarrow gvg^{-1} .$$

As the proof of this theorem would take us too far afield, we omit it. The curious reader is referred to the excellent monograph of Porteous ([88]).

7 The Euclidean case

Because so much of the analysis to be developed on Euclidean space and on more general Riemannian manifolds depends ultimately on the theory of Clifford algebras associated with positive-definite quadratic spaces, it will be very convenient to formulate separately various aspects of the theory for this special class of spaces. Thus, for each $n = 0, 1, \ldots$, let $\mathfrak{A}_n$ ($= \mathfrak{A}_{0,n}$) be the universal Clifford algebra associated with $\mathbb{R}^n$, adopting the convention $\mathfrak{A}_0 = \mathbb{R}$. Formally, each $\mathfrak{A}_n$ is a 2^n-dimensional real associative algebra with identity containing linearly a copy of $\mathbb{R}^n$ so that for any orthonormal basis $\{e_1, \ldots, e_n\}$ of $\mathbb{R}^n$

(7.1)(i) $e_j e_k + e_k e_j = -2\delta_{jk}$ $(1 \le j, k \le n)$,

(7.1)(ii) *the reduced products*

$$e_\alpha = e_{\alpha_1} \cdots e_{\alpha_k} \quad (1 \le \alpha_1 < \cdots < \alpha_k \le n) , \qquad e_\emptyset = 1 ,$$

are a basis for $\mathfrak{A}_n$.

By identifying $\mathbb{R}^n$ with a subspace of $\mathbb{R}^{n+1}$ in an obvious way, we obtain an increasing chain of Clifford algebras

(7.2) $\mathbb{R} \subseteq \mathbb{C} \subseteq \mathbb{H} \subseteq \mathfrak{A}_3 \subseteq \cdots \subseteq \mathfrak{A}_n \subseteq \mathfrak{A}_{n+1} \cdots$.

Let

(7.3) $\mathfrak{A}_n^{(k)} = \mathrm{Span}_{\mathbb{R}} \{ e_{\alpha_1} \cdots e_{\alpha_k} : 1 \le \alpha_1 < \cdots < \alpha_k \le n \}$

be the space of *k-multivectors* spanned by the reduced products e_α of length k in $\mathfrak{A}_n$. Then

(7.4) $\mathfrak{A}_n = \mathfrak{A}_n^{(0)} \oplus \mathfrak{A}_n^{(1)} \oplus \cdots \oplus \mathfrak{A}_n^{(n)}$,

where $\mathfrak{A}_n^{(0)} = \mathbb{R}$ and $\mathfrak{A}_n^{(1)} = \mathbb{R}^n$; the space $\mathfrak{A}_n^{(0)} \oplus \mathfrak{A}_n^{(1)}$ will be frequently thought of as $\mathbb{R}^{n+1}$. If $\{\varepsilon_1, \ldots, \varepsilon_n\}$ is any other orthonormal basis for $\mathbb{R}^n$, then

(7.5) $\varepsilon_j = \sum_k a_{jk} e_k \qquad (1 \le j \le n)$

for some $A = [a_{jk}]$ in $O(n)$, and (7.1)(i) ensures that

(7.6) $\varepsilon_j \varepsilon_k = \sum_{r,s} a_{jr} a_{ks} e_r e_s = \sum_{r \ne s} a_{jr} a_{ks} e_r e_s \qquad (j \ne k)$,

since $[a_{jk}]$ is an orthogonal matrix. Hence $\mathfrak{A}_n^{(2)}$ is well-defined independently of the choice of orthonormal basis. A corresponding calculation shows that this remains true for every $\mathfrak{A}_n^{(k)}$, $k \ge 0$.

Three operators on $\mathfrak{A}_n$ are conveniently defined in terms of the $\mathfrak{A}_n^{(k)}$: for each u in $\mathfrak{A}_n^{(k)}$, $k \ge 0$, set

(7.7) $u' = (-1)^k u$, $u^* = (-1)^{\frac{1}{2} k(k-1)} u$, $\bar{u} = (-1)^{\frac{1}{2} k(k+1)} u$,

and then extend linearly to all of $\mathfrak{A}_n$. These are the principal automorphism, principal anti-automorphism and conjugation respectively, defined earlier, in section 2. Note that $u \to u^*$ and $u \to \bar{u}$ are involutions on $\mathfrak{A}_n$ in the sense of being linear mappings such that

$$(uv)^- = \bar{v} \bar{u} \quad , \quad (\bar{u})^- = u ,$$

and similarly for $u \to u^*$.

As an element of $\mathfrak{A}_n$, each e_j in an orthonormal basis for $\mathbb{R}^n$ will be said to be an *imaginary unit* because $e_j^2 = -1$ in $\mathfrak{A}_n$; furthermore, just as $\mathbb{C}$ and $\mathbb{H}$ can be written as

$$\mathbb{C} = \mathbb{R} \oplus i\mathbb{R} \quad , \quad \mathbb{H} = \mathbb{C} \oplus j\mathbb{C} ,$$

with respect to imaginary units i, j, so $\mathfrak{A}_{n+1}$ can be written as

(7.8) $$\mathfrak{A}_{n+1} = \mathfrak{A}_n \oplus e_{n+1} \mathfrak{A}_n$$

where e_{n+1} is the imaginary unit in $\mathfrak{A}_{n+1}$ which extends an orthonormal basis for $\mathbb{R}^n \subseteq \mathfrak{A}_n$ to one for $\mathbb{R}^{n+1} \subseteq \mathfrak{A}_{n+1}$. The usual basis for $\mathbb{R}^n$ will be specified by calling it the *standard basis*.

The next result exhibits an explicit and very useful *linear* isomorphism between the exterior algebra

(7.9) $$\Lambda^*(\mathbb{R}^n) = \Lambda^0(\mathbb{R}^n) \oplus \Lambda^1(\mathbb{R}^n) \oplus \cdots \oplus \Lambda^n(\mathbb{R}^n)$$

for $\mathbb{R}^n$ and the Clifford algebra $\mathfrak{A}_n$ associated with $\mathbb{R}^n$.

(7.10) Theorem.

The linear extension of the mapping

$$\lambda : e_{\alpha_1} \wedge \cdots \wedge e_{\alpha_k} \longrightarrow e_{\alpha_1} \cdots e_{\alpha_k} \qquad (k \geq 0)$$

to all of $\Lambda^*(\mathbb{R}^n)$ defines a linear isomorphism from $\Lambda^*(\mathbb{R}^n)$ onto $\mathfrak{A}_n$ independently of the choice of orthonormal basis $\{e_j\}$ for $\mathbb{R}^n$.

One consequence of theorem (7.10) is that all the usual Grassmannian geometry associated with $\mathbb{R}^n$ can be found naturally within $\mathfrak{A}_n$. There are many other important consequences as we shall see.

Proof of theorem (7.10). Since the sets of reduced products $e_{\alpha_1} \wedge \cdots \wedge e_{\alpha_k}$ and $e_{\alpha_1} \cdots e_{\alpha_k}$ $(1 \leq \alpha_1 < \cdots < \alpha_k \leq n)$ together with $e_\emptyset$ are bases of $\Lambda^k(\mathbb{R}^n)$ and $\mathfrak{A}_n^{(k)}$ respectively, it is clear that λ extends to a linear isomorphism $\lambda : \Lambda^*(\mathbb{R}^n) \to \mathfrak{A}_n$ for a fixed choice of basis. But, for any other choice of orthonormal basis related to $\{e_j\}$ by (7.5),

(7.11) $$\varepsilon_j \wedge \varepsilon_k = \sum_{r \neq s} a_{jr} a_{ks} e_r \wedge e_s \qquad (j \neq k) ,$$

analogous to (7.6). Hence $\lambda : \Lambda^2(\mathbb{R}^n) \to \mathfrak{A}_n^{(2)}$ is well-defined independently of orthonormal basis. A corresponding proof is valid for every $k \geq 0$, completing the proof. ∎

Now the inner product $u \cdot v$ on $\mathbb{R}^n$ extends naturally to an inner product on $\Lambda^*(\mathbb{R}^n)$. On $\Lambda^k(\mathbb{R}^n)$ it is defined explicitly in a basis-free way by

(7.12) $$(\xi_1 \wedge \cdots \wedge \xi_k , \; \eta_1 \wedge \cdots \wedge \eta_k) = \det(\xi_r \cdot \eta_s)$$

with an extension to $\Lambda^*(\mathbb{R}^n)$ so that $\Lambda^\ell(\mathbb{R}^n)$ is orthogonal to $\Lambda^m(\mathbb{R}^n)$ when $\ell \neq m$. The basis of reduced products $e_{\alpha_1} \wedge \cdots \wedge e_{\alpha_k}$ for $\Lambda^*(\mathbb{R}^n)$ determined by an orthonormal basis $\{e_j\}$ for $\mathbb{R}^n$ clearly is orthonormal with respect to this inner product. Hence $\Lambda^*(\mathbb{R}^n)$ becomes a finite-dimensional Hilbert space in a natural way, and its Hilbert space structure can be transported to $\mathfrak{A}_n$ by λ. By (7.10), therefore, the corresponding inner product on $\mathfrak{A}_n$ is just the usual Euclidean one:

$$(7.13) \quad (u,v) = \left(\sum_\alpha u_\alpha e_\alpha \, , \, \sum_\beta v_\beta e_\beta \right) = \sum_\alpha u_\alpha v_\alpha \, , \qquad u,v \in \mathfrak{A}_n \, ,$$

on $\mathfrak{A}_n$ as a 2^n-dimensional space, and the Hilbert space norm on $\mathfrak{A}_n$ is

$$(7.14) \qquad |u|_\mathfrak{A} = \left| \sum_\alpha u_\alpha e_\alpha \right|_\mathfrak{A} = \left(\sum_\alpha |u_\alpha|^2 \right)^{1/2} .$$

There is an interesting intrinsic description of this Hilbert space structure which brings out clearly its independence of the choice of basis.

(7.15) Definition.

For each $k = 0, 1, \ldots, n$, denote by λ_k the projection $\lambda_k : \mathfrak{A}_n \to \mathfrak{A}_n^{(k)}$ from $\mathfrak{A}_n$ onto its subspace $\mathfrak{A}_n^{(k)}$ of k-multivectors.

Firstly, a wedge and scalar product can be introduced into $\mathfrak{A}_n$ using Clifford multiplication and these λ_k; in fact, when $x \in \mathfrak{A}_n^{(r)}$ and $y \in \mathfrak{A}_n^{(s)}$, set

$$(7.16) \quad x \wedge y = \lambda_{r+s}(xy) \quad (r+s \le n) \quad , \quad x \wedge y = 0 \quad (r+s > n) \, ,$$

and extend linearly; it is related to the wedge product in $\Lambda^*(\mathbb{R}^n)$ by

$$(7.17) \qquad \lambda(\xi \wedge \eta) = \lambda(\xi) \wedge \lambda(\eta) \, , \qquad \xi, \eta \in \Lambda^*(\mathbb{R}^n) \, ,$$

under the linear isomorphism $\lambda : \Lambda^*(\mathbb{R}^n) \to \mathfrak{A}_n$. On the other hand, when

$$(7.18) \qquad\qquad x \cdot y = \lambda_0(\bar{x} y) \, , \qquad x, y \in \mathfrak{A}_n \, ,$$

clearly

$$(7.19) \qquad\qquad u \cdot v = \sum_\alpha u_\alpha v_\alpha = (u,v) \, , \qquad u,v \in \mathfrak{A}_n \, .$$

Consequently, (7.19) coincides with the inner product on $\mathfrak{A}_n$ transported from $\Lambda^*(\mathbb{R}^n)$ by λ. What is more, on restriction to the subspaces $\mathbb{R}^n$ and $\mathbb{R}^{n+1} = \mathbb{R} \oplus \mathbb{R}^n$ of $\mathfrak{A}_n$, the Hilbert space structure induced on $\mathfrak{A}_n$ by $(\cdot,\cdot)$ and $|\cdot|_\mathfrak{A}$ reduces to the standard Euclidean space structure on $\mathbb{R}^n$ and $\mathbb{R}^{n+1}$. Thus there is no reason to distinguish the Hilbert norm $|\cdot|_\mathfrak{A}$ on $\mathfrak{A}_n$ from that of $|\cdot|$ on $\mathbb{R}^n$, and so we shall omit the subscript $\mathfrak{A}$ from $|\cdot|_\mathfrak{A}$. In summary then, through Clifford multiplication $\mathfrak{A}_n$ captures both the algebraic and metric structures on $\Lambda^*(\mathbb{R}^n)$ as well as the Euclidean space structure on $\mathbb{R}^n$. Note also that the norm function $\Delta(u) = \bar{u}u$ coincides with $|u|^2$ when $\Delta(u)$ is real-valued (see (5.14)).

A convenient Banach algebra norm on $\mathfrak{A}_n$ can be introduced by re-garding $\mathfrak{A}_n$ as an algebra of operators on itself.

(7.20) Definition.

For each a in $\mathfrak{A}_n$, let $\|a\|_{\mathfrak{A}}$ be the operator norm of u as an operator on the Hilbert space $\mathfrak{A}_n$; thus

$$\|a\|_{\mathfrak{A}} = \sup\{|au| : |u| = 1\} \ .$$

The norm $|\cdot|$ on $\mathfrak{A}_n$ will be referred to as the Clifford norm, while $\|\cdot\|_{\mathfrak{A}}$ will be called the Clifford operator norm.

Since $\|\cdot\|_{\mathfrak{A}}$ has the submultiplicative property

$$(7.21) \qquad \|ab\|_{\mathfrak{A}} \leq \|a\|_{\mathfrak{A}}\|b\|_{\mathfrak{A}} \ , \qquad a,b \in \mathfrak{A}_n \ ,$$

(unlike $|\cdot|$), $\mathfrak{A}_n$ is a *real* Banach algebra under $\|\cdot\|_{\mathfrak{A}}$ having an identity of norm 1. A detailed description of its Banach algebra structure follows from the next theorem.

(7.22) Theorem.

The principal automorphism and anti-automorphism are isometric on $\mathfrak{A}_n$ in the sense that

$$|u'| = |u^*| = |u| \ , \quad \|a'\|_{\mathfrak{A}} = \|a^*\|_{\mathfrak{A}} = \|a\|_{\mathfrak{A}} \ , \qquad a,u \in \mathfrak{A}_n \ .$$

Furthermore,

    (i) $(au,v) = (u, \bar{a}\,v)$,

    (ii) $|aub| \leq \|a\|_{\mathfrak{A}}|u|\,\|b\|_{\mathfrak{A}}$,

    (iii) $\|\bar{a}\,a\|_{\mathfrak{A}} = \|a\|_{\mathfrak{A}}^2$,

for all a,b, and u in $\mathfrak{A}_n$; in particular,

    (iv) $|a|^2 = \Delta(a) = \|a\|_{\mathfrak{A}}^2$

whenever the 'norm' $\Delta(a) = \bar{a}\,a$ is real-valued.

Proof. The isometries with respect to $|\cdot|$ are clear because both u' and u^* are of the form $\sum_\alpha \pm a_\alpha e_\alpha$ for some choice of signs when $u = \sum_\alpha a_\alpha e_\alpha$. But then

$$\|a'\|_{\mathfrak{A}} = \sup_{|v| \leq 1} |a'u| = \sup_{|v| \leq 1} |au'| = \sup_{|v| \leq 1} |av| = \|a\|_{\mathfrak{A}} \ ;$$

a similar proof shows that $\|a^*\|_{\mathfrak{A}} = \|a\|_{\mathfrak{A}}$. The other properties are proved just as easily. For (i) follows immediately from (7.18) and (7.19), while (ii) follows from the string of inequalities

$$|aub| \leq \|a\|_{\mathfrak{A}}|ub| = \|a\|_{\mathfrak{A}}|b^*u^*| \ .$$

To establish (iii) it is enough to show that

$$\|\bar{a}\,a\|_{\mathfrak{A}} \geq \|a\|_{\mathfrak{A}}^2 \ ,$$

since $\|\bar{a}\|_{\mathfrak{A}} = \|(a')^*\|_{\mathfrak{A}} = \|a\|_{\mathfrak{A}}$. But, by (i),

$$\|\bar{a}\,a\|_{\mathfrak{A}} = \sup |(au, av)| \geq \sup |au|^2 = \|a\|_{\mathfrak{A}}^2 ,$$

taking the supremum over all u, v such that $|u|, |v| \leq 1$. This proves (iii). Finally, the first equality in (iv) has been noted already, and this together with (iii) establishes the second equality, since $\bar{a}\,a$ must be non-negative if it is real. ∎

In general $|a|$ will differ from $\|a\|_{\mathfrak{A}}$, so the following corollary of (7.22) is of considerable interest by itself, but it will be of the utmost importance in dealing with operators on Banach spaces of $\mathfrak{A}_n$-valued functions. Denote by Λ_n the Clifford semi-group

$$(7.23) \qquad \Lambda_n = \{xy \cdots z : x, y, z, \ldots \in \mathbb{R}^{n+1}\}$$

generated multiplicatively in $\mathfrak{A}_n$ by $\mathbb{R}^{n+1}$, and then denote by Γ_n the Clifford group

$$(7.24) \qquad \Gamma_n = \{xy \cdots z : x, y, z, \ldots \in \mathbb{R}^{n+1} \setminus \{0\}\}$$

generated by the non-zero elements in $\mathbb{R}^{n+1}$ (see (5.19), (5.20)). Of course, $\Lambda_n = \Gamma_n \cup \{0\}$.

(7.25) Corollary.
The Clifford norm and Clifford operator norm coincide on Λ_n; more precisely,

$$\|a\|_{\mathfrak{A}} = \Delta(a)^{1/2} = |a| = |x|\,|y| \cdots |z| \qquad (x, y, \ldots, z \in \mathbb{R}^{n+1}) ,$$

holds for any element $a = xy \cdots z$ in Λ_n.

Proof. Since the 'norm' $\Delta(a) = \bar{a}\,a$ is non-negative and multiplicative on Λ_n (see (5.16)), the result follows immediately from (7.22)(iv). ∎

This corollary gives an interesting interpretation to the various spin groups associated with Euclidean space. From (6.8) and (6.12), for instance, we deduce the following.

(7.26) Theorem.
For each $n = 0, 1, \ldots,$

$$\mathrm{Spin}(n+1) = \{a \in \Lambda_n : \|a\|_{\mathfrak{A}} = 1\} ,$$

and

$$\sigma(a) : x \longrightarrow axa^* , \qquad x \in \mathbb{R}^{n+1} \subseteq \mathfrak{A}_n ,$$

defines a two-fold covering $\sigma : \mathrm{Spin}(n+1) \to SO(n+1)$.

Now $O(n)$ can be identified with the subgroup of $SO(n+1)$ fixing the subspace $\mathbb{R}(1, 0, \ldots, 0)$ of $\mathbb{R}^{n+1}$. But this is the space $\mathfrak{A}_n^{(0)}$, thinking of

$\mathbb{R}^{n+1}$ as $\mathfrak{A}_n^{(0)} \oplus \mathfrak{A}_n^{(1)}$. Since $\sigma(a)$ fixes $\mathfrak{A}_n^{(0)}$ if and only if $aa^* = 1$, (6.3) and (6.6) thus give the following.

(7.27) Corollary.

For each $n = 1, 2, \ldots$,

$$\text{Pin}(n) = \{a \in \Lambda_n : aa^* = 1 \ , \ \|a\|_{\mathfrak{A}} = 1\} \ ,$$

while

$$\text{Spin}(n) = \{a \in \Lambda_n \cap \mathfrak{A}_n^+ : aa^* = 1 \ , \ \|a\|_{\mathfrak{A}} = 1\} \ ;$$

furthermore,

$$\sigma(a) : x \longrightarrow axa^* \ , \qquad x \in \mathbb{R}^n \subseteq \mathfrak{A}_n \ ,$$

defines a two-fold covering $\sigma : \text{Pin}(n) \to O(n)$ *as well as* $\sigma : \text{Spin}(n) \to SO(n)$.

It is well-known that the unit sphere in $\mathbb{R}^{n+1}$ has a multiplicative structure if $n = 0, 1$ or 3, when it coincides with $\text{Pin}(1)$, $\text{Spin}(2)$ and $\text{Spin}(3)$ respectively. The results above explain what to do for larger n: one takes the unit sphere in the Clifford semi-group Λ_n, not just the unit sphere in $\mathbb{R}^{n+1}$.

In order to determine precisely the Banach algebra structure of $\mathfrak{A}_n$, recall first that a complex Banach algebra $\mathfrak{B}$ is said to be a C^*-algebra when it has a conjugate-linear involution $a \to \tilde{a}$ satisfying $\|\tilde{a}a\|_{\mathfrak{B}} = \|a\|_{\mathfrak{B}}^2$. For every such $\mathfrak{B}$ there are a complex Hilbert space $\mathcal{H}$ and an algebraic isomorphism $\pi : \mathfrak{B} \to \mathcal{L}(\mathcal{H})$ from $\mathfrak{B}$ into the operator algebra $\mathcal{L}(\mathcal{H})$ so that

(7.28) $\|\pi(a)\|_{\text{op}} = \|a\|_{\mathfrak{B}} \ , \quad \pi(a)^* = \pi(\tilde{a}) \ , \qquad a \in \mathfrak{A}_n \ ,$

where $\| \cdot \|_{\text{op}}$ is the operator norm

(7.29) $\|T\|_{\text{op}} = \sup\{\|T\xi\|_{\mathcal{H}} : \|\xi\|_{\mathcal{H}} \leq 1\}$

on $\mathcal{L}(\mathcal{H})$ and T^* is the operator on $\mathcal{H}$ adjoint to T. By contrast, a real Banach algebra $\mathfrak{B}$ is said to be *real* C^*-algebra when it has a real linear involution $a \to \tilde{a}$ such that $\|\tilde{a}a\|_{\mathfrak{B}} = \|a\|_{\mathfrak{B}}^2$ and $1 + a^*a$ is invertible in $\mathfrak{B}$ for every a in $\mathfrak{B}$. As before there then exist a *real* Hilbert space $\mathcal{H}$ and an algebraic isomorphism $\pi : \mathfrak{B} \to \mathcal{L}(\mathcal{H})$ satisfying (7.28). Conversely, any norm-closed self-adjoint real subalgebra of $\mathcal{L}(\mathcal{H})$ always is a real C^*-algebra whenever $\mathcal{H}$ is a real Hilbert space. Furthermore, each real C^*-algebra $\mathfrak{B}$ is isometrically embedded in a complex C^*-algebra, its complexification $\mathfrak{B}_{\mathbb{C}} = \mathfrak{B} \otimes_{\mathbb{R}} \mathbb{C}$, so that the restriction to $\mathfrak{B}$ of the conjugate-linear involution on $\mathfrak{B}_{\mathbb{C}}$ coincides with the real involution of $\mathfrak{B}$. For instance, when $\mathfrak{B}$ is realized as a subalgebra of $\mathcal{L}(\mathcal{H})$, this $\mathfrak{B}_{\mathbb{C}}$ is just the complex extension of $\mathfrak{B}$ to the complexification $\mathcal{H}_{\mathbb{C}}$ of $\mathcal{H}$.

Theorem (7.22) allows us to apply these C^*-algebra ideas to $\mathfrak{A}_n$, taking for $\mathcal{H}$ the space $\mathfrak{A}_n$ as a Hilbert space under the Clifford norm. In fact, the following theorem is now virtually obvious.

(7.30) Theorem.

Under the Clifford conjugation $a \to \bar{a}$ and the Clifford operator norm, each $\mathfrak{A}_n$ is a real C^*-algebra.

Recall that $\mathfrak{C}_n$ denotes the complex universal Clifford algebra associated with the complex quadratic space $\mathbb{C}^n$ and quadratic form

$$(7.31) \qquad Q_n(z) = z_1^2 + \cdots + z_n^2 , \qquad z = (z_1, \ldots, z_n) \in \mathbb{C}^n .$$

The inner product (7.13) on $\mathfrak{A}_n$ extends to a sesqui-linear inner product

$$(7.32)(\text{i}) \qquad (u, v) = \left(\sum_\alpha u_\alpha e_\alpha , \sum_\beta v_\beta e_\beta \right) = \sum_\alpha u_\alpha \bar{v}_\alpha ,$$

and Hilbert space norm

$$(7.32)(\text{ii}) \qquad |u| = \left| \sum_\alpha u_\alpha e_\alpha \right| = \left(\sum_\alpha |u_\alpha|^2 \right)^{1/2}$$

on $\mathfrak{C}_n$ which coincide with the usual Euclidean ones on $\mathfrak{C}_n$ as a space of complex dimension 2^n. As such $\mathfrak{C}_n$ is just the usual *Hilbert space complexification* of $\mathfrak{A}_n$, and (7.32)(ii) will be said to be the *Clifford norm* on $\mathfrak{C}_n$. Likewise, the Clifford operator norm on $\mathfrak{C}_n$ is the usual operator norm

$$(7.33) \qquad \|a\|_{\mathfrak{C}} = \sup \{ |au| : u \in \mathfrak{C}_n , \ |u| \le 1 \}$$

with respect to which $\mathfrak{C}_n$ is the complexification of $\mathfrak{A}_n$ simply as a Banach algebra, ignoring involutions. Now $\mathfrak{C}_n$ comes equipped with an involution: the Clifford conjugation. But this involution is *complex linear, not conjugate-linear*. So for Banach* algebra purposes, we shall take as involution on $\mathfrak{C}_n$ the conjugate-linear extension $a \to \tilde{a}$ of Clifford conjugation on $\mathfrak{A}_n$. Thus

$$(7.34)(\text{i}) \qquad \tilde{a} = \left(\sum_\alpha u_\alpha e_\alpha \right)^{\sim} = \sum_\alpha \bar{u}_\alpha (-1)^{\frac{1}{2}|\alpha|(|\alpha|+1)} e_\alpha ,$$

whereas

$$(7.34)(\text{ii}) \qquad \bar{a} = \left(\sum_\alpha u_\alpha e_\alpha \right)^{-} = \sum_\alpha u_\alpha (-1)^{\frac{1}{2}|\alpha|(|\alpha|+1)} e_\alpha .$$

The following complex counterpart of theorem (7.30) is also virtually obvious.

(7.35) Theorem.

Under the involution $a \to \tilde{a}$ defined by (7.34)(i) and the Clifford operator norm, each $\mathfrak{C}_n$ is a complex C^*-algebra which is a complexification of the real C^*-algebra $\mathfrak{A}_n$.

By coupling the general structural results obtained earlier for any universal Clifford algebra with those for C^*-algebras, we can now make the structural properties of $\mathfrak{A}_n$ and $\mathfrak{C}_n$ completely transparent. In turn, the significance of all the associated algebra and analysis can then be 'seen' much more clearly. By (3.19), $\mathfrak{A}_n$ is simple unless $n = 4m + 3$, $m \geq 0$, when it splits into the direct sum of two ideals each isomorphic to $\mathfrak{A}_{n-1}$; and, by (3.2), $\mathfrak{C}_n$ is simple unless $n = 2m+1$, $m \geq 0$, when it splits into the direct sum of two ideals each isomorphic to $\mathfrak{C}_{n-1}$. On the other hand, every finite-dimensional real C^*-algebra is the finite direct sum of full matrix rings $M(k, \mathbb{K})$ where $\mathbb{K} = \mathbb{R}, \mathbb{C}$ or $\mathbb{H}$, while every finite-dimensional complex C^*-algebra is the finite direct sum of full matrix rings $M(k, \mathbb{C})$. But every $M(k, \mathbb{K})$ is simple. Hence

$$(7.36)(\text{i}) \qquad \mathfrak{A}_n \cong M(k, \mathbb{K}) \qquad (n \neq 4m + 3) \ ,$$

for some choice of $(k, \mathbb{K})$, while

$$(7.36)(\text{ii}) \qquad \mathfrak{A}_n \cong M(k, \mathbb{K}) \oplus M(k, \mathbb{K}) \qquad (n = 4m + 3) \ ,$$

for the same choice of $(k, \mathbb{K})$ as when $n = 4m + 2$. Of course, these restrictions merely continue the identifications

$$(7.37)(\text{i}) \qquad \mathfrak{A}_0 \cong \mathbb{R} \ , \quad \mathfrak{A}_1 \cong \mathbb{C} \ , \quad \mathfrak{A}_2 \cong \mathbb{H} \ , \quad \mathfrak{A}_3 \cong \mathbb{H} \oplus \mathbb{H} \ ,$$

the first two of which have been used repeatedly already. To describe $\mathfrak{A}_n$, $n > 3$, explicitly we shall show first that

(7.37)(ii)
$$\mathfrak{A}_4 \cong \mathbb{H}^{2 \times 2} \ , \quad \mathfrak{A}_5 \cong \mathbb{C}^{4 \times 4} \ , \quad \mathfrak{A}_6 \cong \mathbb{R}^{8 \times 8} \ , \quad \mathfrak{A}_7 \cong \mathbb{R}^{8 \times 8} \oplus \mathbb{R}^{8 \times 8} \ ,$$

which together with the following fundamental periodicity theorem completely determines the choice of $(k, \mathbb{K})$ for every n.

(7.37)(iii) Theorem.

For each $n \geq 0$ there is a realization of $\mathfrak{A}_{n+8}$ as the algebra

$$\mathfrak{A}_{n+8} = M(16, \mathfrak{A}_n)$$

of all 16×16 matrices having entries from $\mathfrak{A}_n$.

Consequently, when $n = 8m + r$, $0 \leq r < 8$, the choice of $\mathbb{K}$ in (7.36) is given by

(7.38)(i)
$$\mathbb{K} = \mathbb{R} \ (r = 0, 6, 7) \ , \quad \mathbb{K} = \mathbb{C} \ (r = 1, 5) \ , \quad \mathbb{K} = \mathbb{H} \ (r = 2, 3, 4) \ ,$$

and by a dimension count,
(7.38)(ii)
$$k^2 \dim_{\mathbb{R}}(\mathbb{K}) = 2^{8m+r} \ (r \ne 3,7) \ , \ k^2 \dim_{\mathbb{R}}(\mathbb{K}) = 2^{8m+r-1} \ (r = 3,7) \ .$$

The simpler complex analogues of these results were proved earlier in (3.2):

$$(7.39) \qquad \mathbb{C}_{2m} \cong M(2^m, \mathbb{C}) \ , \ \mathbb{C}_{2m+1} \cong M(2^m, \mathbb{C}) \oplus M(2^m, \mathbb{C}) \ .$$

Both (7.37)(ii) and (7.37)(iii) are a consequence of the following more general but weaker version of (2.29).

(7.40) Theorem.

For each $n \ge 0$ there is a realization of $\mathfrak{A}_{n+4}$ as the tensor product

$$\mathfrak{A}_{n+4} \cong \mathfrak{A}_n \otimes_{\mathbb{R}} \mathbb{H}^{2 \times 2} \cong \mathfrak{A}_n \otimes_{\mathbb{R}} \mathfrak{A}_4$$

over $\mathbb{R}$ of $\mathfrak{A}_n$ and $\mathbb{H}^{2 \times 2}$.

By contrast with (2.18), the multiplicative structure on these products is just the usual one on the tensor products of real algebras, namely the $\mathbb{R}$-linear extension of

$$(7.41) \qquad (a_1 \otimes b_1)(a_2 \otimes b_2) = a_1 a_2 \otimes b_1 b_2 \ .$$

Proof of theorem (7.40). Let $\{e_1, \ldots, e_n\}$ be a normalized basis for $\mathfrak{A}_n$, and let $\{i, j, k \ (= ij)\}$ be the usual basis for $\mathbb{H}$. Then

$$\gamma_1 = \begin{bmatrix} 0 & 1 \\ -1 & 0 \end{bmatrix} , \quad \gamma_2 = \begin{bmatrix} 0 & i \\ i & 0 \end{bmatrix} , \quad \gamma_3 = \begin{bmatrix} 0 & j \\ j & 0 \end{bmatrix} , \quad \gamma_4 = \begin{bmatrix} 0 & k \\ k & 0 \end{bmatrix}$$

are elements of $\mathbb{H}^{2 \times 2}$ satisfying

$$\gamma_r \gamma_s + \gamma_s \gamma_r = -2\delta_{rs} I \qquad (1 \le r, s \le 4) \ ;$$

and so, by (3.17),

$$\nu : \mathbb{R}^4 \longrightarrow \mathbb{H}^{2 \times 2} \ , \quad \nu : (x_1, \ldots, x_4) = \sum_{r=1}^{4} x_r \gamma_r$$

extends to an identification of $\mathfrak{A}_4$ with $\mathbb{H}^{2 \times 2}$, since $\dim \mathfrak{A}_4 = 16 = \dim \mathbb{H}^{2 \times 2}$. On the other hand, the product $\gamma_0 = \gamma_1 \cdots \gamma_4$ satisfies

$$\gamma_0^2 = I \ , \quad \gamma_0 \gamma_r + \gamma_r \gamma_0 = 0 \qquad (1 \le r \le 4) \ ;$$

consequently,

$$\eta_r = I \otimes \gamma_r \ (1 \le r \le 4) \ , \quad \eta_{4+j} = e_j \otimes \gamma_0 \ (1 \le j \le n) \ ,$$

are elements of $\mathfrak{A}_n \otimes \mathbb{H}^{2 \times 2}$ such that

$$\eta_k \eta_\ell + \eta_\ell \eta_k = -2\delta_{k\ell} I \qquad (1 \le k, \ell \le n+4) \ ,$$

with respect to the product in (7.41). Hence the linear embedding

$$\nu : \mathbb{R}^{n+4} \longrightarrow \mathfrak{A}_n \otimes \mathbb{H}^{2\times2} \quad, \quad \nu : (x_1, \ldots, x_{n+4}) \longrightarrow \sum_{k=1}^{n+4} x_k \eta_k \; ,$$

extends to an algebra homomorphism π from $\mathfrak{A}_{n+4}$ into $\mathfrak{A}_n \otimes \mathbb{H}^{2\times2}$. By (3.17), π must be an isomorphism except possibly when $n = 4m + 3$. Whatever the value of n, however, every reduced product of the η_k is of the form $e_\alpha \otimes \varepsilon_\beta$ where e_α is a reduced product of the e_j and ε_β is a reduced product of the γ_j; moreover, every tensor product $e_\alpha \otimes \varepsilon_\beta$ arises from such a reduced product. Hence the algebra generated in $\mathfrak{A}_n \otimes \mathbb{H}^{2\times2}$ by $\{\eta_k : 1 \le k \le n + 4\}$ has maximal dimension, and so $\mathfrak{A}_n \otimes \mathbb{H}^{2\times2}$ is a realization of $\mathfrak{A}_{n+4}$. ∎

We can now prove (7.37)(ii). The $\mathfrak{A}_4$-case was proved in (7.40). On the other hand, by (7.37)(i) and (7.40),

$$\mathfrak{A}_5 = \mathbb{C} \otimes \mathbb{H}^{2\times2} \;, \quad \mathfrak{A}_6 = \mathbb{H} \otimes \mathbb{H}^{2\times2} \;, \quad \mathfrak{A}_7 = (\mathbb{H} \otimes \mathbb{H}^{2\times2}) \oplus (\mathbb{H} \otimes \mathbb{H}^{2\times2}) \;,$$

the tensor products being taken over $\mathbb{R}$. But $\mathbb{C} \otimes_\mathbb{R} \mathbb{H}^{2\times2}$ is just the complexification $\mathbb{C}^{4\times4}$ of $\mathbb{H}^{2\times2}$, and $\mathbb{H} \otimes_\mathbb{R} \mathbb{H}^{2\times2}$ is isomorphic to $\mathbb{R}^{8\times8}$. This proves part (ii) of (7.37). As for part (iii), a two-fold application of (7.40) gives

$$\mathfrak{A}_{n+8} \cong \mathfrak{A}_n \otimes_\mathbb{R} \mathbb{H}^{2\times2} \otimes_\mathbb{R} \mathbb{H}^{2\times2}$$

with respect to the usual tensor product (7.41) of algebras. Since

$$\mathbb{H}^{2\times2} \otimes_\mathbb{R} \mathbb{H}^{2\times2} \cong \mathbb{R}^{16\times16} \;,$$

the periodicity theorem follows immediately. ∎

As interesting as these specific characterizations of $\mathfrak{A}_n$ are, the algebraic details should not be allowed to obscure the analytic details. There are really two fundamental ideas that need to be brought out. The first is that $\mathfrak{A}_n$ and $\mathfrak{C}_n$ are Hilbert spaces on which $\mathfrak{A}_n$ and $\mathfrak{C}_n$ are also C^*-algebras with respect to the usual operator norm. The second one is expressed in (7.36) and (7.39): for each $n = 0, 1, \ldots$, there is a finite-dimensional *real* Hilbert space, to be denoted by $\mathfrak{R}_n$, and a finite-dimensional *complex* Hilbert space, the complexification $\mathfrak{R}_n \otimes_\mathbb{R} \mathbb{C}$ of $\mathfrak{R}_n$ to be denoted by $\mathfrak{S}_n$, such that

(7.42)(i) $\mathfrak{A}_n$ *and* $\mathfrak{C}_n$ *can be realized as the C^*-algebras* $\mathcal{L}(\mathfrak{R}_n)$ *(resp.* $\mathcal{L}(\mathfrak{S}_n)$*) when* $n \ne 4\ell + 3$ *(resp.* $n \ne 2\ell + 1$*), while*

(ii) $\mathfrak{A}_n$ *and* $\mathfrak{C}_n$ *split into the direct sum of two ideals isomorphic to the respective C^*-algebras* $\mathcal{L}(\mathfrak{R}_{n-1})$ *and* $\mathcal{L}(\mathfrak{S}_{n-1})$ *in all remaining cases.*

The space $\mathfrak{S}_n$ is usually called the *complex spinor space*, and by analogy we shall call $\mathfrak{R}_n$ the *real spinor space*. Of course

(7.43) $\mathfrak{R}_n \cong \mathbb{K}^k$, $\mathfrak{S}_n \cong \mathbb{C}^{2^m}$ $\left(m = \left[\frac{1}{2}n \right] \right)$,

where $\mathbb{K}$ and k are given by (7.38), but explicit realizations of $\mathfrak{R}_n$ and $\mathfrak{S}_n$ may vary with context. Several such examples will recur throughout the book as prototypes of one of the truly basic ideas underlying all appearances of Clifford algebras and Dirac operators in analysis, that of $\mathfrak{A}_n$-*module* or *Clifford module* as it is more commonly called in the literature. Let $\mathfrak{H}$ be finite-dimensional real or complex Hilbert space and let $\mathcal{L}(\mathfrak{H})$ be the usual C^*-algebra of bounded linear operators on $\mathfrak{H}$, the adjoint of an operator a in $\mathcal{L}(\mathfrak{H})$ being denoted by $\bar{a}$. The norm on $\mathfrak{H}$ will be denoted by $|\cdot|$, the inner product by $(\cdot, \cdot)$, and the operator norm by $\|\cdot\|$; thus

$$(au, v) = (u, \bar{a}v) \quad , \quad \|\bar{a}a\| = \|a\|^2$$

hold for all a in $\mathcal{L}(\mathfrak{H})$ and u, v in $\mathfrak{H}$.

(7.44) Definition.

A finite-dimensional real or complex Hilbert space $\mathfrak{H}$ is said to be an $\mathfrak{A}_n$-module when there exist skew-adjoint operators $e_1, \ldots, e_n$ in $\mathcal{L}(\mathfrak{H})$ such that

$$e_j e_k + e_k e_j = -2\delta_{jk} e_0 \quad (1 \leq j, k \leq n) ,$$

where $e_0 \ (= I)$ is the identity operator.

Much of this section has been devoted to showing that $\mathfrak{R}_n$ and $\mathfrak{A}_n$ are real $\mathfrak{A}_n$-modules, while $\mathfrak{S}_n$ and $\mathfrak{C}_n$ are complex $\mathfrak{A}_n$-modules, for example (see (7.22)(i)). Denote by $\nu : \mathbb{R}^{n+1} \to \mathcal{L}(\mathfrak{H})$ the linear embedding

(7.45) $\nu : (x_0, \ldots, x_n) \longrightarrow \sum_{j=0}^{n} x_j e_j$ $(x \in \mathbb{R}^{n+1})$

of $\mathbb{R}^{n+1}$ into $\mathcal{L}(\mathfrak{H})$. Then by hypothesis

(7.45)' $\overline{\nu(x)}\nu(x) = |x|^2 e_0 = \nu(x)\overline{\nu(x)}$ $(x \in \mathbb{R}^{n+1})$

where $|x| = \left(\sum_j |x_j|^2 \right)^{1/2}$ is the Euclidean length of x; on the other hand,

(7.46) $\|\nu(x)\|^2 = \|\overline{\nu(x)}\nu(x)\| = |x|^2 \|e_0\| = |x|^2$.

Hence ν is an *isometric embedding* of $\mathbb{R}^{n+1}$ into $\mathcal{L}(\mathfrak{H})$; we shall therefore identify $\mathbb{R}^{n+1}$ with a subspace of $\mathcal{L}(\mathfrak{H})$, regarding $e_0, \ldots, e_n$ simply as an orthonormal basis for $\mathbb{R}^{n+1}$, and omitting any mention of ν. With this convention, let $\mathfrak{A}(\mathfrak{H})$ be the real C^*-algebra of $\mathcal{L}(\mathfrak{H})$ generated by $\mathbb{R}^{n+1}$. Now, by the universal property, the restriction of (7.45) to

$(x_1, \ldots, x_n) \to \sum_{j=1}^n x_j e_j$ extends to a homomorphism $\pi : \mathfrak{A}_n \to \mathfrak{A}(\mathfrak{H})$ from $\mathfrak{A}_n$ onto $\mathfrak{A}(\mathfrak{H})$. The structural properties of $\mathfrak{A}_n$ enable us to describe the properties of π.

(7.47) Theorem.
The homomorphism $\pi : \mathfrak{A}_n \to \mathfrak{A}(\mathfrak{H})$ from $\mathfrak{A}_n$ into $\mathcal{L}(\mathfrak{H})$ satisfies the following:

(i) π is always an isomorphism when $n \neq 4\ell + 3$, $\ell \geq 0$;
(ii) π fails to be a homomorphism if and only if $e_1, \ldots, e_n = \pm I$;
(iii) whether or not π is an isomorphism, $\overline{\pi(a)} = \pi(\bar{a})$.

The example of $\mathfrak{A}(\mathfrak{H}) = \mathfrak{H} = \mathbb{H}$, with e_1, e_2, e_3 multiplication by the imaginary units i, j, k, is a case where $e_1 e_2 e_3 = -I$.

Proof of theorem (7.47). Let $\varepsilon_1, \ldots, \varepsilon_n$ be generators for $\mathfrak{A}_n$ such that
$$\varepsilon_j \varepsilon_k + \varepsilon_k \varepsilon_j = -2\delta_{jk} \quad \pi(\varepsilon_j) = e_j \quad (1 \leq j \leq n) \, .$$
Now the kernel of π is a two-sided ideal in $\mathfrak{A}_n$, so π will be an isomorphism if $\mathfrak{A}_n$ is simple, which is the case if $n \neq 4\ell + 3$ or if $n = 4\ell + 3$ and $\varepsilon_1 \cdots \varepsilon_n \notin \mathbb{R}$ (see (3.6),(3.19)). This proves (i) and the necessity half of (ii) because
$$e_1 \cdots e_n = \pi(\varepsilon_1 \cdots \varepsilon_n) = \lambda I$$
where $\lambda = \pm 1$ since
$$(e_1 \cdots e_n)^2 = (-1)^{\frac{1}{2}n(n+1)} = 1$$
when $n = 4\ell + 3$ and $\varepsilon_1 \cdots \varepsilon_n \in \mathbb{R}$. On the other hand,
$$\dim_{\mathbb{R}} \big(\mathfrak{A}(\mathfrak{H}) \big) < 2^n = \dim_{\mathbb{R}} (\mathfrak{A}_n)$$
when $e_1 \cdots e_n = \pm I$, for then $e_n = \mp e_1 \cdots e_{n-1}$. This completes the proof of (ii).

To prove (iii) observe first that
$$(e_{i_1} \cdots e_{i_k})^- = (-e_{i_k}) \cdots (-e_{i_1}) = (-1)^{\frac{1}{2}k(k+1)} e_{i_1} \cdots e_{i_k}$$
$$= (-1)^{\frac{1}{2}k(k+1)} \pi(\varepsilon_{i_1} \cdots \varepsilon_{i_k}) = \pi\big((\varepsilon_{i_1} \cdots \varepsilon_{i_k})^- \big)$$
since each e_j is skew-adjoint. Hence $\pi(\bar{a}) = \overline{\pi(a)}$ holds whenever π is an isomorphism since the reduced products $e_{i_1} \cdots e_{i_k}$ and $\varepsilon_{i_1} \cdots \varepsilon_{i_k}$ form a basis for $\mathfrak{A}(\mathfrak{H})$ and $\mathfrak{A}_n$ respectively. If, however, π is not an isomorphism, then $n = 4\ell + 3$ and $e_1 \cdots e_n = I$, say. Now $E = \frac{1}{2}(I + \varepsilon_1 \cdots \varepsilon_n)$ is a central idempotent in $\mathfrak{A}_n$ (see (3.19)), and
$$\overline{E} = \frac{1}{2}\big(I + (-1)^{\frac{1}{2}n(n+1)} \varepsilon_1 \cdots \varepsilon_n \big) = E$$

because $n = 4\ell + 3$. But, by the definition of E,

$$\pi(\mathfrak{A}_n E) = \pi(\mathfrak{A}_n) \quad , \quad \pi\big(\mathfrak{A}_n(I - E)\big) = 0 \ .$$

Consequently, $\ker \pi = \mathfrak{A}_n(I - E)$ and

$$\pi\big((a - aE)^-\big) = \pi(\bar{a} - \bar{a}E) = 0 \ ,$$

so the kernel of π is closed under conjugation. Hence the property $\pi(\bar{a}) = \overline{\pi(a)}$ is still valid. If $e_1 \cdots e_n = -I$, we set $E = \frac{1}{2}(I - \varepsilon_1 \cdots \varepsilon_n)$. This completes the proof.   ∎

When the $\mathfrak{A}_n$-module $\mathfrak{H}$ is a complex Hilbert space, let $\mathfrak{C}(\mathfrak{H})$ be the complex C^*-algebra generated in $\mathcal{L}(\mathfrak{H})$ by $e_1, \ldots, e_n$. Then, by the universal property, (7.45) extends to a homomorphism $\pi : \mathfrak{C}_n \to \mathfrak{C}(\mathfrak{H})$ from $\mathfrak{C}_n$ onto $\mathfrak{C}(\mathfrak{H})$, and by the previous proof we obtain the complex version of (7.47).

(7.47)′ Theorem.

The homomorphism $\pi : \mathfrak{C}_n \to \mathfrak{C}(\mathfrak{H})$ from $\mathfrak{C}_n$ into $\mathcal{L}(\mathfrak{H})$ satisfies the following:

(i) π is an isomorphism whenever $n = 2m$;

(ii) π fails to be an isomorphism if and only if $e_1 \cdots e_n \in \mathfrak{C}$.

The rigid structure of $\mathfrak{A}_n$ imposes a rigid structure on an $\mathfrak{A}_n$-module. We consider the real case first. As seen already, $\mathfrak{R}_n$ is a $\mathfrak{A}_n$-module with respect say to skew-adjoint operators $e_1, \ldots, e_n$ in $\mathcal{L}(\mathfrak{R}_n)$. But any finite-dimensional real Hilbert space $\mathcal{V}$, $\mathfrak{R}_n \otimes \mathcal{V}$ becomes a Hilbert space under the inner product

(7.48)

$$(\xi, \eta) = \sum_{j,k} (\xi_j, \eta_k)(u_j, v_k) \qquad \left(\xi = \sum_j \xi_j \otimes u_j \ , \ \eta = \sum_k \eta_k \otimes v_k \right) ,$$

the Hilbert–Schmidt inner product derived from those on $\mathfrak{R}_n$ and $\mathcal{V}$; furthermore, the operators $E_1 = e_1 \otimes I, \ldots, E_n = e_n \otimes I$ acting by the identity on $\mathcal{V}$ are clearly skew-adjoint operators on $\mathfrak{R}_n \otimes \mathcal{V}$ such that

(7.49) $$E_j E_k + E_k E_j = -2\delta_{jk}(e_0 \otimes I) \qquad (1 \le j, k \le n) \ .$$

Hence $\mathfrak{R}_n \otimes \mathcal{V}$ also is an $\mathfrak{A}_n$-module; in particular, general structural results for finite-dimensional real C^*-algebras ensure that there is an isometric *-isomorphism

$$\mathcal{L}(\mathfrak{R}_n \otimes \mathcal{V}) \sim \mathcal{L}(\mathfrak{R}_n) \oplus \cdots \oplus \mathcal{L}(\mathfrak{R}_n)$$

of $\dim_{\mathbf{R}}(\mathcal{V})$ copies of $\mathcal{L}(\mathfrak{R}_n)$. More generally, the following holds.

(7.50) Theorem.

If $\mathfrak{H}$ *is a real* $\mathfrak{A}_n$*-module, then, for some integer* $m \geq 1$*, there is an isometric *-isomorphism*

$$\mathcal{L}(\mathfrak{H}) \sim \mathcal{L}(\mathfrak{R}_n) \oplus \cdots \oplus \mathcal{L}(\mathfrak{R}_n)$$

from $\mathcal{L}(\mathfrak{H})$ *onto the direct sum of* m *copies of* $\mathcal{L}(\mathfrak{R}_n)$*.*

In view of the comments prior to (7.50), therefore, every real $\mathfrak{A}_n$-module is isomorphic to $\mathfrak{R}_n \otimes \mathbb{R}^m$ for some $m \geq 1$.

Proof of theorem (7.50). Being a finite-dimensional real C^*-algebra, $\mathcal{L}(\mathfrak{H})$ is isometrically *-isomorphic to a finite direct sum of full matrix rings $M(p, \mathbb{F})$ varying possibly with p and $\mathbb{F} = \mathbb{R}, \mathbb{C}$ or $\mathbb{H}$. On the other hand, by (7.42) and (7.43), $\mathfrak{H}$ is an $M(k, \mathbb{K})$-module where $\mathbb{K}$ and k are given by (7.38). Consequently, $\mathfrak{H}$ is isomorphic to a direct sum of, say, m copies of $\mathbb{K}^k$, and so the algebras $M(p, \mathbb{F})$ all coincide with $M(k, \mathbb{K})$. The theorem now follows from (7.42) and (7.43) again.   ∎

In the case of a complex $\mathfrak{A}_n$-module, the real spinor space $\mathfrak{R}_n$ is replaced by the space $\mathfrak{S}_n$ of complex spinors. Although these results describe in a sense all $\mathfrak{A}_n$-modules, various explicit realizations of them are useful in practice. Let $\mathcal{H}$ be an m-dimensional Hilbert space, real or complex, and let $\{\varepsilon_1, \ldots, \varepsilon_m\}$ be an orthonormal basis for $\mathcal{H}$. The exterior algebra $\Lambda^*(\mathcal{H})$ becomes a Hilbert space extending linearly the inner product

$$(7.51) \qquad (\xi_1 \wedge \cdots \wedge \xi_k \, , \, \eta_1 \wedge \cdots \wedge \eta_k) = \det[\xi_r \cdot \eta_s]$$

on $\Lambda^*(\mathcal{H})$ induced from that on $\mathcal{H}$, just as in the special case of (7.12). To each ε_j there correspond two operators in $\mathcal{L}(\Lambda^*(\mathcal{H}))$: the operator μ_j of exterior multiplication by ε_j, and its Hilbert space adjoint μ_j^* which is interior multiplication by ε_j. Thus μ_j is the linear extension of

$$(7.52)(i) \qquad \mu_j(1) = \varepsilon_j \quad , \quad \mu_j(\xi_1 \wedge \cdots \wedge \xi_k) = \varepsilon_j \wedge \xi_1 \wedge \cdots \wedge \xi_k \, ,$$

while

$$(7.52)(ii) \quad \mu_j^*(\xi_1 \wedge \cdots \wedge \xi_k) = \sum_{\ell=1}^{k} (-1)^{\ell-1} \varepsilon_j \cdot \xi_\ell (\xi_1 \wedge \cdots \wedge \hat{\xi}_\ell \wedge \cdots \wedge \xi_k) \, ,$$

where the hat $\hat{}$ as usual means that the symbol beneath it is to be omitted. By (2.22) or direct calculation,

$$(7.53) \qquad \mu_j \mu_k + \mu_k \mu_j = 0 = \mu_j^* \mu_k^* + \mu_k^* \mu_j^* \quad , \quad \mu_j \mu_k^* + \mu_k^* \mu_j = \delta_{jk} I$$

for all j, k. There are two basic examples, taking $\mathcal{H} = \mathbb{R}^m$ or $\mathbb{C}^m$ and $\{\varepsilon_1, \ldots, \varepsilon_m\}$ the usual orthonormal basis.

(7.54) **(A)**   Define operators $e_1, \ldots, e_n$ on $\Lambda^*(\mathbb{C}^m)$, $n = 2m$, by

(7.54)(i) $e_j = \mu_j - \mu_j^*$, $e_{n-j+1} = i(\mu_j + \mu_j^*)$ $(1 \leq j \leq m)$.

These clearly are skew-adjoint operators on $\Lambda^*(\mathbb{C}^m)$ which by (7.53) satisfy

$$e_j e_k + e_k e_j = -2\delta_{jk} I .$$

Thus $\Lambda^*(\mathbb{C}^m)$ is a complex $\mathfrak{A}_n$-module, $n = 2m$. Since $\dim_{\mathbb{C}}(\Lambda(\mathbb{C}^m)) = 2^m$, the algebra $\mathcal{L}(\Lambda^*(\mathbb{C}^m))$ has complex dimension 2^n. Consequently, by (7.47)′, $\mathcal{L}(\Lambda^*(\mathbb{C}^m))$ is a realization of the complex Clifford algebra $\mathbb{C}_n$, $n = 2m$. Hence $\Lambda^*(\mathbb{C}^m)$ is a realization of the complex spinor space $\mathfrak{S}_n$, $n = 2m$. Now set

(7.54)(ii) $$e_{2m+1} = i^{m-1} e_1 \ldots e_{2m} = (-i) \prod_{j=1}^{m} (\mu_j^* \mu_j - \mu_j \mu_j^*) .$$

(Think of the property $k = ij$ for $\mathbb{H}$.) Then

$$(e_{2m+1})^2 = (-1)^{m-1}(e_1 \cdots e_{2m})^2 = (-1)^{m-1}(-1)^{m(2m-1)} = -1 ,$$

while

$$e_j e_{2m+1} + e_{2m+1} e_j$$
$$= i^{m-1}\big((-1)^j e_1 \cdots \hat{e}_j \cdots e_{2m} + (-1)^{2m-j} e_1 \cdots \hat{e}_j \cdots e_{2m}\big) = 0 .$$

Hence $\Lambda^*(\mathbb{C}^m)$ is also a complex $\mathfrak{A}_{2m+1}$-module, and it can be used to realize the decomposition in (7.42)(ii). Indeed,

$$e_1 \cdots e_{2m} e_{2m+1} = i^{m-1}(-1)^{m(2m-1)} \in \mathbb{C} ,$$

so the embedding $(z_1, \ldots, z_{2m+1}) \to \sum_j z_j e_j$ of $\mathbb{C}^{2m+1}$ into $\mathcal{L}(\Lambda^*(\mathbb{C}^m))$ extends only to a homomorphism, say $S : \mathbb{C}_{2m+1} \to \mathcal{L}(\Lambda^*(\mathbb{C}^m))$ from $\mathbb{C}_{2m+1}$ onto $\mathcal{L}(\Lambda^*(\mathbb{C}^m))$. But $\Lambda^*(\mathbb{C}^m) \oplus \Lambda^*(\mathbb{C}^m)$ also is an $\mathfrak{A}_{2m+1}$-module. It is not difficult to show now that

(7.54)(iii)
$$a \longrightarrow \big(S(a), S(a')\big)   ,   \mathbb{C}_{2m+1} \longrightarrow \mathcal{L}\big(\Lambda^*(\mathbb{C}^m)\big) \oplus \mathcal{L}\big(\Lambda^*(\mathbb{C}^m)\big)$$

defines an isomorphism realizing $\mathbb{C}_{2m+1}$.

(7.54) **(B)**   Define operators $e_1, \ldots, e_n$ on $\Lambda^*(\mathbb{R}^n)$ by

$$e_j = \mu_j - \mu_j^*   (1 \leq j \leq n) .$$

As in the previous example, the e_j are skew-adjoint operators on $\Lambda^*(\mathbb{R}^n)$ which satisfy

$$e_j e_k + e_k e_j = -2\delta_{jk} I   (1 \leq j, k \leq n) .$$

Consequently, $\Lambda^*(\mathbb{R}^n)$ is an $\mathfrak{A}_n$-module. But

$$e_1 \cdots e_n = \prod_{j=1}^{n} (\mu_j - \mu_j^*)$$

is an operator on $\Lambda^*(\mathbb{R}^n)$ such that

$$e_1 \cdots e_n : 1 \longrightarrow \left(\prod_{j=1}^{n} \mu_j - \mu_j^* \right) 1 = \left(\prod_{j=1}^{n} \mu_j \right) 1 = \varepsilon_1 \wedge \cdots \wedge \varepsilon_n \ ;$$

in particular, $e_1 \cdots e_n \notin \mathbb{R}$. Hence the real C^*-algebra $\mathfrak{A}(\Lambda^*(\mathbb{R}^n))$ generated in $\mathcal{L}(\Lambda^*(\mathbb{R}^n))$ by $\mu_j - \bar{\mu}_j$, $1 \leq j \leq n$, is a realization of $\mathfrak{A}_n$ whatever the value of n. Replacing $\mathbb{R}^n$ by $\mathbb{C}^n$ and letting $\mathfrak{C}(\Lambda^*(\mathbb{C}^n))$ be the complex algebra generated in $\mathcal{L}(\Lambda^*(\mathbb{C}^n))$ by $\mu_j - \mu_j^*$, $1 \leq j \leq n$, we see that $\mathfrak{C}(\Lambda^*(\mathbb{C}^n))$ is a realization of $\mathfrak{C}_n$ for every n.

8 Spin(V, Q) as a Lie group

In this section we shall discuss briefly the Lie group and Lie algebra properties of Spin(V, Q), invoking general results except where a result can be established very easily. Throughout the section (V, Q) will be an arbitrary real non-degenerate quadratic space, $\mathrm{End}\, V$ will be the algebra of linear transformations on V, and $GL(V)$ will be the group of invertible linear transformations from V to itself. As an open subset of $\mathrm{End}\, V$, $GL(V)$ inherits an analytic structure with respect to which the group operations are analytic. Thus $GL(V)$ is a Lie group, and its Lie algebra is $\mathrm{End}\, V$ equipped with the bracket operation

$$(8.1) \qquad [X, Y] = XY - YX \ , \qquad X, Y \in \mathrm{End}\, V \ .$$

Locally the analytic structure on $GL(V)$ can be described explicitly using the exponential mapping

$$(8.2) \qquad \exp : \mathrm{End}\, V \longrightarrow GL(V) \quad , \quad \exp X = \sum_{0}^{\infty} \frac{1}{n!} X^n \ ,$$

from the Lie algebra $\mathrm{End}\, V$ into $GL(V)$. For $X \to \exp X$ maps a sufficiently small neighborhood of O in $\mathrm{End}\, V$ bijectively onto a neighborhood of the identity I in $GL(V)$, and translates of this last neighborhood generate a family of charts covering $GL(V)$.

Now let G be a matrix subgroup of $GL(V)$ that is closed in the topology of $GL(V)$. Then there is a unique analytic structure on G with respect to which G is a Lie subgroup of $GL(V)$ having Lie algebra

$$(8.3) \qquad \mathfrak{G} = \left\{ X \in \mathrm{End}\, V : \exp tX \in G \ , \ t \in \mathbb{R} \right\} \ .$$

Furthermore, the exponential mapping $\exp : \mathfrak{G} \to G$ maps some neighborhood of O in $\mathfrak{G}$ bijectively onto a neighborhood of I in G.

(8.4) Examples.

(i) The *special linear group*

$$SL(V) = \big\{ A : A \in GL(V) , \ \det A = 1 \big\}$$

is a Lie subgroup of $GL(V)$ with Lie algebra

$$s\ell(V) = \big\{ X : X \in \operatorname{End} V , \ \operatorname{tr} X = 0 \big\} .$$

(ii) The *orthogonal group*

$$O(V,Q) = \big\{ A : A \in GL(V) , \ Q(Av) = Q(v) , \ v \in V \big\}$$

and *special orthogonal group*

$$SO(V,Q) = \big\{ A : A \in O(V,Q) , \ \det A = 1 \big\}$$

are Lie subgroups of $GL(V)$ whose respective Lie algebras $o(V,Q)$ and $so(V,Q)$ coincide with

(8.5) $\big\{ X \in \operatorname{End} V : B(Xu,v) + B(u,Xv) = 0 , \ u,v \in V \big\} .$

If $SO(V,Q)$ is not connected already, i.e., if Q does not have constant sign on V, the identity component $SO_0(V,Q)$ of $SO(V,Q)$ also is a Lie subgroup of $GL(V)$ whose Lie algebra coincides with (8.5). But the covering group $\widetilde{G}$ of a connected Lie group G always has a canonical Lie group structure such that the associated covering homomorphism $\widetilde{G} \to G$ is analytic. Hence by applying this general result to the various Spin groups we deduce the following.

(8.6) Theorem.

If (V,Q) is a positive-definite or negative-definite quadratic space, then there is a unique analytic structure on $\operatorname{Spin}(V,Q)$ with respect to which it becomes a Lie group such that the two-fold covering homomorphism $\sigma : \operatorname{Spin}(V,Q) \to SO(V,Q)$ is analytic. If (V,Q) is not definite, there is a corresponding result for $\operatorname{Spin}_0(V,Q)$ and $SO_0(V,Q)$.

This analytic structure on $\operatorname{Spin}(V,Q)$ or $\operatorname{Spin}_0(V,Q)$ can now be transported to all the remaining Spin, Pin and Spoin groups so that the corresponding two-fold covering homomorphisms are analytic. For the remainder of this section, however, we shall concentrate on describing more explicitly, though locally, the related analytic structures on the identity components of $\operatorname{Spin}(V,Q)$ and $SO(V,Q)$ using mappings from the respective Lie algebras. Since the Lie algebra of any Lie group depends only on its identity component, $\operatorname{Spin}(V,Q)$ and $\operatorname{Spin}_0(V,Q)$ will have the same Lie algebra whether or not (V,Q) is definite; we shall denote it by $\operatorname{spin}(V,Q)$, and write $\operatorname{spin}(n)$ for the Lie algebra of $\operatorname{Spin}(n)$. Similarly, $SO_0(V,Q)$ and $SO(V,Q)$ will always have the same Lie algebra $so(V,Q)$: the one defined in (8.5). In fact, $\operatorname{spin}(V,Q)$ and $so(V,Q)$

must be isomorphic since $\text{Spin}_0(V, Q)$ is a two-fold covering group of $SO_0(V, Q)$.

Now the Clifford algebra $\mathfrak{A} = \mathfrak{A}(V, Q)$ itself is a Lie algebra with respect to the bracket operation

(8.7) $$[X, Y] = XY - YX \qquad (X, Y \in \mathfrak{A}),$$

defined by Clifford multiplication, and each $X \in \mathfrak{A}$ determines an operator

(8.8) $$d\sigma(X) : \mathfrak{A} \longrightarrow \mathfrak{A} \quad , \quad d\sigma(X) : Y \longrightarrow [X, Y] .$$

On the other hand, the differential of the analytic covering map $\sigma : \text{Spin}(V, Q) \to SO(V, Q)$ determines a Lie algebra isomorphism $d\sigma : \text{spin}(V, Q) \to so(V, Q)$ satisfying

(8.9) $$\sigma(\exp X) = \exp(d\sigma(X)) \qquad (X \in \text{spin}(V, Q)) .$$

The next theorem shows that these two uses of the same notation $d\sigma$ are consistent.

(8.10) Theorem.

The Lie algebra $\text{spin}(V, Q)$ *can be identified with the Lie subalgebra* $\mathfrak{A}^{(2)}(V, Q)$ *of all bivectors in* $\mathfrak{A}(V, Q)$; *furthermore, when* $d\sigma(X)$ *is defined by* (8.8),

(i) $d\sigma(X) \in \text{End } V$ *if* $X \in \mathfrak{A}^{(2)}(V, Q)$, *and*

(ii) $X \to d\sigma(X)$ *is the Lie algebra isomorphism from* $\mathfrak{A}^{(2)}(V, Q)$ *onto* $so(V, Q)$ *satisfying* (8.9).

Proof. Let $\{e_j\}_{j=1}^n$, $n = \dim V$, be a normalized basis for V, and set

$$X_{jk} = \tfrac{1}{2} e_j e_k \qquad (1 \le j, k \le n) .$$

Then $\{X_{jk} : 1 \le j < k \le n\}$ is a basis for $\mathfrak{A}^{(2)}(V, Q)$ and the X_{jk} satisfy the commutation relation

(8.11) $$[X_{jk}, X_{rs}] = Q(e_j)\{\delta_{jr} X_{ks} - \delta_{js} X_{kr}\} - Q(e_k)\{\delta_{kr} X_{js} - \delta_{ks} X_{jr}\} ;$$

in particular, $\mathfrak{A}^{(2)}(V, Q)$ is a Lie subalgebra of $\mathfrak{A}(V, Q)$. On the other hand,

$$\exp(tX_{jk}) = \cos\left(\tfrac{1}{2}t\right) + e_j e_k \sin\left(\tfrac{1}{2}t\right) = e_j\left(e_k \sin\left(\tfrac{1}{2}t\right) - Q(e_j)e_j \cos\left(\tfrac{1}{2}t\right)\right)$$

when $Q(e_j)Q(e_k) = 1$, while

$$\exp(tX_{jk}) = \cosh\left(\tfrac{1}{2}t\right) + e_j e_k \sinh\left(\tfrac{1}{2}t\right) = e_j\left(e_k \sinh\left(\tfrac{1}{2}t\right) - Q(e_j)e_j \cosh\tfrac{1}{2}t\right)$$

when $Q(e_j)Q(e_k) = -1$. Hence, as an even product of elements from $\Sigma_{\mathbf{R}}(V)$, $\exp(tX_{jk})$ always lies in the identity component of $\text{Spin}(V, Q)$,

and so $X_{jk} \in \mathrm{spin}(V, Q)$. Since
$$\dim\big(\mathrm{spin}(V, Q)\big) = \dim\big(so(V, Q)\big) = \tfrac{1}{2}n(n-1) \, ,$$
however, the Lie algebra $\mathfrak{A}^{(2)}(V, Q)$ spanned by these X_{jk} is all of $\mathrm{spin}(V, Q)$.

Now fix X in $\mathfrak{A}^{(2)}(V, Q)$. Then $X = \sum_{j<k} a_{jk} X_{jk}$ where $A = [a_{jk}]$ is a skew-symmetric matrix in $\mathbb{R}^{n \times n}$, and, on V,
$$d\sigma(X)v = \sum_{j<k} a_{jk}[X_{jk}, v] = \sum_{j<k} a_{jk}\big(Q(e_j)E_{jk} - Q(e_k)E_{kj}\big)v$$
where E_{jk} is the linear transformation of V such that
$$E_{jk} : v \longrightarrow v_j e_k \qquad \left(v = \sum_j v_j e_j\right) .$$

Consequently, $X \to d\sigma(X)$ is a linear isomorphism from $\mathfrak{A}^{(2)}(V, Q)$ into $\mathrm{End}\, V$ such that

(8.12) $\qquad d\sigma(X_{jk}) = Q(e_j)E_{jk} - Q(e_k)E_{kj} \qquad (1 \le j, k \le n) .$

Actually, this is a linear isomorphism from $\mathfrak{A}^{(2)}(V, Q)$ into $so(V, Q)$, since
$$B\big(d\sigma(X)u, v\big) + B\big(u, d\sigma(X)v\big) = XB(u, v) - B(u, v)X = 0$$
for all u, v in V (see (8.5)). To check that it is also a Lie algebra isomorphism, observe that
$$\sigma(\exp tX)v = (\exp tX)v(\exp tX)^{-1} \, ,$$
and so
$$\frac{d}{dt}\big(\sigma(\exp tX)v\big)\big|_{t=0} = Xv - vX = [X, v] \, .$$
Thus the operator $d\sigma(X) : v \to [X, v]$ arises from the differential of the Lie group homomorphism $\sigma : \mathrm{Spin}(V, Q) \to SO(V, Q)$. Hence $X \to d\sigma(X)$ is a Lie algebra isomorphism from $\mathrm{spin}(V, Q)$ onto $so(V, Q)$ satisfying (8.9). ∎

The exponential mapping in conjunction with theorem (8.10) gives a mapping from $\mathrm{spin}(V, Q)$ and $so(V, Q)$ into the corresponding Lie groups, and hence determines locally the analytic structure on these groups. But there are similar such mappings which are of considerable interest because of the role they play in the notion of characteristic classes and in the index theorem for Dirac type operators on manifolds. Since the manifolds in question are all Riemannian, we shall develop these ideas solely for the positive-definite case, but many of the constructions can be carried through more generally. So, throughout the remainder of this section, (V, Q) will always denote a positive-definite quadratic space.

Cayley's rational parameterization of the orthogonal group $SO(V, Q)$ is well-known. To describe its principal properties, recall first the following results. If X is a real skew-symmetric $(n \times n)$-matrix, then it cannot always be put in diagonal form, but there always exists an orthogonal matrix S so that SXS^{-1} can be written in block diagonal form

$$(8.13) \qquad SXS^{-1} = \begin{bmatrix} X_1 & & & 0 \\ & \ddots & & \\ & & X_r & \\ 0 & & & 0 \end{bmatrix} \qquad \left(1 \leq r \leq [\tfrac{1}{2}n]\right)$$

where each X_j is a real (2×2)-matrix

$$(8.14) \qquad X_j = \begin{bmatrix} 0 & x_j \\ -x_j & 0 \end{bmatrix} \qquad (1 \leq j \leq r) .$$

On the other hand, let $e_j(t) = e_j(t_1, \dots, t_n)$, $1 \leq j \leq n$, be the *elementary symmetric polynomials*

$$(8.15) \qquad e_1(t) = \sum_j t_j , \quad e_2(t) = \sum_{j<k} t_j t_k, \quad \dots , \quad e_n(t) = t_1 \cdots t_n ,$$

arising from the expansion

$$(8.16) \qquad \prod_{j=1}^{n}(\lambda + t_j) = \lambda^n + e_1(t)\lambda^{n-1} + \cdots + e_n(t) .$$

(8.17) Theorem.
If X is a real skew-symmetric $n \times n$ matrix and SXS^{-1} is its block diagonal form (8.13), then

$$\det(\lambda I + X) = \lambda^n + \sum_{j=1}^{r} e_j(x_1^2, \dots, x_r^2)\lambda^{n-2j} ;$$

in particular, $I + X$ is always invertible.

Proof. Simply observe that

$$\det(\lambda I + X) = \det(\lambda I + SXS^{-1}) = \lambda^{n-2r} \prod_{j=1}^{r}(\lambda^2 + x_j^2) ,$$

and expand this last product. ∎

As the Lie algebra of $SO(V, Q)$ can be identified with the set of all skew-adjoint X in $\operatorname{End} V$ (see (8.5)), theorem (8.17) applies to $so(V, Q)$. Thus

$$A = (I - X)(I + X)^{-1} \qquad (X \in so(V, Q))$$

is well-defined. But

$$(I - X)(I + X)^{-1} = (I + X)^{-1}(I - X) .$$

for such X, so A can be written without ambiguity as

$$(8.18) \qquad A = \frac{I - X}{I + X} .$$

(8.19) Theorem.
  The Cayley mapping

$$X \longrightarrow \frac{I - X}{I + X}$$

is a bijection from $so(V, Q)$ onto the neighborhood

$$\{A \in SO(V, Q) : \det(I + A) \neq 0\}$$

of the identity in $SO(V, Q)$.

The Cayley mapping is not surjective because its range will not contain any A in $SO(V, Q)$ for which $\det(I + A) = 0$, i.e., any A having an eigenvalue -1. Nonetheless, $SO(V, Q)$ can be realized explicitly via the Cayley mapping as a finite union of open sets which prescribe the analytic structure completely. Fix an orthonormal basis $e_1, \ldots, e_n$, $n = \dim V$, for V, and let $\mathcal{J}$ be the finite subset of $SO(V, Q)$ whose matrix representation with respect to this basis is diagonal and has diagonal entries ± 1.

(8.20) Corollary.
  Every A in $SO(V, Q)$ can be written in the form

$$A = J \left(\frac{I - X}{I + X} \right)$$

for some X in $SO(V, Q)$ and J in $\mathcal{J}$. Furthermore, for a given J in $\mathcal{J}$, the choice of X is unique once it exists.

There is a corresponding result for $O(V, Q)$, replacing $\mathcal{J}$ by the larger finite subset of $O(V, Q)$ having the same matrix realization.

Proof of theorem (8.19). When

$$A = \frac{I - X}{I + X} ,$$

then

$$I + A = 2(I + X)^{-1} ,$$

so that $\det(I + A) \neq 0$, and X, A are mutually related by

$$(8.21) \qquad \begin{cases} \text{(i)} & A = (I - X)(I + X)^{-1} = (I + X)^{-1}(I - X) , \\ \text{(ii)} & X = (I - A)(I + A)^{-1} = (I + A)^{-1}(I - A) . \end{cases}$$

Thus the Cayley mapping is a bijection from $so(V, Q)$ into $GL(V, Q)$. But (8.21)(i) ensures that

$$A^t(I + X^t) = I - X^t \; ;$$

where $(\cdot)^t$ denotes the adjoint of an operator in $\operatorname{End} V$ with respect to the inner product B on $V \times V$ associated with Q. Consequently,

$$A^t(I - X) = I + X \; ,$$

since $X \in so(V, Q)$. Thus

$$A^t A = A^t \left(\frac{I - X}{I + X} \right) = I \; ,$$

or, in other words,

$$B(u, v) = B(A^t A u, v) = B(Au, Av) \qquad (u, v \in V) \; .$$

Hence A is an orthogonal transformation on V (see (4.1)). But $\det A = 1$ since

$$\det(I + X) = \det\left((I + X)^t\right) = \det(I - X) \; .$$

Hence $X \to A = (I - X)(I + X)^{-1}$ is a bijection from $so(V, Q)$ into $SO(V, Q)$. ∎

Proof of corollary (8.16). It is enough to show that for each A in $SO(V, Q)$ for which $\det(I + A) = 0$, there exists at least one J in $\mathcal{J}$ such that $\det(I + JA) \neq 0$. But, for any A in $SO(V, Q)$,

$$\sum_{\mathcal{J}} \det(I + JA) = 2^{n-1}(1 + \det A) = 2^n \; ,$$

so at least one term on the left-hand side must be non-zero. If $\det(I + JA) \neq 0$, then

$$JA = \frac{I - X}{I + X} \qquad \left(X \in so(V, Q)\right) \; ;$$

hence

$$A = J \left(\frac{I - X}{I + X} \right) \; ,$$

since $J^2 = I$. ∎

The Cayley mapping defines a mapping from $so(V, Q)$ into $SO(V, Q)$, so naturally one looks for a corresponding mapping $\tau : \operatorname{spin}(V, Q) \to \operatorname{Spin}(V, Q)$ such that

$$(8.22) \qquad \sigma\big(\tau(X)\big) = \frac{I - d\sigma(X)}{I + d\sigma(X)} \qquad \left(X \in \operatorname{spin}(V, Q)\right) \; .$$

The secret to locating this mapping τ is the observation that in the left-hand side of (8.22) the equality

$$(8.23) \qquad \sigma\big(\lambda\tau(X)\big) = \sigma\big(\tau(X)\big)$$

holds for all $\lambda \in \mathbb{R}$, $\lambda > 0$. Hence $\tau(X)$ need not belong to $\mathrm{Spin}(V,Q)$, but only to the half-lines $\{\lambda a : \lambda > 0,\ a \in \mathrm{Spin}(V,Q)\}$ in $\Gamma(V,Q) \cap \mathfrak{A}_+$ through $\mathrm{Spin}(V,Q)$. In a 'letter from Hades', R. Lipschitz pointed out that the Pfaffian is a mapping with this last property. It is conveniently introduced using the wedge product in $\mathfrak{A}(V,Q)$ defined by (7.16). Now for each X in $\mathfrak{A}^{(2)}(V,Q)$ the series

$$(8.24) \quad Pf(X) = I + 2X + \frac{2^2}{2!}X \wedge X + \cdots + \frac{2^m}{m!}X \wedge \cdots \wedge X + \cdots$$

is well-defined, because ultimately it terminates. For instance, (7.16) ensures that

$$Pf\left(\tfrac{1}{2}e_j e_k\right) = I + e_j e_k \qquad (1 \le j < k \le n)$$

whenever $\{e_1, \ldots, e_n\}$ is an orthonormal basis for V; more generally, when

$$(8.25) \qquad X = \sum_{j=1}^{r} x_j X_{2j-1,2j} = \sum_{j=1}^{r} \tfrac{1}{2} x_j e_{2j-1} e_{2j} ,$$

then, in terms of the reduced product basis for $\mathfrak{A}$ determined by this orthonormal basis for V,

$$(8.26) \qquad Pf(X) = I + \sum_{m=1}^{r}\left(\sum_{|\alpha|=2m} x_\alpha e_\alpha\right) = \prod_{j=1}^{r}(I + x_j e_{2j-1} e_{2j})$$

where

$$(8.27)(\mathrm{i}) \quad \alpha = (2j_1 - 1, 2j_1, 2j_2 - 1, \ldots, 2j_m) , \qquad j_1 < j_2 < \cdots < j_m ,$$

and

$$(8.27)(\mathrm{ii}) \qquad x_\alpha = x_{j_1} \cdots x_{j_m} .$$

In view of (8.13), every X in $\mathfrak{A}^{(2)}(V,Q)$ can be written in the form of (8.25) with respect to some orthonormal basis for V (depending on X, of course), and so it is sufficient to take $Pf(X)$ in the form of (8.26), since the definition of $Pf(X)$ clearly is independent of the choice of orthonormal basis. However, a comparison with the Pfaffian as it is usually defined in linear algebra is most readily established by expressing $Pf(X)$ via a reduced product basis for $\mathfrak{A}$ determined by a *fixed* orthonormal basis $\{e_1, \ldots, e_n\}$ for V. In this case, when

$$(8.28)(\mathrm{i}) \qquad X = \sum_{j<k} a_{jk} X_{jk} \qquad (X_{jk} = \tfrac{1}{2} e_j e_k)$$

is an arbitrary element of $\mathfrak{A}^{(2)}(V,Q)$, then

$$(8.28)(\mathrm{ii}) \qquad Pf(X) = I + \sum_{m}\left(\sum_{|\alpha|=2m} x_\alpha e_\alpha\right)$$

where $e_\alpha = e_{\alpha_1} \cdots e_{\alpha_{2m}}$ and

(8.29) $$x_\alpha = \frac{1}{2^m m!} \left(\sum_\sigma \operatorname{sgn}(\sigma) a_{j_1 j_2} \cdots a_{j_{2m-1} j_{2m}} \right)$$

the last sum being taken over all permutations σ of $\{\alpha_1, \ldots, \alpha_{2m}\}$. Classically in linear algebra, the Pfaffian of a skew-symmetric matrix $B = [b_{jk}]$ in $\mathbb{R}^{2m \times 2m}$ is defined by

(8.30) $$pf(B) = \frac{1}{2^m m!} \left(\sum_\sigma \operatorname{sgn}(\sigma) b_{j_1 j_2} \cdots b_{j_{2m-1} j_{2m}} \right)$$

with the sum being taken over all permutations σ of $\{1, 2, \ldots, 2m\}$; it has the fundamental properties
(8.31)

(i) $pf(SBS^{-1}) = (\det S) pf(B)$, (ii) $\det B = \big(pf(B) \big)^2$

for every S in $O(2m)$. The relationship betwen (8.28) and (8.30), as well as the Clifford algebra significance of the properties in (8.31), is easily seen. For if $A = [a_{jk}]$ is the skew-symmetric matrix in $\mathbb{R}^{n \times n}$ derived from (8.28)(i) by requiring that $a_{jk} + a_{kj} = 0$, then to each $\alpha = (\alpha_1, \ldots, \alpha_{2m})$ corresponds the minor B_α of A obtained from the α_1-, ..., α_{2m}-rows and α_1-, ..., α_{2m}-columns of A; for example,

$$B_\alpha = \begin{bmatrix} 0 & a_{12} & a_{14} & a_{1n} \\ -a_{12} & 0 & a_{24} & a_{2n} \\ -a_{14} & \cdot & 0 & a_{2n} \\ -a_{1n} & \cdot & \cdot & 0 \end{bmatrix}$$

when $\alpha = (1, 2, 4, n)$. Every such B_α is a skew-symmetric matrix in $\mathbb{R}^{2m \times 2m}$, and

(8.32) $$Pf(X) = I + \sum_m \left(\sum_{|\alpha|=2m} pf(B_\alpha) e_\alpha \right).$$

In particular, if n is even (say $n = 2N$), then

$$\frac{2^N}{N!} (X \wedge \cdots \wedge X) = pf(A) e_1 \cdots e_{2N}.$$

The $O(n)$-invariance property (8.31)(ii) of the classical Pfaffian follows immediately from this. Indeed, the left-hand side is independent of choice of orthonormal basis $e_1, \ldots, e_{2N}$ for V, while the right-hand side becomes

$$pf(SAS^{-1}) \varepsilon_1 \cdots \varepsilon_{2N} = pf(SAS^{-1}) \det S\, e_1 \cdots e_{2N}$$

under a change of basis

$$e_j \longrightarrow \varepsilon_j = \sum_{k=1}^{2N} b_{jk} e_k \qquad (1 \le j \le 2N)$$

determined by any S in $O(2N)$. Hence

(8.33)(i) $pf(SAS^{-1}) = pf(A) \det S \qquad (S \in O(2N))$,

establishing (8.31)(i). On the other hand, when S is chosen so that SAS^{-1} has block diagonal form

$$\begin{bmatrix} A_1 & & 0 \\ & \ddots & \\ 0 & & A_N \end{bmatrix} \quad , \quad A_j = \begin{bmatrix} 0 & a_j \\ -a_j & 0 \end{bmatrix} ,$$

then

(8.33)(ii)

$$\left(pf(A)\right)^2 = \left(pf(SAS^{-1})\right)^2 = a_1^2 \cdots a_N^2 = \det(SAS^{-1}) = \det A ,$$

establishing (8.31)(ii). This last property arises in an interesting way in the proof of the fundamental mapping property of the Pfaffian.

(8.34) Theorem.

The Pfaffian defines an injection

$$X \longrightarrow Pf(X) = I + \sum_{m \geq 1} \frac{2^m}{m!} (X \wedge \cdots \wedge X)$$

of $\mathrm{spin}(V, Q)$ *into* $\{\lambda a : \lambda > 0, a \in \mathrm{Spin}(V, Q)\}$ *such that*

(i) $\Delta\left(Pf(X)\right) = \det\left(I + d\sigma(X)\right)$, *(ii)* $\sigma\left(Pf(X)\right) = \dfrac{I - d\sigma(X)}{I + d\sigma(X)}$,

for all X in $\mathrm{spin}(V, Q)$; *furthermore, $X \to \sigma(Pf(X))$ is an injection of* $\mathrm{spin}(V, Q)$ *into* $SO(V, Q)$.

There is also an analogue of corollary (8.20). Let $\{e_1, \ldots, e_n\}$ be a fixed orthonormal basis for V. Then any even reduced product $e_{j_1} \cdots e_{j_{2m}}$ of these basis elements is in $\mathrm{Spin}(V, Q)$. Since

$$\sigma(e_\alpha) : e_k \longrightarrow e_\alpha e_k e_\alpha = \begin{cases} (-1)^{\frac{1}{2}|\alpha|} e_k & (k \notin \alpha) , \\ (-1)^{\frac{1}{2}|\alpha|+1} e_k & (k \in \alpha) , \end{cases}$$

for any even reduced product e_α (see (3.4)), it is clear that the matrix realization of $\sigma(e_\alpha)$ as an element of $SO(V, Q)$ is diagonal and has diagonal entries $\{\delta_1, \ldots, \delta_n\}$ where

$$\delta_k = \begin{cases} (-1)^{\frac{1}{2}|\alpha|} & (k \notin \alpha), \\ (-1)^{\frac{1}{2}|\alpha|+1} & (k \in \alpha) . \end{cases}$$

Hence the set $\mathcal{J}$ in corollary (8.20) consists of $\sigma(e_\emptyset)$ $(= I)$ as well as

(8.35) $\{\sigma(e_{j_1} \cdots e_{j_{2m}}) : m \geq 1 , \ 1 \leq j_1 < \cdots < j_{2m} \leq n\}$.

Theorem (8.34) together with that corollary thus gives the following.

(8.36) Corollary.

Every a in Spin(V, Q) *can be written as*

$$a = \left(\frac{1}{\det(I + d\sigma(X))}\right)^{1/2} \left(e_{j_1} \cdots e_{j_{2m}} Pf(X)\right)$$

for some X in spin(V, Q) *and even reduced product $e_{j_1} \cdots e_{j_{2m}}$; in particular, each A in* SO(V, Q) *can be written as*

$$A = J\sigma\left(Pf(X)\right) = J\left(\frac{I - d\sigma(X)}{I + d\sigma(X)}\right)$$

for some X in spin(V, Q) *and J in $\mathcal{J}$.*

Proof of theorem (8.34). Clearly $X \to Pf(X)$ is an injection from spin(V, Q) into $\mathfrak{A}(V, Q)$ since

$$Pf(X) = I + 2X + \sum_{m > 1} \frac{2^m}{m!} X \wedge \cdots \wedge X \; ;$$

in fact, it is an injection from spin(V, Q) into $\mathfrak{A}_+(V, Q)$ because each m-fold wedge product is in $\mathfrak{A}^{(2m)}$. Now fix X and choose an orthonormal basis $\{e_1, \ldots, e_n\}$ for V so that

$$X = \sum_j x_j X_{2j-1, 2j} = \sum_j \tfrac{1}{2} x_j e_{2j-1} e_{2j} \; .$$

Then

(8.37) $$Pf(X) = \prod_j (I + x_j e_{2j-1} e_{2j}) \; ,$$

and so

$$\Delta\left(Pf(X)\right) = 1 + \sum_{j_1 < \cdots < j_m} (x_{j_1} \cdots x_{j_m})^2 = \prod_j (1 + x_j^2) \; .$$

On the other hand,

(8.38) $$d\sigma(X) = \sum_j x_j (E_{2j-1, 2j} - E_{2j, 2j-1}) \; ,$$

where

(8.39) $$E_{2j-1, 2j} : v \longrightarrow v_{2j-1} e_{2j} \qquad \left(v = \sum_j v_j e_j\right) \; .$$

In this case,

$$\det(I + d\sigma(X)) = \prod_j (1 + x_j^2) \; .$$

This establishes (8.34)(i) and proves also that $Pf(X)$ is invertible in $\mathfrak{A}(V, Q)$. To complete the proof, therefore, it is enough to show that

(8.40) $$Pf(X)\xi = \eta Pf(X) \qquad (\xi, \eta \in V)$$

whenever

(8.41) $$(I - d\sigma(X))\xi = (I + d\sigma(X))\eta .$$

For then $Pf(X)$ is in $\Gamma(V, Q) \cap \mathfrak{A}_+$ and

$$\sigma\big(Pf(X)\big) = \frac{I - d\sigma(X)}{I + d\sigma(X)} ;$$

furthermore, since

$$\ker \sigma = \{\lambda I : \lambda \in \mathbb{R} \setminus \{0\}\}$$

(see (5.29)), $Pf(X)$ is uniquely determined among all A in $\Gamma(V, Q) \cap \mathfrak{A}_+$
satisfying

$$\sigma(A) = \frac{I - d\sigma(X)}{I + d\sigma(X)}$$

by the requirement that

$$A = I + \sum_{j=1}^{n} a_j \qquad (a_j \in \mathfrak{A}^{(i)}) ,$$

i.e., the scalar part of A is 1. Hence $X \to \sigma(Pf(X))$ is a bijection from
$\mathrm{spin}(V, Q)$ into $SO(V, Q)$.

To establish (8.40) when (8.41) holds, set $\xi = \sum_{j=1}^{n} \xi_j e_j$ and $\eta = \sum_{j=1}^{n} \eta_j e_j$. Suppose first that n is even, say $n = 2m$. Then, by (8.39),

$$(I - d\sigma(X))\xi = \sum_{j=1}^{m} \{(\xi_{2j-1} + x_j \xi_{2j})e_{2j-1} + (\xi_{2j} - x_j \xi_{2j-1})e_{2j}\} ,$$

and similarly for $(I + d\sigma(X))\eta$. Thus (8.39) is equivalent to

(8.42) $$\left.\begin{cases} \xi_{2j-1} + x_j \xi_{2j} = \eta_{2j-1} - x_j \eta_{2j}, \\ \xi_{2j} - x_j \xi_{2j-1} = \eta_{2j} + x_j \eta_{2j-1} \end{cases}\right\} \qquad (1 \le j \le m) .$$

On the other hand,

$$(1 + x_j e_{2j-1} e_{2j})\xi = (\xi - \xi_{2j-1} e_{2j-1} - \xi_{2j} e_{2j})(1 + x_j e_{2j-1} e_{2j})$$
$$+ \{(\xi_{2j-1} - x_j \xi_{2j})e_{2j-1} + (\xi_{2j} + x_j \xi_{2j-1})e_{2j}\} ,$$

while

$$\eta(1 + x_j e_{2j-1} e_{2j}) = (\eta - \eta_{2j-1} e_{2j-1} - \eta_{2j} e_{2j})(1 + x_j e_{2j-1} e_{2j})$$
$$+ \{(\eta_{2j-1} + x_j \eta_{2j})e_{2j-1} + (\eta_{2j} - x_j \eta_{2j-1})e_{2j}\}$$

since $e_j^2 = -1$. In view of (8.37) and (8.42), an obvious induction argument shows that both $Pf(X)\xi$ and $\eta Pf(X)$ coincide with

(8.43)
$$\sum_{j=1}^{m} \{(\xi_{2j-1} - x_j \xi_{2j})e_{2j-1} + (\xi_{2j} + x_j \xi_{2j-1})e_{2j}\} \prod_{k \ne j} (1 + x_k e_{2k-1} e_{2k}) ,$$

establishing (8.40) when n is even. But when n is odd, say $n = 2m + 1$, and $\xi = \sum_{j=1}^{2m+1} \xi_j e_j$, $\eta = \sum_{j=1}^{2m+1} \eta_j e_j$, property (8.41) is equivalent to (8.42) together with the extra condition $\xi_{2m+1} = \eta_{2m+1}$. In this case both $Pf(X)\xi$ and $\eta Pf(X)$ contain the term

$$\xi_{2m+1} \prod_{j=1}^{m} (1 + x_j e_{2j-1} e_{2j})$$

in addition to (8.43). Nonetheless, (8.40) still holds true. This completes the proof of theorem (8.34). ∎

9 Spin groups as classical Lie groups

Low-dimensional $\mathrm{Spin}_0(p, q)$ groups can be identified with classical Lie groups of matrices having real, complex, or quaternionic entries. In fact, this is often how spin groups arise in the literature without explicit mention of Clifford algebras.

The matrix groups

$$(9.1) \qquad SL(n, \mathbb{F}) = \{g \in \mathbb{F}^{n \times n} : \det g = 1\} \qquad (\mathbb{F} = \mathbb{R}, \mathbb{C}),$$

are completely familiar. To construct the form-preserving classical groups in $\mathbb{F}^{n \times n}$, $\mathbb{F} = \mathbb{R}$, $\mathbb{C}$ or $\mathbb{H}$, first set

$$(9.2) \qquad I_{pq} = \begin{array}{c} \\ p \\ q \end{array} \begin{array}{cc} p \quad\; q \\ \begin{bmatrix} -I_p & 0 \\ 0 & I_q \end{bmatrix} \end{array} \qquad (p + q = n)$$

where I_p is the identity matrix in $\mathbb{R}^{p \times p}$, and let

$$(9.3) \qquad I_{p,q}(v) = [z\ w] I_{pq} \begin{bmatrix} z^* \\ w^* \end{bmatrix}, \qquad v = [\,z\ w\,],$$

be the corresponding (real) quadratic form on $\mathbb{F}^n$, thinking of $\mathbb{F}^n$ as the space $\mathbb{F}^{1 \times n}$ of n-dimensional row vectors; thus in (9.3) z is a p-dimensional row vector, having conjugate transpose z^*, while w is a q-dimensional row vector having conjugate transpose w^*. An $n \times n$ matrix g acting on $\mathbb{F}^{1 \times n}$ by right matrix multiplication $v \to v \cdot g$ is said to *preserve* $I_{p,q}$ when $I_{p,q}(v \cdot g) = I_{p,q}(v)$ for all v in $\mathbb{F}^{1 \times n}$. This will happen if and only if $g I_{pq} g^* = I_{pq}$ where g^* is the conjugate transpose of g, and the set

$$(9.4) \qquad U(p, q; \mathbb{F}) = \{g \in \mathbb{F}^{n \times n} : g I_{pq} g^* = I_{pq}\}$$

of all such $I_{p,q}$-preserving g is a subgroup of $GL_n(\mathbb{F})$. In the case of

$U(p, q; \mathbb{R})$ this group is just a matrix realization for appropriate p, q of the orthogonal group $O(V, Q)$, $\dim V = n$, studied in section 4.

We shall also need to realize some of the low-dimensional spin groups as isomorphic copies of $U(p, q; \mathbb{F})$. For example, when $p = q = n$ and $\mathbb{F} = \mathbb{C}$, let $\tilde{I}_{n,n}$ be the (real) quadratic form defined on $\mathbb{C}^{2n}$ by

$$(9.5) \qquad \tilde{I}_{n,n}(z, w) = [z \; w] \tilde{I}_{nn} \begin{bmatrix} z^* \\ w^* \end{bmatrix} , \qquad z, w \in \mathbb{C}^n ,$$

where

$$(9.6) \qquad \tilde{I}_{nn} = \begin{bmatrix} 0 & -i \\ i & 0 \end{bmatrix} = C^* \begin{bmatrix} -I_n & 0 \\ 0 & I_n \end{bmatrix} C$$

and

$$(9.7) \qquad C = \frac{1}{\sqrt{2}} \begin{bmatrix} I_n & i \\ i & I_n \end{bmatrix} , \qquad i = \sqrt{-1}\, I_n .$$

Simple computations show that $g \to CgC^*$ is an isomorphism from the group

$$(9.8) \qquad \{ g \in \mathbb{C}^{2n \times 2n} : g \tilde{I}_{nn} g^* = \tilde{I}_{nn} \}$$

of all $\tilde{I}_{n,n}$-preserving matrices in $\mathbb{C}^{2n \times 2n}$ onto $U(n, n; \mathbb{C})$.

Finally, to conform with standard notation we set

$$(9.9) \qquad \begin{cases} \text{(i)} & SO(p, q) = \{ g \in U(p, q; \mathbb{R}) : \det g = 1 \} , \\ \text{(ii)} & SU(p, q) = \{ g \in U(p, q; \mathbb{C}) : \det g = 1 \} , \\ \text{(iii)} & Sp(p, q) = U(p, q; \mathbb{H}) . \end{cases}$$

Just as in (9.4), we further set

$$(9.10) \qquad \begin{cases} \text{(i)} & SO(n) = SO(n, 0) = SO(0, n) , \\ \text{(ii)} & SU(n) = SU(n, 0) = SU(0, n) , \\ \text{(iii)} & Sp(n) = Sp(n, 0) = Sp(0, n) . \end{cases}$$

It is well-known that $SU(p, q)$ and $Sp(p, q)$ are connected (semi-simple) Lie groups, which are locally compact, but not compact, when $\min(p, q) > 0$, whereas $SO(n)$, $SU(n)$ and $Sp(n)$ are all compact.

(9.11) Theorem.
Up to isomorphism
    (i) $\mathrm{Spin}(3) = SU(2) = Sp(1)$,
    (ii) $\mathrm{Spin}(4) = SU(2) \times SU(2) = Sp(1) \times Sp(1)$,
    (iii) $\mathrm{Spin}(5) = Sp(2)$,
    (iv) $\mathrm{Spin}_0(2, 1) = SU(1, 1) = SL(2, \mathbb{R})$,
    (v) $\mathrm{Spin}_0(3, 1) = SL(2, \mathbb{C})$,
    (vi) $\mathrm{Spin}_0(4, 1) = Sp(1, 1)$,
    (vii) $\mathrm{Spin}_0(4, 2) = SU(2, 2)$.

Since the Lie algebra of $\mathrm{Spin}_0(p,q)$ is the same as that of $SO_0(p,q)$, all of these isomorphisms are well-known at the Lie algebra level; so theorem (9.11) certainly is true locally. To establish the global isomorphism we proceed systematically, using the same basic construction every time. The appropriate Clifford algebra $\mathfrak{A}$ is realized inside $M(2,\mathbb{F}^{n\times n})$ and the relevant classical matrix group is identified with the group

$$(9.12) \qquad \{g \in \mathfrak{A} : \Delta(g) = 1\}$$

or a subgroup of this group, by relating the norm Δ on $\mathfrak{A}$ with the determinant function or with the matrices I_{nn}, $\tilde{I}_{nn}$ in $M(2,\mathbb{F}^{n\times n})$. The various constructions in (2.29)–(2.38) together with (3.17) become very useful at this stage. As the identification of $\mathbb{F}^{2n\times 2n}$ with $M(2,\mathbb{F}^{n\times n})$ suggests, it will be useful to have a block matrix formulation of the classical groups. Set

$$(9.13) \qquad g = \begin{array}{c} n \\ n \end{array}\!\! \overset{\displaystyle n \quad\; n}{\left[\begin{array}{c|c} a & b \\ \hline c & d \end{array}\right]}, \qquad v = n\overset{\displaystyle n \quad n}{\left[\, z \mid w \,\right]} \; ;$$

then, under right multiplication,

$$(9.14) \qquad g : v \longrightarrow v \cdot g = \left[za + wc \quad zb + wd\right],$$

and the condition $gI_{nn}g^* = I_{nn}$ is equivalent to

$$(9.15) \qquad aa^* - bb^* = I_n \;, \qquad ac^* = bd^* \;, \qquad dd^* - cc^* = I_n \;.$$

Similarly, the condition $g\tilde{I}_{nn}g^* = \tilde{I}_{nn}$ is equivalent to

$$(9.16) \qquad ab^* = ba^* \;, \qquad ad^* - bc^* = I_n \;, \qquad cd^* = dc^* \;.$$

On the other hand, because the relevant classical matrix group is being identified with a subgroup of $\mathfrak{A}$ on which Δ is real-valued, the value of $\Delta(g)$ can be calculated from the expression $\Delta(g) = g\bar{g}$ (see (5.16)).

Proof of theorem (9.11). Since the computations are much the same in each case, most of the details will be omitted.

(i) $\mathrm{Spin}(3) \cong SU(2)$. Under the identification

$$\mathfrak{A}_{0,2} \cong \mathbb{H} = \left\{ Z : Z = \begin{bmatrix} w & z \\ -\bar{z} & \bar{w} \end{bmatrix} \;,\; z,w \in \mathbb{C} \right\}$$

determined by

$$\nu : \mathbb{R}^2 \longrightarrow \mathbb{H} \;, \qquad \nu : (x,y) \longrightarrow \begin{bmatrix} 0 & z \\ -\bar{z} & 0 \end{bmatrix} \;, \qquad z = x + iy \;,$$

it can be shown that

$$Z' = \begin{bmatrix} w & -z \\ \bar{z} & \bar{w} \end{bmatrix} , \quad Z^* = \begin{bmatrix} \bar{w} & z \\ -\bar{z} & w \end{bmatrix} , \quad \overline{Z} = \begin{bmatrix} \bar{w} & -z \\ \bar{z} & w \end{bmatrix} ;$$

consequently,

$$\Delta(Z) = |z|^2 + |w|^2 = \det Z ,$$

and $\mathbf{R}^3$ can be identified with the space of traceless, skew-adjoint matrices

$$X = \begin{bmatrix} ix_1 & x_2 + ix_3 \\ -x_2 + ix_3 & -ix_1 \end{bmatrix}$$

in $\mathfrak{A}_{0,2}$. Thus

$$SU(2) = \left\{ \begin{bmatrix} a & b \\ -\bar{b} & \bar{a} \end{bmatrix} : |a|^2 + |b|^2 = 1 \right\} = \{ g \in \mathfrak{A}_{0,2} : \Delta(g) = 1 \} ,$$

and

$$\sigma : SU(2) \longrightarrow SO(3) \quad , \quad \sigma(g) : X \longrightarrow gXg^*$$

is a two-fold covering.

(ii) $\mathrm{Spin}(4) \cong SU(2) \times SU(2)$. Under the identification

$$\mathfrak{A}_{0,4} \cong \mathbf{H}^{2\times 2} = \left\{ Z : Z = \begin{bmatrix} a & b \\ c & d \end{bmatrix} , \ a,b,c,d \in \mathbf{H} \right\}$$

determined by

$$(9.17) \qquad \nu : \mathbf{R}^4 \longrightarrow \mathbf{H}^{2\times 2} \quad , \quad \nu : (x_0, \ldots, x_3) \longrightarrow \begin{bmatrix} 0 & w \\ -\bar{w} & 0 \end{bmatrix}$$

where $w, \bar{w}$ are the quaternions

$$w = x_0 e_0 + \cdots + x_3 e_3 \quad , \quad \bar{w} = x_0 e_0 - \cdots - x_3 e_3 ,$$

it can be shown that

$$Z' = \begin{bmatrix} a & -b \\ -c & d \end{bmatrix} , \quad Z^* = \begin{bmatrix} \bar{a} & -\bar{c} \\ -\bar{b} & d \end{bmatrix} , \quad \overline{Z} = \begin{bmatrix} \bar{a} & \bar{c} \\ \bar{b} & d \end{bmatrix} ;$$

consequently,

$$(9.18) \qquad \Delta(Z) = \begin{bmatrix} |a|^2 + |b|^2 & a\bar{c} + b\bar{d} \\ c\bar{a} + d\bar{b} & |c|^2 + |d|^2 \end{bmatrix} ,$$

and $\mathbf{R}^4$ can be identified with the space of matrices

$$X = \begin{bmatrix} 0 & x \\ -\bar{x} & 0 \end{bmatrix} \qquad (x \in \mathbf{H}) ,$$

in $\mathfrak{A}_{0,4}$. Thus

$$\mathrm{Spin}(4) = \left\{ \begin{bmatrix} a & 0 \\ 0 & b \end{bmatrix} : |a| = |b| = 1 \right\} \cong SU(2) \times SU(2)$$

and

$$\sigma : SU(2) \times SU(2) \longrightarrow SO(4) \quad , \quad \sigma\left(\begin{bmatrix} a & 0 \\ 0 & b \end{bmatrix} \right) : x \longrightarrow ax\bar{b} ,$$

is a two-fold covering.

(iii) $\text{Spin}(5) \cong Sp(2)$. Under the identification $\mathfrak{A}_{0,4} \cong \mathbb{H}^{2\times 2}$ given in part (ii), $\mathbb{R}^5$ can be identified with the space

$$\left\{ X = \begin{bmatrix} \alpha & w \\ -\bar{w} & \alpha \end{bmatrix} : \alpha \in \mathbb{R} , \ w \in \mathbb{H} \right\} , \qquad \Delta(X) = \alpha^2 + |w|^2 .$$

Now $Sp(2)$ consists of all $g = \begin{bmatrix} a & b \\ c & d \end{bmatrix}$ in $\mathbb{H}^{2\times 2}$ such that

$$|a|^2 + |b|^2 = 1 , \quad a\bar{c} + b\bar{d} = 0 , \quad |c|^2 + |d|^2 = 1 .$$

In view of (9.18), therefore,

$$Sp(2) = \left\{ g \in \mathfrak{A}_{4,0} : \Delta(g) = 1 \right\} ,$$

and

$$\sigma : Sp(2) \longrightarrow SO(5) , \qquad \sigma(g) : X \longrightarrow gXg^*$$

is a two-fold covering.

(iv) $\text{Spin}_0(2,1) \cong SU(1,1) \cong SL(2,\mathbb{R})$. Under the identification

$$\mathfrak{A}_{2,0} \cong \left\{ Z = \begin{bmatrix} z & w \\ \bar{w} & z \end{bmatrix} : z, w \in \mathbb{C} \right\}$$

determined by

$$\nu : \mathbb{R}^{2,0} \longrightarrow \mathbb{C}^{2\times 2} , \qquad \nu : (x,y) \longrightarrow \begin{bmatrix} 0 & z \\ \bar{z} & 0 \end{bmatrix} , \qquad z = x + iy ,$$

it can be shown that

$$Z' = \begin{bmatrix} z & -w \\ -\bar{w} & \bar{z} \end{bmatrix} , \qquad Z^* = \begin{bmatrix} \bar{z} & w \\ \bar{w} & z \end{bmatrix} , \qquad \overline{Z} = \begin{bmatrix} \bar{z} & -w \\ -\bar{w} & z \end{bmatrix} ;$$

consequently,

$$\Delta(Z) = |z|^2 - |w|^2 = \det Z$$

and $\mathbb{R}^{2,1}$ can be identified with the space of self-adjoint matrices of form

$$X = \begin{bmatrix} x_1 & x_2 + ix_3 \\ x_2 - ix_3 & x_1 \end{bmatrix} ,$$

in $\mathfrak{A}_{2,0}$. But then, by (9.15),

$$SU(1,1) = \left\{ \begin{bmatrix} \alpha & \beta \\ \bar{\beta} & \bar{\alpha} \end{bmatrix} : \alpha, \beta \in \mathbb{C} , \ |\alpha|^2 - |\beta|^2 = 1 \right\}$$

$$= \left\{ g \in \mathfrak{A}_{2,0} : \Delta(g) = 1 \right\} ,$$

and so

$$\sigma : SU(1,1) \longrightarrow SO_0(2,1) , \qquad \sigma(g) : X \longrightarrow gXg^*$$

is a two-fold covering.

On the other hand, under the identification

$$\mathfrak{A}_{1,1} \cong \mathbb{R}^{2\times 2} = \left\{ Z : Z = \begin{bmatrix} a & b \\ c & d \end{bmatrix} , \ a, b, c, d \in \mathbb{R} \right\}$$

determined by

$$\nu : \mathbb{R}^{1,1} \longrightarrow \mathbb{R}^{2\times 2} \quad , \quad \nu : (x,y) \longrightarrow \begin{bmatrix} x & -y \\ y & -x \end{bmatrix} \; ,$$

it can be shown that

$$Z' = \begin{bmatrix} d & c \\ b & a \end{bmatrix} \; , \quad Z^* = \begin{bmatrix} a & -c \\ -b & d \end{bmatrix} \; , \quad \overline{Z} = \begin{bmatrix} d & -b \\ -c & a \end{bmatrix} \; ;$$

consequently

$$\Delta(Z) = ad - bc = \det Z \; ,$$

and $\mathbb{R}^{1,2}$ can be identified with the space of self-adjoint matrices

$$X = \begin{bmatrix} x_1 + x_2 & x_3 \\ x_3 & x_1 - x_2 \end{bmatrix} \; ,$$

in $\mathfrak{A}_{1,1}$. Thus

$$SL(2,\mathbb{R}) = \{ g \in \mathfrak{A}_{1,1} : \Delta(g) = 1 \}$$

and

$$\sigma : SL(2,\mathbb{R}) \longrightarrow SO_0(1,2) \quad , \quad \sigma(g) : X \longrightarrow gXg^*$$

is a two-fold covering.

(v) $\mathrm{Spin}_0(3,1) \cong SL(2,\mathbb{C})$. Under the identification

$$\mathfrak{A}_{3,0} \cong \mathbb{C}^{2\times 2} = \left\{ Z : Z = \begin{bmatrix} a & b \\ c & d \end{bmatrix} \; , \; a,b,c,d \in \mathbb{C} \right\}$$

determined by

$$\nu : \mathbb{R}^{3,0} \longrightarrow \mathbb{C}^{2\times 2} \quad , \quad \nu : (x_1, x_2, x_3) \longrightarrow \begin{bmatrix} x_1 & x_2 + ix_3 \\ x_2 - ix_3 & -x_1 \end{bmatrix} \; ,$$

it can be shown that

$$Z' = \begin{bmatrix} \bar{d} & -\bar{c} \\ -\bar{b} & \bar{a} \end{bmatrix} \; , \quad Z^* = \begin{bmatrix} \bar{a} & \bar{c} \\ \bar{b} & \bar{d} \end{bmatrix} \; , \quad \overline{Z} = \begin{bmatrix} d & -b \\ -c & a \end{bmatrix} \; ;$$

consequently,

$$\Delta(Z) = ad - bc = \det Z$$

and $\mathbb{R}^{3,1}$ can be identified with the space of self-adjoint matrices

$$X = \begin{bmatrix} x_1 + x_2 & x_3 + ix_4 \\ x_3 - ix_4 & x_1 - x_2 \end{bmatrix} \; ,$$

in $\mathfrak{A}_{3,0}$. Thus

$$SL(2,\mathbb{C}) = \{ g \in \mathfrak{A}_{3,0} : \Delta(g) = 1 \}$$

and

$$\sigma : SL(2,\mathbb{C}) \longrightarrow SO_0(3,1) \quad , \quad \sigma(g) : X \longrightarrow gXg^*$$

is a two-fold covering.

There is another realization of isomorphism (v) in the form $\mathrm{Spin}_0(1,3) \cong SL(2,\mathbb{C})$, which is of historical importance. Under the identification

$$(9.19) \qquad \mathfrak{A}_{1,3} \cong \left\{ Z = \begin{bmatrix} z & w \\ w' & z' \end{bmatrix} : z, w \in \mathbb{C}^{2\times 2} \right\}$$

determined by

$$(9.20) \qquad \nu : \mathbb{R}^{1,3} \longrightarrow \mathbb{C}^{4\times 4} \quad , \quad \nu : (x_0, x_1, x_2, x_3) \longrightarrow \begin{bmatrix} 0 & w \\ w' & 0 \end{bmatrix}$$

where

$$w = x_0\sigma_0 + \cdots + x_3\sigma_3 \quad , \quad w' = x_0\sigma_0 - \cdots - x_3\sigma_3$$

and $\sigma_0, \ldots, \sigma_3$ are the Pauli matrices, it can be shown that

$$Z' = \begin{bmatrix} z & -w \\ -w' & z' \end{bmatrix} , \quad Z^* = \begin{bmatrix} \bar{z} & w^* \\ \bar{w} & z^* \end{bmatrix} , \quad \overline{Z} = \begin{bmatrix} \bar{z} & -w^* \\ -\bar{w} & z^* \end{bmatrix} ,$$

with respect to the operations on $\mathfrak{A}_{0,3} \cong \mathbb{C}^{2\times 2}$; consequently,

$$\Delta(Z) = \begin{bmatrix} \det z - \det w & \bar{z}w - w^*z' \\ z^*w' - \bar{w}z & \det z - \det w \end{bmatrix} ,$$

and $\mathbb{R}^{1,3}$ can be identified with

$$\left\{ \begin{bmatrix} 0 & w \\ w' & 0 \end{bmatrix} : w \in H(2,\mathbb{C}) \right\}$$

where

$$H(2,\mathbb{C}) = \left\{ \sum_{j=0}^{3} x_j\sigma_j : x_j \in \mathbb{R} \right\}$$

is the space of self-adjoint matrices in $\mathbb{C}^{2\times 2}$. Thus

$$\mathrm{Spin}_0(1,3) = \left\{ \begin{bmatrix} a & 0 \\ 0 & a' \end{bmatrix} : a \in SL(2,\mathbb{C}) \right\} \cong SL(2,\mathbb{C}) ,$$

and

$$\sigma : SL(2,\mathbb{C}) \longrightarrow SO_0(1,3) \quad , \quad \sigma(g) : w \longrightarrow awa^*$$

is a two-fold covering. The Clifford algebra $\mathfrak{A}_{1,3}$ is often called the *Dirac algebra* because the Dirac operator was first introduced by him using the embedding (9.20) of Minkowski space $\mathbb{R}^{1,3}$ into $\mathbb{C}^{4\times 4}$, though of course he did not use the framework of Clifford algebras. This will be discussed in detail in section 2 of chapter II.

(vi) $\mathrm{Spin}_0(4,1) \cong Sp(1,1)$. Under the identification

$$\mathfrak{A}_{4,0} \cong \mathbb{H}^{2\times 2} = \left\{ Z : Z = \begin{bmatrix} a & b \\ c & d \end{bmatrix} , \ a,b,c,d \in \mathbb{H} \right\}$$

determined by

$$\nu : \mathbf{R}^{4,0} \longrightarrow \mathbf{H}^{2\times 2} \quad , \quad \nu : (x_0, \ldots, x_3) \longrightarrow \begin{bmatrix} 0 & w \\ \bar{w} & 0 \end{bmatrix} ,$$

analogous to (9.17), it can be shown that

$$Z' = \begin{bmatrix} a & -b \\ -c & d \end{bmatrix} , \quad Z^* = \begin{bmatrix} \bar{a} & \bar{c} \\ \bar{b} & \bar{d} \end{bmatrix} , \quad \overline{Z} = \begin{bmatrix} \bar{a} & -\bar{c} \\ -\bar{b} & \bar{d} \end{bmatrix} ;$$

consequently,

$$\Delta(Z) = \begin{bmatrix} |a|^2 - |b|^2 & b\bar{d} - a\bar{c} \\ c\bar{a} - d\bar{b} & |d|^2 - |c|^2 \end{bmatrix} ,$$

and $\mathbf{R}^{4,1}$ can be identified with the space of matrices

$$X = \begin{bmatrix} \alpha & x \\ \bar{x} & \alpha \end{bmatrix} \quad (\alpha \in \mathbf{R} , \ x \in \mathbf{H}) ,$$

in $\mathbf{H}^{2\times 2}$. In view of (9.15), therefore,

$$Sp(1,1) = \{ g \in \mathfrak{A}_{4,0} : \Delta(g) = 1 \}$$

and

$$\sigma : Sp(1,1) \longrightarrow SO_0(4,1) \quad , \quad \sigma(g) : X \longrightarrow gXg^*$$

is a two-fold covering.

(vii) $\mathrm{Spin}_0(4,2) \cong SU(2,2)$. Under the identification

$$\mathfrak{A}_{4,1} \cong \mathbf{C}^{4\times 4} = \left\{ Z : Z = \begin{bmatrix} a & b \\ c & d \end{bmatrix} , \ a,b,c,d \in \mathbf{C}^{2\times 2} \right\}$$

determined by

$$\nu : \mathbf{R}^{4,1} \longrightarrow \mathbf{C}^{4\times 4} \quad , \quad \nu : (x_0, \ldots, x_4) \longrightarrow \begin{bmatrix} ix_4 & w \\ \bar{w} & -ix_4 \end{bmatrix} ,$$

it can be shown that

$$Z' = \begin{bmatrix} d' & c' \\ b' & a' \end{bmatrix} , \quad Z^* = \begin{bmatrix} \bar{a} & -\bar{c} \\ -\bar{b} & \bar{d} \end{bmatrix} , \quad \overline{Z} = \begin{bmatrix} d^* & -b^* \\ -c^* & a^* \end{bmatrix}$$

with respect to the operations on $\mathfrak{A}_{3,0} \cong \mathbf{C}^{2\times 2}$; consequently,

$$\Delta(Z) = \begin{bmatrix} ad^* - bc^* & ba^* - ab^* \\ cd^* - dc^* & da^* - cb^* \end{bmatrix} ,$$

and $\mathbf{R}^{4,2}$ can be identified with the space

$$\left\{ X = \begin{bmatrix} z & w \\ \bar{w} & z \end{bmatrix} : z \in \mathbf{C} , \ w \in \mathbf{H} \right\}$$

in $\mathbf{C}^{4\times 4}$. Hence, by (9.16),

$$SU(2,2) = \{ g \in \mathfrak{A}_{4,1} : \Delta(g) = 1 , \ \det g = 1 \} ,$$

and

$$\sigma : SU(2,2) \longrightarrow \mathrm{Spin}_0(4,2) \quad , \quad \sigma(g) : X \longrightarrow gXg^*$$

is a two-fold covering. ∎

Notes and remarks for chapter 1

1. For a good introduction to the general theory of quadratic forms and quadratic spaces, the reader is referred to the monographs of Marcus [79], O'Meara [83], and Porteous [88].

2. The original paper of Clifford on 'geometric algebras' is [23]. The construction of Clifford algebras is scattered throughout the literature; we call particular attention to the classical presentation of Chevalley [22].

3. Many of the structural results of this section may be found in [79].

4. A good, if condensed, presentation of orthogonal transformations may be found in [88]. For a thorough discussion of the conformal group and a proof of theorem (4.16), consult the monograph of Beardon [10].

5. The term 'transformer' has been coined by Takahashi, who gives a good presentation of transformers in [98]. The role of the norm function is discussed in [88] and elsewhere. The term 'spoin' has apparently been introduced by researchers in Clifford analysis; see for example [17]. The spoin construction is briefly discussed in [88].

6. The connectivity of $\text{Spin}(p, q)$ is discussed in some detail in [88]. For a thorough discussion of the connectivity of $SO(p, q)$ the reader is referred to the monograph of Helgason [56].

7. A good reference for the C^*-algebra ideas in this section is the work of Goodearl [50]. For a thorough discussion of the periodicity theorem and the structure of Clifford modules, the topologically sophisticated reader may wish to consult the work of Husemoller [61].

8. The book by Brocker and tom Dieck [19] is a good basic reference for the Euclidean case. The material in this section is also discussed in some detail in [88]. The 'letter from Hades' concerning the Pfaffian is ascribed to Lipschitz and may be found in [75].

9. Extensive calculations, in the spirit of this section, are at the very heart of a recent book by Reese Harvey [54]. Some of the identifications made in this section may also be found in [88].

2

Dirac operators and Clifford analyticity

In this chapter we generalize the classical theory of Hardy spaces of analytic functions to higher dimensions by means of the Euclidean Dirac operator, a first-order, constant-coefficient differential operator having coefficients in $\mathfrak{A}_n$. In many important respects it is the appropriate substitute for the Cauchy–Riemann operators ∂ and $\bar{\partial}$ in higher dimensions. The Dirac operator is distinguished by the fact that it factorizes the Laplacian, so that the functions in its kernel – the so-called *Clifford analytic* functions – are automatically harmonic. We present the basic results in the theory of Clifford analytic functions, emphasizing parallels with classical function theory, and develop a theory of Hardy spaces for such functions for Lipschitz domains in $\mathbb{R}^n$. Along the way there will be important connections with the theory of singular integrals and elliptic boundary value problems on minimally smooth domains.

1 Cauchy–Riemann operators

There are three distinct but equivalent ways of introducing *analytic* complex-valued functions $f = f(z)$ of a complex variable $z = x + iy$:

(1.1)(i) via the complex derivative

$$f'(z_0) = \lim_{z \to z_0} \frac{f(z) - f(z_0)}{z - z_0} \, ,$$

(ii) via power series expansions

$$f(z) = \sum_{n=0}^{\infty} a_n (z - z_0)^n \ ,$$

and

(iii) as solutions of the Cauchy–Riemann equation

$$\bar{\partial}_z f = 0 \ , \quad \bar{\partial}_z = \frac{\partial}{\partial x} + i \frac{\partial}{\partial y} \ .$$

The power and elegance of the theory of analytic functions in all of mathematics and its applications led naturally to the search for similar theories in different settings. In these various settings, however, some of the definitions above may not make sense, as when the division property needed for (1.1)(i) is not well-defined; or a particular definition may be too restrictive and so yield too small a class of 'analytic' functions. In the opposite direction, a definition might be so accommodating that all but the most 'uninteresting' functions are 'analytic.' This is borne out in the simplest of generalizations where the complex field $\mathbb{C}$ is replaced by the quaternionic skew-field $\mathbb{H}$. Indeed, let $f : \mathbb{H} \to \mathbb{H}$ be defined on $\mathbb{H} \sim \mathbb{R}^4$. Then f is said to have a *(left) quaternionic derivative* at w_0 when

$$f'(w_0) = \lim_{h \to 0} \frac{1}{h} \big(f(w_0 + h) - f(w_0) \big)$$

exists. The quotient is well-defined within $\mathbb{H}$ for all f, but the limit exists at most for linear functions, and even then some linear functions fail to have a derivative. Thus this natural generalization of (1.1)(i) yields too small a class of analytic functions. On the other hand, since multiplication in $\mathbb{H}$ is non-commutative, the natural monomial functions of a quaternionic variable are $w \to a_0 w a_1 \cdots a_{r-1} w a_r$ for fixed $a_0, \ldots, a_r$ in $\mathbb{H}$. If $w = t + xi + yj + zk$ is the representation of w in terms of the imaginary units i, j, k, however, the coordinates t, x, y, z themselves are sums

$$t = \tfrac{1}{4}(w - iwi - jwj - kwk) \ , \qquad x = \tfrac{1}{4i}(w - iwi + jwj + kwk) \ ,$$
$$y = \tfrac{1}{4j}(w + iwi - jwj + kwk) \ , \qquad z = \tfrac{1}{4k}(w + iwi + jwj - kwk) \ ,$$

of quaternionic monomials. Consequently, every real polynomial in t, x, y, z would be analytic if (1.1)(ii) were generalized to $\mathbb{H}$ to include quaternionic polynomials. Hence, by contrast, the natural generalization of (1.1)(ii) yields too large a class of analytic functions. This leaves only (1.1)(iii) as a possible source for a definition.

(1.2) Definition.

(Fueter) *Let U be an open set in $\mathbb{H}$ and $f : U \to \mathbb{H}$ a quaternion-valued function defined on U. Then f is said to be left regular, resp. right regular, when*

$$(1.3) \qquad \mathcal{D}_L f = \frac{\partial f}{\partial t} + i\frac{\partial f}{\partial x} + j\frac{\partial f}{\partial y} + k\frac{\partial f}{\partial z} = 0 ,$$

resp.

$$(1.4) \qquad \mathcal{D}_R f = \frac{\partial f}{\partial t} + \frac{\partial f}{\partial x}i + \frac{\partial f}{\partial y}j + \frac{\partial f}{\partial z}k = 0 ,$$

on U.

The fundamental properties

$$ij + ji = jk + kj = ki + ik = 0 , \qquad i^2 = j^2 = k^2 = -1$$

of the imaginary units in $\mathbb{H}$ ensure that

$$(1.5) \qquad \phi(w) = \frac{1}{|w|^4}(t - xi - yj - zk) = \frac{\overline{w}}{|w|^4}$$

is both left- and right-regular, for instance. Intimate connection of $\mathcal{D}_L$ and $\mathcal{D}_R$ with the classical Cauchy–Riemann operators $\overline{\partial}, \partial$ on $\mathbb{C}$ can be established using the properties $k = -ji = ij$ to identify $w \in \mathbb{H}$ with $(\xi, \zeta) \in \mathbb{C}^2$ by

$$w = t + xi + yj + zk = \xi + j\overline{\zeta} = \xi + \zeta j$$

where

$$(1.6) \qquad \xi = t + xi \quad , \quad \zeta = y + zi \in \mathbb{C} ;$$

for then

$$(1.7) \qquad \mathcal{D}_L f = (\overline{\partial}_\xi + j\partial_\zeta)f \quad , \quad \mathcal{D}_R f = f(\overline{\partial}_\xi + \overline{\partial}_\zeta j) .$$

Not surprisingly, therefore, a theory of regular functions, left or right, has been developed following broadly the lines of the corresponding theory for analytic functions, with the function ϕ in (1.5) playing the role of the Cauchy kernel. It is a special case of the theory of Clifford analytic functions to be established shortly. Hence, with analysis on differential manifolds in mind, the approach in (1.1)(iii) via partial differential equations would seem to hold out the best hope for developing a theory which is sufficiently rich and far-reaching, which at the same time is sufficiently restrictive.

Even from the viewpoint of partial differential equations there are several equivalent formulations of analyticity. Let Ω be a open, simply-connected subset of $\mathbb{C}$, and let $f : \Omega \to \mathbb{C}$ be a complex-valued, $\mathcal{C}^1$-function on Ω.

(A) The simplest and most familiar formulation is in terms of the

Cauchy–Riemann equations. If $f = u + iv$, where u, v are real-valued, C^1-functions, then f is analytic on Ω if and only if u, v satisfy

$$(1.8) \qquad \frac{\partial u}{\partial x} = \frac{\partial v}{\partial y} \quad , \quad \frac{\partial u}{\partial y} + \frac{\partial v}{\partial x} = 0$$

throughout Ω.

(B) Equivalently, let

$$(1.9) \qquad \bar{\partial}_z = \frac{\partial}{\partial x} + i\frac{\partial}{\partial y} \quad , \quad \partial_z = \frac{\partial}{\partial x} - i\frac{\partial}{\partial y} \quad , \qquad z = x + iy \, ,$$

be the *Cauchy–Riemann* operators. Then f is analytic on Ω if and only if $\bar{\partial}f = 0$ on Ω, i.e., f is in the kernel of $\bar{\partial}$ on Ω. Since $\bar{\partial}f$ is conjugate to $\partial\bar{f}$, functions in the kernel of ∂ are said to be *conjugate-analytic*; in particular, f is analytic if and only if $\bar{f}$ is conjugate-analytic.

(C) In real terms, define $F : \Omega \to \mathbf{R}^2$ by $F = (v, u)$. Then

$$(1.10) \qquad \operatorname{div} F = \frac{\partial v}{\partial x} + \frac{\partial u}{\partial y} \quad , \quad \operatorname{curl} F = \left(0, 0, \frac{\partial u}{\partial x} - \frac{\partial v}{\partial y}\right) .$$

Consequently, f is analytic if and only if $\operatorname{div} F = 0$ and $\operatorname{curl} F = 0$ on Ω; by thinking of F as a 1-form on $\Omega \subseteq \mathbf{R}^2$, we can also express this last pair of equations by saying that F is in the kernel of the Hodge–deRham (d, d^*)-system on 1-forms on Ω.

(D) Since a solution on Ω of $\operatorname{curl} F = 0$ and $\operatorname{div} F = 0$ is precisely the gradient $F = \nabla\Phi$ of a real-valued harmonic function Φ, $f = u + iv$ is analytic on Ω if and only if $u = \frac{\partial\Phi}{\partial y}$ and $v = \frac{\partial\Phi}{\partial x}$, i.e., $f = i\partial\Phi$, where Φ is a real-valued harmonic function on Ω.

(E) The Cauchy–Riemann operators themselves can be expressed in real form by identifying $\mathbf{C}$ with the subalgebra

$$\left\{ \begin{bmatrix} a & -b \\ b & a \end{bmatrix} : a, b \in \mathbf{R} \right\}$$

of $M(2, \mathbf{R})$, and $f = u + iv$ with $f = \begin{bmatrix} u & -v \\ v & u \end{bmatrix}$. For then

$$(1.11) \qquad \bar{\partial}f = \left(\begin{bmatrix} 1 & 0 \\ 0 & 1 \end{bmatrix} \frac{\partial}{\partial x} + \begin{bmatrix} 0 & -1 \\ 1 & 0 \end{bmatrix} \frac{\partial}{\partial y} \right) \begin{bmatrix} u & -v \\ v & u \end{bmatrix} ,$$

while

$$(1.12) \qquad \partial f = \left(\begin{bmatrix} 1 & 0 \\ 0 & 1 \end{bmatrix} \frac{\partial}{\partial x} - \begin{bmatrix} 0 & -1 \\ 1 & 0 \end{bmatrix} \frac{\partial}{\partial y} \right) \begin{bmatrix} u & -v \\ v & u \end{bmatrix} .$$

Hence in all cases an analytic function arises as a vector-valued or matrix-valued solution of a particular first-order, constant-coefficient, elliptic system of partial differential equations, and in every case the *components* of these solutions are harmonic real-valued functions. This last property is expressed most vividly either by regarding the *first-order system* as arising from a *factorization* of a *second-order operator* in the sense that

(1.13) $$\partial\bar{\partial} = \bar{\partial}\partial = \Delta ,$$

or by *characterizing* the solutions as gradients of harmonic functions. Whatever the interpretation, however, the role of the first-order system is to single out a particular subspace, the space of analytic functions on Ω, from the space of all harmonic functions on Ω. The importance of analytic function theory stems from the richness of the properties enjoyed by the functions in this subspace, but not necessarily enjoyed by all harmonic functions. The *Cauchy integral* property is one such example: if f is analytic on Ω and γ is a positively oriented simple closed curve in Ω, then

(1.14) $$f(z) = \frac{1}{2\pi i} \int_{\gamma} \frac{f(w)}{w - z}\, dw$$

for every z inside γ; conversely, if (1.14) holds for all such γ and z, then f is analytic. Many of the fundamental properties of analytic functions follow quickly from this *reproducing kernel property*. The distinction between analytic and harmonic functions is also apparent in their subharmonicity properties.

(1.15) Definition.
   Let Ω be an open connected set in $\mathbb{R}^n$ and $F : \Omega \to \mathbb{R}$ a real-valued function on Ω having continuous second-order derivatives in the set $\Omega' = \{x \in \Omega : F(x) > 0\}$. Then F is said to be subharmonic on Ω if $\Delta F \geq 0$ on Ω'.

   By the mean-value property for harmonic functions, a function $F : x \to |f(x)|^p$ will be subharmonic whenever $p \geq 1$ and f is a (possibly vector-valued) harmonic function on Ω ($\subseteq \mathbb{R}^n$); as the example of $f(x_1, \ldots, x_n) = x_1$ shows, the subharmonicity of $x \to |f(x)|^p$ may fail for harmonic functions when $p < 1$. For analytic functions, however, the subharmonicity is still valid.

(1.16) Theorem.
   If f is analytic in a simply connected open set Ω in $\mathbb{C}$, then both $\log|f|$ and $|f|^p$, $p > 0$, are subharmonic on Ω.

Proof. In both cases the proof makes use of the factorization $\Delta = \bar{\partial}\partial$ of the Laplacian by the Cauchy–Riemann operators. Let $\phi = \partial f$. Then

$$\partial \bar{f} = \bar{\partial} f = \bar{\partial} \phi = 0 \,, \qquad \bar{\partial}\bar{f} = \bar{\phi} \,.$$

Now $\log |f| > 0$ holds if and only if $|f| > 1$; consequently,

$$\Omega' = \{z : \log|f| > 0\} \subseteq \Omega'' = \{z : |f| \neq 0\} \,.$$

On Ω'',

$$\Delta|f| = \bar{\partial}\partial \log\left(f^{\frac{1}{2}}\bar{f}^{\frac{1}{2}}\right) = \bar{\partial}\left(\frac{1}{|f|}\partial(f^{\frac{1}{2}}\bar{f}^{\frac{1}{2}})\right)$$

$$= \bar{\partial}[\phi\bar{f}^{\frac{1}{2}}/(|f|f^{\frac{1}{2}})] = \bar{\partial}(\phi/f) = 0 \,.$$

So $\log|f|$ actually is harmonic on Ω'', hence subharmonic on Ω. On the other hand, for $p > 0$,

$$\Delta|f|^p = \bar{\partial}\partial(f^{\frac{1}{2}p}\bar{f}^{\frac{1}{2}p}) = \tfrac{1}{2}p\bar{\partial}(f^{\frac{1}{2}(p-2)}\phi\bar{f}^{\frac{1}{2}p})$$

$$= \tfrac{1}{2}pf^{\frac{1}{2}(p-2)}\phi(\bar{\partial}\bar{f}^{\frac{1}{2}p}) = \tfrac{1}{4}p^2|f|^{p-2}|\phi|^2 \geq 0$$

on Ω''. Hence $|f|^p$, $p > 0$, also is subharmonic on Ω.   ∎

In summary then, the Cauchy–Riemann operators $(\bar{\partial}, \partial)$ belong to the class of differential operators ð which on Euclidean space have the following properties:

(1.17)(i) each ð is a first-order system of constant coefficient, linear, homogeneous elliptic differential operators

(ii) every function in the kernel of ð has harmonic components,

(iii) for each ð there is a reproducing kernel reproducing solutions of ð$f = 0$ from their boundary values.

The Fueter operators belong to this class also. For if $\overline{\mathcal{D}}_L$ and $\overline{\mathcal{D}}_R$ are the operators

$$(1.18) \qquad \overline{\mathcal{D}}_L f = \frac{\partial f}{\partial t} - i\frac{\partial f}{\partial x} - j\frac{\partial f}{\partial y} - k\frac{\partial f}{\partial z}$$

and

$$(1.19) \qquad \overline{\mathcal{D}}_R f = \frac{\partial f}{\partial t} - \frac{\partial f}{\partial x}i - \frac{\partial f}{\partial y}j - \frac{\partial f}{\partial z}k$$

conjugate to $\mathcal{D}_L$ and $\mathcal{D}_R$ respectively, then the pairs $(\mathcal{D}_L, \overline{\mathcal{D}}_L)$, $(\mathcal{D}_R, \overline{\mathcal{D}}_R)$ factor the Laplacian on $\mathbb{R}^4$ in the sense that

$$(1.20) \qquad \mathcal{D}_L\overline{\mathcal{D}}_L = \overline{\mathcal{D}}_L\mathcal{D}_L = \Delta \,, \qquad \mathcal{D}_R\overline{\mathcal{D}}_R = \overline{\mathcal{D}}_R\mathcal{D}_R = \Delta \,.$$

A function in the kernel of any of these four operators will thus have harmonic components. Consequently, all four have properties (1.17)(i),(ii). But, provided one accepts the existence of a Cauchy integral theory for these operators (as one can), they will also have the third of the properties in (1.17). Hence the Fueter operators $\mathcal{D}_L$, $\overline{\mathcal{D}}_L$, $\mathcal{D}_R$, and $\overline{\mathcal{D}}_R$ all belong to the same class as the Cauchy–Riemann operators $\overline{\partial}$ and ∂. The search for a function theory analogous to analytic function theory, but valid on more general manifolds, thus comes down to the following fundamental though vague problems.

I. *On Euclidean space describe other first-order systems having the three properties in (1.17), and classify as far as possible all such systems.*

II. *Again on Euclidean space, develop the function theory associated with systems satisfying (1.17).*

III. *By allowing non-constant coefficient and non-homogeneous systems in (1.17) develop the corresponding theory whenever possible on more general Riemannian manifolds, for instance on oriented Riemannian manifolds with or without boundary.*

In view of their connection with (linear) partial differential equations, it is not surprising that these problems are connected also with many of the most fundamental aspects of analysis on manifolds.

2 Dirac operators past and present

In the last section we saw the crucial role of the factorization $\Delta = \overline{\partial}\partial$ in complex function theory. The Fueter operators $\mathcal{D}$ and $\overline{\mathcal{D}}$ yielded an analogous factorization of the Laplacian in $\mathbb{R}^4$. In general, the term 'Dirac operator' is used to refer to any first-order operator which factorizes the 'Laplacian' for a given quadratic space of arbitrary signature. In this section we shall make these concepts precise and put them in some historical perspective.

In 1928, Dirac introduced a first-order linear operator in order to express the square root of the wave operator

$$\Box = \frac{\partial^2}{\partial x_0^2} - \left(\frac{\partial^2}{\partial x_1^2} + \frac{\partial^2}{\partial x_2^2} + \frac{\partial^2}{\partial x_3^2} \right) \ .$$

Although Dirac did not explicitly use the Clifford algebra framework, he constructed this first order operator using the Dirac algebra which (as we have seen in chapter 1, section 9, example (v)) is a particular realization of the algebra $\mathfrak{A}_{1,3}$. We briefly review the construction of the Dirac algebra here, using essentially the same method that Dirac did. Let $\sigma_0, \sigma_1, \sigma_2, \sigma_3$ denote the Pauli matrices, introduced in chapter 1, (2.3). We recall that

$$\sigma_0^2 = \sigma_1^2 = \sigma_2^2 = \sigma_3^2 = I \quad , \quad \sigma_j \sigma_k = -i\sigma_\ell$$

when $\{j, k, \ell\}$ is any cyclic permutation of $\{1, 2, 3\}$. We use the Pauli matrices to construct the Dirac γ-matrices

$$\gamma_0 = \begin{bmatrix} 0 & \sigma_0 \\ \sigma_0 & 0 \end{bmatrix} \quad, \quad \gamma_j = \begin{bmatrix} 0 & \sigma_j \\ -\sigma_j & 0 \end{bmatrix} \quad, \qquad j = 1, 2, 3 \ .$$

The γ-matrices satisfy

$$\gamma_0^2 = I \ , \quad \gamma_1^2 = \gamma_2^2 = \gamma_3^3 = -I \ , \quad \gamma_j \gamma_k = -\gamma_k \gamma_j \ , \qquad j \neq k$$

and are easily seen to generate the algebra $\mathfrak{A}_{1,3}$ as a real subalgebra of $\mathbb{C}^{4 \times 4}$. Consequently, as a differential operator on, say, $\mathbb{C}^4$-valued (i.e., spinor-valued) functions on an open set $U \subseteq \mathbb{R}^4$,

$$(2.1) \qquad D = \gamma_0 \frac{\partial}{\partial x_0} + \gamma_1 \frac{\partial}{\partial x_1} + \gamma_2 \frac{\partial}{\partial x_2} + \gamma_3 \frac{\partial}{\partial x_3}$$

has the desired property: $D^2 = \square$.

Not long after Dirac's original construction, Brauer and Weyl generalized it to any arbitrary finite-dimensional quadratic space of arbitrary signature. With this generalization in mind, we proceed to a general definition of the Dirac operator associated to a given quadratic space.

Let $\mathfrak{A} = \mathfrak{A}(V, Q)$ be the universal Clifford algebra associated with a real non-degenerate quadratic space (V, Q). Then the space $C^\infty(\Omega, \mathfrak{A})$ of smooth $\mathfrak{A}$-valued functions on any open set Ω in V is an $\mathfrak{A}$-module under pointwise multiplication. In particular, by identifying V as usual with a subspace of $\mathfrak{A}$, we can regard any X in V as an element of $\mathfrak{A}$, and hence as a multiplier on $C^\infty(\Omega, \mathfrak{A})$. On the other hand, each X in V gives rise by parallel translation to a vector field ∂_X

$$\partial_X f(v) = \frac{d}{dt} f(v + tX)\big|_{t=0} \ , \qquad v \in \Omega \ ,$$

acting on smooth functions on Ω, whether scalar- or vector-valued. It is well-known that $X \to \partial_X$ is linear in X in the sense that

$$(2.2)(i) \qquad \partial_{\alpha X + \beta Y} = \alpha \partial_X + \beta \partial_Y \ , \qquad \alpha, \beta \in \mathbb{R} \ , \ X, Y \in V \ ,$$

while each ∂_X acts linearly in the sense that

$$(2.2)(ii) \qquad \partial_X(\alpha f + \beta g) = \alpha \partial_X f + \beta \partial_X g \ , \qquad \alpha, \beta \in \mathbb{R} \ .$$

Now let $\{e_1, \ldots, e_n\}$ be a normalized basis for V, and let $\partial_1, \ldots, \partial_n$ be the corresponding vector fields.

(2.3) Definition.

The Dirac operator D associated with (V, Q) is the first-order differential operator $D = \sum_{j=1}^n Q(e_j) e_j \partial_j$ on $C^\infty(\Omega, \mathfrak{A})$ whose coefficients are the generators $e_1, \ldots, e_n$ of $\mathfrak{A}(V, Q)$ acting by pointwise multiplication on $C^\infty(\Omega, \mathfrak{A})$. The Laplacian Δ_Q is the second-order, constant-coefficient differential operator $\Delta_Q = \sum_{j=1}^n Q(e_j)(\partial_j)^2$.

Just as

$$a^2 = \left(\sum_{j=1}^{n} a_j e_j\right)^2 = -\left(\sum_{j=1}^{n} a_j^2 Q(e_j)\right) = -Q(a) , \qquad a \in V ,$$

is the basic Clifford property of $\mathfrak{A}$, so

$$(2.4) \qquad D^2 = \left(\sum_{j=1}^{n} Q(e_j) e_j \partial_j\right)^2 = -\left(\sum_{j=1}^{n} Q(e_j) \partial_j^2\right) = -\Delta_Q$$

is the basic property of the Dirac operator relating D to Δ_Q. Since Δ_Q has constant scalar-valued coefficients, it is defined on any smooth function, whether scalar- or vector-valued. Thus the basic property of the Dirac operator realizes D as a *linear first-order* differential operator which is the square root of the constant coefficient second order differential operator, $-\Delta_Q$. But the linearity of D can only be achieved at the expense of introducing operator-valued coefficients. Nonetheless, Δ_Q acts separately on each real-valued component of any f in the domain of D. For instance, by taking the set $\{e_\alpha\}$ of all reduced products $e_\alpha = e_{\alpha_1} \cdots e_{\alpha_k}$, $0 \le |\alpha| \le n$, as a basis for $\mathfrak{A}$, we obtain the following.

(2.5) Theorem.
  If $f = \sum_\alpha f_\alpha(x) e_\alpha$ is a solution of $Df = 0$ in $C^\infty(\Omega, \mathfrak{A})$ with each f_α real-valued, then $\Delta_Q f_\alpha = 0$.

Proof. By (2.4)

$$= D^2 f = -\sum_\alpha (\Delta_Q f_\alpha) e_\alpha .$$

Hence $\Delta_Q f_\alpha = 0$ since the e_α are linearly independent. ∎

Although defined here in terms of a fixed normalized basis of V, both D and Δ_Q are independent of the particular choice of basis.

(2.6) Theorem.
  Let $\{e_j\}$, $\{\varepsilon_j\}$ be normalized bases of V such that $Q(e_j) = Q(\varepsilon_j)$, and let $\{\partial_j\}$, $\{\delta_j\}$ be the associated vector fields. Then

$$\sum_j Q(e_j) e_j \partial_j = \sum_j Q(\varepsilon_j) \varepsilon_j \delta_j \quad , \quad \sum_j Q(e_j) \partial_j^2 = \sum_j Q(\varepsilon_j) \delta_j^2 .$$

Proof. Fix A in $O(V, Q)$ so that $\varepsilon_j = A e_j$, $1 \le j \le n$. Then $A \sim [a_{jk}]$ where

$$\varepsilon_j = \sum_k a_{jk} e_k \quad , \quad \sum_j Q(\varepsilon_j) a_{jk} a_{j\ell} = Q(e_k) \delta_{k\ell} ,$$

and so

$$\delta_j = \partial_{Ae_j} = \sum_\ell a_{j\ell} \partial_\ell ,$$

using the linearity property of ∂_X in (2.2)(i). Consequently,

$$\sum_j Q(\varepsilon_j)\varepsilon_j\delta_j = \sum_{k,\ell}\left(\sum_j Q(\varepsilon_j)a_{jk}a_{j\ell}\right)e_k\partial_\ell = \sum_k Q(e_k)e_k\partial_k .$$

Similarly,

$$\sum_j Q(\varepsilon_j)\delta_j^2 = \sum_{k,\ell}\left(\sum_j Q(\varepsilon_j)a_{jk}a_{j\ell}\right)\partial_k\partial_\ell = \sum_k Q(e_k)\partial_k^2 ,$$

completing the proof. ∎

More generally, of course, the Dirac operator D associated with (V, Q) may be viewed as an operator on $C^\infty(\Omega, \mathfrak{H})$ where $\mathfrak{H}$ is any finite-dimensional $\mathfrak{A}$-module upon which $\mathfrak{A}$ acts on the left. For example, Dirac's original operator (2.1) is precisely the Dirac operator for $\mathbb{R}^{1,3}$ acting on functions taking values in the $\mathfrak{A}_{1,3}$-module $\mathbb{C}^4$. We shall refer to the Dirac operator for $\mathbb{R}^n$ ($= \mathbb{R}^{0,n}$) acting on $C^\infty(\Omega, \mathfrak{H})$, where $\mathfrak{H}$ is an $\mathfrak{A}_n$-module, as the *standard Euclidean Dirac operator*.

Now let us consider the connection between the standard Dirac operator and the Cauchy–Riemann operators ∂ and $\bar\partial$. Under the identification of the Clifford algebra $\mathfrak{A}_2$ ($\cong \mathbb{H}$) given in chapter 1, theorem (9.11)(i), the Dirac operator D is easily seen to be the off-diagonal operator

$$D = \begin{bmatrix} 0 & \bar\partial \\ -\partial & 0 \end{bmatrix} , \quad \bar\partial = \frac{\partial}{\partial x_1} + i\frac{\partial}{\partial x_2} , \quad \partial = \frac{\partial}{\partial x_1} - i\frac{\partial}{\partial x_2} ,$$

while every $f \in C^\infty(\Omega, \mathfrak{A}_2)$ may be expressed as

$$f = \begin{bmatrix} f_1 & -\bar{f}_2 \\ f_2 & \bar{f}_1 \end{bmatrix} , \quad f_1, f_2 \in C^\infty(\Omega, \mathbb{C}) .$$

In this realization, the algebra $\mathfrak{A}_2$ splits into the sum of its diagonal part $\mathfrak{A}_2^+ \sim \mathbb{C}$ and its off-diagonal part $\mathfrak{A}_2^- \sim \mathbb{C}$. When restricted to functions having values in $\mathfrak{A}_2^+$, D coincides with ∂, i.e.,

$$\begin{bmatrix} 0 & \bar\partial \\ -\partial & 0 \end{bmatrix}\begin{bmatrix} f_1 & 0 \\ 0 & \bar{f}_1 \end{bmatrix} = \begin{bmatrix} 0 & \bar\partial f_1 \\ \partial f_1 & 0 \end{bmatrix} ,$$

while the restriction of D to $\mathfrak{A}_2^-$ is just $\bar\partial$:

$$\begin{bmatrix} 0 & \bar\partial \\ -\partial & 0 \end{bmatrix}\begin{bmatrix} 0 & -\bar{f}_2 \\ f_2 & 0 \end{bmatrix} = \begin{bmatrix} \bar\partial f_2 & 0 \\ 0 & \partial\bar{f}_2 \end{bmatrix} .$$

Thus $\partial, \bar\partial$ arise by restricting the standard Dirac operator D to the components $\mathfrak{A}_2^+$, $\mathfrak{A}_2^-$, respectively, in the $\mathbb{Z}(2)$-grading of $\mathfrak{A}_2$ induced by the principal automorphism. For this reason, ∂ and $\bar{\partial}$ are often called *graded Dirac operators*; we shall discuss these in chapter 4.

3 Clifford analyticity

For Euclidean space $\mathbb{R}^n$, the standard Dirac D-operator acting on functions which take values in an $\mathfrak{A}_n$-module $\mathfrak{H}$ gives rise to a natural class of differential operators satisfying (1.17). The examples of the previous section show the great richness of this class. The field of 'Clifford analysis' has been particularly concerned with the specific case of $\mathfrak{H} = \mathfrak{A}_n$; but many important classical operators are obtained by restricting the action of D to functions taking values in a subspace of $\mathfrak{H}$.

In the present chapter, we will concern ourselves with the basic connections between Dirac operators, classical complex function theory, and classical Hardy space theory. With this in mind we shall fix certain notations to be used in this chapter once for all. We suppose that $\mathfrak{H}$ is an $\mathfrak{A}_n$-module, as defined in chapter 1, (7.44). Then the Dirac D-operator and its adjoint $\overline{D}$ are the first-order systems of differential operators on $\mathcal{C}^\infty(\Omega, \mathfrak{H})$ defined by

$$(3.1) \qquad D = \sum_{j=1}^n e_j \frac{\partial}{\partial x_j} \quad , \quad \overline{D} = \sum_{j=1}^n \bar{e}_j \frac{\partial}{\partial x_j}$$

where Ω is any open connected set in $\mathbb{R}^n$.

(3.2) Definition.
A function f in $\mathcal{C}^\infty(\Omega, \mathfrak{H})$ is said to be Clifford analytic on Ω when $Df = 0$ on Ω.

Since $D\overline{D} = \Delta_n$, where Δ_n is the Laplacian on $\mathbb{R}^n$, the real-valued components of any Clifford analytic function are always harmonic. On the other hand, if Φ is any real-valued harmonic function on Ω and

$$f = \overline{D}\Phi = \sum_{j=1}^n \bar{e}_j \frac{\partial \Phi}{\partial x_j}$$

(thinking of $\mathbb{R} \subseteq \mathfrak{A}_n$), then

$$Df = D\overline{D}\Phi = \Delta_n \Phi = 0 .$$

Thus f is Clifford analytic on Ω; so there are many $\mathfrak{A}_n$-valued Clifford analytic functions on any open set in $\mathbb{R}^n$. In particular, when Γ_n is the fundamental solution for Δ_n on $\mathbb{R}^n \setminus \{0\}$ defined by

(3.3)
$$\Gamma_2(x) = \frac{1}{2\pi} \log |x| \quad , \quad \Gamma_n(x) = \frac{1}{\omega_n(2-n)} |x|^{-(n-2)} \qquad (n > 2) ,$$

where ω_n is the surface area of the unit sphere in $\mathbb{R}^n$, the function

(3.4)
$$\phi(x) = \overline{D}\Gamma_n(x) = \frac{1}{\omega_n} \frac{x}{|x|^n} , \qquad x \neq 0 ,$$

is Clifford analytic everywhere away from the origin. Later this function ϕ will play the role of Cauchy kernel for Clifford analytic functions.

In the special case $\mathfrak{H} = \mathfrak{A}_n$, one should perhaps call any solution of $Df = 0$ a *left* Clifford analytic function by contrast with the solutions of $\sum_{j=1}^{n} \frac{\partial f}{\partial x_j} e_j = 0$ which might then be called *right* Clifford analytic functions. However, conjugation in $\mathfrak{A}_n$ ensures that

$$\left(\sum_j e_j \frac{\partial f}{\partial x_j} \right)^{-} = -\left(\sum_j \frac{\partial \bar{f}}{\partial x_j} e_j \right)$$

where the conjugate $\bar{f}$ of f has the obvious meaning $(\bar{f})(x) = \overline{f(x)}$; thus f will be left Clifford analytic if and only if its conjugate $\bar{f}$ is right Clifford analytic. Since

$$\overline{\phi(x)} = \frac{1}{\omega_n} \frac{x}{|x|^n} = -\phi(x) ,$$

a function can be both left and right Clifford analytic without, however, being a constant function. Thus there exist non-trivial *two-sided* Clifford analytic functions. More generally, the definition of D could also have been given with respect to *right* $\mathfrak{A}_n$-modules $\mathfrak{H}$. For our purposes, it is sufficient to view the action of D as an action on the left, bearing in mind that there is a corresponding theory for action on the right.

There are formal resemblances between D and the classical Cauchy–Riemann operators. Let $\{e_\alpha\}$ be the basis of $\mathfrak{A}_n$ of all reduced products $e_\alpha = e_{\alpha_1} \cdots e_{\alpha_k}$ and $f = \sum_\alpha f_\alpha(x) e_\alpha$ the corresponding decomposition of any f in $C^\infty(\Omega, \mathfrak{A}_n)$ as a sum having smooth real-valued functions f_α as coefficients. Then

(3.5)
$$Df = \sum_\alpha \left(\sum_{j=1}^{n} \frac{\partial f_\alpha}{\partial x_j} e_j e_\alpha \right) .$$

Now in reduced form each product $e_j e_\alpha$ will contain $|\alpha| + 1$ or $|\alpha| - 1$ terms according as $j \notin \alpha$ or $j \in \alpha$. Thus

(3.6)
$$Df = \sum_\beta F_\beta(x) e_\beta$$

where each function F_β is of the form

$$(3.7) \qquad F_\beta(x) = \sum_{j=1}^{n} \pm \frac{\partial f_\alpha}{\partial x_j} \ ,$$

for some choice of signs $\pm$, and for each fixed β the $\alpha = \alpha(j)$ is chosen so that $\alpha \cup \{j\} = \beta$ if $j \in \beta$ or $\alpha = \beta \cup \{j\}$ if $j \notin \beta$. Hence $Df = 0$ if and only if the 2^n real-valued functions f_α satisfy the 2^n first-order differential equations determined by $F_\beta = 0$. These functions F_β are the analogue of (1.10), but they are unlikely to have the same practical value unless $f_\alpha = 0$ for a large number of α, i.e., unless f takes values in some fixed subspace of $\mathfrak{A}_n$.

The realization of D in complex terms analogous to (1.9) comes from the case $\mathfrak{H} = \mathfrak{S}_n$, $n = 2m$, with $\mathfrak{S}_n = \Lambda^*(\mathbb{C}^m)$ and

$$(3.8) \qquad e_j = \mu_j - \mu_j^* \ , \quad e_{m+j} = i(\mu_j + \mu_j^*) \ , \qquad 1 \le j \le m \ ,$$

(see chapter 1 (7.54)(**A**)).

(3.9) Theorem.
   When $\mathfrak{H} = \Lambda^*(\mathbb{C}^m)$ and $e_1, \ldots, e_{2m}$ are defined by (3.8), the corresponding standard Dirac operator D is given by

$$D = \sum_{j=1}^{m} (\mu_j - \mu_j^*) \frac{\partial}{\partial x_j} + \sum_{j=1}^{m} i(\mu_j + \mu_j^*) \frac{\partial}{\partial y_j} = \sum_{j=1}^{m} (\mu_j \bar{\partial}_j - \mu_j^* \partial_j)$$

where

$$\bar{\partial}_j = \frac{\partial}{\partial x_j} + i \frac{\partial}{\partial y_j} \ , \quad \partial_j = \frac{\partial}{\partial x_j} - i \frac{\partial}{\partial y_j} \ , \qquad 1 \le j \le m \ .$$

In the very important case when $\mathfrak{H} = \Lambda^*(\mathbb{R}^n)$ and $e_j = \mu_j - \mu_j^*$ (see chapter 1 (7.54)(**B**)), D reduces to the 'div-curl' system of (1.10).

(3.10) Theorem.
When $\mathfrak{H} = \Lambda^*(\mathbb{R}^n)$ and $e_j = \mu_j - \mu_j^*$, the corresponding standard Dirac operator is given by

$$D = \sum_{j=1}^{n} \mu_j \frac{\partial}{\partial x_j} - \sum_{j=1}^{n} \mu_j^* \frac{\partial}{\partial x_j} \ .$$

Since

$$(3.11) \qquad \sum_{j=1}^{n} \mu_j \frac{\partial}{\partial x_j} = d \ , \quad \sum_{j=1}^{n} \mu_j^* \frac{\partial}{\partial x_j} = d^*$$

are just the respective exterior and interior derivatives, we see that D

can be realized as $d - d^*$ on forms; hence on k-forms, say, D coincides with the Hodge–deRham (d, d^*)-system.

We show now that the standard Dirac operator D on $\mathfrak{H}$-valued functions has all the properties in (1.17). Property (1.17)(ii), for instance, is an immediate consequence of the basic factorization property $D^2 = -\Delta_n$, as noted earlier. The next result completes the proof for property (1.17)(i).

(3.12) Theorem.

The standard Dirac operator is a determined elliptic operator on $\mathfrak{H}$-valued functions in the sense that the symbol mapping

$$\sigma_\lambda : \xi \longrightarrow \left(\sum_{j=1}^n \lambda_j e_j \right) \xi , \qquad \xi \in \mathfrak{H} ,$$

is invertible, i.e., surjective as well as injective, as a linear transformation on $\mathfrak{H}$ for every non-zero $\lambda = (\lambda_1, \dots, \lambda_n)$ in $\mathbb{R}^n$.

Proof. Since $Q(\sum_j \lambda_j e_j) = |\lambda|^2$, the element $\sum_j \lambda_j e_j$ will be invertible in $\mathcal{L}(\mathfrak{H})$ when $\lambda \neq 0$. Hence σ_λ will be an invertible operator for such λ. ∎

To establish (1.17)(iii) let M be an n-dimensional, compact, oriented C^∞-manifold with boundary ∂M contained in Ω. For $1 \leq j \leq n$, let $d\hat{x}_j$ be the $(n-1)$-form obtained by setting

$$(3.13) \qquad d\hat{x}_j = dx_1 \wedge \cdots \wedge \widehat{dx_j} \wedge \cdots \wedge dx_n ,$$

and then let $d\sigma(x)$ be the $\mathcal{L}(\mathfrak{H})$-valued $(n-1)$-form given by

$$(3.14) \qquad d\sigma(x) = \sum_{j=1}^n (-1)^{j-1} e_j \, d\hat{x}_j .$$

Thus to each f in $C^\infty(\Omega, \mathfrak{H})$ there corresponds an $\mathfrak{H}$-valued $(n-1)$-form

$$(3.15) \qquad \omega = d\sigma(x) f(x) = \sum_{j=1}^n (-1)^{j-1} e_j f(x) \, d\hat{x}_j$$

having exterior derivative

$$(3.16) \qquad d\omega = \sum_{j=1}^n (-1)^{j-1} e_j \frac{\partial f}{\partial x_j} \, dx_j \wedge d\hat{x}_j = Df(x) \, dV(x)$$

on Ω where $dV(x) = dx_1 \wedge \cdots \wedge dx_n$ is the volume element on Ω. Now for each $x \in \partial M$, let $\eta(x) = \sum_{j=1}^n \eta_j e_j$ be the outer unit normal to ∂M at x. Then

$$(3.17) \qquad (-1)^{j-1} d\hat{x}_j = \eta_j(x) \, dS(x)$$

where $dS(x)$ is the scalar element of surface area on ∂M. Consequently, on ∂M,

(3.18) $\qquad\qquad \omega = d\sigma(x)f(x) = \eta(x)f(x)\,dS(x)\ .$

The analogue for $\mathfrak{H}$-valued functions on $\mathbf{R}^n$ of the classical Borel–Pompieu result for complex-valued functions on $\mathbf{C}$ now follows by a standard use of Stokes' theorem.

(3.19) Theorem.
  If M is a compact, n-dimensional, oriented C^∞-manifold in Ω, then for each f in $C^\infty(\Omega, \mathfrak{H})$

$$f(z) = \frac{1}{\omega_n} \int_{\partial M} \frac{\bar{x} - \bar{z}}{|x - z|^n}\, d\sigma(x)f(x) - \frac{1}{\omega_n} \int_M \frac{\bar{x} - \bar{z}}{|x - z|^n}\, Df(x)\,dV(x)$$

when z is an interior point of M, whereas

$$\int_{\partial M} \frac{\bar{x} - \bar{z}}{|x - z|^n}\, d\sigma(x)f(x) = \int_M \frac{\bar{x} - \bar{z}}{|x - z|^n}\, Df(x)\,dV(x)$$

whenever z lies outside $M \cup \partial M$.

(3.20) Corollary.
  (Cauchy Integral Theorem) *If f is a Clifford analytic function on Ω, then*

$$f(z) = \frac{1}{\omega_n} \int_{\partial M} \frac{\bar{x} - \bar{z}}{|x - z|^n}\, d\sigma(x)f(x)$$

for each z in the interior of M.

Hence on $\mathfrak{H}$-valued functions the standard Dirac operator has all the properties in (1.17). In proving theorem (3.19) and related results it will be useful to separate out some 'integration by parts' formulæ. The hypotheses will be the same as those for (3.19).

(3.21) Lemma.
  Let ϕ, f be smooth scalar-valued functions on Ω. Then

$$\int_M \left\{ \phi(x)\frac{\partial f}{\partial x_j} + \frac{\partial \phi}{\partial x_j} f(x) \right\} dV(x) = \int_{\partial M} \phi(x)f(x)\eta_j(x)\,dS(x)\ ,$$

for each j, $1 \le j \le n$.

Proof. Since the exterior derivative of the $(n-1)$-form

$$\omega = (-1)^{j-1}\phi(x)f(x)\,d\hat{x}_j = \phi(x)f(x)\eta_j(x)\,dS(x)$$

is given by

$$d\omega = \frac{\partial}{\partial x_j}\big(\phi(x)f(x)\big)\,dV(x) = \left\{ \phi(x)\frac{\partial f}{\partial x_j} + \frac{\partial \phi}{\partial x_j} f(x) \right\} dV(x)\ ,$$

the result follows immediately from Stokes' theorem.   ■

Applying lemma (3.21) componentwise we obtain the corresponding $\mathfrak{H}$-valued version:

$$(3.22) \quad \int_M \left\{ \phi(x) \left(\sum_{j=1}^n e_j \frac{\partial f}{\partial x_j} \right) + \left(\sum_{j=1}^n \frac{\partial \phi}{\partial x_j} e_j \right) f(x) \right\} dV(x)$$
$$= \int_{\partial M} \phi(x) \eta(x) f(x) \, dS(x)$$

for all ϕ in $\mathcal{C}^\infty(\Omega, \mathfrak{A}_n)$ and f in $\mathcal{C}^\infty(\Omega, \mathfrak{H})$.

Proof of theorem (3.19). Suppose first that z is an interior point of M, and for all sufficiently small $\varepsilon > 0$ let $M_\varepsilon = M \setminus B_\varepsilon(z)$ where $B_\varepsilon(z)$ is the open ball of radius ε centered at z. Then, with the notation of (3.3) and (3.4), the function

$$\phi(x - z) = \overline{D}\Gamma_n(x - z) = \frac{1}{\omega_n} \frac{\bar{x} - \bar{z}}{|x - z|^n}$$

is a (two-sided) Clifford analytic function on every M_ε. Applying (3.22) to M_ε with this choice of ϕ, we thus obtain

$$\int_{M_\varepsilon} \phi(x - z) \, Df(x) \, dV(x) = \int_{\partial M_\varepsilon} \phi(x - z) \eta(x) f(x) \, dS(x$$
$$(3.23) \qquad = \left(\int_{\partial M} - \int_{\Sigma_\varepsilon(z)} \right) \phi(x - z) \eta(x) f(x) \, dS(x)$$

where $\Sigma_\varepsilon(z)$ is the sphere of radius ε centered at z. But

$$\omega_n \phi(x - z) \eta(x) = \left(\frac{\bar{x} - \bar{z}}{|x - z|^n} \right) \left(\frac{x - z}{|x - z|} \right) = \frac{-(x - z)^2}{|x - z|^{n+1}} = \frac{1}{|x - z|^{n-1}}$$

on $\Sigma_\varepsilon(z)$. Consequently,

$$\lim_{\varepsilon \to 0} \int_{\Sigma_\varepsilon(z)} \phi(x - z) \eta(x) f(x) \, dS(x)$$
$$= \lim_{\varepsilon \to 0} \left(\frac{1}{\omega_n \varepsilon^{n-1}} \int_{\Sigma_\varepsilon(z)} f(x) \, dS(x) \right) = f(z) \ .$$

Letting $\varepsilon \to 0$ throughout (3.23), we get the generalized Borel–Pompieu result for interior points. For exterior points it is sufficient to repeat the previous proof without bothering to exclude $B_\varepsilon(z)$.   ■

Analogues for Clifford analytic functions of familiar results in analytic function theory, especially those depending on the Cauchy integral theorem, can now be established.

(3.24) Theorem.

(Mean-value theorem) *Let $B_r(z_0)$ be the ball of radius r centered at z_0 which is contained in Ω, and let $|B_r(z_0)|$ be its volume. Then*

$$f(z_0) = \frac{1}{|B_r(z_0)|} \int_{B_r(z_0)} f(x) \, dV(x)$$

for any function f which is Clifford analytic in Ω.

Proof. By the Cauchy integral theorem applied to $B_r(z_0)$,

$$f(z_0) = \frac{1}{\omega_n} \int_{\Sigma_r(z_0)} \frac{\bar{x} - \bar{z}_0}{|x - z_0|^n} \, \eta(x) f(x) \, dS(x)$$

$$= \frac{1}{r^n \omega_n} \int_{\Sigma_r(z_0)} (\bar{x} - \bar{z}_0) \eta(x) f(x) \, dS(x)$$

$$= \frac{n}{r^n \omega_n} \int_{B_r(z_0)} f(x) \, dV(x) \ .$$

This completes the proof since $|B_r(z_0)| = \frac{1}{n} r^n \omega_n$. ∎

(3.25) Theorem.

(Cauchy theorem) *Let M be a compact, n-dimensional, oriented C^∞-manifold in Ω. Then*

(3.26) $$\int_{\partial M} \eta(x) f(x) \, dS(x) = 0$$

for every Clifford analytic function f in $C^\infty(\Omega, \mathfrak{H})$.

Proof. Take $\phi \equiv 1$ in (3.22). ∎

Conversely, much the same proof as in the classical case gives the following.

(3.27) Theorem.

(Morera's theorem) *If f is a continuous $\mathfrak{H}$-valued function on Ω such that (3.26) holds for every compact, n-dimensional, oriented C^∞-manifold M in Ω, then f is Clifford analytic on Ω.*

As in chapter 1, section 7 we shall denote the norm on $\mathfrak{H}$ by $|\cdot|$, the inner product by $(\cdot, \cdot)$ and the operator norm in $\mathcal{L}(\mathfrak{H})$ by $\|\cdot\|$. We fix an orthonormal basis $\{v_\alpha\}$ for $\mathfrak{H}$; for each u in $\mathfrak{H}$ we denote by u_α the projection (u, v_α) of u onto v_α. When f is an $\mathfrak{H}$-valued function, we define the component function f_α by setting $f_\alpha(x) = (f(x))_\alpha$.

(3.28) Theorem.

(Maximum Modulus Principle) *Let f be a function which is Clifford analytic on a connected open set Ω. Then*

$$(3.29) \qquad\qquad |f(\omega)| \leq |f(\omega_0)| \,, \qquad \omega \in \Omega \,,$$

holds for some ω_0 in Ω if and only if f is a constant function.

(3.30) Corollary.

If Ω is a bounded open set in $\mathbb{R}^n$, then

$$\sup_{x \in \Omega} |f(x)| = \sup_{x \in \partial\Omega} |f(x)|$$

whenever f is Clifford analytic on Ω and continuous on $\partial\Omega \cup \Omega$.

Proof of theorem (3.28). Clearly (3.29) must hold for every ω_0 in Ω if f is a constant function. Conversely, suppose that (3.29) holds for some ω_0 in Ω and set

$$\Omega_0 = \left\{ \omega \in \Omega : |f(\omega)| = |f(\omega_0)| \right\} \,.$$

Then we claim that $\Omega = \Omega_0$. For the continuity of $\omega \to |f(\omega)|$ ensures that Ω_0 is closed in Ω. On the other hand, just as in the classical proof, the mean-value theorem ensures that Ω_0 is open in Ω. Hence $\Omega = \Omega_0$, since Ω is assumed to be connected. In this case

$$0 = \Delta_n |f(\omega)|^2 = \Delta_n \left(\sum_\alpha f_\alpha(\omega) \overline{f_\alpha(\omega)} \right)$$

$$= \sum_{\alpha,j} \left| \frac{\partial f_\alpha}{\partial \omega_j}(\omega) \right|^2$$

throughout Ω because each f_α is harmonic. Consequently,

$$\frac{\partial f_\alpha}{\partial \omega_j}(\omega) = 0 \,, \qquad \omega \in \Omega \,, \ 1 \leq j \leq n \,,$$

for each α, and so f is constant on Ω. ∎

(3.31) Theorem.

(Weierstrass) *Let Ω be an open set in $\mathbb{R}^n$. Then the space $\mathcal{A}(\Omega, \mathfrak{H})$ of all Clifford analytic functions on Ω is sequentially complete under the topology of uniform convergence on compact sets.*

Proof. Choose a sequence $\{K_m\}$ of compact sets K_m in Ω such that

(i) K_k is a n-dimensional, oriented C^∞-manifold in Ω,
(ii) $\bigcup K_k = \Omega$,

and let $\{f_m\}$ be a sequence of functions in $\mathcal{A}(\Omega, \mathfrak{H})$ which converges uniformly on each K_k. By the Cauchy integral theorem,

$$f_n(z) = \frac{1}{\omega_n} \int_{\partial K_k} \frac{\bar{x} - \bar{z}}{|x - z|^n} \, d\sigma(x) f_m(x)$$

for each z in the interior of K_k and each k. By differentiation under the integral, we easily see that $\{f_m\}$ is convergent also in the space $\mathcal{C}^\infty(\Omega, \mathfrak{H})$ of smooth, $\mathfrak{H}$-valued functions on Ω, i.e., each sequence $\{\partial^\beta f_m\}$, $|\beta| \geq 0$, of partial derivatives converges uniformly on compact sets of Ω. Since $\mathcal{C}^\infty(\Omega, \mathfrak{H})$ is sequentially complete, this ensures the existence of f in $\mathcal{C}^\infty(\Omega, \mathfrak{H})$ such that

$$\lim_{m \to \infty} \partial^\beta f_m(x) = \partial^\beta f(x)$$

for each x in Ω and multi-index β. Hence $Df = 0$ on Ω, completing the proof. ∎

To describe the subharmonicity properties of Clifford analytic functions the following criterion, a Kato-type inequality, will be useful.

(3.32) Theorem.
  Fix $p < 1$ and let f be any Clifford analytic function on Ω. Then $x \to |f(x)|^p$ is subharmonic on Ω if

$$(3.33) \qquad \left| \sum_{j=1}^{n} \nu_j \frac{\partial f}{\partial x_j}(x) \right|^2 \leq \frac{1}{2-p} \left(\sum_{j=1}^{n} \left| \frac{\partial f}{\partial x_j}(x) \right|^2 \right)$$

holds for all x in Ω and $\nu = \sum_{j=1}^{n} \nu_j e_j$ in the unit sphere Σ_{n-1} in $\mathbb{R}^n$.

Since the components of any Clifford analytic function are harmonic, the mean-value property (or the Cauchy integral theorem) ensures that $x \to |f(x)|^p$ is sub-harmonic for $p \geq 1$. So theorem (3.32) deals with the range of values of p of most interest.

Proof of theorem (3.32). If $f(x) \neq 0$, then

$$\frac{\partial}{\partial x_j} |f(x)|^p = \frac{\partial}{\partial x_j} (f(x), f(x))^{\frac{1}{2}p} = p|f(x)|^{p-2} \left(f(x), \frac{\partial f}{\partial x_j} \right)$$

and

$$\frac{\partial^2}{\partial x_j^2} |f(x)|^p = p(p-2)|f(x)|^{p-4} \left(f(x), \frac{\partial f}{\partial x_j} \right)^2$$

$$+ p|f(x)|^{p-2} \left\{ \left| \frac{\partial f}{\partial x_j}(x) \right|^2 + \left(f(x), \frac{\partial^2 f}{\partial x_j^2} \right) \right\}.$$

But f has harmonic components. Consequently,

$$\Delta_n |f(x)|^p = p(2-p)|f(x)|^{p-4}$$

$$\times \left\{ \left(\frac{1}{2-p}\right)|f(x)|^2 \left(\sum_{j=1}^{n} \left|\frac{\partial f}{\partial x_j}(x)\right|^2\right) - \sum_{j=1}^{n} \left(f(x), \frac{\partial f}{\partial x_j}(x)\right)^2 \right\}$$

whenever $f(x) \neq 0$. Hence $x \to |f(x)|^p$ is subharmonic on Ω if and only if

$$(3.34) \qquad \sum_{j=1}^{n} \left(f(x), \frac{\partial f}{\partial x_j}(x)\right)^2 \leq \frac{1}{2-p}|f(x)|^2 \left(\sum_{j=1}^{n} \left|\frac{\partial f}{\partial x_j}(x)\right|^2\right).$$

But

$$\left(\sum_{j=1}^{n} \left(f(x), \frac{\partial f}{\partial x_j}(x)\right)^2\right)^{1/2} = \max \sum_{j=1}^{n} \nu_j \left(f(x), \frac{\partial f}{\partial x_j}(x)\right)$$

$$\leq \max |f(x)| \left|\sum_{j=1}^{n} \nu_j \frac{\partial f}{\partial x_j}(x)\right|,$$

the maximum in both cases being taken over all $\nu = \sum_j \nu_j e_j$ in Σ_{n-1}. Hence $x \to |f(x)|^p$ will be subharmonic on Ω whenever (3.33) holds for all $x \in \Omega$ and $\nu \in \Sigma_{n-1}$. ∎

In the special case $\mathfrak{H} = \mathfrak{A}_n$, we obtain an analogue of theorem (1.16) for $\mathfrak{A}_n$-valued Clifford analytic functions which varies with dimension. Recall (see chapter 1, (7.14)) that the Hilbert space norm on $\mathfrak{A}_n$ with respect to which the basis of reduced products $\{e_\alpha\}$ is an orthonormal basis is denoted by $|\cdot|$. Then we have the following.

(3.35) Theorem.
   Let Ω be any open set in $\mathbb{R}^n$, $n \geq 2$. Then
   (i) whenever $p \geq \frac{n-2}{n-1}$ and whenever f is an $\mathfrak{A}_n$-valued Clifford analytic function on Ω, $x \to |f(x)|^p$ is subharmonic on Ω, but
   (ii) if $\frac{n-2}{n-1} > p > 0$, there is an $\mathfrak{A}_n$-valued Clifford analytic function f such that $x \to |f(x)|^p$ is not subharmonic anywhere on its domain of definition.

Proof. We prove first the positive result. In view of criterion (3.33), it is enough to show that

$$(3.36) \qquad \left|\sum_{j=1}^{n} \nu_j \frac{\partial f}{\partial x_j}(x)\right|^2 \leq \left(\frac{n-1}{n}\right) \sum_{j=1}^{n} \left|\frac{\partial f}{\partial x_j}(x)\right|^2, \qquad x \in B_r,$$

is satisfied for all $\nu = (\nu_1, \ldots, \nu_n)$ in Σ_{n-1} whenever f is a function which is Clifford analytic in a fixed ball B_r of radius r centered at the origin. For then (3.36) continues to hold throughout all of the open set Ω because D commutes with translation. But in B_r, any orthonormal basis of $\mathbb{R}^n$ can be used to define the Dirac operator; in particular, for each fixed $\nu = (\nu_1, \ldots, \nu_n)$ in Σ_{n-1}, an orthonormal basis $\{\varepsilon_1, \ldots, \varepsilon_n\}$ with $\varepsilon_1 = \nu_1 e_1 + \cdots + \nu_n e_n$ can be used. Let $\delta_1, \ldots, \delta_n$ be the corresponding vector fields. Then, by (2.2)(i),

$$\delta_1 = \sum_{k=1}^n \nu_k \frac{\partial}{\partial x_k} \quad , \quad \delta_j = \sum_{k=1}^n a_{jk} \frac{\partial}{\partial x_k} \qquad (j > 1) \,,$$

where all the ν and $a_j = (a_{j1}, \ldots, a_{jn})$ are mutually orthogonal vectors in Σ_{n-1}. With respect to these vector fields, (3.36) becomes

$$|\delta_1 f(x)| \leq \left(\frac{n-1}{n}\right) \sum_{j=1}^n |\delta_j f(x)|^2 \,, \qquad x \in B_r \,.$$

Hence it is enough to establish (3.36) with $\nu = (1, 0, \ldots, 0)$, thinking of $\frac{\partial}{\partial x_1}, \ldots, \frac{\partial}{\partial x_n}$ as the vector fields on $\mathbb{R}^n$ determined by an *arbitrary* orthonormal basis $\{e_1, \ldots, e_n\}$ for $\mathbb{R}^n$. But the Cauchy–Riemann type equations obtained by setting (3.7) equal to zero still hold in the same form

$$(3.37) \qquad F_\beta(x) = \sum_{j=1}^n \pm \frac{\partial f_\alpha}{\partial x_j}(x) = 0 \,, \qquad \alpha = \alpha(j) \,, \ x \in B_r \,,$$

provided both the vector fields and basis of reduced products are defined with respect to this new basis $\{e_1, \ldots, e_n\}$.

Now, by the Cauchy-Schwarz inequality,

$$|c_1|^2 = \left| \sum_{j=2}^n c_j \right|^2 \leq (n-1) \sum_{j=2}^n |c_j|^2$$

whenever $c_1, \ldots, c_n$ are real or complex numbers such that $c_1 + \cdots + c_n = 0$, hence

$$(3.38) \qquad |c_1|^2 \leq \left(\frac{n-1}{n}\right) \sum_{j=1}^n |c_j|^2$$

for all such $c_1, \ldots, c_n$. Applying this inequality to each of the differential equations arising from (3.37), taking c_1 in each instance to be the term $\frac{\partial}{\partial x_1}(f_\alpha(x))$, and then summing over all 2^n equations, we deduce that

$$\left| \frac{\partial f}{\partial x_j}(x) \right|^2 \leq \left(\frac{n-1}{n}\right) \sum_{j=1}^n \left| \frac{\partial f}{\partial x_j}(x) \right|^2 \,, \qquad x \in \Omega \,,$$

holds for all Clifford analytic functions f on Ω. This proves the first half of the theorem.

To prove the second half, set

$$f(x) = \frac{x}{|x|^n} , \qquad x \neq 0 .$$

Then $|f(x)| = |x|^{-(n-1)}$, and

$$\Delta |f(x)|^p = p(n-1)\big[(n-1)p - n + 2\big]|x|^{-(n-1)p-2} , \qquad x \neq 0 ,$$

which will be negative when $p < \frac{n-2}{n-1}$.   ∎

In chapter 1, section 7, it was shown that every $\mathfrak{A}_n$-module $\mathfrak{H}$ is isomorphic to $\mathfrak{R}_n \otimes V$ for some finite-dimensional Hilbert space V, where $\mathfrak{R}_n$ is the spinor module (see the remark following (7.50) in chapter 1). The action of $\mathfrak{A}_n$ on $\mathfrak{R}_n \otimes V$ is the action induced by that of $\mathfrak{A}_n$ upon $\mathfrak{R}_n$. Moreover, $\mathfrak{R}_n$ has an explicit realization as a minimal left ideal in the algebra $\mathfrak{A}_n$. Consequently, we see that theorem (3.35)(i) continues to hold true when $\mathfrak{A}_n$ is replaced by $\mathfrak{R}_n$; still more generally, we have the following.

(3.39) Corollary.

Let Ω be any open set in $\mathbb{R}^n$, $n \geq 2$. Suppose that $\mathfrak{H}$ is an $\mathfrak{A}_n$-module and let $p \geq \frac{n-2}{n-1}$. Then for every Clifford analytic f in $C^\infty(\Omega, \mathfrak{H})$, $x \to |f(x)|^p$ is subharmonic on Ω.

4 Spaces of analytic functions

Let Ω be an open connected set in $\mathbb{C}$ having boundary $\partial\Omega$, and let $\mathcal{A}(\Omega)$ be the space of all complex-valued analytic functions on Ω. Because of the Weierstrass property, this space is complete under the topology of uniform convergence on compact sets, but two families of complete normed subspaces of $\mathcal{A}(\Omega)$ – the Hardy and Bergman spaces – are of much greater analytic interest, the first in connection with L^p-boundary regularity results and the second in representation theory for the group of holomorphic automorphisms of Ω.

Classical H^p theory – the study of spaces of functions analytic in the unit disk $D = \{z \in \mathbb{C} : |z| < 1\}$ or the upper half-plane $\mathbb{C}_+ = \{z \in \mathbb{C} : \operatorname{Im} z > 0\}$ – started in 1915 when G.H. Hardy considered the mean value

$$(4.1) \qquad M_p(F, r) = \left(\frac{1}{2\pi} \int_0^{2\pi} |f(re^{i\theta})|^p \, d\theta\right)^{1/p} , \qquad 0 < p < \infty ,$$

on the circle $|z| = r$, $0 < r < 1$, of a function F analytic in D. He showed that $M_p(F, r)$ behaves in a similar manner to the maximum modulus

(4.2) $\displaystyle\sup_{|z|\leq r} |F(z)| = \sup_{0\leq\theta\leq 2\pi} |F(re^{i\theta})| = M_\infty(F, r)\,,$ $0 < r < 1\,;$

more precisely,

(4.3)(i) $M_p(F, r)$ *is a strictly monotonic increasing function of* r *unless* F *is constant,*

(4.3)(ii) $\log M_p(F, r)$ *is a convex function of* $\log r$, *i.e.,*

$$M_p(F, r) \leq \big[M_p(F, r_0)\big]^\theta \big[M_p(F, r_1)\big]^{1-\theta}$$

whenever $\log r = \theta \log r_0 + (1 - \theta) \log r_1$.

For M_∞, property (4.3)(i) is just the maximum modulus theorem, while (4.3)(ii) is the Hadamard three-circles theorem. Shortly afterwards, a systematic study was begun of the space $H^p(D)$, $0 < p < \infty$, of all analytic functions F on D for which

(4.4) $\displaystyle \|F\|_{H^p(D)} = \sup_{0<r<1} \left(\frac{1}{2\pi} \int_0^{2\pi} |F(re^{i\theta})|^p\, d\theta \right)^{1/p}$ $(0 < p < \infty)\,,$

is finite. The name Hardy space was attached to $H^p(D)$ by F. Riesz. Ever since, Hardy H^p theory 'has proved a fruitful field of investigation, requiring the most subtle techniques of real-variable and complex variable theory' to quote Flett's introduction to Hardy's work on the subject. Indeed, as the maximum modulus result in (4.3)(i) suggests, $H^p(D)$-theory is intimately conected with $L^p(\partial D)$-boundary value theory, i.e., with function-theory on the unit circle $\mathbb{T} = \{z \in \mathbb{C} : |z| = 1\}$, hence with boundary regularity results for the Cauchy–Riemann operator $\bar{\partial}$.

Completely analogous to (4.4), there is a Hardy $H^p(\mathbb{C}_+)$-space of analytic functions F on $\mathbb{C}_+$ for which

(4.5) $\displaystyle \|F\|_{H^p(\mathbb{C}_+)} = \sup_{t>0} \left(\int_{-\infty}^\infty |F(x + it)|^p\, dx \right)^{1/p}$ $(0 < p < \infty)\,,$

is finite. In this case the corresponding boundary value theory is concerned with function theory on $\mathbb{R}$, identified with the real axis in $\mathbb{C}$. These Hardy spaces on different domains are closely related, however. Let

(4.6) $\displaystyle \Phi(z) = \frac{i - z}{z + i}\,,$ $z \in \mathbb{C}_+\,,$

be the conformal transformation from $\mathbb{C}_+$ onto D, and for each analytic function G on D define F on $\mathbb{C}_+$ by

(4.7) $\displaystyle F(z) = (z + i)^{2/p} G\Big(\frac{i - z}{z + i}\Big)\,,$ $z \in \mathbb{C}_+\,,$

choosing an analytic branch of $(z + i)^{2/p}$ on $\mathbb{C}_+$, p fixed. Then F is an analytic function that is in $H^p(\mathbb{C}_+)$ if and only if G is in $H^p(D)$. For the time being then we restrict attention to the Hardy spaces $H^p(\mathbb{C}_+)$, $p > 0$, and recall largely without proof their basic properties, with the potential for generalization to the corresponding spaces of Clifford analytic functions in mind.

Every F in $H^p(\mathbb{C}_+)$ has boundary values in $L^p(\mathbb{R})$ in the following sense. Let $\Gamma_\alpha(x)$ be the cone

$$(4.8) \qquad \Gamma_\alpha(x) = \big\{ z \in \mathbb{C}_+ : |\Re z - x| < \alpha \Im z \big\}$$

in $\mathbb{C}_+$ based at $x \in \mathbb{R}$ with aperture α, $\alpha > 0$. This cone defines a 'non-tangential approach region' to x. We say that a function F on $\mathbb{C}_+$ has a *non-tangential limit* at x if

$$(4.9) \qquad \lim_{z \to x}\text{-n.t.}\, F(z) = \lim_{\substack{z \to x \\ z \in \Gamma_\alpha(x)}} F(z)$$

exists as z approaches x through values in $\Gamma_\alpha(x)$. The following result follows straightforwardly from properties of harmonic functions and weak*-compactness results.

(4.10) Theorem.

(Fatou) *Suppose* $F \in H^p(\mathbb{C}_+)$, $p > 1$. *Then there is a function* f *in* $L^p(\mathbb{R})$ *such that*

(i) $\lim_{z \to x}\text{-n.t.}\, F(z) = f(x)$ *exists for almost all x in* $\mathbb{R}$,

(ii) $\lim_{t \to 0} \int_{-\infty}^{\infty} |F(x + it) - f(x)|^p \, dx = 0$.

The function F in theorem (4.10) can be recovered from its boundary function f by taking the Cauchy integral of f. More generally, suppose f is any real- or complex-valued function in $L^p(\mathbb{R})$, $p \geq 1$. The *Cauchy integral* of f is the function

$$(4.11) \qquad \mathcal{C}f(z) = \frac{1}{2\pi i} \int_{-\infty}^{\infty} \frac{f(t)}{t - z}\, dt \,, \qquad z \in \mathbb{C} \setminus \mathbb{R} \,.$$

Clearly $\mathcal{C}f$ is analytic on $\mathbb{C}_+$, and it is easily seen that

$$(4.12) \quad \mathcal{C}f(z) = \tfrac{1}{2}\big((P_t * f)(x) + i(Q_t * f)(x) \big) \,, \qquad z = x + it \in \mathbb{C}_+ \,,$$

where

$$(4.13) \qquad P_t(x) = \frac{1}{\pi}\Big(\frac{t}{x^2 + t^2} \Big) = \Re\Big(\frac{-1}{i\pi z} \Big)$$

is the *Poisson kernel* for $\mathbb{R}$ and $P_t * f$ the *Poisson integral* of f, while

$$(4.14) \qquad Q_t(x) = \frac{1}{\pi}\Big(\frac{x}{x^2 + t^2} \Big) = \Im\Big(\frac{-1}{i\pi z} \Big)$$

is the *conjugate Poisson kernel* for $\mathbb{R}$ and $Q_t * f$ the *conjugate Poisson integral* of f. Basic properties of $\mathcal{C}f$ can be summarized as follows.

(4.15) Theorem.

If $f \in L^p(\mathbb{R})$, $p > 1$ then Cf is in $H^p(\mathbb{C}_+)$ and

$$\|Cf\|_{H^p} \leq C_p \|f\|_{L^p} \ ;$$

in particular,

$$\lim_{t \to 0}(P_t * f)(x) = f(x) \quad , \quad \lim_{t \to 0}(Q_t * f)(x) = \mathcal{H}f(x)$$

where $\mathcal{H}$ is the singular integral operator defined by

$$\mathcal{H}f(x) = p.v. \ \frac{1}{\pi} \int_{-\infty}^{\infty} \frac{f(y)}{x - y} \, dy \ .$$

In view of (4.10), this function $\mathcal{H}f$, the *Hilbert transform* of f, satisfies

$$(4.16) \qquad \|\mathcal{H}f\|_p \leq C_p \|f\|_p \ , \qquad f \in L^p(\mathbb{R}) \ ,$$

for $1 < p < \infty$. On the other hand, when the Fourier transform of f is defined by

$$(4.17) \qquad \mathcal{F}f(s) = \widehat{f}(s) = \int_{-\infty}^{\infty} e^{-ixs} f(x) \, dx \ , \qquad s \in \mathbb{R} \ ,$$

then

$$(4.18) \qquad (\mathcal{H}f)^{\wedge}(s) = -i \operatorname{sgn} s \widehat{f}(s) \ ,$$

and so

$$(4.19) \qquad \mathcal{F}\big((I - i\mathcal{H})f\big) = 2\chi_{(-\infty, 0]} \mathcal{F}f \ .$$

From theorems (4.10) and (4.15) we thus derive a *boundary characterization* of $H^p(\mathbb{C}_+)$.

(4.20) Theorem.

For a function f in $L^p(\mathbb{R})$, $1 < p < \infty$, the following assertions are equivalent:

(i) f *is the almost-everywhere non-tangential limit of a function*
   F *in* $H^p(\mathbb{C}_+)$,

(ii) $(I - i\mathcal{H})f = 0$,

(iii) the Fourier transform $\widehat{f}$ of f has support in $[0, \infty)$.

Exactly the same result as (4.10) can be established, replacing the *analytic* function F with a *harmonic* function F satisfying the same norm condition (4.5). It is only for $p \leq 1$, therefore, that the distinction between analyticity and harmonicity becomes apparent, and it is the extra subharmonicity property (1.16) of analytic functions that can be made responsible for this distinction.

(4.21) Theorem.

Suppose $F \in H^p(\mathbb{C}_+)$, $0 < p \leq 1$. Then there is a function f in $L^p(\mathbb{R})$ such that

(i) $\lim_{\substack{z \to x}} \text{-n.t.}\, F(z) = f(x)$ *exists for almost all x in $\mathbb{R}$,*

(ii) $\lim_{y \to 0} \int_{-\infty}^{\infty} |F(x + iy) - f(x)|^p \, dx = 0$.

Proof. (Sketch) Choose any $q > 1$ and define Φ on $\mathbb{C}_+$ by $\Phi(z) = |F(z)|^{p/q}$. Then Φ is non-negative and subharmonic on $\mathbb{C}_+$, while

$$\sup_{t>0} \int_{-\infty}^{\infty} |\Phi(x + it)|^q \, dx = \sup_{t>0} \int_{-\infty}^{\infty} |F(x + it)|^p \, dx < \infty \, .$$

In this case, Φ has a least-harmonic majorant Ψ which is the Poisson integral $P_t * f$ of a function $f \in L^q(\mathbb{R})$, since $q > 1$; furthermore, Ψ will have non-tangential limits almost everywhere on $\mathbb{R}$. In turn this ensures the almost everywhere non-tangential boundedness of Φ, and hence the existence almost everywhere of non-tangential limits for F. The second result in (4.21) follows using dominated convergence and the Hardy–Littlewood maximal function.   ∎

There is a boundary characterization of $H^1(\mathbb{C}_+)$ analogous to theorem (4.20).

(4.22) Theorem.

For a function f in $L^1(\mathbb{R})$ the following assertions are equivalent:

(i) *f is the almost-everywhere non-tangential limit of a function F in $H^1(\mathbb{C}_+)$,*

(ii) *$\mathcal{H}f$ is in $L^1(\mathbb{R})$ and $(I - i\mathcal{H})f = 0$,*

(iii) *$\mathcal{H}f$ is in $L^1(\mathbb{R})$ and the Fourier transform $\hat{f}$ of f vanishes on $(-\infty, 0]$.*

The difference in the precise form of theorems (4.20) and (4.22) arises solely from the failure of inequality (4.16) when $p = 1$. The Hilbert transform is merely of weak type (1,1) in the sense that

$$\text{meas}\left\{ x \in \mathbb{R} : |\mathcal{H}f(x)| > \lambda \right\} \leq \frac{C_1}{\lambda} \|f\|_1 \, , \qquad \lambda > 0 \, ,$$

holds for all f in $L^1(\mathbb{R})$.

(4.23) Definition.

Denote by $H^p(\partial\mathbb{C}_+)$, $1 \leq p < \infty$, the subspace of complex-valued functions in $L^p(\mathbb{R})$ satisfying any one, and hence all, of the three conditions in (4.20) when $1 < p < \infty$, and in (4.22) when $p = 1$.

Combining the previous results we now obtain the following.

(4.24) Theorem.
   The space $H^p(\partial\mathbb{C}_+)$, $1 \le p < \infty$, is a closed subspace of $L^p(\mathbb{R})$ and the Cauchy integral $f \to Cf$ is an isometric isomorphism from $H^p(\partial\mathbb{C}_+)$ onto $H^p(\mathbb{C}_+)$, extending each f on $\partial\mathbb{C}_+$ analytically to a function $F = Cf$ on $\mathbb{C}_+$ such that

$$\lim_{z \to x}\text{-n.t.}\, F(z) = f(x) .$$

   There is an entirely analogous boundary characterization for Hardy spaces in the unit disk D, replacing $H^p(\mathbb{C}_+)$ by $H^p(D)$ and $H^p(\partial\mathbb{C}_+)$ by $H^p(\mathbb{T})$. Since this is completely classical we omit the details.

   Two other characterizations of Hardy spaces, illustrating important properties of analytic functions, are of importance. The first exploits the fact that an analytic function is essentially determined by its harmonic real part, and has led to an extensively developed real H^p theory whose avoidance of $\bar{\partial}$ and analytic function theory has made it applicable in widely varying settings. Here, however, it will be thought of as an expression of the *analytic* over-determinedness of the Cauchy–Riemann operator $\bar{\partial}$, in contrast to the algebraic determinedness of $\bar{\partial}$ as expressed by ellipticity in terms of its symbol (see theorem (3.12)).

(4.25) Definition.
For $1 < p < \infty$, let $\Re H^p(\mathbb{R})$ be the space $\Re L^p(\mathbb{R})$ of all real-valued L^p-functions on $\mathbb{R}$, whereas, for $p = 1$, let $\Re H^1(\mathbb{R})$ be the space of all real-valued L^1-functions f such that $\mathcal{H}f$ is also an L^1-function.

   These $\Re H^p(\mathbb{R})$ are Banach spaces with respect to the norms

(4.26)(i) $\qquad \|f\|_{\Re H^p} = \left(\int_{-\infty}^{\infty} |f(x)|^p \, dx \right)^{1/p} \qquad (1 < p < \infty) ,$

and

(4.26)(ii) $\qquad \|f\|_{\Re H^1} = \int_{-\infty}^{\infty} \left(|f(x)| + |\mathcal{H}f(x)| \right) dx \qquad (p = 1) .$

There are corresponding spaces $\Re H^p(\mathbb{R})$, $0 < p < 1$, defined in terms of so-called atomic or molecular decompositions. In view of theorems (4.20) and (4.22), $\Re H^p(\mathbb{R})$ consists of the functions (or distributions when $p < 1$) which are the real part of the boundary values of functions in $H^p(\mathbb{C}_+)$; hence,

(4.27) $\qquad \Re H^p(\mathbb{R}) = \{ f : f = \Re g , \; g \in H^p(\partial\mathbb{C}_+) \} ,$

as the notation was chosen to suggest. Since $\Im g = \mathcal{H}f$ when $f = \Re g$, we deduce

(4.28) Theorem.
    *The Cauchy integral $f \rightarrow Cf$ is a linear isomorphism from $\Re H^p(\mathbb{R})$,
$1 \leq p < \infty$, onto $H^p(\mathbb{C}_+)$.*

For $p < 1$ there is an analogous result. The second characterization
will later facilitate the introduction of Hardy spaces on more general
domains in $\mathbb{C}$.

(4.29) Definition.
    *Let F be any real or complex-valued continuous function on $\mathbb{C}_+$.
Then the non-tangential maximal function $N(F)$ of F is defined on $\mathbb{R} \sim
\partial \mathbb{C}_+$ by*

$$N(F)(x) = \sup_{z \in \Gamma_\alpha(x)} |F(z)| , \qquad x \in \mathbb{R} .$$

It would be more precise to write $N_\alpha(F)$ instead of $N(F)$, since the
definition depends on the aperture of each Γ_α, but usually in applications
one of the first results to be established is independence of α. Now

$$(4.30) \qquad \int_{-\infty}^{\infty} |F(x+it)|^p \, dx \leq \int_{-\infty}^{\infty} N(F)(x)^p \, dx , \qquad 0 < p < \infty ,$$

for any $t > 0$ and aperture α; so an analytic function F will clearly be
in $H^p(\mathbb{C}_+)$, $0 < p < \infty$, whenever its non-tangential maximal function
$N(F)$ is in $L^p(\mathbb{R})$. What is not so clear is whether F is in $H^p(\mathbb{C}_+)$ when
only the non-tangential maximal function $N(U)$ of its real part $U = \Re F$
is known to be in $L^p(\mathbb{R})$, though for $1 < p < \infty$ this can be deduced
from (4.16) since the imaginary part $V = \Im F$ of F is given by $V =
\mathcal{H}U$, again an expression of the over-determinedness property of analytic
functions. The whole problem was completely clarified by Burkholder–
Gundy–Silverstein, completing well-known results of Hardy–Littlewood.

(4.31) Theorem.
    *An analytic function F is in $H^p(\mathbb{C}_+)$, $0 < p < \infty$, if and only if
the non-tangential maximal function $N(U)$ of its harmonic real part U
is in $L^p(\mathbb{R})$; furthermore, there exist constants A_p and B_p such that*

$$A_p \|F\|_{H^p} \leq \left(\int_{-\infty}^{\infty} N(U)^p(x) \, dx \right)^{1/p} \leq B_p \|F\|_{H^p} .$$

By analogy with $\Re H^p(\mathbb{R})$, let $\Re H^p(\mathbb{C}_+)$ be the space of real-valued
harmonic functions U on $\mathbb{C}_+$ for which

$$(4.32) \qquad \|U\|_{\Re H^p} = \left(\int_{-\infty}^{\infty} N(U)^p(x) \, dx \right)^{1/p} \qquad (0 < p < \infty)$$

is finite. Then for each U in $\Re H^p(\mathbb{C}_+)$ there is a harmonic conjugate V (uniquely specified by its behavior at infinity) so that $U + iV$ is in $H^p(\mathbb{C}_+)$ and $U \to U + iV$ is a linear isomorphism from $\Re H^p(\mathbb{C}_+)$ onto $H^p(\mathbb{C}_+)$.

To summarize, the Hardy space $H^p(\mathbb{C}_+)$, $1 \leq p < \infty$, is a Banach space of complex-valued functions in the kernel of a first-order, *determined*, elliptic operator, the Cauchy–Riemann operator $\bar{\partial}$; but the functions in $H^p(\mathbb{C}_+)$ are *over-determined* in the sense that there are spaces $\Re H^p(\mathbb{R})$ and $\Re H^p(\mathbb{C}_+)$ of *real-valued* functions together with linear isomorphisms from these spaces onto $H^p(\mathbb{C}_+)$. Diagrammatically,

(4.33)

$$
\begin{array}{ccccccc}
\Re H^p(\mathbb{C}_+) & \longrightarrow & H^p(\mathbb{C}_+) & & U = P_t * f & \longrightarrow & U + iV \\
\uparrow & \nearrow & \uparrow & & \uparrow & \nearrow^{Cf} & \uparrow Cg \\
\Re H^p(\mathbb{R}) & \longrightarrow & H^p(\partial\mathbb{C}_+) & & f & \longrightarrow & g = f + i\mathcal{H}f
\end{array}
$$

The existence of $\Re H^p(\mathbb{C}_+)$ is a reflection in part of the fact that the solutions of $\bar{\partial}$ have harmonic components, whereas the existence of $\Re H^p(\mathbb{R})$ and $H^p(\mathbb{R})$ is a boundary regularity result for $\bar{\partial}$.

Hardy spaces defined in terms of the non-tangential maximal function can be associated with many other domains in $\mathbb{C}$. In his thesis, Carlos Kenig [63] completely developed the theory of Hardy spaces for domains in $\mathbb{C}$ having minimally smooth boundary, i.e., Lipschitz domains. The balance of this section is devoted to a summary of his main results (presented without proof), with emphasis upon the parallels with the classical theory. To begin, let ϕ be a real-valued function on $\mathbb{R}$, which is (at least) Lipschitz, and let

(4.34)
$$ M_\phi = \{ Z = x + it : t > \phi(x) \} $$

be the open set in $\mathbb{C}$ above the graph

(4.35)
$$ \partial M_\phi = \{ X = x + i\phi(x) : x \in \mathbb{R} \} $$

of ϕ. Without loss of generality, we assume $\phi(0) = 0$; then ∂M_ϕ contains the origin and $i \in M_\phi$. Equally, the open set below the graph of ϕ could be considered, and when there is occasion to do so it will be denoted by M_ϕ^-, with M_ϕ^+ then being used to distinguish (4.34) from M_ϕ^-. Being Lipschitz, ϕ has a derivative ϕ' almost everywhere on $\mathbb{R}$. The usual Lebesgue L^p spaces $L^p(\partial M_\phi)$ of real- or complex-valued functions f on ∂M_ϕ are defined with respect to arc-length measure; thus

(4.36)
$$ \|f\|_p = \left(\int_{\partial M_\phi} |f(X)|^p \, dX \right)^{1/p} = \left(\int_{-\infty}^{\infty} |f(x)|^p \big(1 + |\phi'(x)|^2\big)^{1/2} \, dx \right)^{1/p} $$

where consistently we shall write $f(x)$ instead of $f(x, \phi(x))$ when $X = (x, \phi(x))$. Now the region M_ϕ is simply connected and so conformally equivalent to $\mathbb{C}_+$. Let $\Phi : \mathbb{C}_+ \to M_\phi$ be the conformal mapping which is uniquely specified by requiring that $\Phi(i) = i$, $\Phi(\infty) = \infty$. Then $F \to G = F \circ \Phi$ and $G \to F = G \circ \Phi^{-1}$ identify the analytic functions on M_ϕ with those on $\mathbb{C}_+$. On the other hand, the following boundary properties of Φ are (essentially) classical.

(4.37) Theorem.
 Let $\Phi : \mathbb{C}_+ \to M_\phi$ be a conformal mapping. Then

 (i) *Φ extends to a topological isomorphism from $\mathbb{C}_+ \cup \mathbb{R}$ onto $M_\phi \cup \partial M_\phi$,*
 (ii) *on $\mathbb{R}$, $\Phi'(x)$ exists almost everywhere and is locally integrable,*
 (iii) *sets of measure 0 in $\mathbb{R}$ correspond to sets of measure 0 in ∂M_ϕ, and vice versa,*
 (iv) *both Φ and its inverse Φ^{-1} preserve angles at almost every boundary point.*

 For any α, $0 < \alpha < \arctan(1/\|\phi'\|_\infty)$, and $X = (x, \phi(x))$ on ∂M_ϕ, let

$$(4.38) \qquad \Gamma_\alpha(X) = \Big\{ Z \in \mathbb{C} : |\Re Z - z| < \tan \alpha (\Im Z - \phi(x)) \Big\}$$

be the vertical cone at X having aperture α; the choice of α ensures that each $\Gamma_\alpha(X)$ lies wholly in M_ϕ. Thus the non-tangential maximal function $N(F)$,

$$(4.39) \qquad N(F)(X) = \sup\{|F(Z)| : Z \in \Gamma_\alpha(X)\} \,,$$

is well-defined for any continuous real or complex-valued function F on M_ϕ.

(4.40) Definition.
 The Hardy space $H^p(M_\phi)$, $0 < p < \infty$, consists of all analytic functions F on M_ϕ for which

$$\|F\|_{H^p} = \left(\int_{\partial M_\phi} N(F)(X)^p \, dX \right)^{1/p}$$
$$= \left(\int_{-\infty}^{\infty} N(F)(x)^p \big(1 + |\phi'(x)|^2\big)^{1/2} \, dx \right)^{1/p}$$

is finite.

 This definition does not depend on the choice of aperture α, $0 <$

$\alpha < \arctan(1/\|\phi'\|_\infty)$. In view of (4.30) and the Burkholder–Gundy–Silverstein theorem (see (4.31)), these Hardy spaces coincide with the classical ones on $\mathbb{C}_+$ when $\phi \equiv 0$. Thus, at this stage, the principal interest is in seeing how much $H^p(M_\phi)$ differs from $H^p(\mathbb{C}_+)$ and how much they are alike as ϕ varies, or, equivalently, as the conformal transformation $\Phi : \mathbb{C}_+ \to M_\phi$ varies.

First, there is an isomorphism between $H^p(M_\phi)$ and $H^p(\mathbb{C}_+)$ analogous to the one that (4.6) determines between $H^p(D)$ and $H^p(\mathbb{C}_+)$.

(4.41) Theorem.
The conformal mapping $\Phi : \mathbb{C}_+ \to M_\phi$ determines an analytic function g on $\mathbb{C}_+$ so that

(i) $\lim_{z \to x} \text{-n.t.}\ g(z) = |\Phi'(x)|$ *on* $\mathbb{R}$,

(ii) F *is in* $H^p(M_\phi)$ *if and only if* $(e^g)^{1/p} F \circ \Phi$ *is in* $H^p(\mathbb{C}_+)$, $0 < p < \infty$.

By introducing weighted Hardy spaces on $\mathbb{C}_+$, we can make theorem (4.41) more explicit.

(4.42) Theorem.
Fix p, $0 < p < \infty$, and let F be an analytic function on M_ϕ. Then F is in $H^p(M_\phi)$ if and only if

$$\sup_{t>0} \int_{-\infty}^\infty |F(\Phi(x+it))|^p |\Phi'(x)|\, dx$$

is finite; furthermore, there exist constants A_p, B_p so that

$$A_p \|F\|_{H^p} \le \sup_{t>0} \left(\int_{-\infty}^\infty |F(\Phi(x+it))|^p |\Phi'(x)|\, dx \right)^{1/p} \le B_p \|F\|_{H^p}$$

with respect to the norm of F in $H^p(M_\phi)$.

Finally in this string of positive comparisons between $H^p(M_\phi)$ and $H^p(\mathbb{C}_+)$, there is a Burkholder–Gundy–Silverstein theorem.

(4.43) Theorem.
An analytic function F is in $H^p(M_\phi)$, $0 < p < \infty$, if and only if the non-tangential maximal function $N(U)$ of its harmonic real part U is in $L^p(\partial M_\phi)$; furthermore, there exist constants A_p and B_p such that

$$A_p \|F\|_{H^p} \le \left(\int_{\partial M_\phi} N(U)(X)^p\, dX \right)^{1/p} \le B_p \|F\|_{H^p}.$$

Just as before therefore, we can use the space $\Re H^p(M_\phi)$ of all real-valued functions U on M_ϕ for which

$$\|U\|_{\Re H^p} = \left(\int_{\partial M_\phi} N(U)(X)^p \, dX\right)^{1/p}$$

is finite as an expression of the analytic over-determinedness of $H^p(M_\phi)$. The boundary value expressions of this analytic over-determinedness, however, are where we shall find the most important divergence in the properties of $H^p(M_\phi)$ and $H^p(\mathbb{C}_+)$. In view of theorems (4.41) and (4.42) together with (4.37)(iii), (iv), for instance, every F in $H^p(M_\phi)$ has non-tangential boundary values almost everywhere on ∂M_ϕ. So the space $\Re H^p(\partial M_\phi)$ of all these boundary values is well-defined, and certainly will be a subspace of $\Re L^p(\partial M_\phi)$ when $1 \leq p < \infty$. The key question is whether $\Re H^p(\partial M_\phi)$ and $\Re L^p(\partial M_\phi)$ coincide, a question which is answered as follows.

Suppose that ω is a non-negative locally integrable function on $\mathbb{R}$. The function ω is said to belong to the Muckenhoupt class A_q, where $1 \leq q < \infty$, provided that

$$(4.44) \qquad \sup_I \left(|I|^{-1}\int_I \omega(y)\,dy\right)\left(|I|^{-1}\int_I \omega(y)^{-1/(q-1)}\,dy\right)^{q-1} < \infty$$

where the supremum is taken over all intervals I, and for $q = 1$ the second factor on the left-hand side of (4.44) is replaced by the L^∞ norm of ω^{-1} on I. Now, let $p_0 = \inf\{q : |\Phi| \in A_q\}$. Then we have $1 \leq p_0 < \infty$, and the next result follows.

(4.45) Theorem.
     Whenever $p > p_0$, $\Re H^p(\partial M_\phi) = \Re L^p(\partial M_\phi)$. On the other hand, if $1 < p \leq p_0$, $\Re H^p(\partial M_\phi)$ is a proper subspace of $\Re L^p(\partial M_\phi)$.

We recall that for $1 < p < \infty$, $\omega \in A_p$ if and only if the following weighted norm inequality obtains for the Hilbert transform $\mathcal{H}$:

$$(4.46) \qquad \int_\mathbb{R} |\mathcal{H}f(x)|^p \omega(x)\,dx \leq c \int_\mathbb{R} |f(x)|^p \omega(x)\,dx$$

where c is a constant independent of f. The proof of the second part of theorem (4.45) relies upon this alternative characterization of A_p. In particular, we can reformulate theorem (4.45) as follows.

(4.47) Corollary.
     For $1 < p < \infty$, $\Re H^p(\partial M_\phi) = \Re L^p(\partial M_\phi)$ if and only if the weighted norm inequality (4.46) obtains, with $\omega = |\Phi|$.

To conclude this section, we shall give a simple example of a Lipschitz domain M_ϕ for which the critical exponent p_0 is strictly greater than 1. Let M_ϕ denote the wedge

$$M_\phi = \left\{ z \in \mathbb{C} : \left| \arg z - \frac{\pi}{2} \right| < \alpha \right\}$$

where $0 < \alpha < \pi/2$. The conformal map $\Phi : \mathbb{C}_+ \to M_\phi$ is given by

$$\Phi(z) = ie^{-i\alpha} z^{2\alpha/\pi}$$

so that $|\Phi(x)| = |x|^{2\alpha/\pi}$ for $x \in \mathbb{R}$. A simple calculation shows that $|\Phi| \in A_p$, $1 < p < \infty$, if and only if $p > 1 + \frac{2\alpha}{\pi}$. Letting α range over $(0, \pi/2)$, we obtain for each $p \in (1,2)$ a domain M_ϕ for which the equality $\Re H^p(\partial M_\phi) = \Re L^p(\partial M_\phi)$ fails.

5 Spaces of Clifford analytic functions I: the upper half-space

Hardy spaces and Bergman spaces in $\mathcal{A}(\Omega, \mathfrak{H})$, $\Omega \subseteq \mathbb{R}^n$, can be introduced exactly as in the classical case, replacing analyticity in the usual sense by Clifford analyticity. The classical theory, however, fails to carry over intact, both in principle and in technique, to this more general setting. For instance, the weaker subharmonicity property (3.39) of $\mathcal{A}(\Omega, \mathfrak{H})$ means that boundary regularity results can be established only for $p > \frac{n-2}{n-1}$ rather than for all $p > 0$ as in the classical case. More crucially, the absence of a higher-dimensional Riemann mapping theorem prevents us from moving freely from one simply connected set Ω to another by conformal mapping, forcing the introduction of new techniques to establish corresponding results for Clifford analytic functions. The availability of the Cauchy integral, though, more than makes up for this deficiency. In this section, we will develop Hardy space theory for the upper half-space $\Omega = \mathbb{R}^n_+$. In the following two sections we will consider the Cauchy integral operator and Hardy space theory on Lipschitz domains.

For $p > 0$, the Hardy space $H^p(\mathbb{R}^n_+, \mathfrak{H})$ of Clifford analytic functions is defined to be the space of all Clifford analytic F in $\mathbb{R}^n_+$ satisfying

$$(5.1) \qquad \|F\|_{H^p(\mathbb{R}^n_+)} = \sup_{t>0} \left(\int_{\mathbb{R}^{n-1}} |F(x+it)|^p \, dx \right)^{1/p} < \infty \, .$$

The boundary-value theory for $H^p(\mathbb{R}^n_+, \mathfrak{H})$ is concerned with functions defined on $\mathbb{R}^{n-1} = \partial \mathbb{R}^n_+$. The existence of non-tangential boundary values when $p > 1$ follows as in the proof of theorem (4.10) from properties of Poisson integrals in $\mathbb{R}^n$. The analogue of theorem (4.21) is valid only for $\frac{n-2}{n-1} < p \leq 1$, as we shall see. We begin with some notation.

By analogy to (4.8), for $\alpha > 0$ we define, for $x \in \mathbb{R}^{n-1}$,

$$(5.2) \qquad \Gamma_\alpha(x) = \left\{ z = (y,t) \in \mathbb{R}_+^n : |y - x| < \alpha t \right\} ;$$

this is the cone based at $x \in \mathbb{R}^{n-1}$ with aperture α. As before, we say that a function F on $\mathbb{R}_+^n$ has a *non-tangential limit at* x if

$$(5.3) \qquad \lim_{z \to x}\text{-n.t.}\, F(z) = \lim_{\substack{z \to x \\ z \in \Gamma_\alpha(x)}} F(z)$$

exists as z approaches x from above and non-tangentially. By analogy to the classical case, we shall show that for $p > \frac{n-2}{n-1}$, every $H^p(\mathbb{R}_+^n, \mathfrak{H})$ function has almost-everywhere non-tangential limits. Then, for $p \geq 1$, we shall give a boundary integral characterization of $H^p(\mathbb{R}_+^n, \mathfrak{H})$, whereby the Hardy space may be identified with a subspace of $L^p(\mathbb{R}^{n-1}, \mathfrak{H})$ on the boundary. Finally, for $p > 1$, we shall show that, in the special case of $\mathfrak{H} = \mathfrak{A}_n$, $H^p(\mathbb{R}_+^n, \mathfrak{A}_n)$ is effectively isomorphic to $L^p(\mathbb{R}^{n-1}, \mathfrak{A}_{n-1})$, while $H^1(\mathbb{R}_+^n \mathfrak{A}_n)$ is isomorphic to a subspace of $L^1(\mathbb{R}^{n-1}, \mathfrak{A}_{n-1})$.

We begin with the following analogue of theorems (4.10) and (4.21):

(5.4) Theorem.
   Suppose $F \in H^p(\mathbb{R}_+^n, \mathfrak{H})$, $p > \frac{n-2}{n-1}$. Then there is a function $f \in L^p(\mathbb{R}^{n-1}, \mathfrak{H})$ such that

(i) $\lim_{z \to x}\text{-n.t.}\, F(z) = f(x)$ *exists for almost all* $x \in \mathbb{R}^{n-1}$,

(ii) $\lim_{t \to 0} \int_{\mathbb{R}^{n-1}} |F(x,t) - f(x)|^p \, dx = 0$.

Proof. The proof is a minor modification of the proof of theorem (4.21); we sketch the argument here. Let $p_0 = \frac{n-2}{n-1}$ and define Φ on $\mathbb{R}_+^n$ by $\Phi(z) = |F(z)|^{p_0}$. Then Φ is non-negative and subharmonic on $\mathbb{R}_+^n$, so that, in particular, there exists a least-harmonic majorant Ψ satisfying $\Phi(z) \leq \Psi(z)$ and

$$(5.5)$$

$$\sup_{t>0} \int_{\mathbb{R}^{n-1}} \Psi(x,t)^{p/p_0} \, dx = \sup_{t>0} \int_{\mathbb{R}^{n-1}} \Phi(x,t)^{p/p_0} \, dx = \|F\|_{H^p(\mathbb{R}_+^n)} \ .$$

Since $p/p_0 > 1$, $\Psi(x,t)$ is the Poisson integral $P_t * g$ of a function $g \in L^{p/p_0}(\mathbb{R}^{n-1})$. Thus Ψ will have almost-everywhere non-tangential limits on $\mathbb{R}^{n-1}$, and will be non-tangentially bounded almost everywhere. This assures the existence of almost-everywhere non-tangential limits for F, and (ii) follows from dominated convergence and arguments involving Poisson integrals and the Hardy–Littlewood maximal function, as in the classical case. $\blacksquare$

   As we shall demonstrate shortly, for $p \geq 1$, F may be recovered from its boundary function f by taking the Cauchy integral of f. In general, suppose $p \geq 1$ and $f \in L^p(\mathbb{R}^{n-1}, \mathfrak{H})$. We define the *Cauchy integral of* f on $\mathbb{R}^{n-1}$ by setting

(5.6)
$$Cf(z) = \frac{1}{\omega_n} \int_{\mathbb{R}^{n-1}} \frac{z-u}{|z-u|^n} \, d\sigma(u) f(u) = \frac{1}{\omega_n} \int_{\mathbb{R}^{n-1}} \frac{u-z}{|u-z|^n} \, e_n f(u) \, du$$

for $z = (x,t) \in \mathbb{R}^n \setminus \mathbb{R}^{n-1}$. Cf is clearly analytic on $\mathbb{R}_+^n$, and

(5.7)
$$Cf(z) = \frac{1}{2}(P_t * f)(x) + \frac{1}{2}\sum_{j=1}^{n-1} e_j e_n (Q_t^{(j)} * f)(x)$$

where, for $x \in \mathbb{R}^{n-1}$,

(5.8)
$$P_t(x) = \frac{2}{\omega_n} \frac{t}{[t^2 + |x|^2]^{n/2}} = \frac{2}{\omega_n} \frac{1}{t^{n-1}} \frac{1}{[1 + |x/t|^2]^{n/2}}$$

is the *Poisson kernel* for $\mathbb{R}^{n-1}$, and, for $1 \leq j \leq n-1$,

(5.9)
$$Q_t^{(j)}(x) = \frac{2}{\omega_n} \frac{x_j}{[t^2 + |x|^2]^{n/2}} = \frac{2}{\omega_n} \frac{1}{t^{n-1}} \frac{x_j/t}{[1 + |x/t|^2]^{n/2}}$$

is the jth *conjugate Poisson kernel* for $\mathbb{R}^{n-1}$. A computation of Fourier transforms shows that

(5.10) $\quad \widehat{P}_t(\xi) = e^{-t|\xi|} \quad, \quad \widehat{Q}_t(\xi) = -ie^{-t|\xi|} \xi_j |\xi|^{-1} \qquad (1 \leq j \leq n-1)\,,$

where the Fourier transform is defined according to the normalization

(5.11)
$$\hat{g}(\xi) = \int_{\mathbb{R}^{n-1}} e^{-ix\cdot\xi} g(x)\, dx\,.$$

We easily verify that $Q_t^{(j)} = R_j P_t$, where R_j is the jth *Riesz transform*, given by

(5.12)
$$R_j g(x) = \frac{2}{\omega_n} \int_{\mathbb{R}^{n-1}} \frac{x_j - u_j}{|x - u|^n} g(u)\, du\,,$$

in view of the fact that

(5.13)
$$(R_j g)^\wedge(\xi) = -i\xi_j |\xi|^{-1} \hat{g}(\xi)\,.$$

The Riesz transforms are the natural higher-dimensional analogue of the Hilbert transform; they are bounded operators on $L^p(\mathbb{R}^{n-1})$ for $1 < p < \infty$, and they map $L^1(\mathbb{R}^{n-1})$ boundedly into weak-$L^1(\mathbb{R}^{n-1})$. By analogy to our earlier work, we define the operator

(5.14)
$$\mathcal{H} = \sum_{j=1}^{n-1} e_j R_j$$

so that (5.7) becomes

(5.15) $\quad Cf(z) = \left\{ P_t * \frac{1}{2}(I + \mathcal{H}e_n)f \right\}(x)\,, \qquad z = (x,t) \in \mathbb{R}_+^n\,.$

By virtue of (5.15), the following result is a straightforward consequence of the properties of the Poisson kernel.

(5.16) Theorem.
   Suppose that either (i) $1 < p < \infty$ and $f \in L^p(\mathbb{R}^{n-1}, \mathfrak{H})$, or (ii) $p = 1$ and $f, \mathcal{H}f \in L^1(\mathbb{R}^{n-1}, \mathfrak{H})$. Then $Cf \in H^p(\mathbb{R}^n_+, \mathfrak{H})$, and

$$\lim_{z \to x}\text{-n.t.}\, Cf(z) = \tfrac{1}{2}(I + \mathcal{H}e_n)f(x)$$

for almost all $x \in \mathbb{R}^{n-1}$.

   By analogy to definition (4.29), we introduce the non-tangential maximal function:

(5.17) Definition.
   Let F be any $\mathfrak{H}$-valued function on $\mathbb{R}^n_+$. Then the non-tangential maximal function $N(F)$ of F is defined, for $x \in \mathbb{R}^n$, by

$$N(F)(x) = \sup_{z \in \Gamma_\alpha(x)} |F(z)| \ .$$

   It is easily seen, as before, that for any $t > 0$ and for any aperture α,

$$(5.18) \qquad \int_{\mathbb{R}^{n-1}} |F(x,t)|^p\, dx \le \int_{\mathbb{R}^{n-1}} N(F)(x)^p\, dx \ , \qquad 0 < p < \infty \ .$$

Consequently, if F is Clifford analytic in $\mathbb{R}^n_+$ and if $N(F) \in L^p(\mathbb{R}^{n-1})$, then $F \in H^p(\mathbb{R}^n_+, \mathfrak{H})$. In fact, we obtain the following Burkholder–Gundy–Silverstein result.

(5.19) Theorem.
   Let $\frac{n-2}{n-1} < p < \infty$. A Clifford analytic function F is in $H^p(\mathbb{R}^n_+, \mathfrak{H})$ if and only if the non-tangential maximal function $N(F)$ is in $L^p(\mathbb{R}^{n-1})$; furthermore, there exist constants A_p and B_p such that

$$A_p\|F\|_{H^p(\mathbb{R}^n_+)} \le \left(\int_{\mathbb{R}^{n-1}} N(F)^p(x)\, dx \right)^{1/p} \le B_p\|F\|_{H^p(\mathbb{R}^n_+)}$$

Proof. It clearly suffices to show that the H^p-norm of F controls the L^p-norm of $N(F)$. Let Φ, Ψ, g be as in the proof of theorem (5.4). Then, for $x \in \mathbb{R}^{n-1}$,

$$N(F)(x) \le \left[N(\Psi)(x)\right]^{1/p_0} \le K\left[M(g)(x)\right]^{1/p_0}$$

where K is a constant depending on n, and M denotes the Hardy–

Littlewood maximal function. Thus

$$\int_{\mathbb{R}^{n-1}} N(F)^p(x)\,dx \le K \int_{\mathbb{R}^{n-1}} \left[M(g)(x) \right]^{p/p_0} dx$$

$$\le K \int_{\mathbb{R}^{n-1}} |g(x)|^{p/p_0}\,dx$$

$$= K\|F\|^p_{H^p(\mathbb{R}^n_+)}$$

where K is a constant, different at each occurrence, depending only upon n and p. This completes the proof. ∎

The characterization of $H^p(\mathbb{R}^n_+, \mathfrak{H})$ given by theorem (5.19) plays a crucial role in the proof of the following boundary integral characterization for $p \ge 1$.

(5.20) Theorem.
    Let $1 \le p < \infty$ and suppose $F \in H^p(\mathbb{R}^n_+, \mathfrak{H})$. Then $F = Cf$, where f is the almost-everywhere non-tangential limit of F given by theorem (5.4).

Proof. Since F is harmonic, F must be the Poisson extension of f, i.e., for all $z = (x, t) \in \mathbb{R}^n_+$, $F(z) = P_t * f(x)$. Standard estimates for Poisson integrals ensure the existence of a constant $L = L_{n,p} > 0$ such that

$$(5.21) \qquad \sup_{x \in \mathbb{R}^{n-1}} |F(x, t)| \le L\|F\|_{H^p(\mathbb{R}^n_+)} t^{(1-n)/p} \ .$$

Now, fix $z \in (x, t) \in \mathbb{R}^n_+$, and let $R > 10|z| > t > \varepsilon > 0$. Defining

$$(5.22) \qquad \Sigma_{\varepsilon,R} = \{ \zeta = (\xi, s) \in \mathbb{R}^n_+ : |\zeta| = R \ , \ x > \varepsilon \} \ ,$$

$$(5.23) \qquad D_{\varepsilon,R} = \{ \zeta = (\xi, s) \in \mathbb{R}^n_+ : |\zeta| \le R \ , \ s = \varepsilon \} \ ,$$

we see that z is inside $S_{\varepsilon,R} = \Sigma_{\varepsilon,R} \cup D_{\varepsilon,R}$. By the Cauchy integral formula,

$$(5.24) \quad F(z) = \frac{1}{\omega_n} \int_{S_{\varepsilon,R}} \frac{z - \zeta}{|z - \zeta|^n} \eta(\zeta) F(\zeta)\,dS(\zeta)$$

$$= \frac{1}{\omega_n} \int_{D_{\varepsilon,R}} \frac{\zeta - z}{|\zeta - z|^n} e_n F(\zeta)\,dS(\zeta) + \frac{1}{\omega_n} \int_{\Sigma_{\varepsilon,R}} \frac{z - \zeta}{|z - \zeta|^n} \eta(\zeta) F(\zeta)\,dS(\zeta) \ .$$

We will use (5.21) to show that the second integral tends to 0 as $R \to \infty$.

To begin, we have

$$\left| \frac{1}{\omega_n} \int_{\Sigma_{\varepsilon,R}} \frac{z-\zeta}{|z-\zeta|^n} \eta(\zeta) F(\zeta)\, dS(\zeta) \right| \leq \frac{1}{\omega_n} \int_{\Sigma_{\varepsilon,R}} |z-\zeta|^{1-n} |F(\zeta)|\, dS(\zeta)$$

(5.25)

$$\leq \frac{1}{\omega_n} \left(\frac{10}{9} \right)^{n-1} R^{1-n} \|F\|_{H^p(\mathbb{R}_+^n)} \int_{\Sigma_{\varepsilon,R}} s^{(1-n)/p}\, dS(\zeta)$$

where we have used (5.21) together with the fact that, for $\zeta \in \Sigma_{\varepsilon,R}$, $\frac{9}{10}R \leq |z-\zeta| \leq 11R$. Integration in polar coordinates yields the estimate

$$(5.26) \quad \int_{\Sigma_{\varepsilon,R}} s^{(1-n)/p}\, dS(\zeta) \leq L R^{n-1} \int_{\arcsin \varepsilon/R}^{\pi/2} (R \sin \theta)^{(1-n)/p}\, d\theta$$

$$\leq L \times \begin{cases} \varepsilon^{(p+1-n)/p} R^{n-2}, & 1 \leq p < n-1 \\ \log \dfrac{\pi R}{2\varepsilon}, & p = n-1 \\ 1, & p \geq n. \end{cases}$$

Combining (5.25) and (5.26), we have

$$(5.27) \quad \left| \frac{1}{\omega_n} \int_{\Sigma_{\varepsilon,R}} \frac{z-\zeta}{|z-\zeta|^n} \eta(\zeta) F(\zeta)\, dS(\zeta) \right|$$

$$\leq L \|F\|_{H^p(\mathbb{R}_+^n)} \cdot \begin{cases} \varepsilon^{(p+1-n)/p} R^{-1}, & 1 \leq p < n-1 \\ R^{1-n} \log \dfrac{\pi R}{2\varepsilon}, & p = n-1 \\ R^{1-n}, & p \geq n. \end{cases}$$

In any event, both sides of (5.27) tend to 0 as $R \to \infty$. Thus, taking the limit as $R \to \infty$ in (5.24) yields

$$(5.28) \quad F(z) = \frac{1}{\omega_n} \int_{\mathbb{R}^{n-1}} \frac{\xi - x + (\varepsilon - t)e_n}{\big[|\xi - x|^2 + (\varepsilon - t)^2\big]^{n/2}} e_n F(\xi, \varepsilon)\, d\xi$$

for all ε with $0 < \varepsilon < t$. It is easy to see that

$$(5.29) \quad \begin{aligned} F(x,t) &= \{ P_{t-\varepsilon} * \tfrac{1}{2}(I + \mathcal{H}e_n)(P_\varepsilon * f) \}(x) \\ &= \{ P_t * \tfrac{1}{2}(I + \mathcal{H}e_n) f \}(x) = Cf(z), \end{aligned}$$

where we have used the semigroup property of the Poisson kernel, together with the fact that $\mathcal{H}e_n$ commutes with $P_\varepsilon *$. ∎

Theorem (5.20) says, in effect, that, for $1 < p < \infty$, $H^p(\mathbb{R}_+^n, \mathfrak{H})$ is precisely the image of $L^p(\mathbb{R}^{n-1}, \mathfrak{H})$ under the Cauchy integral operator, while $H^1(\mathbb{R}_+^n, \mathfrak{H})$ is the image under C of the set of all $f \in L^1(\mathbb{R}^{n-1}, \mathfrak{H})$ for which $\mathcal{H}f \in L^1(\mathbb{R}^{n-1}, \mathfrak{H})$. Moreover, for $1 \leq p < \infty$, $f \in L^p(\mathbb{R}^{n-1}, \mathfrak{H})$ arises as the non-tangential boundary value of an $H^p(\mathbb{R}_+^n, \mathfrak{H})$-function if and only if $f = \mathcal{H}e_n f \in L^p(\mathbb{R}^{n-1}, \mathfrak{H})$.

We consider the special case $\mathfrak{H} = \mathfrak{A}_n$. Recall the direct-sum decomposition $\mathfrak{A}_n = \mathfrak{A}_{n-1} \oplus e_n \mathfrak{A}_{n-1}$; since e_n is the inner unit normal on $\partial \mathbb{R}^n_+ = \mathbb{R}^{n-1}$, we may view this as a decomposition of the algebra into its 'tangential' and 'normal' components on $\partial \mathbb{R}^n_+$. Writing $f = f_{\mathrm{Tan}} + e_n f_{\mathrm{Nor}}$, with $f_{\mathrm{Tan}}, f_{\mathrm{Nor}}$ taking values in $\mathfrak{A}_{n-1}$, we easily see that $f = \mathcal{H}e_n f$ if and only if

(5.30) $$ f_{\mathrm{Tan}} = -\mathcal{H}f_{\mathrm{Nor}} \quad , \quad f_{\mathrm{Nor}} = -\mathcal{H}f_{\mathrm{Tan}} \; . $$

Since $\mathcal{H}^2 = I$, the two equations in (5.30) are equivalent. In particular, f is *determined* by *either* f_{Nor} *or* f_{Tan}. By analogy to definition (4.25), we make the following.

(5.31) Definition.
     For $1 < p < \infty$, *let* $\operatorname{Tan} H^p(\mathbb{R}^{n-1})$ *denote the space* $L^p(\mathbb{R}^{n-1}, \mathfrak{A}_{n-1})$, *whereas, for* $p = 1$, *let* $\operatorname{Tan} H^1(\mathbb{R}^{n-1})$ *denote the space of all* $f \in L^1(\mathbb{R}^{n-1}, \mathfrak{A}_{n-1})$ *such that* $\mathcal{H}f \in L^1(\mathbb{R}^{n-1}, \mathfrak{A}_{n-1})$.

     The spaces $\operatorname{Tan} H^p(\mathbb{R}^{n-1})$ are Banach spaces with respect to the norms

(5.32)(i) $$ \|f\|_{\operatorname{Tan} H^p} = \left(\int_{\mathbb{R}^{n-1}} |f(x)|^p_{\mathfrak{A}_{n-1}} \, dx \right)^{1/p} \quad (1 < p < \infty) \, , $$

(5.32)(ii) $$ \|f\|_{\operatorname{Tan} H^1} = \int_{\mathbb{R}^{n-1}} \left(|f(x)|_{\mathfrak{A}_{n-1}} + |\mathcal{H}f(x)|_{\mathfrak{A}_{n-1}} \right) dx \; . $$

These are the higher-dimensional generalizations of the spaces $\Re H^p(\mathbb{R})$, and they consist precisely of the tangential components of the boundary values of $H^p(\mathbb{R}^n_+, \mathfrak{A}_n)$-functions. As in the classical case, we obtain the following result.

(5.33) Theorem.
     The Cauchy integral is a linear isomorphism from $\operatorname{Tan} H^p(\mathbb{R}^{n-1})$ *onto* $H^p(\mathbb{R}^n_+, \mathfrak{A}_n)$, *for* $1 \le p < \infty$.

These results for $\mathfrak{A}_n$ clearly extend to any $\mathfrak{A}_n$-module $\mathfrak{H}$ which admits a splitting of the form $\mathfrak{H} = \mathfrak{H}_0 \oplus e_n \mathfrak{H}_0$, with $\mathfrak{H}_0$ an $\mathfrak{A}_{n-1}$-module.

6 Cauchy integrals and Hilbert transforms on Lipschitz domains

In the complex plane the principal-value singular integral operator associated to the Cauchy integral, the Hilbert transform, has long played a central role in analytic function theory. In the last section we saw that the operator $\mathcal{H}$, which is effectively an embedding of the Riesz transforms into $\mathfrak{A}_{n-1}$, played a correspondingly central role in the development of H^p theory for $\mathbb{R}^n_+$. In section 4, we discussed the theory of H^p spaces in Lipschitz domains in $\mathbb{C}$, as developed by Kenig. An alternative aproach to this theory makes use of the Cauchy integral and Hilbert transform for Lipschitz curves. Calderón, Coifman, McIntosh, and Meyer, working in the late 1970s and early 1980s, obtained the crucial L^p estimates for the Hilbert transform on Lipschitz curves, which form the groundwork of this alternate approach.

By analogy to the classical case, we would like to develop a theory of H^p spaces of Clifford analytic functions on Lipschitz domains in $\mathbb{R}^n$. In this context, conformal mapping is no longer available as a tool in the development of the theory, and so we must rely on an understanding of the Hilbert transform defined on the boundary of Lipschitz domains in $\mathbb{R}^n$. In this section, we shall prove the L^p-boundedness of the Hilbert transform for Lipschitz domains in $\mathbb{R}^n$. Our proof makes use of the seminal ideas of Coifman, McIntosh, and Meyer in the Clifford algebra setting, in a way that reveals in interesting and crucial ways the central role of the semigroup Λ_n (see chapter 1, (7.23) and (7.25)) in estimating Clifford-valued singular integrals. We note that the results of this section are easily extended to other L^p-spaces using standard techniques; the L^2-estimates are at the very heart of the theory, and so we concentrate our efforts there.

We begin with some definitions.

(6.1) Definition.

Let $M \subseteq \mathbb{R}^n$ be a bounded domain, and let $\beta > 0$. We say that M is a Lipschitz domain with constant less than or equal to β if and only if for every $Q \in \partial M$ there is a ball B centered at Q, a coordinate system $(x_1, \ldots, x_{n-1}, t) = (x, t)$ with origin at Q, and a function $\phi : \mathbb{R}^{n-1} \to \mathbb{R}$ such that $\phi(0) = 0$ and

(6.2)(i) $|\phi(z) - \phi(y)| \leq \beta|x - y|$ for all $x, y \in \mathbb{R}^{n-1}$;

(6.2)(ii) $M \cap B = \{(x, t) : t > \phi(x)\} \cap B$.

If, in addition, each function ϕ is an element of $C^m(\mathbb{R}^{n-1}, \mathbb{R})$ for some fixed positive integer m, we say that M is a C^m-domain. The Hilbert transform for M is defined for $\mathfrak{H}$-valued functions f on ∂M and for $Z \in \partial M$ by

(6.3) $\mathcal{H}_M f(Z) = \dfrac{2}{\omega_n} \text{p.v.} \displaystyle\int_{\partial M} \dfrac{Q - Z}{|Q - Z|^n} \, d\sigma(Q) f(Q)$

$$= \frac{2}{\omega_n} \lim_{\varepsilon \to 0} \int_{\partial M \setminus B_\varepsilon(Z)} \frac{Q - Z}{|Q - Z|^n} \, d\sigma(Q) f(Q)$$

where, for $\varepsilon > 0$ and $Z \in \partial M$, $B_\varepsilon(Z)$ is the surface ball of radius ε centered at Z.

The main result of this section is the following.

(6.4) Theorem.
    Let M be a Lipschitz domain in $\mathbb{R}^n$. Then $\mathcal{H}_M f(Z)$ exists pointwise almost everywhere and in $L^2(\partial M, \mathfrak{H})$ for each $f \in L^2(\partial M, \mathfrak{H})$. Furthermore

$$(6.5) \qquad \int_{\partial M} |\mathcal{H}_M f(Z)|^2 \, dS(Z) \leq C \int_{\partial M} |f(Z)|^2 \, dS(Z) ,$$

where C depends only upon n and M.

The existence of the limit (6.3) pointwise almost everywhere is a local result, in the sense that the convergence is only in question near Z. In local coordinates, a Lipschitz domain is the region above the graph of a Lipschitz function; thus the existence of the limit (6.3) is an immediate consequence of the existence of the limit in the special case where M is precisely the region above a Lipschitz graph. In addition, by using a sufficiently fine partition of unity on ∂M, we can derive the norm estimate (6.5) for the general case from the corresponding estimate for special Lipschitz domains. Hence it is certainly enough to prove the following.

(6.6) Theorem.
    Let M be the open set in $\mathbb{R}^n$ above the graph $\{(x, \phi(x)) : x \in \mathbb{R}^{n-1}\}$ of a Lipschitz function ϕ. Then $\mathcal{H}_M f$ exists pointwise almost everywhere on ∂M and in $L^2(\partial M, \mathfrak{H})$, for each $f \in L^2(\partial M, \mathfrak{H})$; furthermore
(6.7)
$$\left(\int_{\partial M} |\mathcal{H}_M f(X)|^2 \, dS(X) \right)^{1/2} \leq C(1 + \|\nabla \phi\|_\infty)^5 \left(\int_{\partial M} |f(X)|^2 \, dS(X) \right)^{1/2}$$
where C depends only on n.

In the passage from the special to the general case precise control on the operator norm of $\mathcal{H}_M$ will be lost because of the use of a partition of unity. Nonetheless, this norm will still depend only on n and M.

For the balance of this section we shall assume unless otherwise noted that M is a special Lipschitz domain as described in the statement of theorem (6.6). We pause to note that, if $Z = (z, \phi(z)) \in \partial M$, then

(6.8) $\mathcal{H}_M f(Z)$

$$= \frac{2}{\omega_n} \lim_{\varepsilon \to 0} \int_{|z-x|>\varepsilon} \frac{z - x + e_n(\phi(z) - \phi(x))}{\left[|z - x|^2 + (\phi(z) - \phi(x))^2\right]^{n/2}}$$
$$\bullet (e_n - D\phi(x)) f(x) \, dx$$

$$= \frac{2}{\omega_n} \lim_{\varepsilon \to 0} \int_{|x-z|>\varepsilon} \frac{(x - z)e_n + (\phi(z) - \phi(x))}{\left[|x - z|^2 + (\phi(z) - \phi(x))^2\right]^{n/2}} g(x) \, dx \ ,$$

where $f(x)$ is used to denote $f(x, \phi(x))$ and $g(x) = -(1 - De_n\phi(x))f(x)$. When $\phi \equiv 0$, so that $M = \mathbb{R}^n_+$, it is easily seen that $\mathcal{H}_M$ is simply $\mathcal{H}e_n$, where $\mathcal{H}$ is given by (5.14). In order to estimate (6.8), it suffices to consider the related operator $\mathcal{H}_\phi$, given by

(6.9)

$$\mathcal{H}_\phi g(Z) = \frac{2}{\omega_n} \lim_{\varepsilon \to 0} \int_{|x-z|>\varepsilon} \frac{x - z + (\phi(z) - \phi(x))}{\left[|x - z|^2 + (\phi(z) - \phi(x))^2\right]^{n/2}} g(x) \, dx$$

in view of the fact that $e_1e_n, e_2e_n, \ldots, e_{n-1}e_n$ generate a subalgebra of $\mathfrak{A}_n$ that is isomorphic to $\mathfrak{A}_{n-1}$. When $\phi \equiv 0$, $\mathcal{H}_\phi$ is just $-\mathcal{H}$.

In order to estimate $\mathcal{H}$, we shall make use of the Fourier transform; consequently we shall consider $\mathfrak{A}_n$ as a subspace of the complex Clifford algebra $\mathfrak{C}_n$ in the natural way, and we shall assume without loss of generality that $\mathfrak{H}$ is actually a complex Hilbert space. If $\alpha = \sum \alpha_B e_B \in \mathfrak{C}_n$, then we define

(6.10)
$$|\alpha|_\mathfrak{C} = \left(\sum |\alpha_B|^2\right)^{1/2} .$$

The definition of the Clifford operator norm $\| \cdot \|_\mathfrak{A}$ given in chapter 1, (7.20), extends naturally to $\mathfrak{C}_n$ and will be denoted by $\| \cdot \|_\mathfrak{C}$. In this section only, we extend conjugation from $\mathfrak{A}_n$ to $\mathfrak{C}_n$ in a non-standard way, by defining, for $\alpha \in \mathfrak{C}_n$,

(6.11)
$$\bar{\alpha} = \sum \bar{\alpha}_B \bar{e}_B \ ;$$

with this definition, we have $\overline{(\alpha\beta)} = \bar{\beta}\bar{\alpha}$, as before. This notion of conjugation corresponds to the adjoint in $\mathcal{L}(\mathfrak{H})$. Recall that the 'norm' function $\Delta : \mathfrak{A}_n \to \mathfrak{A}_n$ is defined by

(6.12)
$$\Delta(\alpha) = \bar{\alpha}\alpha \ , \qquad \alpha \in \mathfrak{A}_n$$

(see chapter 1, (5.7)). We extend Δ to $\mathfrak{C}_n$ by using (6.12) together with our non-standard conjugation in $\mathfrak{C}_n$. Note that, for $\alpha \in \mathfrak{C}_n$, the scalar part of $\Delta(\alpha)$ is given by

(6.13)
$$\Delta(\alpha)_\phi = \sum |\alpha_B|^2 = |\alpha|_\mathfrak{C}^2 .$$

We shall need to use the following lemma.

(6.14) Lemma.
  If $\alpha \in \Lambda_n$, $\beta \in \mathfrak{C}_n$, then $|\alpha\beta|^2_{\mathfrak{C}} = |\beta\alpha|^2_{\mathfrak{C}} = \Delta(\alpha)|\beta|^2_{\mathfrak{C}} = \|\alpha\beta\|^2_{\mathfrak{C}}$.

Proof. We have

$$|\alpha\beta|^2_{\mathfrak{C}} = \Delta(\alpha\beta)_\phi = (\bar{\beta}\bar{\alpha}\alpha\beta)_\phi = (\bar{\beta}\Delta(\alpha)\beta)_\phi = \Delta(\alpha)\|\beta\|^2_{\mathfrak{C}} \ ,$$
$$|\beta\alpha|^2_{\mathfrak{C}} = |\bar{\alpha}\bar{\beta}|^2_{\mathfrak{C}} = \Delta(\bar{\alpha})|\bar{\beta}|^2_{\mathfrak{C}} = \Delta(\alpha)|\beta|^2_{\mathfrak{C}} \ .$$

The equality of $|\alpha\beta|^2_{\mathfrak{C}}$ and $\|\alpha\beta\|^2_{\mathfrak{C}}$ now follows as in theorem (7.22) of chapter 1. ∎

We now obtain an important integral representation formula for the operator $\mathcal{H}$. The symbol of this operator under Fourier transformation is given by

$$(6.15) \qquad (\mathcal{H}g)^\wedge(\xi) = -i\frac{\xi}{|\xi|}\hat{g}(\xi) \qquad (\xi \in \mathbb{R}^{n-1}) \ .$$

If we let P_t, Q_t denote the operators

$$(6.16) \quad P_t = (I - t^2\Delta)^{-1} = (I + t^2D^2)^{-1} \quad , \quad Q_t = tDP_t \qquad (t \in \mathbb{R})$$

we see that P_t, Q_t are given by the Fourier multipliers

$$(6.17) \qquad \hat{p}_{|t|}(\xi) = (1 + t^2|\xi|^2)^{-1} \quad , \quad \hat{q}_t(\xi) = it\xi\hat{p}_{|t|}(\xi)$$

and, moreover, that

$$(6.18) \quad \hat{p}_{|t|}(\xi) \pm i\hat{q}_t(\xi) = (1 \mp t\xi)^{-1} \quad , \quad P_t \pm iQ_t = (I \mp itD)^{-1} \ .$$

In view of the identity

$$(6.19) \qquad \frac{1}{|\xi|} = \frac{1}{\pi}\int_{-\infty}^{\infty} \frac{1}{1 + t^2|\xi|^2} \, dt \ , \qquad \xi \neq 0$$

we easily obtain

$$(6.20) \qquad -i\frac{\xi}{|\xi|} = \frac{1}{\pi}\int_{-\infty}^{\infty} \frac{-it\xi}{1 + t^2|\xi|^2}\frac{dt}{t} = -\frac{1}{\pi}\int_{-\infty}^{\infty}\hat{q}_t(\xi)\frac{dt}{t} \ .$$

Using the shorthand $R_t = P_t + iQ_t$, it easily follows that

$$(6.21) \qquad \mathcal{H}g = \frac{i}{\pi}\,\text{p.v.}\int_{-\infty}^{\infty} R_tg\frac{dt}{t} = \frac{i}{\pi}\lim_{\varepsilon\to 0}\int_{\varepsilon<|t|<1/\varepsilon} R_tg\frac{dt}{t}$$

and, equivalently, that

$$(6.22) \qquad \mathcal{H}g = \frac{i}{\pi}\int_{0}^{\infty}(R_t - R_{-t})g\frac{dt}{t} \ .$$

This is only the simplest case of the following remarkable representation theorem.

(6.23) Theorem.

   If ϕ is any real-valued Lipschitz function on $\mathbb{R}^{n-1}$, then

$$(6.24) \qquad \mathcal{H}_\phi = \frac{1}{\pi i} \text{ p.v.} \int_{-\infty}^{\infty} e^{i\phi/t} R_t e^{-i\phi/t} \frac{dt}{t} \ .$$

Proof of theorem (6.23). We may write the right-hand side of (6.24) as $S + V$, where S is the scalar operator given by

$$(6.25) \qquad Sg(x) = \frac{1}{\pi i} \text{ p.v.} \int_{-\infty}^{\infty} e^{i\phi/t} P_t e^{-i\phi/t} g(x) \frac{dt}{t}$$

and V is the vector operator given by

$$(6.26) \qquad Vg(x) = \frac{1}{\pi} \text{ p.v.} \int_{-\infty}^{\infty} e^{i\phi/t} Q_t e^{-i\phi/t} g(x) \frac{dt}{t} \ .$$

S and V are integral operators; we shall evaluate their kernels. P_t is, in fact, a Bessel potential operator of order 2, given by convolution with the dilate $p_{n-1,|t|} = |t|^{1-n} p_{n-1}(\cdot t^{-1})$, where p_{n-1} is a radial function given by

$$(6.27) \quad p_{n-1}(x) = B_{n-1}(r) = \frac{1}{4\pi\Gamma(1/2)} \int_0^{\infty} e^{-\pi r^2/t} e^{-t/4\pi} t^{(2-n)/2} \frac{dt}{t}$$

$(r = |x|, n \geq 2)$. The functions B_n satisfy the recurrence

$$(6.28) \qquad B_{n+2}(r) = -\frac{1}{2\pi r} \frac{\partial}{\partial r} B_n(r) \quad , \quad B_1(r) = \frac{1}{2} e^{-r} \ .$$

If we write $B_{n-1,|t|}(r) = p_{n-1,|t|}(x)$, it is easily seen that

$$(6.29) \quad B_{2k+1,|t|}(r) = \frac{1}{2|t|} \left(-\frac{1}{2\pi r} \frac{\partial}{\partial r} \right)^k e^{-r/|t|} \ , \qquad k = 0, 1, 2, \dots \ .$$

Now, if $n - 1 = 2k + 1$ for some $k \in \{0, 1, 2, \dots\}$, we have

$(6.30) \ Sg(x)$

$$= \int_{\mathbb{R}^{2k+1}} \left\{ \text{p.v.} \int_{-\infty}^{\infty} \frac{1}{\pi i} \exp\left(i \frac{\phi(x) - \phi(y)}{t} \right) B_{2k+1,|t|}(|x - y|) \frac{dt}{t} \right\} g(y) \, dy$$

so that the kernel of S is given by

$$(6.31) \qquad \frac{1}{2\pi i} \left(-\frac{1}{2\pi r} \frac{\partial}{\partial r} \right)^k \int_{-\infty}^{\infty} \exp\left(-\left(\frac{r}{|t|} - \frac{i\alpha}{t} \right) \right) \frac{dt}{t|t|} \ ,$$

where we let $r = |x-y|$, $\alpha = (\phi(x) - \phi(y))$. A straightforward calculation shows that (6.31) equals

$$(6.32) \qquad \frac{1}{2\pi i} \left(-\frac{1}{2\pi r} \frac{\partial}{\partial r} \right)^k \frac{2i\alpha}{r^2 + \alpha^2} = \frac{k!}{\pi^{k+1}} \frac{\alpha}{(r^2 + \alpha^2)^{k+1}}$$

$$= \frac{k!}{\pi^{k+1}} \frac{\phi(x) - \phi(y)}{\left[|x - y|^2 + (\phi(x) - \phi(y))^2 \right]^{k+1}}$$

$$= \frac{2}{\omega_n} \frac{\phi(x) - \phi(y)}{\left[|x - y|^2 + \left(\phi(x) - \phi(y) \right)^2 \right]^{n/2}} \, ,$$

which is precisely the scalar part of the kernel of $\mathcal{H}_\phi$. To calculate the kernel of $\mathcal{V}$, we write $\mathcal{V} = \sum_{j=1}^{n-1} e_j V_j$, and note that the kernel of V_j is then

$$(6.33) \qquad \mathrm{p.v.} \int_{-\infty}^{\infty} \frac{1}{\pi} \exp\left(i \frac{\phi(x) - \phi(y)}{t} \right) \frac{\partial}{\partial z_j} \, B_{2k+1,|t|}(|x - y|) \, \frac{dt}{t} \, ,$$

where $z = x - y$. It is not difficult to see that (6.33) equals

$$(6.34) \qquad \frac{\partial}{\partial z_j} \left(\frac{k!}{\pi^{k+1}} \frac{\alpha}{(|z|^2 + \alpha^2)^{k+1}} \right)$$

where α is regarded as a constant with respect to z; i.e., we obtain

$$(6.35) \qquad \frac{2}{\omega_n} \frac{y_j - x_j}{\left[|x - y|^2 + \left(\phi(x) - \phi(y) \right)^2 \right]^{n/2}} \, .$$

Summing on j, we see that the kernel of $\mathcal{V}$ is precisely the vector part of the kernel of $\mathcal{H}_\phi$.

If $n - 1$ is even, we make use of the easily verifiable fact that

$$(6.36) \qquad p_{n-1}(x) = \int_{\mathbf{R}} p_n(x, s) \, ds \qquad (n \geq 2) \, ,$$

together with our earlier calculations, to show once again that $\mathcal{H}_\phi = \mathcal{S} + \mathcal{V}$. ∎

The theorem admits a still more amazing and useful corollary.

(6.37) Corollary.
Let ϕ be any real-valued Lipschitz function on $\mathbf{R}^{n-1}$ and let μ denote the operator of multiplication by $D\phi$. Then

$$(6.38) \qquad \mathcal{H}_\phi = \frac{1}{\pi i} \mathrm{p.v.} \int_{-\infty}^{\infty} (I - \mu - itD)^{-1} \, \frac{dt}{t} \, .$$

Proof. Note that

$$
\begin{aligned}
e^{i\phi/t} R_t e^{-i\phi/t} &= \left[e^{i\phi/t} (I - itD) e^{-i\phi/t} \right]^{-1} \\
&= \left[I - it e^{i\phi/t} D e^{-i\phi/t} \right]^{-1} \\
&= \left\{ I - it e^{i\phi/t} \left[-i\mu/t (e^{-i\phi/t}) + e^{-i\phi/t} D \right] \right\}^{-1} \\
&= (I - \mu - itD)^{-1} \, ;
\end{aligned}
$$

we have made crucial use of the face that $e^{-i\phi/t}$ is scalar-valued, since D does not, in general, obey the product rule. Then (6.38) follows from (6.24). ∎

We now make use of the representation (6.38) to prove theorem

(6.6). At this stage, we will assume that μ is the operator of multiplication by a function – which we shall also call μ – which maps $\mathbb{R}^{n-1}$-valued functions to functions with values in the semigroup $\Lambda_n \subseteq \mathfrak{A}_n$. We shall also assume that $\|\Delta(\mu)\|_\infty < 1$, so that the norm of μ as a multiplier is less than 1. Then we may write

$$(I - \mu - itD)^{-1} = (I - itD)^{-1}[I - \mu(I - itD)^{-1}]^{-1}$$
$$= R_t[I - \mu R_t]^{-1} .$$

Since the symbol of the operator R_t is $\mathbb{R}^{n-1}$-valued, we easily see that, as an operator on, for example, $L^2(\mathbb{R}^{n-1}, \mathfrak{H})$, μR_t has norm less than 1. Thus we may write

$$(6.39) \qquad (I - \mu - itD)^{-1} = R_t \sum_{k=0}^{\infty} (\mu R_t)^k .$$

Now note that

$$(6.40) \quad T_\mu = \frac{1}{\pi i} \, \text{p.v.} \int_{-\infty}^{\infty} (I - \mu - itD)^{-1} \frac{dt}{t}$$

$$= \frac{1}{\pi i} \int_0^\infty \left[(I - \mu - itD)^{-1} - (I - \mu + itD)^{-1} \right] \frac{dt}{t}$$

$$= \sum_{k=0}^{\infty} \frac{1}{\pi i} \int_0^\infty \left\{ R_t(\mu R_t)^k - R_{-t}(\mu R_{-t})^k \right\} \frac{dt}{t}$$

provided that the series converges. In turn, we have
(6.41)

$$T_\mu = \sum_{k=0}^{\infty} \sum_{s=0}^{k} \frac{1}{\pi} \int_0^\infty \left\{ (R_t\mu)^{k-s} Q_t(\mu P_t)^s + (R_{-t}\mu)^{k-s} Q_t(\mu P_t)^s \right\} \frac{dt}{t} .$$

Let $\|\mu\|_{\text{op}} = \|\Delta(\mu)\|_\infty^{1/2}$ denote the norm of μ as an operator on $L^2(\mathbb{R}^{n-1}; \mathfrak{H})$. Then, for $f \in L^2(\mathbb{R}^{n-1}, \mathfrak{H})$,

$$(6.42) \qquad \|T_\mu f\|_2 \leq \sum_{k=0}^{\infty} \sum_{s=0}^{k} \frac{2}{\pi} \|\mu\|_{\text{op}}^{k-2} \left\{ \int_0^\infty \|Q_t(\mu P_t)^s f\|_2^2 \frac{dt}{t} \right\}^{1/2} .$$

We prove the following.

(6.43) Theorem.

There is a constant c_n depending only upon n such that, for all $\mu \in L^\infty(\mathbb{R}^{n-1}, \Lambda_n)$, $f \in L^2(\mathbb{R}^{n-1}, \mathfrak{H})$,

$$(6.44) \quad \left\{ \int_0^\infty \|Q_t(\mu P_t)^s f\|_2^2 \frac{dt}{t} \right\}^{1/2} \leq c_n(1+s)\|\mu\|_{\text{op}}^s \|f\|_2 , \qquad s \geq 0 .$$

Proof. We write Q_t in terms of its components as $Q_t = \sum_{j=1}^{n-1} e_j Q_{j,t}$.

Then
(6.45)

$$Q_{j,t}(\mu P_t)^s = (P_t\mu)^s Q_{j,t} + \sum_{\ell=0}^{s-1}(P_t\mu)^{s-\ell-1}(Q_{j,t}\mu P_t - P_t\mu Q_{j,t})(\mu P_t)^\ell .$$

We introduce the following function spaces and norms. A measurable function $F : \mathbb{R}^n_+ \to \mathfrak{H}$ is said to belong to $L^{p,2}$ if and only if

$$\|F\|_{p,2} = \left\{ \int_{\mathbb{R}^{n-1}} \|F(x,\cdot)\|_{L^p(\mathbb{R}_+,\frac{dt}{t},\mathfrak{H})} \, dx \right\}^{1/2}$$

is finite; we are concerned with $p = 2$ and $p = \infty$. The following norm estimates are easily verified using Plancherel's theorem:

(6.46)
$$\begin{cases} \text{(i)} & \|P_t\mu\|_{L^{2,2}\to L^{2,2}} \le \|\mu\|_{\text{op}} , \\ \text{(ii)} & \|Q_t\|_{L^2\to L^{2,2}} = \dfrac{1}{\sqrt{2}} . \end{cases}$$

We claim, moreover, that there is a dimensional constant c_n such that

(6.47)
$$\|(\mu P_t)^\ell\|_{L^2\to L^{\infty,2}} \le c_n\|\mu\|_{\text{op}}^\ell .$$

For, suppose that $f \in L^2(\mathbb{R}^{n-1},\mathfrak{H})$. Then
(6.48) $|(\mu P_t)^\ell f(x)|$

$$\le \int_{\mathbb{R}^{n-1}} \|\mu(x)p_t(x-y_1)\mu(y_1)p_t(y_1-y_2)\bullet$$
$$\cdots \bullet p_t(y_{\ell-1}-y_\ell)f(y_\ell)\|_{\mathbb{C}} \, dy_1 \cdots dy_\ell$$

$$\le \|\mu\|_{\text{op}}^\ell p_t * \cdots * p_t * |f|(x)$$

$$\le \|\mu\|_{\text{op}}^\ell M(|f|)(x) ,$$

where, in the last inequality, M denotes the Hardy–Littlewood maximal operator, and we have made use of the fact that p_t is a radially decreasing, positive function with $\|p_t\|_1 = 1$. As a consequence of (6.48), (6.47) holds with c_n equal to the norm of M on $L^2(\mathbb{R}^{n-1},\mathbb{R})$.

Finally, we claim that there is a dimensional constant c_n such that
(6.49) $$\|Q_{j,t}\mu P_t - P_t\mu Q_{j,t}\|_{L^{\infty,2}\to L^{2,2}} \le c_n\|\mu\|_{\text{op}} .$$
To see this, suppose for the moment that $F \in L^{\infty,2}(\mathbb{R}^n_+)$ is scalar-valued and that μ is a *real-valued* function in $\text{BMO}(\mathbb{R}^{n-1})$. Then, by the product rule for D_j,

$$Q_{j,t}(\mu P_t F) = P_t t D_j(\mu P_t F) = P_t\big([t D_j\mu] \cdot [P_t F]\big) + P_t\mu Q_{j,t}F .$$

We estimate the norm of $P_t([tD_j\mu] \cdot [P_tF])$ on $L^{2,2}(\mathbf{R}_+^n, \mathbf{C})$ by duality:

$$(6.50) \quad \|P_t([tD_j\mu] \cdot [P_tF])\|_{L^{2,2}}$$

$$= \sup_{\|G\|_{2,2}=1} \left| \iint_{\mathbf{R}_+^n} G \cdot P_t([tD_j\mu] \cdot [P_tF]) \frac{dxdt}{t} \right|$$

$$= \sup_{\|G\|_{2,2}=1} \left| \int_{\mathbf{R}^{n-1}} \mu\left(D_j \int_0^\infty [P_tG] \cdot [P_tF] \frac{dt}{t} \right) dx \right|.$$

As a consequence of the tent space theory of Coifman, Meyer, and Stein,

$$D_j \int_0^\infty [P_tG] \cdot [P_tF] \frac{dt}{t} \in H_{at}^1 ,$$

the atomic H^1-space predual to BMO, and its H_{at}^1-norm is bounded by $c_n\|F\|_{\infty,2}\|G\|_{2,2}$, where c_n is a dimensional constant. Returning, then, to the general case in which $\mu \in L^\infty(\mathbf{R}^{n-1}, \Lambda_n)$, we easily obtain the estimate (6.49) by applying this 'scalar' argument on components.

The estimate (6.44) now follows from (6.45)–(6.47) and (6.49). ∎

Now, combining (6.42) and (6.44), we see that, if $\|\mu\|_{op} < 1$, then

$$\|T_\mu\|_{L^2 \to L^2} \leq c_n \sum_{k=0}^\infty \|\mu\|_{op}^k \sum_{s=0}^k (1+s)$$

$$= c_n \sum_{k=0}^\infty \frac{(k+2)(k+1)}{2} \|\mu\|_{op}^k$$

from which the following is immediate.

(6.51) Corollary.

Let $\mu \in L^\infty(\mathbf{R}^{n-1}, \Lambda_n)$ and suppose $\|\mu\|_{op} < 1$. Then the operator T_μ, defined by (6.40), satisfies

$$(6.52) \qquad \|T_\mu\|_{L^2 \to L^2} \leq c_n(1 - \|\mu\|_{op})^{-3}$$

where c_n is a purely dimensional constant.

Theorem (6.6) is now a relatively easy consequence of corollary (6.51). Suppose $\mu \in L^\infty(\mathbf{R}^{n-1}, \mathbf{R}^n)$, set $B = 1 - \mu$, let $s > 0$, and make the change of variables $y = (st)^{-1}$ in (6.40). Then

$$(6.53)$$

$$T_\mu = \frac{s}{\pi} \lim_{\varepsilon \to 0} \int_{s\varepsilon < |y| < s/\varepsilon} (D + iysB)^{-1} \, dy = \frac{s}{\pi} \text{ p.v.} \int_{-\infty}^\infty (D + iysB)^{-1} \, dy .$$

By corollary (6.51), when $\|1 - sB\|_{op} < 1$, we have

$$(6.54) \qquad \|T_\mu\|_{L^2 \to L^2} \leq 8c_n s(1 - \|1 - sB\|_{op}^2)^{-3} .$$

Now note that, since $1 - sB \in L^\infty(\mathbf{R}^{n-1}, \Lambda_n)$,

$$(6.55) \qquad \|1 - sB\|_{op}^2 = (1 - s)^2 \|\mu\|_{op}^2 .$$

Choosing $s = (1 + \|\mu\|_{\mathrm{op}}^2)^{-1}$, we obtain

(6.56) $$\|1 - sB\|_{\mathrm{op}}^2 = 1 - (1 + \|\mu\|_{\mathrm{op}}^2)^{-1} < 1$$

so that, by (6.54),

(6.57) $$\|T_\mu\|_{L^2 \to L^2} \le 8c_n (1 + \|\mu\|_{\mathrm{op}}^2)^2 \le C(1 + \|\mu\|_{\mathrm{op}})^4$$

where C depends only on n. The estimate (6.7), and hence theorem (6.6), follow from (6.57).   ∎

7 Spaces of Clifford analytic functions II: Lipschitz domains

In this section we develop a theory of Hardy spaces of Clifford analytic functions for Lipschitz domains in $\mathbb{R}^n$. We shall focus our attention on the case of special Lipschitz domains, as in section 6, and we continue in the notation established there. We begin with some terminology.

Recalling that ϕ is a Lipschitz function with constant $\beta = \|\nabla\phi\|_\infty$, we fix a constant $\beta_0 > \beta$ and define, for $Q = (x, \phi(x)) \in \partial M$, the cone

$$V(Q) = \left\{ (z,t) : z \in \mathbb{R}^{n-1} , \ t \in \mathbb{R} , \ (t - \phi(x)) > \beta_0 |z - x| \right\} \subseteq M .$$

For any $\mathfrak{H}$-valued function u defined in M, we define the non-tangential maximal function $N(u)$ of u at Q by setting

$$N(u)(Q) = \sup_{Y \in V(Q)} |u(Y)| ;$$

the non-tangential limit of u at Q, when it exists, is given by

$$u^+(Q) = \lim_{Y \to Q}\text{-n.t.}\, u(Y) = \lim_{\substack{Y \to Q \\ Y \in V(Q)}} u(Y) .$$

The following result is basic to all that we shall do in this section.

(7.1) Theorem.

Let $1 < p < \infty$. Then, for all $f \in L^p(\partial M, \mathfrak{H})$, the Cauchy integral

(7.2) $$C_M f(Y) = \frac{1}{\omega_n} \int_{\partial M} \frac{Y - Q}{|Y - Q|^n} \eta(Q) f(Q) \, dS(Q)$$

has non-tangential maximal function in $L^p(\partial M)$. Moreover,

(i) $\|N(C_M f)\|_p \le C_p \|f\|_p$,

(ii) $C_M f$ has non-tangential limits for almost all $Q \in \partial M$, and $(C_M f)^+(Q) = \frac{1}{2} f(Q) + \frac{1}{2} \mathcal{H}_M f(Q)$ almost everywhere.

The proof of theorem (7.1) follows from theorem (6.4) by application of the standard techniques of Calderón–Zygmund theory, so we shall omit it. We pause to note a very important corollary.

For $f \in L^p(\partial M, \mathfrak{H})$, the single- and double-layer potentials of f are defined, for $Y \in M$, by

$$(7.3) \qquad \mathcal{S}_M f(Y) = \frac{1}{\omega_n(2-n)} \int_{\partial M} \frac{1}{|Y-Q|^{n-2}} \, f(Q) \, dS(Q) \,,$$

$$(7.4) \qquad \mathcal{K}_M f(Y) = \frac{1}{\omega_n} \int_{\partial M} \frac{(Y-Q) \cdot \eta(Q)}{|Y-Q|^n} \, f(Q) \, dS(Q) \,.$$

A straightforward calculation reveals that $C_M f = -D[\mathcal{S}_M(\eta f)]$; moreover, the 'scalar part' of the kernel of C_M is precisely the negative of the kernel of $\mathcal{K}_M$. In view of these relationships, the following corollary is immediate.

(7.5) Corollary.
    Let $1 < p < \infty$. Then, for all $f \in L^p(\partial M, \mathfrak{H})$,

(i) $\|N(\mathcal{K}_M f)\|_p$, $\|N(D\mathcal{S}_M f)\|_p \le C_p \|f\|_p$,

(ii) $(\mathcal{K}_M f)^+(Q) = (\frac{1}{2}I - K_M)f(Q)$ for almost all $Q \in \partial M$, where

$$K_M f(Q) = \frac{1}{\omega_n} \int_{\partial M} \frac{(X-Q) \cdot \eta(X)}{|X-Q|^n} \, f(X) \, dS(X) \,,$$

(iii) $(D\mathcal{S}_M f)^+(Q) = \frac{1}{2}\eta(Q)f(Q) + S_M f(Q)$ for almost all $Q \in \partial M$, where

$$S_M f(Q) = \frac{1}{\omega_n} \int \frac{Q-X}{|Q-X|^n} \, f(X) \, dS(X) \,.$$

Corollary (7.5) is absolutely crucial to the solution of Dirichlet and Neumann problems for the Laplacian by the method of layer potentials in Lipschitz domains. Indeed, this result was the starting point in the work of Verchota, Dahlberg, and Kenig on the solution of elliptic boundary value problems in minimally smooth domains.

In this section, we shall use theorem (7.1) together with a result of Verchota on the Dirichlet problem for the Laplacian to give a boundary integral characterization of the spaces

$$(7.6)$$
$$H^p(M, \mathfrak{H}) = \{ F \in C^\infty(M, \mathfrak{H}) : DF = 0 \text{ in } M \,, \ N(F) \in L^p(\partial M, \mathbb{R}) \}$$

for an appropriate range of p. We shall make $H^p(M)$ a normed space $(1 \le p \le \infty)$ by defining

$$(7.7) \qquad\qquad \|F\|_{H^p(M)} = \|N(F)\|_p \,.$$

By theorem (5.19), when $M = \mathbb{R}_+^n$, this is consistent with the previous definition.

    The fundamental result on the solvability of the Dirichlet problem, which we state without proof, is as follows.

(7.8) Theorem.
  (Verchota) *There exists $\varepsilon(\beta)$, $0 < \varepsilon(\beta) \leq 1$, such that $\frac{1}{2}I + K_M$ is invertible on $L^p(\partial M, \mathfrak{H})$ provided that $2 - \varepsilon(\beta) < p < \infty$. Moreover, for this same range of p, there corresponds to each $f \in L^p(\partial M, \mathfrak{H})$ a unique function u such that*

  (i) u is harmonic in Ω,
  (ii) $N(u) \in L^p(\partial M, \mathbb{R})$,
  (iii) $u^+(Q) = f(Q)$ for almost all $Q \in \partial M$.

  Furthermore, u is given by $K_M(\frac{1}{2}I + K_M)^{-1}f$ almost everywhere on ∂M, and, if M is C^1, we may take $\varepsilon(\beta) = 1$.

  Now, if $2 - \varepsilon(\beta) < p < \infty$ and $F \in H^p(M, \mathfrak{H})$, it follows from basic results on harmonic measure that F^+ exists almost everywhere and belongs to $L^p(\partial M, \mathfrak{H})$, with $F = K_M(T_M F^+)$, $T_M = (\frac{1}{2}I + K_M)^{-1}$. In particular, (7.4) may be used to deduce estimates on the growth of F at infinity; with such estimates, we can show that $C_M F^+ = F$. But, in fact, we can prove a stronger result using corollary (3.39). Let p_0 denote the larger of 1, $\frac{n-2}{n-1}(2 - \varepsilon(\beta))$. Then we have the following.

(7.9) Theorem.
  Let $p > p_0$ and suppose $F \in H^p(M, \mathfrak{H})$. Then F^+ exists almost everywhere and belongs to $L^p(\partial M, \mathfrak{H})$, and $F = CF^+$ in M. Moreover, when $p_0 = 1$ the result is also true for $p = 1$.

Proof. Suppose that $r > 0$ and $Q = (x, \phi(x)) \in \partial M$, and define
$$(7.10) \qquad M_r = \left\{(x,t) : x \in \mathbb{R}^{n-1}, \ t \in \mathbb{R}, \ t > \phi(x) + r\right\},$$
which is the 'upper half-space' above the graph of the Lipschitz function $\phi + r$ (which has constant β). Fix $X \in M$, and let $R > 10|X| > r$. Define
$$(7.11)$$
$$\Sigma_{r,R} = \left\{Y \in \mathbb{R}^n : |Y| = R, \ Y \in M_r\right\}, \quad D_{r,R} = \left\{Y \in \partial M_r : |Y| \leq R\right\}.$$
For all r sufficiently small, X is inside $S_{r,R} = \Sigma_{r,R} \cup D_{r,R}$. By the Cauchy integral formula,
$$(7.12) \qquad F(X) = \frac{1}{\omega_n} \int_{S_{r,R}} \frac{X-Y}{|X-Y|^n} d\sigma(Y) F(Y)$$
$$= \frac{1}{\omega_n} \int_{D_{r,R}} \frac{X-Y}{|X-Y|^n} d\sigma(Y) F(Y)$$
$$+ \frac{1}{\omega_n} \int_{\Sigma_{r,R}} \frac{X-Y}{|X-Y|^n} d\sigma(Y) F(Y) .$$

We wish to show that the second integral tends to 0 as R tends to ∞. To do this, we first estimate the size of $F(Y)$ for $Y \in \Sigma_{r,R}$.

Since $p > p_0 \geq \frac{n-2}{n-1}(2 - \varepsilon(\beta))$, we may choose q with $p/(2 - \varepsilon(\beta)) > q > \frac{n-2}{n-1}$. Then, by corollary (3.39), $|F|^q$ is subharmonic in M, so it has a least-harmonic majorant Ψ with $N(\Psi) \in L^{p/q}$. Since $p/q > 2 - \varepsilon(\beta)$, it can be seen that Ψ^+ exists almost everywhere, $\|\Psi^+\|_{p/q} \leq \|N(\Psi)\|_{p/q}$, and $\Psi = \mathcal{K}_M(T_M\Psi^+)$ in M. Moreover, it follows that F^+ exists almost everywhere and belongs to $L^p(\partial M, \mathfrak{H})$.

Let $\delta > 0$. We can find a compactly supported function $g_\delta \in L^p(\partial M, \mathbb{R})$ such that $\|T_M\Psi^+ - g_\delta\|_{p/q} < \delta$. Now choose R so large that the distance from $\Sigma_{r,R}$ to $\operatorname{supp} g_\delta$ is $\geq R/2$. Then, for $Y \in \Sigma_{r,R}$,

$$(7.13) \quad \left| \Psi(Y) - \frac{1}{\omega_n} \int_{\partial M} \frac{(Y - Z) \cdot \eta(Z)}{|Y - Z|^n} g_\delta(Z) \, dS(Z) \right|$$

$$\leq \delta \left(\int_{|Y-Z|>r} |Y - Z|^{(1-n)\lambda} \, dS(Z) \right)^{1/\lambda} = k_r \delta$$

where $\frac{1}{\lambda} + \frac{q}{p} = 1$ and k_r is a constant depending upon r (as well as n, p, q); in the estimate (7.13) we use the fact that, for $Z \in \partial\Omega$, $|Y - Z| \geq r$. Moreover

$$\left| \frac{1}{\omega_n} \int_{\partial\Omega} \frac{(Y - Z) \cdot \eta(Z)}{|Y - Z|^n} g_\delta(Z) \, dS(Z) \right|$$

$$(7.14) \quad \leq \frac{1}{\omega_n} \left(\int_{\operatorname{supp} g_\delta} |Y - Z|^{(1-n)\lambda} \, dS(Z) \right)^{1/\lambda} \|g_\delta\|_{p/q}$$

$$\leq k R^{(1-n)q/p} \|g_\delta\|_{p/q}$$

where k is a constant independent of R (but depending upon n, p, q); the estimate (7.14) uses the fact that, for $Z \in \operatorname{supp} g_\delta$, $|Y - Z| \sim R$. Combining (7.13) and (7.14), we obtain, for $Y \in \Sigma_{r,R}$,

$$(7.15) \quad |F(Y)| \leq \Psi(Y)^{p/q} \leq \left[k_r \delta + k R^{(1-n)q/p}(\delta + \|T\Psi^+\|_{p/q}) \right]^{p/q}.$$

But, in fact, what we have shown is that, for all $\delta > 0$, there exists $R_0 > 0$ such that $R > R_0$ and $Y \in \Sigma_{r,R}$ imply (7.15). Returning to (7.12), we have, for $R > R_0$,

$$(7.16) \quad \left| \frac{1}{\omega_n} \int_{\Sigma_{r,R}} \frac{X - Y}{|X - Y|^n} \, d\sigma(Y) F(Y) \right|$$

$$\leq \frac{1}{\omega_n} \int_{\Sigma_{r,R}} |X - Y|^{(1-n)} |F(Y)| \, dS(Y)$$

$$\leq k_n \left[k_r \delta + k R^{(1-n)q/p}(\delta + \|T\Psi^+\|_{p/q}) \right]^{p/q},$$

where k_n is a constant depending only upon n. Note that the second inequality in (7.16) follows from the fact that $R > 10|X|$; thus for $Y \in \Sigma_{r,R}$, $|X - Y| \sim R$. Fixing δ and letting $R \to \infty$, we obtain

$$(7.17) \qquad \left| F(X) - \frac{1}{\omega_n} \int_{\partial M_r} \frac{X - Y}{|X - Y|^n} d\sigma(Y) F(Y) \right| \leq k_n k_r \delta \, ,$$

combining (7.12) and (7.16). Since $\delta > 0$ is arbitrary, it follows that for all $X \in M_r$, $F(X)$ may be represented by its Cauchy integral on ∂M_r. In parametric form, this is

$$(7.18) \quad F(X) = \frac{1}{\omega_n} \int_{\mathbb{R}^{n-1}} \frac{x - y + e_n(t - \phi(y) - r)}{\left[|x - y|^2 + (t - \phi(y) - r)^2 \right]^{n/2}} \cdot$$
$$\left(\nabla \phi(y) - e_n \right) F\left(y, \phi(y) + r \right) dy$$

letting $X = (x, t)$, $Y = (y, \phi(y) + r) \in \partial M_r$.

For every $X \in M$, we can find $r_0 > 0$ such that $X \in M_r$ for all $r \leq r_0$. For this range of rs, we obtain

$$(7.19) \quad \left| \frac{x - y + e_n(t - \phi(y) - r)}{\left[|x - y|^2 + (t - \phi(y) - r)^2 \right]^{n/2}} \left(\nabla \phi(y) - e_n \right) F\left(y, \phi(y) + r \right) \right|$$

$$\leq \left[|x - y|^2 + (t - \phi(y) - r_0)^2 \right]^{(1-n)/2} (1 + \beta) N(F)(y) \, ;$$

moreover, if $\frac{1}{p'} + \frac{1}{p} = 1$, Hölder's inequality gives

$$(7.20) \quad \int_{\mathbb{R}^{n-1}} \left[|x - y|^2 + (t - \phi(y) - r_0)^2 \right]^{(1-n)/2} (1 + \beta) N(F)(y) \, dy$$

$$\leq (1 + \beta) \| N(F) \|_p \left\{ \int_{\mathbb{R}^{n-1}} \left[|x - y|^2 + (t - \phi(y) - r_0)^2 \right]^{(1-n)p'/2} dy \right\}^{1/p'}$$

which is finite (for $p = 1$, the third factor on the right hand side of (7.20) is replaced by the appropriate L^∞ norm). Consequently, the Lebesgue dominated convergence theorem allows us to take the limit as $r \to 0+$ under the integral sign in (7.18), obtaining $F = CF^+$. ∎

(7.21) Corollary.

Let $p > p_0$ and suppose $f \in L^p(\partial M, \mathfrak{H})$. Then there exists $F \in H^p(M, \mathfrak{H})$ such that $f = F^+$ almost everywhere on ∂M, if and only if $\mathcal{H}_M f = f$. If $p = p_0 = 1$, then given $f \in L^1(\partial M, \mathfrak{H})$ there exists $F \in H^1(M, \mathfrak{H})$ with $F^+ = f$ almost everywhere, if and only if $\mathcal{H}_M f \in L^1(\partial M, \mathfrak{H})$ and $\mathcal{H}_M f = f$.

Proof. For $p > p_0$, this characterization of the boundary values of $H^p(M, \mathfrak{H})$-functions is immediate from theorems (7.1) and (7.9). If $p_0 = 1$ and $f \in L^1(\partial M, \mathfrak{H})$ satisfies $\mathcal{H}_M f \in L^1(\partial M, \mathfrak{H})$, standard techniques of Calderón–Zygmund theory show that $N(C_M f) \in L^1(\partial M, \mathbb{R})$ and that theorem (7.1)(ii) continues to hold. Then, once again, the desired result follows from theorem (7.9). ∎

By analogy to the classical case, we denote by $H^p(\partial M, \mathfrak{H})$ the space of nontangential boundary values of $H^p(M, \mathfrak{H})$. Then corollary (7.21) together with theorem (7.9) yields the following.

(7.22) Corollary.
  For $p > p_0$, $H^p(\partial M, \mathfrak{H})$ is the subspace of $L^p(\partial M, \mathfrak{H})$ consisting of those functions f for which $\mathcal{H}_M f = f$. If $p_0 = 1$, then $H^1(\partial M, \mathfrak{H})$ is the subspace of $L^p(\partial M, \mathfrak{H})$ consisting of those functions f for which $\mathcal{H}_M f \in L^1(\partial M, \mathfrak{H})$ and $f = \mathcal{H}_M f$.

Conspicuously absent from this discussion is any decomposition of $H^p(\partial M, \mathfrak{H})$ into 'real' and 'imaginary' (or, as in section 5, 'normal' and 'tangential') parts. The analytic over-determinedness of the operator D should in some way be reflected in the structure of $H^p(\partial M, \mathfrak{H})$. Except in the special case $M = \mathbb{R}^n_+$, $\mathfrak{H} = \mathfrak{A}_n$, the following conjecture remains open.

(7.23) Conjecture.
  For every Clifford module $\mathfrak{H}$ there exists $\eta \in \mathbb{R}^n$ with $\eta^2 = -1$, and a subspace $\mathfrak{H}_0$ of $\mathfrak{H}$, such that
  (i) $\mathfrak{H}$ admits the splitting $\mathfrak{H} = \mathfrak{H}_0 \oplus \eta \mathfrak{H}_0$,
  (ii) the Cauchy integral operator C_M is an isomorphism from $L^p(\partial M, \mathfrak{H}_0)$ onto $H^p(M, \mathfrak{H})$ for all $p > p_0$,
  (iii) if Tan denotes the projection of $\mathfrak{H}$ onto $\mathfrak{H}_0$, then the boundary operator mapping $F \to \mathrm{Tan}(F^+)$ is continuous from $H^p(M, \mathfrak{H})$ onto $L^p(\partial M, \mathfrak{H}_0)$ for all $p > p_0$.

Notes and remarks for chapter 2

1. In their classic text, Stein and Weiss present a very fine discussion of generalizing $\bar{\partial}$ to elliptic systems in $\mathbb{R}^n$ (see [96]); the reader is referred to their work for an excellent presentation of generalized Cauchy–Riemann systems. The work of Fueter dates back to a series of papers in the 1930s, beginning with [39] and [40]. The latter-day

descendants of Fueter are well represented by the monograph of Brackx, Delanghe, and Sommen [17]; for historical perspective on the development of Fueter's ideas, see the references and comments there.

2. In his famous 1928 paper on the spinning electron ([33]), the first 'Dirac operator' was introduced as a square root of the wave operator. His definition was in the setting of Minkowski space, but it was soon extended to any finite-dimensional quadratic space of arbitrary signature by Brauer and Weyl ([18]). The relationship between the standard Dirac operator and the Cauchy–Riemann $\partial, \bar{\partial}$-operators foreshadows the idea of graded Dirac operators, which we discuss in chapter 4; see also the recent monograph of Roe [90].

3. Brackx, Delanghe, and Sommen have written a very detailed monograph ([17]) which fully develops the parallels between classical complex function theory and Clifford analytic function theory. For the subharmonicity results, we are clearly indebted to Stein and Weiss; see, for example [95] and [96].

4. There are several readily accessible accounts of classical Hardy space theory; here we mention particularly the monographs of Duren [34], Garnett [41], and Koosis [70], and the excellent survey of Coifman and Weiss [26]. The results of Kenig on Hardy spaces for Lipschitz domains in the plane are found in [63]; see also [64].

5. The monograph [96] is a good general reference for this section. The arguments in theorems (5.4) and (5.19) are based upon ideas from the paper of C. Fefferman and Stein [36]. The proof of the representation theorem (5.20) follows classical ideas set forth in [70]. The idea of normal–tangential decomposition is inspired in part by the work of Korányi and Vági for Riesz systems on balls (see [71]).

6. The main theorem of this section is a generalization of the celebrated Cauchy integral theorem of Calderón, Coifman, McIntosh, and Meyer ([20], [24]). The theorem was first proved for small Lipschitz constants by Murray ([82]) and extended to all special Lipschitz domains by McIntosh ([80]); the proof given here is essentially due to McIntosh. The proof of theorem (6.23) is due to Murray and has not been previously published.

7. The results of Verchota first appeared in his doctoral dissertation, and are published in [101]; a convenient summary may be found in the excellent survey paper [64]. The results on harmonic measure needed for the proof of theorem (7.9) are due to Dahlberg [28].

3

Representations of Spin (V, Q)

Discussion of further applications of the theory of Dirac operators and Clifford algebras now begins. The style of exposition will change somewhat, with fewer details being given than before, so that greater demands are placed upon the reader. A wider variety of topics can then be covered. In this chapter we shall discuss the representation theory for the group $\mathrm{Spin}(V, Q)$, concentrating almost entirely on the case of the compact group $\mathrm{Spin}(n)$ which arises when (V, Q) is an n-dimensional (real) positive- or negative-definite quadratic space (see also chapter 5). In the non-compact case, the characterization of the irreducible unitary representations is only just being discovered. But in the compact case these representations are all finite-dimensional and their 'parameterization' has been known for many years, thanks to the effort of Cartan, Weyl *et al.* This will be given in section 2 after some routine preliminaries have been disposed of in section 1. For detailed applications to analysis, however, various explicit realizations of these representations are needed. Representations of $O(n)$, hence $SO(n)$, and of $\mathrm{Spin}(n)$, on spaces of harmonic polynomials on $\mathbb{R}^n$ are well-known, and their role in such singular integrals as Riesz transforms is well-understood. Realizations of more general representations of $O(n)$ on harmonic polynomials on the space $\mathbb{R}^{r \times n}$ of real matrices are far less widely known; their use in singular integrals has hardly begun. In sections 4 and 5 we give a fairly detailed account of some aspects of the theory of polynomials of matrix argument, using it as a vehicle for presenting some aspects of polynomial invariant theory. While not perhaps directly related to Spin

groups, invariant theory has become important in research centered on
the Atiyah–Singer index theorem which is directly related to the Dirac
operator. The fact that much research on the theory of polynomials of
matrix argument remains to be done was further motivation for including a discussion of them. The standard theory for polynomials on $\mathbb{R}^n$ is
given in streamlined form in section 3.

1 Elements of representation theory

In this introductory section we shall review some of the basic ideas in
the representation theory of groups. Fuller details can be found in any
of the excellent treatments of this theory (see Notes and remarks).

Let G be a group and $\mathcal{V}$ a vector space over $\mathbb{F}$, where $\mathbb{F} = \mathbb{R}, \mathbb{C}$ or
$\mathbb{H}$. A representation of G is a pair $(\mathcal{V}, \tau)$ in which τ is a homomorphism
from G into the group $\mathrm{Aut}(\mathcal{V})$ of invertible $\mathbb{F}$-linear transformations on
$\mathcal{V}$. Thus $\tau(g)$, $\tau(g)^{-1}$ are $\mathbb{F}$-linear operators on $\mathcal{V}$ such that

$$(1.1) \qquad \tau(g_1 g_2) = \tau(g_1)\tau(g_2) , \qquad \tau(g^{-1}) = \tau(g)^{-1}$$

for all g_1, g_2 and g in G. When G is a topological group and $\mathcal{V}$ is a
topological space, we shall assume that $\tau : G \to \mathrm{Aut}(\mathcal{V})$ is continuous
in the sense that $g \to \tau(g)v$ is continuous from G into $\mathcal{V}$ for each v
in $\mathcal{V}$. In practice, it will often be convenient to think and speak of $\mathcal{V}$
as simply a *G-module*, with $v \to gv$, $v \in \mathcal{V}$ being the operator on $\mathcal{V}$
representing g when $\mathcal{V}$ is a left G-module and $v \to vg^{-1}$ in the right G-
module case. Whatever the point of view, however, the representation
is said to be *finite-* or *infinite-dimensional* according as $\mathcal{V}$ is finite- or
infinite-dimensional.

Let $\mathcal{V}$ be a G-module, say a left G-module. A subspace $\mathcal{U}$ in $\mathcal{V}$ which
is *G-invariant* in the sense that $gu \in \mathcal{U}$ for all g in G and u in $\mathcal{U}$, is called
a *submodule* of $\mathcal{V}$ or a *subrepresentation*. If $\mathcal{V}$ is finite-dimensional it is
said to be *irreducible* when it contains no submodules other than $\{0\}$
or $\mathcal{V}$; otherwise, it is said to be *reducible*. In the infinite-dimensional
case an irreducible module $\mathcal{V}$ is one containing no closed submodules
other than $\{0\}$ or $\mathcal{V}$. Now let $\mathcal{V}'$ be a second (left) G-module for the
same choice of $\mathbb{F}$. An $\mathbb{F}$-linear transformation $\mathcal{S} : \mathcal{V} \to \mathcal{V}'$ is said to be
equivariant if

$$(1.2) \qquad \mathcal{S}(gv) = g(\mathcal{S}v) \qquad (g \in G , \ v \in \mathcal{V}) ,$$

and $\mathcal{V}, \mathcal{V}'$ are said to be *equivalent G-modules* when there is a linear isomorphism from $\mathcal{V}$ onto $\mathcal{V}'$ which is equivariant; in the infinite-

dimensional case all $\mathbb{F}$-linear transformations are assumed to be continuous.

A simple result due to Schur is very useful when dealing with finite-dimensional representations.

(1.3) Theorem.

(Schur's Lemma) *Let* V, V' *be irreducible, finite-dimensional G-modules. Then*

(i) *every equivariant mapping* $S : V \to V'$ *is either* 0 *or an isomorphism,*

(ii) *every equivariant mapping* $S : V \to V$ *is of the form* $Sv = \lambda v$ *on* V *for some* $\lambda \in \mathbb{C}$ *when* V *is a complex module.*

By far the most important topology on a G-module arises from an inner product.

(1.4) Definition.

Let V *be a, say, left complex G-module. Then an Hermitian inner product* $(\cdot, \cdot)$ *for* V *is said to be G-invariant when* $(gu, gv) = (u, v)$ *for all* u, v *in* V *and* g *in* G.

In representation-theoretic terms therefore, a representation (V, τ) will be *unitary* where there is an inner product $(\cdot, \cdot)$ for V which is G-invariant in the sense that $(\tau(g)u, \tau(g)v) = (u, v)$. But suppose V is G-module having some inner product $\langle \cdot, \cdot \rangle$ which is not necessarily G-invariant. Then, provided G is compact so that the integral

$$(1.5) \qquad (u, v) = \int_G \langle gu, gv \rangle \, dg \qquad (u, v \in V)$$

is well defined, (1.5) defines a G-invariant Hermitian inner product for V. Compactness of G ensures also that every irreducible unitary representation of G is finite-dimensional, and that complex G-modules can be written as the direct sum of irreducible G-submodules. Thus from the outset in the case of compact groups we can assume that every complex G-module comes with a G-invariant inner product structure; but if G is non-compact, construction of such inner products is very difficult indeed. In fact, when (V, Q) is not positive- or negative-definite, few constructions of $\text{Spin}(V, Q)$-invariant inner products are known. This prompts the following definition.

(1.6) Definition.

Let G *be a compact group. We denote by* $\mathcal{M}_{\mathbb{C}}(G)$ *the set of all equivalence classes of irreducible unitary representations of* G, *two such repre-*

sentations $(\mathcal{V}, \tau)$ and $(\mathcal{V}', \tau')$ being equivalent when $\mathcal{V}, \mathcal{V}'$ are equivalent as G-modules.

The *sum*

(1.7) $(\mathcal{V}_1 \oplus \mathcal{V}_2, \tau_1 \oplus \tau_2)$, $(\tau_1 \oplus \tau_2)g : v_1 + v_2 \to \tau_1(g)v_1 + \tau_2(g)v_2$

of unitary representations of any group G, whether compact or not, is again a unitary representation. On the other hand, under the *Hilbert–Schmidt inner product*

(1.8) $(u, v) = \sum_{j,k} (\xi_j, \xi_k')(\eta_j, \eta_k')$ $(\xi_j, \xi_k' \in \mathcal{V}_1 \, , \, \eta_j, \eta_k' \in \mathcal{V}_2)$

of $u = \sum_j \xi_j \otimes \eta_j$ and $v = \sum_k \xi_k' \otimes \eta_k'$, defined with respect to the inner products on $\mathcal{V}_1$ and $\mathcal{V}_2$, the tensor product $\mathcal{V}_1 \otimes \mathcal{V}_2$ is a Hilbert space and the *tensor product*

(1.9) $(\mathcal{V}_1 \otimes \mathcal{V}_2, \tau_1 \otimes \tau_2)$, $(\tau_1 \otimes \tau_2)g : v_1 \otimes v_2 \to \tau_1(g)v_1 \otimes \tau_2(g)v_2$

also is a unitary representation of G. In general, of course, neither $(\mathcal{V}_1 \oplus \mathcal{V}_2, \tau_1 \oplus \tau_2)$ nor $(\mathcal{V}_1 \otimes \mathcal{V}_2, \tau_1 \otimes \tau_2)$ will be irreducible even if $(\mathcal{V}_i, \tau_i)$ are, a point which will become very important later (see for instance, section 2 of this chapter and section 4 of chapter 4).

A unitary representation of a Lie group G is frequently studied by looking at the representation $(\mathcal{V}, d\tau)$ of the Lie algebra $\mathfrak{G}$ of G that it defines. This brings algebraic ideas to the fore, particularly the structural properties of $\mathfrak{G}$, in contrast to the analytic ideas we shall be emphasizing after the early part of section 2. Let $(\mathcal{V}, \tau)$ be a finite-dimensional representation of a Lie group G, compact or not. Then $t \to \tau(\exp tX)v$ is differentiable as a function of t for each X in $\mathfrak{G}$ and $V \in \mathcal{V}$, and

(1.10) $d\tau(X)v = \dfrac{d}{dt}\tau(\exp tX)v\big|_{t=0}$

defines a representation of $\mathfrak{G}$ on $\mathcal{V}$, the so-called *derived representation* of $\mathfrak{G}$.

To complete the preliminaries we shall study the interconnections among real, complex, and quaternionic representations of a *compact* group G. Let $\mathcal{V}$ be a complex G-module. A *real structure* on $\mathcal{V}$ is a conjugate-linear mapping $\mathcal{J}$ on $\mathcal{V}$ such that

$$\mathcal{J}^2 = I \quad , \quad \mathcal{J}(gv) = g\mathcal{J}(v) \qquad (g \in G \, , \, v \in \mathcal{V})$$

while a *quaternionic structure* on $\mathcal{V}$ is a conjugate-linear mapping $\mathcal{J}$ on $\mathcal{V}$ satisfying

$$\mathcal{J}^2 = -I \quad , \quad \mathcal{J}(gv) = g\mathcal{J}(v) \qquad (g \in G \, , \, v \in \mathcal{V}) \, .$$

In both cases $\mathcal{J}$ is called a *structure map*. A unitary representation

$(\mathcal{V}, \tau)$ of G is said to be of *real* or *quaternionic type* according as $\mathcal{V}$ admits a real or quaternionic structure; if $(\mathcal{V}, \tau)$ is not of real or quaternionic type we shall say it is of complex type. In the real case, let

$(1.11)_{\mathbb{R}}$ $\mathcal{V}_+ = \{v \in \mathcal{V} : \mathcal{J}(v) = v\}$, $\mathcal{V}_- = \{v \in \mathcal{V} : \mathcal{J}(v) + v = 0\}$

be the ± 1-eigenspaces of $\mathcal{J}$; whereas, in the quaternionic case, let

$(1.11)_{\mathbf{H}}$ $\mathcal{V}_+ = \{v \in \mathcal{V} : \mathcal{J}(v) = iv\}$, $\mathcal{V}_- = \{v \in \mathcal{V} : \mathcal{J}(v) + iv = 0\}$

be the $\pm i$-eigenspaces of $\mathcal{J}$. Then, for $(1.11)_{\mathbb{R}}$, $(\mathcal{V}_+, \tau)$ and $(\mathcal{V}_-, \tau)$ are real representations of G, which are obviously equivalent under the equivariant mapping $v \to iv$. Similarly, $(\mathcal{V}_+, \tau)$ and $(\mathcal{V}_-, \tau)$ are equivalent quaternionic representations of G for $(1.11)_{\mathbf{H}}$. Now let $\mathcal{S} : \mathcal{V} \to \mathcal{V}'$ be an equivariant mapping from one complex G-module onto a second one $\mathcal{V}'$. If $\mathcal{J}$ is real structure on $\mathcal{V}$, then clearly $\mathcal{J}' = \mathcal{S} \circ \mathcal{J} \circ \mathcal{S}^{-1}$ is a real structure on $\mathcal{V}'$; a similar result holds for quaternionic structures. Thus *real type* (resp. *quaternionic type*) is a property shared by all elements of an equivalence class in $\mathcal{M}_{\mathbb{C}}(G)$. We shall denote by $\mathcal{M}_{\mathbb{R}}(G)$ (resp. $\mathcal{M}_{\mathbf{H}}(G)$) those equivalence classes in $\mathcal{M}_{\mathbb{C}}(G)$ all of whose elements are of real type (resp. quaternionic type). This emphasis on real, complex, and quaternionic types arises because the classification of the real spinor space $\mathfrak{R}_n$ given in section 7 of chapter 1 will be important in the next section. In this connection note that

$(1.12)(i)$ $\mathcal{J}_1 \otimes \mathcal{J}_2$ *is a real structure on* $\mathcal{V}_1 \otimes \mathcal{V}_2$ *when* $\mathcal{J}_1, \mathcal{J}_2$ *are real structures on* $\mathcal{V}_1$ *and* $\mathcal{V}_2$ *respectively, or when both are quaternionic structures,*

$(1.12)(ii)$ $\mathcal{J}_1 \otimes \mathcal{J}_2$ *is a quaternionic structure on* $\mathcal{V}_1 \otimes \mathcal{V}_2$ *when* $\mathcal{J}_1$ *is a real structure on* $\mathcal{V}_1$ *and* $\mathcal{J}_2$ *is a quaternionic structure on* $\mathcal{V}_2$, *or vice versa.*

2 Signature, fundamental representations

As $\mathrm{Spin}(n)$ is a compact group, being topologically isomorphic to a sphere, as well as a Lie group, we can study the sets $\mathcal{M}_{\mathbf{F}}(\mathrm{Spin}(n))$, $n \geq 3$, of equivalence classes of irreducible $\mathrm{Spin}(n)$-modules, $\mathbf{F} = \mathbb{R}$, $\mathbb{C}$, or $\mathbf{H}$. There are really two different topics to pursue, however. The first seeks a 'labeling' of the elements of $\mathcal{M}_{\mathbb{C}}(\mathrm{Spin}(n))$ by their *signature* along with a description of the generators of $\mathcal{M}_{\mathbb{C}}(\mathrm{Spin}(n))$, the so-called *fundamental representations*, from which all others can be built by Cartan composition. This largely is an algebraic problem whose solution leads easily to characterizations of $\mathcal{M}_{\mathbb{R}}(\mathrm{Spin}(n))$ and $\mathcal{M}_{\mathbf{H}}(\mathrm{Spin}(n))$. The second seeks specific examples of members of a particular equivalence class of irreducible representations. There will be many such examples for each equivalence class, and the most convenient choice usually varies with context. Since these examples often arise in kernels of Spin(n)-invariant differential or integral operators, this second problem is largely an analytic one, as we shall see throughout the remaining chapters. In the present section the principal focus will be on the first problem, although the relation of the second one with $\mathfrak{A}_n$-modules will be brought out.

From a theoretical standpoint, the most convenient realization of Spin(n) is the one given in corollary (7.27) in chapter 1 where it is identified with the multiplicative group

$$(2.1) \qquad \mathrm{Spin}(n) = \left\{ a \in \Lambda_n \cap \mathfrak{A}_n^+ : aa^* = 1 \ , \ \|a\|_{\mathfrak{A}} = 1 \right\}$$

of *even* elements in the Clifford semigroup

$$(2.1)' \qquad \Lambda_n = \left\{ xy \cdots z : x, y, \ldots, z \in \mathbb{R}^{n+1} \right\} \ .$$

In this construction

$$(2.2) \qquad \sigma(a) : x \longrightarrow axa^* \qquad (x \in \mathbb{R}^n \subseteq \mathfrak{A}_n)$$

defines a two-fold covering homomorphism $\sigma : \mathrm{Spin}(n) \to SO(n)$. Two groups, the maximal torus T_n and Weyl group W_n, associated with Spin(n) have a particularly important role to play. Fix the standard basis $\{e_1, \ldots, e_n\}$ for $\mathbb{R}^n$ regarded as imaginary units in $\mathfrak{A}_n$, and set

$$X_j = \tfrac{1}{2} e_j e_{n-j+1}, \qquad \left(1 \le j \le p = \left[\tfrac{1}{2} n \right] \right) \ .$$

Then

$$X_j^2 = -\tfrac{1}{4} \ , \qquad X_j X_k = X_k X_j \quad (j \ne k) \ ;$$

consequently,

$$\exp \theta_j X_j = \cos \tfrac{1}{2} \theta_j + (e_j e_{n-j+1}) \sin \tfrac{1}{2} \theta_j \qquad \text{(DeMoivre)}$$

and

$$T_n = \left\{ \prod_{j=1}^{p} \exp \theta_j X_j : 0 \le \theta_j \le 4\pi \right\}$$

is an abelian subgroup of Spin(n). It is not hard to see that T_n actually is a *maximal* abelian subgroup; since such subgroups are unique up to conjugation in Spin(n), we can speak of *the* maximal torus T_n. But it must be observed that

$$(2.3) \qquad \beta : (\theta_1, \ldots, \theta_p) \longrightarrow \prod_{j=1}^{p} \exp \theta_j X_j$$

is only a homomorphism from $\mathbb{R}^p/(4\pi\mathbb{Z})^p$ onto T_n, not an isomorphism:

the kernel of β consists of $(\theta_1, \ldots, \theta_p)$ with $\theta_j = 0$ or 2π and $\theta_1 + \cdots + \theta_p \equiv 0 \pmod{4\pi}$. Now $\sigma(\exp \theta_j X_j)$ fixes every e_k except for

(2.4)(i) $\sigma(\exp \theta_j X_j) : e_j \longrightarrow (\cos \theta_j)e_j + (\sin \theta_j)e_{n-j+1}$

and

(2.4)(ii) $\sigma(\exp \theta_j X_j) : e_{n-j+1} \longrightarrow (-\sin \theta_j)e_j + (\cos \theta_j)e_{n-j+1}$.

Thus $\sigma(\exp \theta_j X_j)$ acts on $\mathbf{R}^n$ solely as a rotation through θ_j in the $(j, n-j+1)$-plane, and so in matrix form

$$\sigma\left(\prod_{j=1}^{p} \exp \theta_j X_j\right) = \begin{bmatrix} \cos \theta_1 & 0 & \cdots & 0 & \sin \theta_1 \\ 0 & \cos \theta_2 & \cdots & \sin \theta_2 & 0 \\ \vdots & \vdots & & \vdots & \vdots \\ 0 & -\sin \theta_2 & \cdots & \cos \theta_2 & 0 \\ -\sin \theta_1 & 0 & \cdots & 0 & \cos \theta_1 \end{bmatrix}.$$

The group T_n is used to 'label' the set $\mathcal{M}_{\mathbf{C}}(\mathrm{Spin}(n))$. Let $(\mathcal{V}, \tau)$ be an irreducible unitary representation of $\mathrm{Spin}(n)$. Its restriction to the abelian subgroup T_n will be a direct sum of one-dimensional representations of T_n each of the form

$$\prod_{j=1}^{p} \exp \theta_j X_j \longrightarrow e^{i(m_1\theta_1 + \cdots + m_p\theta_p)}$$

for some choice of $(m_1, \ldots, m_p)$ in $\frac{1}{2}\mathbf{Z}$; such integers m_j are said to be *integer* or *half-integer* according as m_j or $2m_j$ is in $\mathbf{Z}$. Not all choices of $(m_1, \ldots, m_p)$ can occur, however, because

(2.3)' $e^{i(m_1\theta_1 + \cdots + m_p\theta_p)} = 1$ $\big((\theta_1, \ldots, \theta_p) \in \ker \beta\big)$.

To calculate the effect of this restriction, set $\theta_j = 2\pi\varepsilon_j$, so that $\varepsilon_j = 0$ or 1 and $\varepsilon_1 + \cdots + \varepsilon_p$ is an even integer. But then, by (2.3)', $m_1\varepsilon_1 + \cdots + m_p\varepsilon_p$ must be an integer for all possible such $(\varepsilon_1, \ldots, \varepsilon_p)$. Consequently, each p-tuple $(m_1, \ldots, m_p)$ consists entirely of integers or entirely of non-zero half-integers. We call such a p-tuple a *weight* of the representation. In linear algebra terms this is all very natural. For let

$$V_\nu = \left\{ v \in \mathcal{V} : \tau\left(\prod_{j=1}^{p} \exp \theta_j X_j\right)v = e^{i(m_1\theta_1 + \cdots + m_p\theta_p)}v, \right.$$

$$\left. (\theta_1, \ldots, \theta_p) \in \mathbf{R}^p/4\pi\mathbf{Z}^p \right\}$$

be the weight subspace of $\mathcal{V}$ corresponding to the weight $\nu = (m_1, \ldots, m_p)$; it is just the common eigenspace in $\mathcal{V}$ of the abelian group of operators $\tau(t)$, $t \in T$. Thus

(2.5) $$\mathcal{V} = \oplus \sum_{\nu \in S_\tau} V_\nu \ ,$$

the sum being taken over the set S_τ of all weights σ for which V_σ is a non-trivial subspace of $\mathcal{V}$. There is a natural simple ordering on S_τ, the lexicographic ordering, in which $\nu = (m_1, \ldots, m_p) > \nu' = (m'_1, \ldots, m'_p)$ if the first non-zero difference of $m_j - m'_j$ is positive. Since equivalent representations will have equivalent weight subspace decompositions, this ordered set S_τ will be the same for all representations equivalent to $(\mathcal{V}, \tau)$.

Associated with the maximal torus T_n in Spin(n) is the Weyl group W_n. Let

$$N(T_n) = \left\{ g \in \text{Spin}(n) \ , \ g T_n g^{-1} = T_n \right\}$$

be the *normalizer* of T_n in Spin(n). Then the finite group $W_n = N(T_n)/T_n$ is called the *Weyl group* of Spin(n). Its importance in the present context comes from its action as a finite group of symmetries on S_τ. When $n = 2p + 1$ this group consists of all permutations of $(m_1, \ldots, m_p)$, together with an arbitrary number of changes of sign; when $n = 2p$, the group still allows all permutations, but allows only an even number of changes of sign. From general results it also follows that, if ν is a weight for $(\mathcal{V}, \tau)$, then so is every Weyl group translate $\omega(\nu)$, $\omega \in W_n$. If we now apply these ideas to (2.5), we see that among all ν in S_τ there is a unique highest one $(m_1, \ldots, m_p)$ with respect to the lexicographic ordering, and it will have the form

(2.6)
$$m_1 \geq \cdots \geq m_p \geq 0 \ (n = 2p + 1) \quad , \ m_1 \geq \cdots \geq m_{p-1} \geq |m_p| \ (n = 2p) \ .$$

This unique weight is called the *highest weight* for $(\mathcal{V}, \tau)$. A fundamental result of Cartan as applied to Spin(n) relates the highest weights to $\mathcal{M}_{\mathbb{C}}(\text{Spin}(n))$.

(2.7) Theorem.

  (i) The weight subspace corresponding to the highest weight of an irreducible unitary representation of Spin(n) *is one-dimensional.*

  (ii) Every p-tuple $(m_1, \ldots, m_p)$ consisting entirely of integers or entirely of non-zero half-integers which satisfy (2.6) is the highest weight of exactly one irreducible unitary representation of Spin(n) *(up to equivalence).*

Following standard terminology, we shall call the highest weight

$(m_1, \ldots, m_p)$ for $(\mathcal{V}, \tau)$ the *signature* of $(\mathcal{V}, \tau)$, and any ϕ in $\mathcal{V}$ such that $\mathbb{C}\phi$ is the weight subspace of $\mathcal{V}$ corresponding to this highest weight will be called a *highest weight vector*. Since equivalent representations clearly have the same signature (and conversely), theorem (2.7) provides the desired labeling of $\mathcal{M}_{\mathbb{C}}(\mathrm{Spin}(n))$.

(2.8) Theorem.

The mapping $(\mathcal{V}, \tau) \to$ *signature* $(\mathcal{V}, \tau)$ *extends to a 1-1 mapping from* $\mathcal{M}_{\mathbb{C}}(\mathrm{Spin}(n))$ *onto the set of all p-tuples* $(m_1, \ldots, m_p)$, $p = \left[\frac{1}{2}n\right]$, *each consisting entirely of integers* m_j *or entirely of non-zero half-integers* m_j *which satisfy* (2.6).

Construction of (reasonably) explicit realizations of representations of $\mathrm{Spin}(n)$ having given signature can be accomplished inside appropriate Clifford modules. As might be expected from the structural properties of these modules and from (2.6), the explicit realizations frequently depend on whether $n = 2p$ or $n = 2p + 1$. Suppose first that $n = 2p$. Then the complex spinor space $\mathfrak{S}_n$ can be realized as the complex Hilbert space $\Lambda^*(\mathbb{C}^p)$ with

(2.9) $\qquad e_j = \mu_j - \mu_j^* \quad , \quad e_{2p-j+1} = i(\mu_j + \mu_j^*) \qquad (1 \le j \le p) \,,$

the imaginary units in $\mathcal{L}(\Lambda^*(\mathbb{C}^p)) \sim \mathfrak{C}_n$ determined by the usual orthonormal basis $\{\varepsilon_1, \ldots, \varepsilon_p\}$ for $\mathbb{C}^p$ (see chapter 1, $(7.54)(\mathbf{A})$); thus

$$\exp \theta_j X_j = \cos \tfrac{1}{2}\theta_j + i \sin \tfrac{1}{2}\theta_j (\mu_j \mu_j^* - \mu_j^* \mu_j) \,,$$

since $\mu_j^2 = (\mu_j^*)^2 = 0$. Let

(2.10)(i) $\qquad \Lambda_+^*(\mathbb{C}^p) = \{\xi_1 \wedge \cdots \wedge \xi_{2k} : \xi_j \in \mathbb{C}^p \,,\ k \ge 0\}$

and

(2.10)(ii) $\qquad \Lambda_-^*(\mathbb{C}^p) = \{\xi_1 \wedge \cdots \wedge \xi_{2k+1} : \xi_j \in \mathbb{C}^p \,,\ k \ge 0\}$

be the subspaces of $\Lambda^*(\mathbb{C}^p)$ consisting respectively of all even and all odd wedge products of elements from $\mathbb{C}^p$. Then $\Lambda_+^*(\mathbb{C}^p)$ and $\Lambda_-^*(\mathbb{C}^p)$ are $\mathrm{Spin}(n)$-modules since

(2.11) $\qquad \mu_j, \mu_j^* : \Lambda_{\pm}^*(\mathbb{C}^p) \longrightarrow \Lambda_{\mp}^*(\mathbb{C}^p) \,.$

Furthermore, since $\bar{a} = a^*$ for any a in the even part $\mathfrak{A}_n^+$ of $\mathfrak{A}_n$, theorem (7.35) in chapter 1 ensures that

(2.12)(i) $\qquad S(a) : v \longrightarrow S(a)v \qquad \left(v \in \Lambda_+^*(\mathbb{C}^p)\right)$

defines a unitary representation of $\mathrm{Spin}(n)$ on $\Lambda_+^*(\mathbb{C}^p)$ where S is the isomorphism from $\mathfrak{C}_n$ onto $\mathcal{L}(\Lambda^*(\mathbb{C}^p))$, $n = 2p$, extending the embedding $(x_1, \ldots, x_n) \to \sum_j x_j e_j$ defined by (2.9); similarly,

(2.12)(ii) $\qquad S(a) : v \longrightarrow S(a)v \qquad \left(v \in \Lambda_-^*(\mathbb{C}^p)\right)$

defines a corresponding unitary representation of Spin(n) on $\Lambda_-^*(\mathbb{C}^p)$. Now let

(2.13) $s_+ = (\frac{1}{2}, \ldots, \frac{1}{2}, \frac{1}{2})$, $s_- = (\frac{1}{2}, \ldots, \frac{1}{2}, -\frac{1}{2})$

be p-tuples of non-zero half-integers satisfying (2.6).

(2.14) Theorem.
   When $n = 2p$, the subspaces $\Lambda_+^*(\mathbb{C}^p)$ and $\Lambda_-^*(\mathbb{C}^p)$ of $\Lambda^*(\mathbb{C}^p)$ are irreducible Spin(n)-modules under the representations S; furthermore, the signature of $\Lambda_+^*(\mathbb{C}^p)$ is s_+ or s_- according as p is even or odd, whereas the signature of $\Lambda_-^*(\mathbb{C}^p)$ is s_+ or s_- according as p is odd or even.

Proof. Since $\mathfrak{A}_n^+$ is the linear span of Spin(n), the irreducibility of $\Lambda_+^*(\mathbb{C}^p)$ and $\Lambda_-^*(\mathbb{C}^p)$ as Spin(n)-modules will follow immediately once their irreducibility as $\mathfrak{A}_n^+$-modules has been established. But $\mathfrak{A}_{2p}^+$ is isomorphic to $\mathfrak{A}_{2p-1}$. Consequently, $\Lambda_\pm^*(\mathbb{C}^p)$ are complex $\mathfrak{A}_{2p-1}$-modules having dimension 2^{p-1}, the same as that of the complex spinor space $\mathfrak{S}_{2p-2}$ which is an irreducible $\mathfrak{A}_{2p-1}$-module. Hence $\Lambda_\pm(\mathbb{C}^p)$ are irreducible Spin(n)-modules.

   To determine their signature fix a basis element $\varepsilon_{i_1} \wedge \cdots \wedge \varepsilon_{i_k}$ in $\Lambda^*(\mathbb{C}^p)$. Then
$\exp \theta_j X_j : \varepsilon_{i_1} \wedge \cdots \wedge \varepsilon_{i_k}$

$$\rightarrow \left(\cos \tfrac{1}{2}\theta_j + i \sin \tfrac{1}{2}\theta_j \mu_j \mu_j^*\right) \varepsilon_{i_1} \wedge \cdots \wedge \varepsilon_{i_k} = e^{\frac{1}{2} i \theta_j} \varepsilon_{i_1} \wedge \cdots \wedge \varepsilon_{i_k}$$

when $\{i_1, \ldots, i_k\}$ contains j, whereas
$\exp \theta_j X_j : \varepsilon_{i_1} \wedge \cdots \wedge \varepsilon_{i_k}$

$$\rightarrow \left(\cos \tfrac{1}{2}\theta_j - i \sin \tfrac{1}{2}\theta_j \mu_j^* \mu_j\right) \varepsilon_{i_1} \wedge \cdots \wedge \varepsilon_{i_k} = e^{-\frac{1}{2} i \theta_j} \varepsilon_{i_1} \wedge \cdots \wedge \varepsilon_{i_k}$$

when $\{i_1, \ldots, i_k\}$ does not contain j. Consequently, $\mathbb{C}\varepsilon_{i_1} \wedge \cdots \wedge \varepsilon_{i_k}$ is a weight space for T_n corresponding to the weight $(m_1, \ldots, m_p)$ where

$$m_j = \cdot\tfrac{1}{2} \;\; (j \in \{i_1, \ldots, i_k\}) \;\;,\;\; m_j = -\tfrac{1}{2} \;\; (j \notin \{i_1, \ldots, i_k\}) \;.$$

In particular, the highest weight for $\Lambda_+^*(\mathbb{C}^p)$ will be $(\frac{1}{2}, \ldots, \frac{1}{2}, \frac{1}{2})$ or $(\frac{1}{2}, \ldots, \frac{1}{2}, -\frac{1}{2})$ depending on whether p is even or odd; similarly, the highest weight for $\Lambda_-^*(\mathbb{C}^p)$ will be $(\frac{1}{2}, \ldots, \frac{1}{2}, \frac{1}{2})$ or $(\frac{1}{2}, \ldots, \frac{1}{2}, -\frac{1}{2})$ depending on whether p is odd or even. In both cases, $\varepsilon_1 \wedge \cdots \wedge \varepsilon_p$ or $\varepsilon_1 \wedge \cdots \wedge \varepsilon_{p-1}$ will be respective highest weight vectors. This completes the proof.   ■
   Now let

(2.15)(i) $d_r = (\underbrace{1, \ldots, 1}_{r}, 0, \ldots) \qquad (r < p)$

and

(2.15)(ii) $\qquad d_p^+ = (1, \ldots, 1, 1) \qquad, \qquad d_p^- = (1, \ldots, 1, -1)$

be p-tuples of integers satisfying (2.6). Realizations of representations of $\mathrm{Spin}(n)$, $n = 2p$, having these as signature are obtained by extending σ to all of $\mathfrak{A}_n$ and thence to $\mathfrak{C}_n$ after complexifying $\mathfrak{A}_n$. More precisely, define $\sigma(a)$ on all of $\mathfrak{A}_n$ by

(2.16) $\qquad \sigma(a) : u \longrightarrow aua^* \qquad (a \in \mathrm{Spin}(n)\ ,\ u \in \mathfrak{A}_n)\ .$

It clearly extends to a representation of $\mathrm{Spin}(n)$ on the complexification $\mathfrak{C}_n$ of $\mathfrak{A}_n$. We shall denote this extension by σ also. Under the sesquilinear inner product on $\mathfrak{C}_n$ defined by (7.32)(i) in chapter 1, $(\mathfrak{C}_n, \sigma)$ is a unitary representation of $\mathrm{Spin}(n)$. But it is highly reducible. Indeed,

(2.17) $\sigma(a) : xy \cdots z \to (axa^*)(aya^*) \cdots (aza^*)$

$$= \big(\sigma(a)x\big)\big(\sigma(a)y\big) \cdots \big(\sigma(a)z\big) \qquad (x, y, \ldots, z \in \mathbb{R}^n)\ ,$$

so each $\sigma(a)$ actually defines a mapping $\sigma(a) : \mathfrak{A}_n^{(k)} \to \mathfrak{A}_n^{(k)}$ of the space $\mathfrak{A}_n^{(k)}$ of k-multivectors into itself (see chapter 1, (7.3)). Consequently, $\mathfrak{C}_n$ decomposes under σ into the sum

(2.18) $\qquad \mathfrak{C}_n = \mathfrak{C}_n^{(0)} \oplus \mathfrak{C}_n^{(1)} \oplus \cdots \oplus \mathfrak{C}_n^{(n)}$

of $\mathrm{Spin}(n)$-modules $\mathfrak{C}_n^{(k)}$ where

(2.18)′ $\qquad \mathfrak{C}_n^{(k)} = \mathrm{span}_{\mathbb{C}}\{ e_{\alpha_1} \cdots e_{\alpha_k} : 1 \le \alpha_1 < \cdots < \alpha_k \le n \}$

is the complexification of $\mathfrak{A}_n^{(k)}$. Since $\mathfrak{C}_n$ is *linearly* isomorphic to $\Lambda^*(\mathbb{C}^n)$ (see chapter 1, (7.10)), the representation $(\mathfrak{C}_n, \sigma)$ can be identified with a representation on an exterior algebra just as $(\mathfrak{S}_n^{\pm}, S)$ were. In fact, $\mathfrak{C}_n^{(1)} = \mathbb{C}^n$ and the action of $\sigma(a)$ on $\mathfrak{C}_n^{(1)}$, $a \in \mathrm{Spin}(n)$, is just the *canonical* action of $\sigma(a)$ on $\mathbb{C}^n$ as an element of $SO(n)$. This last action extends to any exterior power $\Lambda^k(\mathbb{C}^n)$ setting

(2.17)′ $\qquad \sigma(a) : \xi_1 \wedge \cdots \wedge \xi_k \longrightarrow \sigma(a)\xi_1 \wedge \cdots \wedge \sigma(a)\xi_k \qquad (\xi_j \in \mathbb{C}^n)\ .$

In view of (2.17), the linear isomorphism $\lambda : \Lambda^k(\mathbb{C}^n) \to \mathfrak{C}_n^{(k)}$ defined by (7.10) in chapter 1, is clearly $\mathrm{Spin}(n)$-equivariant. Hence we can and shall use $(\mathfrak{C}_n, \sigma)$, $(\Lambda^*(\mathbb{C}^n), \sigma)$ interchangeably.

(2.19) Theorem.
When $n = 2p$, every space $\mathfrak{C}_n^{(k)}$ of k-multivectors in $\mathfrak{C}_n$ other than $\mathfrak{C}_n^{(p)}$ is an irreducible $\mathrm{Spin}(n)$-module under the unitary representation σ; furthermore, the signature of $(\mathfrak{C}_n^{(k)}, \sigma)$ is d_k if $k < p$, and d_{n-k} if $k > p$. The reducible $\mathrm{Spin}(n)$-module $\mathfrak{C}_n^{(p)}$ decomposes into two irreducible

Spin(n)-*modules, one having signature* d_p^+, *the other having signature* d_p^-.

Proof. Define a new orthonormal basis for $\mathbf{C}^n$ ($\subseteq \mathfrak{C}_n$) by setting

$$\delta_j = \frac{1}{\sqrt{2}}(e_j - ie_{n-j+1}), \quad \delta_{n-j+1} = \frac{1}{\sqrt{2}}(e_j + ie_{n-j+1}) \quad (j = 1, \ldots, p).$$

Now $\sigma(\exp \theta_j X_j)$ fixes every δ_k except for $k = j$ and $n - j + 1$ where, by (2.4),

(2.20)
$$\sigma(\exp \theta_j X_j) : \delta_j \longrightarrow e^{i\theta_j}\delta_j, \quad \sigma(\exp \theta_j X_j) : \delta_{n-j+1} \longrightarrow e^{-i\theta_j}\delta_{n-j+1}.$$

Contrary to what one might expect, however, the products $\delta_{\alpha_1} \cdots \delta_{\alpha_k}$ do not form a basis for $\mathfrak{C}_n^{(k)}$; indeed, such a product need not lie in $\mathfrak{C}_n^{(k)}$ as the example

$$\delta_1 \delta_n = \tfrac{1}{2}(e_1 - ie_n)(e_1 + ie_n) = 1 + ie_1 e_n$$

shows. (The problem arises because the change of basis $\{e_1, \ldots, e_n\} \to \{\delta_1, \ldots, \delta_n\}$ comes from a unitary transformation, not an orthogonal one.) To overcome this difficulty, observe that

$$\delta_1 \wedge \delta_n = \tfrac{1}{2}(e_1 - ie_n) \wedge (e_1 + ie_n) = ie_1 e_n,$$

whereas

$$\delta_1 \wedge \delta_2 = \tfrac{1}{2}(e_1 - ie_n) \wedge (e_2 - ie_{n-1}) = \delta_1 \delta_2,$$

using the wedge product on $\mathfrak{A}_n$ defined by (7.15) in chapter I; and, crucially, both of these belong to $\mathfrak{C}_n^{(2)}$. In fact, more generally,

$$\{\delta_{\alpha_1} \wedge \cdots \wedge \delta_{\alpha_k} : 1 \leq \alpha_1 \wedge \cdots \wedge \alpha_k \leq n\}$$

defines a basis for $\mathfrak{C}_n^{(k)}$ each element of which is a weight vector because

(2.20)' $\quad \sigma\left(\prod_{j=1}^{p} \exp \theta_j X_j\right) : \delta_{\alpha_1} \wedge \cdots \wedge \delta_{\alpha_k} \longrightarrow \left(\prod_{j=1}^{p} e^{im_j \theta_j}\right)\delta_{\alpha_1} \wedge \cdots \wedge \delta_{\alpha_k}$

with $m_j = 0$ if $\{\alpha_1, \ldots, \alpha_k\}$ contains neither j nor $n-j+1$, and $m_j = -1$, 0, or 1 otherwise. This restricts the set of possible weights for $(\mathfrak{C}_n^{(k)}, \sigma)$.

To determine irreducibility of $(\mathfrak{C}_n^{(k)}, \sigma)$, we return to the basis for $\mathfrak{C}_n^{(k)}$ of reduced products $e_\alpha = e_{\alpha_1} \cdots e_{\alpha_k}$ (see (2.18)'). Let $\mathcal{U}$ be a Spin(n)-submodule of $\mathfrak{C}_n^{(k)}$. If $\mathcal{U} \neq \{0\}$, there exists $u = \sum_{|\alpha|=k} \lambda_\alpha e_\alpha$ in $\mathcal{U}$ such that $\lambda_\beta \neq 0$ for at least one β, $|\beta| = k$. By using induction on the number of non-zero coefficients λ_β, we shall attempt to prove that $\mathcal{U}$ contains every reduced product e_α, $|\alpha| = k$. When this occurs $\mathcal{U}$ will coincide with $\mathfrak{C}_n^{(k)}$, and $(\mathfrak{C}_n^{(k)}, \sigma)$ will be irreducible. So suppose first that only one λ_β is non-zero, say $\beta = (\beta_1, \ldots, \beta_k)$. Then $e_\beta \in \mathcal{U}$. But, up to sign, any e_α can be written as $\sigma(a)e_\beta$ for some a in Spin(n) choosing the appropriate permutation $\sigma(a)$ of the basis $\{e_1, \ldots, e_n\}$ of $\mathbb{R}^n$ and, in the case of an odd permutation, substituting $-e_1$ for e_1. Suppose next that at least two coefficients $\lambda_\beta, \lambda_\gamma$ are non-zero. Set $\beta = \{\beta_1, \ldots, \beta_k\}$ and $\gamma = \{\gamma_1, \ldots, \gamma_k\}$. If $2k < n$, there exist integers r, s such that $r \in \beta$ but $r \notin \gamma$, and $s \notin \beta \cup \gamma$. Choose $\sigma(a)$ in $SO(n)$ so that $\sigma(a)$ fixes every e_j except for

$$\sigma(a) : e_r \longrightarrow -e_r \quad , \quad \sigma(a) : e_s \longrightarrow -e_s .$$

Then $u + \sigma(a)u \in \mathcal{U}$ has fewer non-vanishing coefficients than u, but it still has at least one. The obvious induction argument now shows that $(\mathfrak{C}_n^{(k)}, \sigma)$ is irreducible provided $k < p$. Since the highest weight for $(\mathfrak{C}_n^{(k)}, \sigma)$ will be $(\underbrace{1, \ldots, 1}_{k}, 0, \ldots)$ corresponding to the highest weight

vector $\delta_1 \wedge \cdots \wedge \delta_k$ $(= \delta_1 \cdots \delta_k)$, its signature is d_k. Although this proof fails for $\mathfrak{C}_n^{(p)}$, simple modifications of it show that $(\mathfrak{C}_n^{(p)}, \sigma)$ is an irreducible unitary representation of $\text{Pin}(n)$, using the covering homomorphism $\sigma : \text{Pin}(n) \to O(n)$ (see chapter 1, (7.27)).

Proof of the remaining irreducibility results requires an idea that will be exceedingly important later on. We shall define a linear mapping $\eta : \mathfrak{C}_n \to \mathfrak{C}_n$ satisfying

(2.21)
$$\begin{cases} \text{(i)} \quad \eta(\sigma(a)u) = \sigma(a)\eta(u) \quad , \quad \text{(ii)} \quad \eta^2 = I \quad , \quad \eta(bu) = b'\eta(u) , \\ \text{(iii)} \quad \eta \text{ maps } \mathfrak{C}_n^{(k)} \text{ onto } \mathfrak{C}_n^{(n-k)} \quad (0 \le k \le n) \end{cases}$$

for all $a \in \text{Spin}(n)$, $b \in \mathfrak{A}_n$, and $u \in \mathfrak{C}_n$. By property (i), η is an equivariant mapping from the irreducible $\text{Spin}(n)$-module $\mathfrak{C}_n^{(k)}$ onto $\mathfrak{C}_n^{(n-k)}$, $0 \le k < p$, and η will split $\mathfrak{C}_n^{(p)}$ into two $\text{Spin}(n)$-modules

$$(\mathfrak{C}_n^{(p)})_+ = \{u \in \mathfrak{C}_n^{(p)} : \eta(u) = u\} \ , \quad (\mathfrak{C}_n^{(p)})_- = \{u \in \mathfrak{C}_n^{(p)} : \eta(u) + u = 0\} \ ,$$

the ± 1 eigenspaces of η in $\mathfrak{C}_n^{(p)}$. Consequently, $\mathfrak{C}_n^{(n-k)}$ will be irreducible and have signature d_k, $k < p$, leaving only the corresponding results for $(\mathfrak{C}_n^{(p)})_\pm$ to be proved. The latter requires the precise definition of η:

(2.22) $\eta(u) = i^p(e_1 \cdots e_n)u \quad (u \in \mathfrak{C}_n , \ n = 2p) .$

The proof of (2.21) is a simple matter; for

$$(i^p e_1 \cdots e_n)^2 = (-1)^{\frac{1}{2}n}(-1)^{\frac{1}{2}n(n+1)} = 1$$

while

$$(e_1 \cdots e_n)e_j = -e_j(e_1 \cdots e_n) \quad (1 \le j < n) ,$$

since n is even (see chapter 1, (3.4)). Properties (i) and (ii) of (2.21) are thus clear. On the other hand,

$$\eta(e_{\alpha_1} \cdots e_{\alpha_k}) = (\pm 1)e_{\beta_1} \cdots e_{\beta_{n-k}}$$

where $\{\beta_1,\ldots,\beta_{n-k}\} = \{1,\ldots,n\}\setminus\{\alpha_1,\ldots,\alpha_k\}$. This proves $(2.21)(\text{iii})$.
Now let $\mathcal{U}$ be a $\text{Spin}(n)$-submodule of $(\mathbb{C}_n^{(p)})_+$, $\mathcal{U} \neq \{0\}$, and set
$$\mathcal{V} = \{u + \sigma(e_1)v : u, v \in \mathcal{U}\} = \mathcal{U} + \sigma(e_1)\mathcal{U} .$$
Since
$$\text{Pin}(n) = \{a \in \Lambda_n : aa^* = 1 , \ \|a\|_{\mathfrak{A}} = 1\} = \{a, e_1 b : a, b \in \text{Spin}(n)\} ,$$
$\mathcal{V}$ is a non-trivial $\text{Pin}(n)$-submodule of $\mathbb{C}_n^{(p)}$; consequently, $\mathcal{V} = \mathbb{C}_n^{(p)}$
because $(\mathbb{C}_n^{(p)}, \sigma)$ is an irreducible representation of $\text{Pin}(n)$. But, by
$(2.21)(\text{ii})$,
$$\eta\big(\sigma(e_1)u\big) = -\sigma(e_1)\eta(u) \qquad (u \in \mathbb{C}_n) ,$$
so
$$\mathcal{U} \subseteq (\mathbb{C}_n^{(p)})_+ \ , \quad \sigma(e_1)\mathcal{U} \subseteq (\mathbb{C}_n^{(p)})_- \ , \quad \mathcal{U} \oplus \sigma(e_1)\mathcal{U} = \mathbb{C}_n^{(p)} .$$
Hence $\mathcal{U} = (\mathbb{C}_n^{(p)})_+$, establishing irreducibility; similarly, $(\mathbb{C}_n^{(p)})_-$ is an
irreducible $\text{Spin}(n)$-module. Now
$$\sigma\left(\prod_{j=1}^p \exp\theta_j X_j\right) : \delta_1 \wedge \cdots \wedge \delta_p \longrightarrow e^{i(\theta_1+\cdots+\theta_{p-1}+\theta_p)}\delta_1 \wedge \cdots \wedge \delta_p .$$
But if $\delta_1 \wedge \cdots \wedge \delta_p = u_+ + u_-$ with u_+, u_- elements of $(\mathbb{C}_n^{(p)})_+$, $(\mathbb{C}_n^{(p)})_-$
respectively, then
$$\sigma\left(\prod_{j=1}^p \exp\theta_j X_j\right) : u_\pm \longrightarrow e^{i(\theta_1+\cdots+\theta_{p-1}+\theta_p)}u_\pm .$$
In view of $(2.20)'$, however, this can occur if and only if $u_\pm$ are constant
multiples of $\delta_1 \wedge \cdots \wedge \delta_p$. Thus
$$\eta(\delta_1 \wedge \cdots \wedge \delta_p) = \text{const.} \ \delta_1 \wedge \cdots \wedge \delta_p .$$
To determine the constant, observe that
$$\begin{aligned}
\eta(\delta_1 \wedge \cdots \wedge \delta_p) &= 2^{-\frac{1}{2}p}\eta\big((e_1 - ie_n) \cdots (e_p - ie_{p+1})\big) \\
&= 2^{-\frac{1}{2}p}\eta(e_1 \cdots e_p + \cdots + (-i)^p e_n \cdots e_{p+1}) \\
&= 2^{-\frac{1}{2}p} i^p e_1 \cdots e_n(e_1 \cdots e_p + \cdots + (-i)^p e_n \cdots e_{p+1}) \\
&= 2^{-\frac{1}{2}p}(-1)^p\big((-i)^p e_n \cdots e_{p+1} + \cdots + e_1 \cdots e_p\big) \\
&= (-1)^p \delta_1 \wedge \cdots \wedge \delta_p .
\end{aligned}$$
Hence $(\mathbb{C}_n^{(p)})_+$ has signature d_p^+ if p is even, while $(\mathbb{C}_n^{(p)})_-$ will have
signature d_p^+ if p is odd. Similarly, by using $\delta_1 \wedge \cdots \wedge \delta_{p-1} \wedge \delta_{p+1}$ we see
that $(\mathbb{C}_n^{(p)})_+$ has signature d_p^- when p is odd and $(\mathbb{C}_n^{(p)})_-$ has signature
d_p^- when p is even. This completes the proof.   ∎

Hints of the future geometric importance of the η-mapping defined by

(2.22) can be gleaned from an orthogonal change of basis $\{e_1, \cdots, e_n\} \rightarrow$ $\{\varepsilon_1, \cdots, \varepsilon_n\}$, $\varepsilon_j = \sum_k a_{jk} e_k$. For

$$\varepsilon_1 \cdots \varepsilon_n = (\det A) e_1 \cdots e_n \qquad \left(A = [a_{jk}] \right) .$$

Consequently, each orthonormal basis of $\mathbb{R}^n$ having the same *orientation* as the standard basis defines the same η; if it has the opposite orientation, then η is replaced by $-\eta$. Of course, from an algebraic point of view, the latter would simply reverse the roles of the $\pm$ eigenspaces. It will also be instructive to have η realized on $\Lambda^*(\mathbb{C}^n)$, an $\mathfrak{A}_n$-module, isomorphic to $\mathfrak{C}_n$. Now by (7.54)(**B**) in chapter 1, $e_j = \mu_j - \mu_j^*$, $1 \le j \le n$. Consequently

$$(2.23) \qquad \eta(u) = i^p \left(\prod_{j=1}^n \mu_j - \mu_j^* \right) u \qquad \left(u \in \Lambda^*(\mathbb{C}^{2p}) \right) .$$

The η-mapping is also well-defined on the complex spinor space $\mathfrak{S}_n$, $n = 2p$, where it becomes
(2.23)$'$

$$\eta(u) = i^p (e_1 \cdots e_n) u = \left(\prod_{j=1}^p \mu_j^* \mu_j - \mu_j \mu_j^* \right) u \qquad \left(u \in \mathfrak{S}_n \sim \Lambda^*(\mathbb{C}^p) \right) .$$

Just as in the case of (2.21)(i),(ii), we see that
(2.24)
$$(i) \quad \eta\big(S(a) u \big) = S(a) \eta(u) \quad , \quad (ii) \quad \eta^2 = I \quad , \quad \eta(bu) = b' \eta(u) ,$$

for all $a \in \mathrm{Spin}(n)$, $b \in \mathfrak{A}_n$, and $u \in \mathfrak{S}_n$. Since $\eta = (-1)^k$ on $\Lambda^k(\mathbb{C}^p)$, the ± 1-eigenspaces $\mathfrak{S}_n^{\pm}$ of η in $\mathfrak{S}_n$ are now given by

$$(2.24)' \qquad \mathfrak{S}_n^+ = \Lambda_+^*(\mathbb{C}^p) \quad , \quad \mathfrak{S}_n^- = \Lambda_-^*(\mathbb{C}^p) .$$

Hence the decomposition of the complex spinor space $\mathfrak{S}_n$ into its irreducible $\mathrm{Spin}(n)$-submodules arises from this eigenspace decomposition exactly as the decomposition of the 'middle-space' $\mathfrak{C}_n^{(p)}$ of the complex Clifford algebra $\mathfrak{C}_n$ did, further confirming the algebraic importance of η.

Now let $n = 2p + 1$, and let
$$(2.25) \qquad s = (\tfrac{1}{2}, \ldots, \tfrac{1}{2}, \tfrac{1}{2}) \quad , \quad d_r = (\underbrace{1, \ldots, 1}_{r}, 0, \ldots) \qquad (1 \le r \le p)$$

be p tuples of non-zero half-integers and integers respectively satisfying (2.6). The construction of irreducible unitary representations of $\mathrm{Spin}(n)$ having these as signature is similar to but simpler than the ones in the even-dimensional case, so we shall be brief. Using the ideas in (7.54)(**A**) in chapter 1, let $\mathcal{V} = \Lambda^*(\mathbb{C}^p)$ and let $S : \mathfrak{C}_n \longrightarrow \mathcal{L}(\Lambda^*(\mathbb{C}^p))$, $n = 2p + 1$, be the *homomorphism* obtained by extending to $\mathfrak{C}_n$ the linear embedding

$(x_1, \ldots, x_n) \rightarrow \sum_j x_j e_j$ of $\mathbf{R}^n$, $n = 2p + 1$, in $\mathcal{L}(\Lambda^*(\mathbf{C}^p))$ where

(2.26) $e_j = \mu_j - \mu_j^*$, $e_{n-j+1} = i(\mu_j + \mu_j^*)$ $(1 \le j \le p)$

and

(2.26)' $$e_{p+1} = (-i)\left(\prod_{j=1}^{p} \mu_j^* \mu_j - \mu_j \mu_j^*\right)$$

are defined with respect to the usual basis $\{\varepsilon_1, \ldots, \varepsilon_p\}$ for $\mathbf{C}^p$. The restriction to $\mathbf{C}_{2p}$ of this *homomorphism* $S : \mathbf{C}_{2p+1} \rightarrow \mathcal{L}(\Lambda^*(\mathbf{C}^p))$ coincides with the *isomorphism* $S : \mathbf{C}_{2p} \rightarrow \mathcal{L}(\Lambda^*(\mathbf{C}^p))$ used earlier because (2.26) coincides with (2.9). Consequently, the representation (2.12) is just the restriction to Spin(2p) of the representation

$$S(a) : v \longrightarrow S(a)v \qquad (v \in \Lambda^*(\mathbf{C}^p))$$

of Spin(2p+1), regarding Spin(2p) as a subgroup of Spin(2p+1). Hence use of the same symbol S in both cases is both consistent and deliberate. Unlike the Spin(2p) case, however, it defines an irreducible representation of Spin(2p + 1).

(2.27) Theorem.
    When $n = 2p + 1$, $\Lambda^*(\mathbf{C}^p)$ is an irreducible Spin(n)-module under the unitary representation S; its signature is s.

Proof. The restriction to Spin(2p + 1) of

$$S : \mathbf{C}_{2p+1} \longrightarrow \mathcal{L}(\Lambda^*(\mathbf{C}^p))$$

is easily seen to be unitary. Now let $\mathcal{U}$ be an irreducible Spin(2p + 1)-submodule of $\Lambda^*(\mathbf{C}^p)$. Since $\mathcal{U}$ must also be a Spin(2p)-module, $\mathcal{U}$ contains at least one of $\Lambda_+^*(\mathbf{C}^p)$ and $\Lambda_-^*(\mathbf{C}^p)$ in view of (2.14). But $\mu_j : \Lambda^k(\mathbf{C}^p) \rightarrow \Lambda^{k+1}(\mathbf{C}^p)$ while

$$\left(\prod_{j=1}^{p} \mu_j^* \mu_j - \mu_j \mu_j^*\right) : u \longrightarrow (-1)^k u \qquad (u \in \Lambda^k(\mathbf{C}^p)) \, ,$$

i.e., e_{p+1} preserves $\Lambda^*(\mathbf{C}^p)$. Consequently

$$e_j e_{p+1} : \Lambda_{\pm}^*(\mathbf{C}^p) \longrightarrow \Lambda_{\mp}^*(\mathbf{C}^p) \qquad (1 \le j \le n) \, .$$

Hence $\mathcal{U}$ contains both of $\Lambda_+^*(\mathbf{C}^p)$, $\Lambda_-^*(\mathbf{C}^p)$, establishing irreducibility. It has highest weight s with corresponding highest weight vectors $\varepsilon_1 \wedge \cdots \wedge \varepsilon_p$. ∎

    Now let $\mathcal{V}$ be the complex Clifford algebra $\mathbf{C}_n$ and let σ be the extension of (2.16) to $\mathbf{C}_n$ as before. Then the proof in (2.19) of the irreducibility of $(\mathbf{C}_{2p}^{(k)}, \sigma)$, $k < p$, carries over to $(\mathbf{C}_{2p+1}^{(k)}, \sigma)$, $k \le p$. This proves the following.

(2.28) Theorem.

When $n = 2p + 1$, every space $\mathfrak{C}_n^{(k)}$ of k-multivectors in $\mathfrak{C}_n$ is an irreducible Spin(n)-module under the unitary representation σ; its signature is d_k if $k \leq p$, and d_{n-k} if $k > p$.

Just as in the even-dimensional case, the representation $(\mathfrak{C}_n^{(k)}, \sigma)$ of Spin(n), $n = 2p + 1$, is equivalent to $(\Lambda^k(\mathbb{C}^n), \sigma)$ under the equivariant mapping $\lambda : \Lambda^*(\mathbb{C}^n) \to \mathfrak{C}_n$. All other irreducible unitary representations of Spin(n) can now be built up (to within equivalence) by a general process known as *Cartan composition* (see (2.29)). Let (V, τ), (V', τ') be irreducible unitary representations of Spin(n) having respective signatures $(m_1, \ldots, m_p)$, $(m'_1, \ldots, m'_p)$, $p = \left[\frac{1}{2}n\right]$, with corresponding highest weight vectors ϕ, ϕ'. As we showed in the previous section, the tensor product $(V \otimes V', \tau \otimes \tau')$ of these representations is again a unitary representation of Spin(n), but in general it is highly reducible, and decomposing $V \otimes V'$ into its irreducible Spin(n)-submodules is an extremely difficult task. One submodule is canonically defined, however. For let $V \boxtimes V'$ be the irreducible Spin(n)-submodule of $V \otimes V'$ generated by the tensor product $\phi \otimes \phi'$ of the highest weight vectors ϕ, ϕ'. Then

$$(\tau \otimes \tau')\left(\prod_{j=1}^p \exp \theta_j X_j\right) : \phi \otimes \phi' \longrightarrow \left(\prod_{j=1}^p e^{i(m_j + m'_j)\theta_j}\right) \phi \otimes \phi' \, ,$$

so $(m_1 + m'_1, \ldots, m_p + m'_p)$ is a weight for $V \boxtimes V'$; in fact, this must be the highest weight for $V \boxtimes V'$ or for any other irreducible submodule of $V \otimes V'$ since a weight for such a submodule must be the sum of a weight for V and one for V'. But a general result of Cartan ensures that $V \boxtimes V'$ is the only irreducible submodule of $V \otimes V'$ having highest weight $(m_1 + m'_1, \ldots, m_p + m'_p)$, the sum of those for V and V'. Hence $V \boxtimes V'$ is frequently called the *highest weight submodule* of $V \otimes V'$ or *Cartan composition* of V, V'. This brings us to the following basic result, making (2.7)(ii) completely explicit. For concreteness it is stated in terms of representations on exterior products.

(2.29) Theorem.

Whether $n = 2p$ or $2p + 1$, there are p irreducible unitary representations

(i) $\left(\Lambda^k(\mathbb{C}^n), \sigma\right)(1 \leq k \leq p - 2)$, $\left(\Lambda^*_+(\mathbb{C}^p), S\right)$, $\left(\Lambda^*_-(\mathbb{C}^p), S\right)(n = 2p)$,

(ii) $\left(\Lambda^k(\mathbb{C}^n), \sigma\right)(1 \leq k \leq p - 1)$, $\left(\Lambda^*(\mathbb{C}^p), S\right)(n = 2p + 1)$,

of Spin(n) *which via successive Cartan compositions generate all other irreducible unitary representations of* Spin(n) *up to equivalence.*

These p representations are called the *fundamental representations* of Spin(n).

Proof of (2.29). In (2.29)(i) the representation $(\Lambda^{p-1}(\mathbb{C}^n), \sigma)$ is missing from those in (2.19). But $(\Lambda^{p-1}(\mathbb{C}^n), \sigma)$ has the same signature d_{p-1} as the Cartan composition

$$\left(\Lambda_+^*(\mathbb{C}^p), S\right) \boxtimes \left(\Lambda_-^*(\mathbb{C}^p), S\right) \qquad (n = 2p) ,$$

and so is equivalent to it. On the other hand, the representations $((\mathbb{C}_n^{(p)})_+, \sigma)$ and $((\mathbb{C}_n^p)_-, \sigma)$ in (2.19) can be omitted from (2.29)(i) because

$$\left((\mathbb{C}_n^{(p)})_+, \sigma\right) \quad , \quad (\mathfrak{S}_n^+, S) \boxtimes (\mathfrak{S}_n^+, S) \qquad (n = 2p)$$

have the same signature d_p^+, while

$$\left(\mathbb{C}_n^{(p)})_-, \sigma\right) \quad , \quad (\mathfrak{S}_n^-, S) \boxtimes (\mathfrak{S}_n^-, S) \qquad (n = 2p)$$

have the same signature d_p^-. Similary, $(\Lambda^p(\mathbb{C}^n), \sigma)$ can be omitted from the representations of Spin($2p + 1$) in (2.28) because it has the same signature d_p as

$$\left(\Lambda^*(\mathbb{C}^p), S\right) \boxtimes \left(\Lambda^*(\mathbb{C}^p), S\right) \qquad (n = 2p + 1) .$$

Now fix a p-tuple $(m_1, \ldots, m_p)$ of integers or non-zero half-integers satisfying (2.6). It is clear that non-negative integers $\alpha_1, \ldots, \alpha_p$ can be chosen so that

$$(m_1, \ldots, m_p) = \alpha_1 d_1 + \cdots + \alpha_{p-2} d_{p-2} + \alpha_{p-1} s_+ + \alpha_p s_-$$

when $n = 2p$, and

$$(m_1, \ldots, m_p) = \alpha_1 d_1 + \cdots + \alpha_{p-1} d_{p-1} + \alpha_p s$$

when $n = 2p + 1$. Hence every irreducible unitary representation of Spin(n) having signature $(m_1, \ldots, m_p)$ coincides up to equivalence with iterated Cartan compositions of the representations in (2.29). ∎

Theorem (2.29) has a number of important consequences. Consider first the Hilbert space $\mathfrak{H} = \Lambda^*(\mathbb{C}^p) \otimes \Lambda^*(\mathbb{C}^p)$ under the Hilbert–Schmidt norm. Let $e_1, \ldots, e_n$ be skew-adjoint operators on $\Lambda^*(\mathbb{C}^p)$ with respect to which it becomes a complex $\mathfrak{A}_n$-module, $n = 2p$ or $2p + 1$, (see (7.54)(**A**) in chapter 1). Then $\mathfrak{H}$ is a complex $\mathfrak{A}_n$-module with respect to the skew-adjoint operators $E_1 = e_1 \otimes I, \ldots, E_n = e_n \otimes I$ acting by the identity on the second term in $\mathfrak{H}$. On the other hand, $(\mathfrak{H}, S \otimes S)$ is a unitary representation of Spin(n) which, as noted several times already, is highly reducible; in fact, by the Cartan composition argument used in the proof of (2.29), it contains irreducible Spin(n)-modules having respective signatures d_{p-1}, d_p^+ and d_p^- when $n = 2p$, and d_p when $n = 2p + 1$. Still more is true, however.

(2.30) Theorem.
 The tensor product $\mathfrak{H} = \Lambda^*(\mathbb{C}^p) \otimes \Lambda^*(\mathbb{C}^p)$ is a $\mathrm{Spin}(n)$-module under the unitary representation $S \otimes S$, whether $n = 2p$ or $2p + 1$; it contains irreducible $\mathrm{Spin}(n)$-submodules having respective signatures

$$(i) \quad d_k \ (1 \leq k \leq p - 1) \ , \ d_p^+, d_p^- \ (n = 2p)$$

$$(ii) \quad d_k \ (1 \leq k \leq p) \ (n = 2p + 1) \ .$$

Proof. For any finite-dimensional Hilbert space $\mathcal{H}$ with inner product $\langle \cdot, \cdot \rangle$, the linear extension of

$$T_{\xi \otimes \eta} : \xi \otimes \eta : u \longrightarrow \langle u, \eta \rangle \xi \qquad (u \in \mathcal{H})$$

is an isomorphism from $\mathcal{H} \otimes \mathcal{H}$ onto $\mathcal{L}(\mathcal{H})$. But if τ is a unitary representation of $\mathrm{Spin}(n)$ on $\mathcal{H}$, then $a : T \to \tau(a)T\tau(a)^{-1}$ is a representation of $\mathrm{Spin}(n)$ such that

$$T_{\tau(a)\xi \otimes \tau(a)\eta} = \tau(a)(T_{\xi \otimes \eta})\tau(a)^{-1} \qquad \big(a \in \mathrm{Spin}(n)\big) \ .$$

Hence the linear extension of $\xi \otimes \eta \to T_{\xi \otimes \eta}$ is an equivariant mapping from $\mathcal{H} \otimes \mathcal{H}$ onto $\mathcal{L}(\mathcal{H})$. Let us apply this to the unitary representation $(\Lambda^*(\mathbb{C}^p), S)$ of $\mathrm{Spin}(n)$, $n = 2p$. Since $\mathcal{L}(\Lambda^*(\mathbb{C}^p))$ is simply a realization of $\mathfrak{C}_{2p}$, it is easily checked that

$$(2.31) \qquad \big(\Lambda^*(\mathbb{C}^p) \otimes \Lambda^*(\mathbb{C}^p), S \otimes S\big) \ , \qquad (\mathfrak{C}_{2p}, \sigma)$$

are equivalent representations of $\mathrm{Spin}(2p)$. The first part of (2.30) now follows immediately from (2.19). On the other hand, $\mathrm{Spin}(2p+1)$ can be realized inside $\mathfrak{A}_{2p}$ (see chapter 1, (7.26)), and (2.31) still gives equivalent representations of this realization of $\mathrm{Spin}(2p+1)$. But $\mathfrak{C}_{2p}^{(2k)} \oplus \mathfrak{C}_{2p}^{(2k+1)}$ is an irreducible $\mathrm{Spin}(2p + 1)$-module of $\mathfrak{C}_{2p}$ having signature d_k. Part (ii) of (2.30) follows at once, completing the proof. ∎

Combining (2.30) with (2.29) we obtain yet another basic result.

(2.32) Theorem.
 The m-fold tensor product

$$\mathfrak{H} = \Lambda^*(\mathbb{C}^p) \otimes \cdots \otimes \Lambda^*(\mathbb{C}^p)$$

is a $\mathrm{Spin}(n)$-module under the unitary representation $S \otimes \cdots \otimes S$, whether $n = 2p$ or $2p + 1$; furthermore, m can be chosen so that $\mathfrak{H}$ contains an irreducible $\mathrm{Spin}(n)$-submodule having prescribed signature $(m_1, \ldots, m_p)$.

The problem with (2.32) is that a particular sub-module of $\mathfrak{H}$ is not easy to recognize; other more explicit realizations are frequently needed.

We can use (2.29) to study also the representations of $SO(n)$ and $O(n)$. Because of the two-fold covering homomorphism $\sigma : \mathrm{Spin}(n) \to SO(n)$, there exists for each g in $SO(n)$ an a_g in $\mathrm{Spin}(n)$ such that $\sigma(\pm a_g) = g$. Consequently, if $(\mathcal{V}, \tau)$ is a representation of $\mathrm{Spin}(n)$, then $g \to \tau(\pm a_g)$ defines a representation of $SO(n)$, but it will be *single-valued* if and only if $\tau(I) = \tau(-I)$; otherwise it will be a *double-valued* representation of $SO(n)$. The representations of $\mathrm{Spin}(n)$ giving single-valued representations of $SO(n)$ are easily characterized.

(2.33) Theorem.

An irreducible unitary representation of $\mathrm{Spin}(n)$ *descends to a single-valued representation of* $SO(n)$ *if and only if its signature* $(m_1, \ldots, m_p)$, $p = [\frac{1}{2}n]$, *consists of integers satisfying* (2.6).

Proof. Suppose first that $\tau(I) = \tau(-I)$. Under the mapping $\beta : \mathbb{R}^p/(4\pi\mathbb{Z})^p \to T_n$,

$$\beta(\theta_1, \ldots, \theta_p) = \prod_{j=1}^{p} \exp \theta_j X_j = -I$$

if and only if $\theta_j = 2\pi\varepsilon_j$, $\varepsilon_j = 0$ or 1, and $\varepsilon_1 + \cdots + \varepsilon_p$ is an odd integer. Thus

$$e^{2\pi i(m_1\varepsilon_1 + \cdots + m_p\varepsilon_p)} = 1$$

for all such ε_j when $\tau(I) = \tau(-I)$, and so each m_j must be an integer. Conversely, if the signature of $(\mathcal{V}, \tau)$ consists entirely of integers, it can be built up from successive Cartan compositions of those representations having signature

$$d_k \ (k < p) \,, \ d_p^{\pm} \ (n = 2p) \quad ; \quad d_k \ (k \le p) \ (n = 2p + 1) \,.$$

But each of these has been realized as an irreducible $\mathrm{Spin}(n)$-submodule of $(\Lambda^*(\mathbb{C}^n), \sigma)$, which virtually by definition descends to a single-valued representation of $SO(n)$. As any Cartan composition of them will also be single-valued, the proof is complete. ∎

Each of the irreducible representations of $SO(n)$ on $\Lambda^k(\mathbb{C}^n)$, $2k < n$, defined by (2.17)′ extends to a unitary representation of $O(n)$. This has profound consequences for realizing the single-valued representations of $SO(n)$. For the restriction $2k < n$ ensures that the signature of Cartan compositions of $(\Lambda^*(\mathbb{C}^n), \sigma)$, $2k < n$, has the form
(2.34)

$$(m_1, \ldots, m_{p-1}, 0) \ (n = 2p) \quad , \quad (m_1, \ldots, m_p) \ (n = 2p + 1) \,,$$

where, by (2.6), each m_j is an integer such that

(2.35) $$m_1 \geq m_2 \geq \cdots \geq 0 .$$

Any finite set $(m_1, m_2, \ldots)$ of integers satisfying (2.35) is called a *partition* (of the number $m_1 + m_2 + \cdots$). Now a *polynomial representation* of the (real) general linear group $GL(n)$ $(= GL(n, \mathbb{R}))$ is, roughly speaking, a finite-dimensional representation of $GL(n)$ on a space of polynomial functions in several variables, and a fundamental result relates partitions to such polynomial representations.

(2.36) Theorem.

(Cartan–Schur–Weyl) *There is a natural 1–1 correspondence between (equivalence classes of) irreducible polynomial representations of $GL(n)$ and the partitions $(m_1, m_2, \ldots)$ where at most n of the m_j are non-zero.*

In a sense which can be made precise, the partition $(m_1, m_2, \ldots)$ is the *signature* of the representation. Since $SO(n)$ and $O(n)$ are subgroups of $GL(n)$, we should expect to find irreducible single-valued representations of $SO(n)$ having signature (2.34) by restricting to $SO(n)$ the polynomial representation of $GL(n)$ having the same signature. The irreducible $SO(n)$-submodule will then be singled out by imposing *harmonicity* conditions on the polynomials. This will be carried out in the remaining sections of this chapter. The question studiously avoided so far, however, is whether these polynomial functions are real- or complex-valued. To answer this we must know whether a representation of $\mathrm{Spin}(n)$ has real, complex or quaternionic type.

(2.38) Theorem.

The equivalence classes in $\mathcal{M}_{\mathbb{C}}(\mathrm{Spin}(n))$ having signature d_k, $1 \leq 2k < n$, are all of real type, while the type of all those having signature $s_\pm$ when $n = 2p$, or s when $n = 2p + 1$ is

(i) real if $n = 8\ell$, $8\ell + 1$ and $8\ell + 7$, $\ell \geq 0$,

(ii) complex if $n = 4\ell + 2$, $\ell \geq 0$, and

(iii) quaternionic if $n = 8\ell + 3$, $8\ell + 4$, and $8\ell + 5$, $\ell \geq 0$.

The type of any Cartan composition of these representations is determined by (1.12); in particular, those having signature (2.34) are all of real type because they are Cartan compositions of the d_k, $2k < n$. Hence real- or complex-valued polynomials can be used interchangeably to realize the representations with signature (2.34). That the d_k are of real type is clear, since (2.17) defines a representation of $\mathrm{Spin}(n)$ on $\mathfrak{A}_n^{(k)}$ and on $\Lambda^*(\mathbb{R}^n)$ which is just the real part under complex conjugation of the corresponding representation on $\mathbb{C}_n^{(k)}$ and on $\Lambda^*(\mathbb{C}^n)$. The proof of the remainder of (2.38) uses (7.38) in chapter 1; details can be found elsewhere (see Notes and remarks).

3 Class 1 representations

The most familiar representations of $O(n)$, hence of $SO(n)$ and Spin(n), are on spaces of harmonic polynomials on $\mathbb{R}^n$, $n \geq 3$. These are explicit realizations of the irreducible representations having signature $(m, 0, \ldots, 0)$, $m \geq 0$, but their importance in harmonic analysis can be traced to the fact that there are harmonic polynomials which are unchanged by any rotation fixing some direction in $\mathbb{R}^n$. Such polynomials are often called *axial polynomials*. Group-theoretically, this is the class 1 property and it is the aspect on which we shall concentrate.

Let $\mathcal{P}(\mathbb{R}^n)$ be the algebra of real- or complex-valued polynomial functions on $\mathbb{R}^n$ equipped with the positive-definite inner product defined by

$$(3.1) \qquad \langle P, Q \rangle = P\left(\frac{\partial}{\partial z}\right)\overline{Q}(z)\Big|_{z=0} \,,$$

the so-called 'Fischer inner product'. Under matrix multiplication on the right, $\mathcal{P}(\mathbb{R}^n)$ is a $GL(n)(= GL(n,\mathbb{R}))$-module on which

$$(3.2) \qquad \big(\rho(g)f\big)(z) = f(zg) \qquad (z \in \mathbb{R}^n \,,\ g \in GL(n))$$

is a representation of $GL(n)$. This clearly is a continuous representation of $GL(n)$, but it will only be unitary when restricted to $O(n)$ since

$$(3.3) \qquad \langle \rho(g)P, Q \rangle = \langle P, \rho(g')Q \rangle \qquad (g \in GL(n)) \,,$$

where g' is the transpose of g. The decomposition of $\mathcal{P}(\mathbb{R}^n)$ into its irreducible $GL(n)$-submodules is well-known.

(3.4) Theorem.
For each m the space $\mathcal{P}_m(\mathbb{R}^n)$ of degree m homogeneous polynomials is an irreducible $GL(n)$-submodule of $\mathcal{P}(\mathbb{R}^n)$, and

$$(3.4)' \qquad \mathcal{P}(\mathbb{R}^n) = \bigoplus \sum_0^\infty \mathcal{P}_m(\mathbb{R}^n)$$

is the decomposition of $\mathcal{P}(\mathbb{R}^n)$ into its irreducible $GL(n)$-submodules.

This decomposition can be achieved explicitly using Taylor series expansions. Here we shall use the notation

$$(3.5) \qquad z \cdot \frac{\partial}{\partial \zeta} = \sum_{j=1}^n z_j \frac{\partial}{\partial \zeta_j} \qquad (z \in \mathbb{R}^n)$$

to express the derivative of a function f in the 'direction' z as $(z \cdot \frac{\partial}{\partial \zeta})f$; in particular, the formal Taylor series of f is given by

$$(3.6) \qquad f(\zeta + z) = \sum_{m=0}^{\infty} \frac{1}{m!} \left(z \cdot \frac{\partial}{\partial \zeta}\right)^m f(\zeta) \qquad (z, \zeta \in \mathbf{R}^n) .$$

Since $(z \cdot \frac{\partial}{\partial \zeta})^m f \big|_{\zeta=0}$ is always a degree m homogeneous polynomial (or the zero polynomial), the Taylor series

$$(3.7) \qquad P(z) = \sum_{m=0}^{\infty} \frac{1}{m!} \left(z \cdot \frac{\partial}{\partial \zeta}\right)^m P \bigg|_{\zeta=0} \qquad (z \in \mathbf{R}^n)$$

makes (3.4)' completely explicit. The dimension

$$(3.8) \qquad \dim \mathcal{P}_m(\mathbf{R}^n) = \frac{(n+m-1)!}{m!(n-1)!} = \binom{n+m-1}{m}$$

of each space $\mathcal{P}_m(\mathbf{R}^n)$ can be expressed in a corresponding way in terms of the formal power series

$$(3.9) \qquad (1-t)^{-n} = \sum_{m=0}^{\infty} \left(\dim \mathcal{P}_m(\mathbf{R}^n)\right) t^m .$$

Representations of $O(n)$ now arise by restricting attention to the space $\mathcal{H}(\mathbf{R}^n)$ of harmonic polynomials in $\mathcal{P}(\mathbf{R}^n)$. Indeed, the representation (3.2) commutes with the Laplacian on $\mathbf{R}^n$ in the sense that

$$(3.10) \qquad \Delta\big(\rho(k)f\big) = \rho(k)(\Delta f) \qquad \big(f \in \mathcal{C}^{\infty}(\mathbf{R}^n)\big)$$

if (and only if) k is in $O(n)$; thus $\rho(k)$ preserves both $\mathcal{H}(\mathbf{R}^n)$ and its subspace $\mathcal{H}_m(\mathbf{R}^n)$ of degree m homogeneous harmonic polynomials. Hence $\mathcal{H}(\mathbf{R}^n)$ is an $O(n)$-module on which each $\rho(k)$ is a unitary operator with respect to the Fischer inner product (3.1) as we deduced earlier from (3.3). There is, however, a very useful inner product on $\mathcal{H}(\mathbf{R}^n)$ for which the unitarity of $\rho(k)$ is clearer. Set

$$(3.11) \qquad (P, Q) = \int_{\mathbf{R}^n} P(z) \overline{Q(z)}\, e^{-\frac{1}{2}|z|^2}\, dz \qquad \big(P, Q \in \mathcal{H}(\mathbf{R}^n)\big) .$$

Simply by changing variables in the integral, we see that

$$\big(\rho(k)P, \rho(k)Q\big) = (P, Q) \qquad \big(k \in O(n)\big) .$$

The quantitative relation between (3.11) and the Fischer inner product is easily calculated.

(3.12) Lemma.
  On $\mathcal{H}(\mathbf{R}^n)$

$$\int_{\mathbf{R}^n} P(z) \overline{Q(z)}\, e^{-\frac{1}{2}|z|^2}\, dz = 2\pi^{\frac{1}{2}n} P\left(\frac{\partial}{\partial z}\right) \overline{Q(z)} \bigg|_{z=0} .$$

Proof. Since $P(\frac{\partial}{\partial z})Q$ is harmonic, the mean-value property for harmonic functions ensures that

$$\int_{\mathbf{R}^n} P\Big(\frac{\partial}{\partial z}\Big)\overline{Q(z)}\, e^{-\frac{1}{2}|z|^2}\, dz$$

$$= \int_0^\infty e^{-\frac{1}{2}r^2} r^{n-1} \Big(\int_{\Sigma_{n-1}} P\Big(\frac{\partial}{\partial z}\Big)\overline{Q(z)}\Big|_{z=r\omega}\, d\omega\Big)\, dr$$

$$= P\Big(\frac{\partial}{\partial z}\Big)\overline{Q(z)}\Big|_{z=0} \Big(|\Sigma_{n-1}| \int_0^\infty e^{-\frac{1}{2}r^2} r^{n-1}\, dr\Big)$$

$$= 2\pi^{\frac{1}{2}n} \langle P, Q\rangle \ .$$

On the other hand,

$$\int_{\mathbf{R}^n} P\Big(\frac{\partial}{\partial z}\Big)\overline{Q(z)}\, e^{-\frac{1}{2}|z|^2}\, dz = \int_{\mathbf{R}^n} \overline{Q}(z) P\Big(-\frac{\partial}{\partial z}\Big) \big(e^{-\frac{1}{2}|z|^2}\big)\, dz \ .$$

But

$$e^{-\frac{1}{2}|z|^2} = \frac{1}{(2\pi)^{n/2}} \int_{\mathbf{R}^n} e^{is\cdot z} e^{-\frac{1}{2}|s|^2}\, ds \ .$$

Consequently, by Hecke's identity,

$$P\Big(-\frac{\partial}{\partial z}\Big) e^{-\frac{1}{2}|z|^2} = \frac{1}{(2\pi)^{n/2}} \int_{\mathbf{R}^n} P(is) e^{-\frac{1}{2}|s|^2} e^{is\cdot z}\, ds = P(z) e^{-\frac{1}{2}|z|^2} \ ,$$

since P is a harmonic polynomial. This establishes the relationship between the two inner products on $\mathcal{H}(\mathbf{R}^n)$. ∎

(3.12)′ Remark.

Notice that the proof of (3.12) has actually shown that

$$\int_{\mathbf{R}^n} P(z)\overline{Q(z)}\, e^{-\frac{1}{2}|z|^2}\, dz$$

$$= \int_{\mathbf{R}^n} P\Big(\frac{\partial}{\partial z}\Big)\overline{Q(z)}\, e^{-\frac{1}{2}|z|^2}\, dz$$

$$= 2\pi^{\frac{1}{2}n} P\Big(\frac{\partial}{\partial z}\Big)\overline{Q(z)}\Big|_{z=0} \ .$$

As the homogeneous Taylor polynomial $(z\cdot\frac{\partial}{\partial \zeta})^m f\big|_{\zeta=0}$ of any harmonic function lies in $\mathcal{H}_m(\mathbf{R}^n)$, the Taylor series expansion (3.7) decomposes $\mathcal{H}(\mathbf{R}^n)$ into the orthogonal direct sum of its $O(n)$-submodules $\mathcal{H}_m(\mathbf{R}^n)$. More precisely, we have the following.

(3.13) Theorem.

The decomposition of $\mathcal{H}(\mathbb{R}^n)$, $n \geq 3$, into its irreducible $O(n)$-sub-modules is given by

$$(3.13)' \qquad \mathcal{H}(\mathbb{R}^n) = \bigoplus \sum_{m=0}^{\infty} \mathcal{H}_m(\mathbb{R}^n) \ ;$$

furthermore, $(\mathcal{H}_m(\mathbb{R}^n), \rho)$ is a realization of the irreducible unitary representation of $SO(n)$ having signature $(m, 0, \ldots, 0)$.

Proof. Since $P(\frac{\partial}{\partial z})\overline{Q(z)} = 0$ when $\deg P > \deg Q$, it is clear that $(3.13)'$ is an orthogonal direct sum. It is also well-known that each $\mathcal{H}_m(\mathbb{R}^n)$ is an irreducible $O(n)$- (and $SO(n)$-) module. Thus we have to identify its signature. Now when $\mathcal{H}(\mathbb{R}^n)$ consists of complex-valued polynomials, the function

$$\Phi_\alpha(z) = (z_1 + iz_n)^{\alpha_1}(z_2 + iz_{n-1})^{\alpha_2}\cdots \qquad \bigl(z = (z_1, \ldots, z_n)\bigr)$$

lies in $\mathcal{H}_m(\mathbb{R}^n)$ for each $\alpha = (\alpha_1, \ldots, \alpha_n)$, $|\alpha| = m$, and

$$\rho(d)\Phi_\alpha = \bigl(e^{i\alpha_1\theta_1}e^{i\alpha_2\theta_2}\cdots\bigr)\Phi_\alpha$$

when d is the block diagonal matrix

$$\begin{bmatrix} \cos\theta_1 & 0 & \cdots & 0 & \sin\theta_1 \\ 0 & \cos\theta_2 & \cdots & \sin\theta_2 & 0 \\ \vdots & \vdots & & \vdots & \vdots \\ 0 & -\sin\theta_2 & \cdots & \cos\theta_2 & 0 \\ -\sin\theta_1 & 0 & \cdots & 0 & \cos\theta_1 \end{bmatrix} .$$

Consequently,

$$\rho(d) : (z_1 + iz_n)^m \longrightarrow e^{im\theta_1}(z_1 + iz_n)^m \ ;$$

and so $\Phi_{(m,0,\ldots,0)}$ is a highest weight vector for $\mathcal{H}_m(\mathbb{R}^n)$ having weight $(m, 0, \ldots, 0)$. Taking the real part of $\Phi_{(m,0,\ldots,0)}$, we obtain the same result for real-valued harmonic polynomials because $\rho(k)f$ is real-valued whenever f itself is real-valued. ∎

The decomposition of $\mathcal{P}_m(\mathbb{R}^n)$ into its irreducible $O(n)$-submodules is given by the well-known 'separation of variables theorem':

$$(3.14)$$

$$\mathcal{P}_m(\mathbb{R}^n) = \mathcal{H}_m(\mathbb{R}^n) \oplus |z|^2\mathcal{H}_{m-2}(\mathbb{R}^n) \oplus \cdots \oplus \begin{cases} |z|^m\mathcal{H}_0(\mathbb{R}^n) & (m \text{ even}), \\ |z|^{m-1}\mathcal{H}_1(\mathbb{R}^n) & (m \text{ odd}); \end{cases}$$

in particular,

$$(3.15) \qquad \dim\mathcal{H}_m(\mathbb{R}^n) = \dim\mathcal{P}_m(\mathbb{R}^n) - \dim\mathcal{P}_{m-2}(\mathbb{R}^n)$$

$$= \frac{(m + n - 2)!(n + 2m - 1)}{(n-1)!m!} .$$

Hence

(3.16)
$$\frac{1 - t^2}{(1 - t)^n} = \sum_{m=0}^{\infty} (\dim \mathcal{H}_m(\mathbb{R}^n)) t^n \ .$$

A study of the class 1 property of $(\mathcal{H}_m(\mathbb{R}^n), \rho)$ requires a detailed knowledge of ultra-spherical (or Gegenbauer) polynomials C_m^{λ}, $\Re\lambda > 0$. These are conveniently defined by the generating function

(3.17) $(1 - 2rt + r^2)^{-\lambda} = \sum_{m=0}^{\infty} C_m^{\lambda}(t) r^m$ $(|r| < |t \pm (t^2 - 1)^{1/2}|)$

or, equally, as

(3.17)′ $C_m^{\lambda}(t) = \dfrac{\Gamma(2\lambda + m)}{m! \Gamma(2\lambda)} F\!\left(-m, 2\lambda + m; \lambda; \tfrac{1}{2}(1 - t)\right)$

where

(3.18) $F(\alpha, \beta; \gamma, t) = \sum_{k=0}^{\infty} \dfrac{(\alpha)_k (\beta)_k}{(\gamma)_k} \dfrac{1}{k!} t^k$

is the usual $_2F_1$-hypergeometric function defined in terms of the monomials t^k, $k \geq 0$, and the standard Pochhammer symbol

(3.18)′ $(\alpha)_k = \alpha(\alpha + 1) \ \cdots \ (\alpha + k - 1) = \Gamma(\alpha + k)/\Gamma(\alpha) \ .$

Since $F(a, b; c; 0) = 1$, this normalizes C_m^{λ} so that

$$C_m^{\lambda}(1) = \frac{\Gamma(2\lambda + m)}{m! \Gamma(2\lambda)} = \binom{m + 2\lambda - 1}{m}$$

which also follows easily from (3.17). Together with quadratic transformations for hypergeometric functions these ensure that

(3.19) $C_{2m}^{\lambda}(t) = (-1)^m \dfrac{\Gamma(\lambda + m)}{m! \Gamma(\lambda)} F\!\left(-m, m + \lambda; \tfrac{1}{2}; t^2\right) ,$

while

(3.19)′ $C_{2m+1}^{\lambda}(t) = (-1)^m \dfrac{\Gamma(\lambda + m + 1)}{m! \Gamma(\lambda)} 2t F\!\left(-m, m + \lambda + 1; \tfrac{3}{2}; t^2\right) .$

From (3.17) and (3.19) it now follows that

(3.20) $C_m^{\lambda}(-t) = (-1)^m C_m^{\lambda}(t) \ .$

Property (3.19) enables us to associate two polynomials to each C_m^{λ}, $\Re\lambda > 0$, one a polynomial on $\mathbb{R}^n$, the other a polynomial on $\mathbb{R}^n \times \mathbb{R}^n$. Set

(3.21)
$$\phi_m^{(\lambda)}(z) = |z|^m C_m^{\lambda}\!\left(\frac{y}{|z|}\right) \Big/ C_m^{\lambda}(1) = |z|^m C_m^{\lambda}(\omega_z \cdot \mathbf{1}) \Big/ C_m^{\lambda}(1) \quad (z = x + y\mathbf{1})$$

and

(3.21)′ $Z_m^{(\lambda)}(z, \zeta) = |z|^m |\zeta|^m C_m^{\lambda}(\omega_z \cdot \omega_\zeta) \Big/ C_m^{\lambda}(1) \quad (z, \zeta \in \mathbb{R}^n) ,$

where $\mathbf{1} = (0, \dots, 0, 1)$ is the North Pole on Σ_{n-1} and $z = |z|\omega_z$, $\zeta = |\zeta|\omega_\zeta$ are the spherical polar representations of z and ζ respectively. In this case

$$\phi_m^{(\lambda)}(z) = Z_m^{(\lambda)}(z, \mathbf{1}) \qquad (z \in \mathbb{R}^n) \;;$$

furthermore, by (3.17),

$$(3.22) \qquad |1 - rz|^{-2\lambda} = \sum_{m=0}^{\infty} \binom{m + 2\lambda - 1}{m} \phi_m^{(\lambda)}(z) r^m$$

and

$$(3.22)' \qquad |\zeta - rz|^{-2\lambda} = \sum_{m=0}^{\infty} \binom{m + 2\lambda - 1}{m} Z_m^{(\lambda)}(z, \zeta) |\zeta|^{2(\lambda - m)} r^m \;,$$

while by (3.19) there exist constants $a_{mj}^{(\lambda)}$ so that

$$(3.23) \qquad \phi_m^{(\lambda)}(z) = \sum_{j=0}^{[\frac{1}{2}m]} a_{mj}^{(\lambda)} |z|^{2j} y^{m-2j} \qquad (z = x + y\mathbf{1})$$

and

$$(3.23)' \qquad Z_m^{(\lambda)}(z, \zeta) = \sum_{j=0}^{[\frac{1}{2}m]} a_{mj}^{(\lambda)} |z|^{2j} |\zeta|^{2j} (z \cdot \zeta)^{m-2j} \qquad (z, \zeta \in \mathbb{R}^n) \;.$$

Consequently, $\phi_m^{(\lambda)}$ is a polynomial in $\mathcal{P}_m(\mathbb{R}^n)$ which is left fixed by any rotation of $\mathbb{R}^n$ around the y-axis; similarly, $Z_m^{(\lambda)}$ is a polynomial in $\mathcal{P}_m(\mathbb{R}^n) \otimes \mathcal{P}_m(\mathbb{R}^n)$ which is fixed by any simultaneous rotation of z and ζ. To express this group-theoretically, set $\xi = \mathbf{1}k \; (k \in O(n))$. Then

$$\omega_{zk^{-1}} \cdot \mathbf{1} = \omega_z \cdot \mathbf{1}k = \omega_z \cdot \xi \qquad \left(k \in O(n) \;, \; z \in \mathbb{R}^n\right) \;,$$

and so

$$\left(\rho(k)\phi_m^{(\lambda)}\right)(z) = \phi_m^{(\lambda)}(zk^{-1}) = |z|^m C_m^\lambda(\omega_z \cdot \xi) / C_m^\lambda(1) \;.$$

Hence $\rho(k)\phi_m^{(\lambda)} = \phi_m^{(\lambda)}$ if and only if k belongs to the subgroup $M \; (\sim O(n-1))$ fixing $\mathbf{1}$. On the other hand,

$$(3.24) \qquad Z_m^{(\lambda)}(zk, \zeta k) = Z_m^{(\lambda)}(z, \zeta) \qquad (z, \zeta \in \mathbb{R}^n)$$

for all k in $O(n)$. Thus $Z_m^{(\lambda)}$ is an $O(n)$-invariant polynomial in $\mathcal{P}_m(\mathbb{R}^n) \otimes \mathcal{P}_m(\mathbb{R}^n)$, while $\phi_m^{(\lambda)}$ is an $O(n-1)$-invariant polynomial in $\mathcal{P}_m(\mathbb{R}^n)$. These last two properties foreshadow one way polynomial invariant theory will enter into the theory.

The crucial link with harmonic polynomials will follow from the next result in what is essentially a special case of Maxwell's theory of poles.

(3.25) Theorem.
  For each λ, $\Re\lambda > 0$,

$$\left(z \cdot \frac{\partial}{\partial \zeta}\right)^m |\zeta|^{-2\lambda} = \gamma_m^\lambda |\zeta|^{-2(\lambda+m)} Z_m^{(\lambda)}(z, \zeta)$$

where

$$\gamma_m^\lambda = (-1)^m \frac{\Gamma(2\lambda + m)}{\Gamma(2\lambda)} .$$

When $\lambda = \frac{1}{2}(n - 2)$, the function $\zeta \to |\zeta|^{-2\lambda}$ is the fundamental solution of the Laplacian on $\mathbb{R}^n$ (up to a constant); in particular, it is a harmonic function on $\mathbb{R}^n \setminus \{0\}$. But then $(z \cdot \frac{\partial}{\partial \zeta})^m |\zeta|^{-(n-2)}$ will be harmonic in both z and ζ. Hence

$$(z, \zeta) \longrightarrow |\zeta|^{-2(\lambda+m)} Z_m^{(\lambda)}(z, \zeta) \qquad (\lambda = \tfrac{1}{2}(n - 2))$$

is harmonic separately in z and ζ, which in turn shows that

$$z \longrightarrow \phi_m^{(\lambda)}(z) = Z_m^{(\lambda)}(z, 1) \qquad (\lambda = \tfrac{1}{2}(n - 2))$$

is harmonic. Since $Z_m^{(\lambda)}(z, \zeta) = Z_m^{(\lambda)}(\zeta, z)$, it follows that $Z_m^{(\lambda)}$ is harmonic separately in both variables. This proves the following.

(3.26) Corollary.
  When $\lambda = \frac{1}{2}(n - 2)$, $\phi_m^{(\lambda)}$ *is in* $\mathcal{H}_m(\mathbb{R}^n)$ *and* $Z_m^{(\lambda)}$ *is in* $\mathcal{H}_m(\mathbb{R}^n) \otimes \mathcal{H}_m(\mathbb{R}^n)$.

Proof of theorem (3.25). We expand $1/|z - \zeta|^{2\lambda}$ as an infinite series in two ways. First, there is a Taylor series expansion

$$\frac{1}{|z - \zeta|^{2\lambda}} = \sum_{m=0}^{\infty} (-1)^m \frac{1}{m!} \left(z \cdot \frac{\partial}{\partial \zeta}\right)^m \left(\frac{1}{|\zeta|^{2\lambda}}\right) ,$$

while, by (3.17),

$$\frac{1}{|z - \zeta|^{2\lambda}} = \frac{1}{|\zeta|^{2\lambda}} \sum_{m=0}^{\infty} \left(\frac{|z|}{|\zeta|}\right)^m C_m^\lambda(\omega_z \cdot \omega_\zeta) ,$$

both series converging for all z sufficiently small in relation to ζ. Hence, by homogeneity,

$$(-1)^m \frac{1}{m!} \left(z \cdot \frac{\partial}{\partial \zeta}\right)^m \left(\frac{1}{|\zeta|^{2\lambda}}\right) = \left(\frac{|z|^m}{|\zeta|^{2\lambda+m}}\right) C_m^\lambda(\omega_z \cdot \omega_\zeta)$$

$$= C_m^\lambda(1) \frac{1}{|\zeta|^{2(\lambda+m)}} Z_m^{(\lambda)}(z, \zeta) ,$$

without restriction on z and ζ.   ∎

We can now turn to a detailed study of the 'class 1 property' of $(\mathcal{H}_m(\mathbb{R}^n), \rho)$.

(3.27) Theorem.

(Class 1 property) *When* $\lambda = \frac{1}{2}(n-2)$, $\phi_m^{(\lambda)}$ *is in* $\mathcal{H}_m(\mathbb{R}^n)$ *and, apart from constant multiples, it is the only element in* $\mathcal{H}_m(\mathbb{R}^n)$ *satisfying* $\rho(k)\phi = \phi$ *for all* k *in the subgroup* M *of* $O(n)$ *fixing* **1**.

Rather than prove this well-known result directly, it will be very instructive to exhibit explicitly the decomposition of $\mathcal{H}_m(\mathbb{R}^n)$ into its M-invariant submodules. The class 1 property says merely that the space $\mathbb{R}\phi_m^{(\lambda)}$ (or $\mathbb{C}\phi_m^{(\lambda)}$) *occurs in this decomposition*, and that it is the *only space in the decomposition on which every* k *in* M *acts as the identity transformation*.

The corresponding subgroup G_0 of $GL(n)$ fixing **1** will be isomorphic to $GL(n-1)$, and the decomposition of $\mathcal{P}_m(\mathbb{R}^n)$ into its G_0-invariant submodules is clearly given by

$$(3.28) \qquad \mathcal{P}_m(\mathbb{R}^n) = \sum_{j=0}^{m} y^{n-j}\mathcal{P}_j(\mathbb{R}^{n-1}) \qquad (\mathbb{R}^n = \mathbb{R}^{n-1} + \mathbb{R}\mathbf{1}) ,$$

since every P in $\mathcal{P}_m(\mathbb{R}^n)$ can be written uniquely in the form

$$(3.29) \quad P(z) = y^m P_0(x) + y^{m-1}P_1(x) + \cdots + P_m(x) \qquad (z = x + y\mathbf{1})$$

with P_j in $\mathcal{P}_j(\mathbb{R}^{n-1})$. Clearly $\mathbb{R}y^n$ (or $\mathbb{C}y^n$) is the only subspace in (3.28) on which every h in G_0 acts as the identity transformation. The analogous result for harmonic polynomials replaces the G_0-invariants y^{n-j} in (3.29) with the M-invariant polynomials $\phi_{m-j}^{(\lambda+j)}$, $\lambda = \frac{1}{2}(n-2)$.

(3.30) Theorem.

When $\lambda = \frac{1}{2}(n-2)$ *the polynomial* $Q = \phi_{m-j}^{(\lambda+j)}P_j$, $P_j \in \mathcal{H}_j(\mathbb{R}^{n-1})$, *lies in* $\mathcal{H}_m(\mathbb{R}^n)$, *and every* P *in* $\mathcal{H}_m(\mathbb{R}^n)$ *can be written uniquely in the form*

$$P(z) = \sum_{j=0}^{m} \phi_{m-j}^{(\lambda+j)}(z)P_j(x) \qquad (z = x + y\mathbf{1})$$

with P_j *in* $\mathcal{H}_j(\mathbb{R}^n)$.

Since each $\phi_{m-j}^{(\lambda+j)}$ is M-invariant,

$$(3.31) \qquad\qquad \mathcal{H}_m(\mathbb{R}^n) = \sum_{j=0}^{m} \phi_{m-j}^{(\lambda+j)}\mathcal{H}_j(\mathbb{R}^{n-1})$$

decomposes $\mathcal{H}_m(\mathbb{R}^n)$ into irreducible M-modules. The class 1 property of $\mathcal{H}_m(\mathbb{R}^n)$ now follows immediately.

Proof of (3.30). As a simple calculation shows, the function

$$z = x + y\mathbf{1} \longrightarrow |1 - rz|^{-2(\lambda+j)}h(x) \qquad \left(\lambda = \tfrac{1}{2}(n-2)\right)$$

is harmonic for each r and h in $\mathcal{H}_j(\mathbb{R}^{n-1})$. On the other hand,

$$|1 - rz|^{-2(\lambda+j)} = \sum_{k=0}^{\infty} \phi_k^{(\lambda+j)}(z) r^k .$$

Thus $h \to \phi_{m-j}^{(\lambda+j)} h$ embeds $\mathcal{H}_j(\mathbb{R}^{n-1})$ in $\mathcal{H}_m(\mathbb{R}^n)$; furthermore,

$$\phi_{m-j}^{(\lambda+j)}\mathcal{H}_j(\mathbb{R}^{n-1}) \cap \phi_{m-k}^{(\lambda+k)}\mathcal{H}_k(\mathbb{R}^{n-1}) = \{0\}$$

when $j \neq k$. But, by (3.15) and (3.28),

$$\dim \mathcal{H}_m(\mathbb{R}^n) = \dim \mathcal{P}_m(\mathbb{R}^n) - \dim \mathcal{P}_{m-2}(\mathbb{R}^n)$$

$$= \sum_{j=0}^{m} \mathcal{P}_j(\mathbb{R}^{n-1}) - \sum_{j=2}^{m} \mathcal{P}_{j-2}(\mathbb{R}^{n-1}) = \sum_{j=0}^{m} \dim \mathcal{H}_j(\mathbb{R}^{n-1}) .$$

Hence the direct sum $\sum_{j=0}^{m} \phi_{m-j}^{(\lambda+j)}\mathcal{H}_j(\mathbb{R}^{n-1})$ is all of $\mathcal{H}_m(\mathbb{R}^n)$, completing the proof. ∎

Finally in this section we shall derive two important consequences of the class 1 property of $\mathcal{H}_m(\mathbb{R}^n)$: the reproducing kernel property of $Z_m^{(\lambda)}$ and the integral representation of $P_m^{(\lambda)}$.

(3.32) Theorem.
When $\lambda = \frac{1}{2}(n-2)$,

$$\int_{\mathbb{R}^n} Z_m^{(\lambda)}(z,\zeta)\overline{P(\zeta)}\, e^{-\frac{1}{2}|\zeta|^2}\, d\zeta = (\phi_m^{(\lambda)}, \phi_m^{(\lambda)})P(z) \qquad (z \in \mathbb{R}^n)$$

for all P in $\mathcal{H}_m(\mathbb{R}^n)$ where

$$(\phi_m^{(\lambda)}, \phi_m^{(\lambda)}) = \int_{\mathbb{R}^n} |\phi_m^{(\lambda)}(z)|^2 e^{-\frac{1}{2}|z|^2}\, dz .$$

The actual numerical value of $(\phi_m^{(\lambda)}, \phi_m^{(\lambda)})$ can be calculated using (3.12), (3.18), (3.20) and (3.23).

Proof. Set

$$Q(z) = \int_{\mathbb{R}^n} Z_m^{(\lambda)}(z,\zeta)\overline{P(\zeta)}\, e^{-\frac{1}{2}|\zeta|^2}\, d\zeta .$$

Then, by (3.26), Q is a polynomial in $\mathcal{H}_m(\mathbb{R}^n)$, while the $O(n)$-invariance of $Z_m^{(\lambda)}$ ensures that

$$\int_{\mathbb{R}^n} Z_m^{(\lambda)}(z,\zeta)\overline{P(\zeta k)}\, e^{-\frac{1}{2}|\zeta|^2}\, d\zeta = Q(zk) \qquad (k \in O(n)) .$$

Hence

$$P \longrightarrow \int_{\mathbb{R}^n} Z_m^{(\lambda)}(z,\zeta)\overline{P(\zeta)}\, e^{-\frac{1}{2}|\zeta|^2}\, d\zeta$$

is an $O(n)$-equivariant mapping on $\mathcal{H}_m(\mathbb{R}^n)$, and so, by Schur's lemma (see (1.3)),

$$\int_{\mathbb{R}^n} Z_m^{(\lambda)}(z,\zeta)\overline{P(\zeta)}\, e^{-\frac{1}{2}|\zeta|^2}\, d\zeta = c_{n,m}P(z) \qquad (P \in \mathcal{H}_m(\mathbb{R}^n)) \ .$$

To evaluate the constant, take $P = \phi_m^{(\lambda)}$ and $z = 1$. Then

$$c_{n,m}\phi_m^{(\lambda)}(1) = \int_{\mathbb{R}^n} |\phi_m^{(\lambda)}(\zeta)|^2\, e^{-\frac{1}{2}|\zeta|^2}\, d\zeta = (\phi_m^{(\lambda)}, \phi_m^{(\lambda)}) \ ,$$

completing the proof. ∎

(3.33) Theorem.

(Laplace type representation) *When* $\lambda = \frac{1}{2}(n-2)$, *the Gegenbauer polynomial* C_m^λ *is given by*

$$C_m^\lambda(t) = C_m^\lambda(1)\left(\int_{\Sigma_{n-2}} f\big((1-t^2)^{1/2}\omega_0 + t\mathbf{1}\big)\, d\omega_0\right) \qquad (|t| < 1)$$

where f *is any polynomial in* $\mathcal{H}_m(\mathbb{R}^n)$ *such that* $f(1) = 1$ *and* $d\omega_0$ *is normalized rotation-invariant measure on the unit sphere* Σ_{n-2} *in* $\mathbb{R}^{n-1}$.

Proof. Given such an f, set

$$P(z) = \int_M f(xk + y\mathbf{1})\, dk \qquad (z = x + y\mathbf{1}) \ .$$

Then P is an M-invariant polynomial in $\mathcal{H}_m(\mathbb{R}^n)$ and $P(1) = 1$. The class 1 property thus ensures that $P = \phi_m^{(\lambda)}$. But $C_m^\lambda(\omega \cdot 1) = C_m^\lambda(1)\phi_m^{(\lambda)}(\omega)$ on Σ_{n-1}; consequently,

$$C_m^\lambda(\omega \cdot 1) = C_m^\lambda(1)\left(\int_M f(\xi k + t\mathbf{1})\, dk\right) \qquad (\omega = \xi + t\mathbf{1}) \ .$$

Now each ω in Σ_{n-1} can be written uniquely as $\omega = (1-t^2)^{1/2}\omega_0 + t\mathbf{1}$ for some ω_0 in Σ_{n-2}, and in such a coordinatization

$$\int_M f(\xi k + t\mathbf{1})\, dk = \int_{\Sigma_{n-2}} f\big((1-t^2)^{1/2}\omega_0 + t\mathbf{1}\big)\, d\omega_0 \ .$$

From this the integral representation of C_m^λ follows immediately. ∎

4 Polynomials of matrix argument

From an analytic point of view, one of the most convenient ways of realizing those representations of $\mathrm{Spin}(n)$ that descend to single-valued representations of $SO(n)$ is on polynomials of matrix argument. In this setting, a direct generalization of the one studied in the previous section, it is pointless and unnatural, however, to bring in $\mathrm{Spin}(n)$, and so for the most part we shall avoid any mention of spin groups. In fact the (real) general linear groups as much as orthogonal groups are the ones of principal interest in this section which is devoted to setting out some of the basic ideas associated with polynomial functions on the space $\mathbb{R}^{r \times n}$, $r \leq n$, of $r \times n$ real matrices – the polynomials we shall speak of as having matrix argument. In concrete terms, these are just the polynomials in the coordinate functions $z \to z_{jk}$ $(1 \leq j \leq r, 1 \leq k \leq n)$ on $\mathbb{R}^{r \times n}$. The space of all polynomial functions, whether real- or complex-valued, will be denoted by $\mathcal{P}(\mathbb{R}^{r \times n})$, and if $\frac{\partial}{\partial z} = [\frac{\partial}{\partial z_{jk}}]$ denotes the $r \times n$ matrix of partial derivatives with respect to these coordinate functions, then the Fischer inner product on $\mathcal{P}(\mathbb{R}^{r \times n})$ is given as before by

(4.1) $$\langle P, Q \rangle = P\Big(\frac{\partial}{\partial z}\Big)\overline{Q}(z)\Big|_{z=0}.$$

As $\mathbb{R}^{r \times n}$ is a left $GL(r)$-module and a right $GL(n)$-module under matrix multiplication, there are corresponding representations
(4.2)
$$\big(\lambda(h)f\big)(z) = f(h^{-1}z) \,, \quad \big(\rho(k)f\big)(z) = f(zk) \quad \big(h \in GL(r) \,, \ k \in GL(n)\big)$$
on $\mathcal{P}(\mathbb{R}^{r \times n})$ which are still well-defined on all of $\mathcal{C}^{\infty}(\mathbb{R}^{r \times n})$.

Now $GL(1)$ is the multiplicative group $\mathbb{R}_*$ of non-zero real numbers acting on $\mathbb{R}^n$ by scalar multiplication $x \to \alpha z$, so a polynomial P in $\mathcal{P}(\mathbb{R}^n)$ is homogeneous of degree m when $P(\alpha z) = \alpha^m P(z)$, i.e., when $P(\alpha z) = \chi_m(\alpha)P(z)$ where χ_m is the character $\chi_m : \alpha \to \alpha^m$ of $\mathbb{R}_*$. Consequently, the decomposition (3.4)$'$ of $\mathcal{P}(\mathbb{R}^n)$ into its $GL(n)$-irreducible submodules $\mathcal{P}_m(\mathbb{R}^n)$, as well of $\mathcal{H}(\mathbb{R}^n)$ into its irreducible $O(n)$-submodules $\mathcal{H}_m(\mathbb{R}^n)$, is intimately tied in with the characters of $GL(1)$ which extend to functions on all of $\mathbb{R}$. But what seems like algebraic overkill for $\mathbb{R}^n$ is exactly the right notion to use for $\mathbb{R}^{r \times n}$: left multiplication by $GL(r)$ and various of its subgroups is used to specify refined homogeneity conditions on $\mathcal{P}(\mathbb{R}^{r \times n})$. Indeed, left multiplication is an operation on rows, so GL_r or its subgroups can be used to specify *row-homogeneity*. But here, for simplicity, we shall consider only the case of row-homogeneity with respect to $GL(r)$, replacing the characters $\chi_m : \alpha \to \alpha^m$, $m \geq 0$, of $GL(1)$ by the characters $h \to (\det h)^m$, $m \geq 0$, of $GL(r)$. These are the characters which extend to all of $\mathbb{R}^{r \times r}$.

(4.3) Definition.

A polynomial P in $\mathcal{P}(\mathbb{R}^{r \times n})$ is said to be determinantally homogeneous of degree m when

$$P(hz) = (\det h)^m P(z) \qquad (z \in \mathbb{R}^{r \times n})$$

for all h in $\mathbb{R}^{r \times r}$.

To construct examples of such functions denote by $\Delta_{k_1 \cdots k_r}$

$$(4.4) \quad \Delta_{k_1 \cdots k_r}(z) = \det \begin{vmatrix} z_{1k_1} & \cdots & z_{1k_r} \\ \vdots & & \vdots \\ z_{rk_1} & \cdots & z_{rk_r} \end{vmatrix} \quad (1 \leq k_1 < \cdots < k_r \leq n) ,$$

the determinantal minor formed from the $k_1, \ldots, k_r$ columns of $x \in \mathbb{R}^{r \times n}$. Clearly

$$(4.4)' \quad \Delta_{k_1 \cdots k_r}(hz) = (\det h)\Delta_{k_1 \cdots k_r}(z) \quad \left(h \in GL(r) \right) .$$

Hence when homogeneity in $\mathcal{P}(\mathbb{R}^{r \times n})$ is determined by $GL(r)$, these minors $z \to \Delta_{k_1 \cdots k_r}(z)$ will play exactly the role that the coordinate functions $z \to z_k$ do for $\mathbb{R}^n$. For instance, any degree m homogeneous polynomial in these minors will be determinantally homogeneous of degree m. Before embarking on a discussion of the general theory however, it is instructive to look at a special case. Set

$$(4.5) \quad \mathcal{W} = \mathrm{span}_{\mathbb{C}}\{\Delta_{k_1 \cdots k_r} : 1 \leq k_1 < \cdots < k_r \leq n\} .$$

This is a $GL(n)$-submodule of $\mathcal{P}(\mathbb{R}^{r \times n})$ consisting of polyomials which are determinantally homogeneous of degree 1. On the other hand, since the $\Delta_{k_1 \cdots k_r}$ are *linearly* independent, $\mathcal{W}$ has dimension $\binom{n}{r}$, the same as that of $\Lambda^r(\mathbb{C}^n)$ on which the fundamental representation of $SO(n)$ having signature $d_r = (\underbrace{1, \ldots, 1}_{r}, 0, \ldots 0)$ is realized (see (2.17)). In fact,

$(\mathcal{W}, \rho)$ is just another realization of this same fundamental representation, and so instead of using tensor products of $\Lambda^r(\mathbb{C}^n)$ to realize the representations having signature $(\underbrace{m, \ldots, m}_{r}, 0, \ldots, 0)$, we can use degree m determinantally homogeneous polynomials on $\mathbb{R}^{r \times n}$. Irreducible representations of $O(n)$ are then obtained by restricting attention to harmonic polynomials, just as was done in the previous section. These are analytically more appealing realizations of representations of $GL(n)$ and $O(n)$, but a brief excursion into classical polynomial invariant theory is needed to develop the details.

Since its beginnings in the middle of the nineteenth century, invariant theory has suffered a number of well-publicized deaths and rebirths. In its most recent re-incarnation following the ever-increasing understanding of Hermann Weyl's book on the subject, invariant theory seems to have permeated all of mathematics from combinatorics and statistics to differential equations, characteristic classes and index theory. In fact, wherever there is a natural action of a group expressed, say, through various symmetries, invariant-theoretic ideas seem to play a basic role. Their importance in representation theory should come as no surprise therefore; we have seen it already in studying class 1 representations, for instance, and the classical Pfaffian is a very important example of an $SO(n)$-invariant which is not an $O(n)$-invariant (see chapter 1, (8.31)).

The notion of an invariant is a very simple one. Let K be any group, finite or not, Lie or not, and let $\mathcal{V}$ be a (right) K-module over $\mathbb{R}$ or $\mathbb{C}$ which may be finite or infinite-dimensional. Then an element v in $\mathcal{V}$ is said to be a K-invariant when $v \cdot k = v$ for all k in K, i.e., when K fixes v. For instance, if V is a finite-dimensional (right) K-module, the algebra $\mathcal{P}(V)$ of polynomial functions on V becomes a K-module under the right regular representation

(4.6) $\qquad \big(\rho(k)P\big)(v) = P(v \cdot k) \qquad (v \in V \ , \ k \in K) ,$

and a K-invariant in $\mathcal{P}(V)$ is simply a polynomial P satisfying $\rho(k)P = P, k \in K$. Clearly the set of all such invariant polynomials is a subalgebra, $\mathcal{P}(V)^K$, of $\mathcal{P}(V)$. For varying choices of V and K, $\mathcal{P}(V)^K$ will be of paramount importance in this chapter. In just the same way, if V is a finite-dimensional left H-module and

(4.7) $\qquad \big(\lambda(h)P\big)(v) = P(h^{-1}v) \qquad (v \in V \ , \ h \in H) ,$

is the left regular representation of H on $\mathcal{P}(V)$, an H-invariant in $\mathcal{P}(V)$ is a polynomial P satisfying $\lambda(h)P = P, h \in H$; the set $\mathcal{P}(V)^H$ of all such H-invariants is a subalgebra of $\mathcal{P}(V)$. When V is a left H-module as well as a right K-module, we can also consider the algebra $\mathcal{P}(V)^{H \times K}$ of $(H \times K)$-invariants.

Invariant polynomials are often used in symbols defining invariant pseudo-differential operators, and this is a particularly important analytic use of invariant theory. For example, let V be a finite-dimensional real Hilbert space that is, say, a right K-module. Then

(4.8) $\qquad \big(\rho(k)f\big)(\zeta) = f(\zeta k) \quad , \quad \big((\rho \otimes \rho)(k)\phi\big)(z,\zeta) = \phi(zk, \zeta k)$

define representations of K on $C^\infty(V)$ and $C^\infty(V \times V)$ respectively. But K also acts contragrediently on V by $v \rightarrow v(k^{-1})'$ where k' is the adjoint of k with respect to the inner product $(\cdot, \cdot)$ on $V \times V$; in addition, the inner product is a K-invariant in the sense that

(4.9) $\qquad \big(z, k, s(k^{-1})'\big) = (z, s) \qquad (z, s \in V)$

holds for all k. More generally, let P be any polynomial on $V \times V$ satisfying the invariance condition

(4.9)' $\qquad P\big(zk, s(k^{-1})'\big) = P(z, s) \qquad (z, s \in V)$

for all k. By thinking of $P(z, \frac{\partial}{\partial \zeta})$ as a finite order differential operator whose coefficients are polynomials in z, we obtain a differential operator

$$(4.10) \qquad P\left(z, \frac{\partial}{\partial \zeta}\right) : f(\zeta) \longrightarrow F(z, \zeta) = P\left(z, \frac{\partial}{\partial \zeta}\right) f(\zeta)$$

from $\mathcal{C}^\infty(V)$ into $\mathcal{C}^\infty(V \times V)$. Indeed, any P in $\mathcal{P}(V \times V)$ could be used to define such an operator, but the crucial point of condition (4.9)$'$ is to ensure that $P(z, \frac{\partial}{\partial \zeta})$ is an invariant differential operator as made precise by the following theorem.

(4.11) Theorem.
Let (4.10) be the differential operator determined by a polynomial P on $V \times V$ satisfying (4.9)$'$. Then

$$P\left(z, \frac{\partial}{\partial \zeta}\right) : f(\zeta k) \longrightarrow F(zk, \zeta k) \qquad (k \in K) ,$$

i.e., $P(z, \frac{\partial}{\partial \zeta})$ is an equivariant operator from $(\mathcal{C}^\infty(V), \rho)$ into $(\mathcal{C}^\infty(V \times V), \rho \otimes \rho)$.

Proof. Essentially the result is just a sophisticated version of the chain rule in multi-variable calculus, but use of the invariance condition (4.9)$'$ is brought out more clearly by interpreting $P(z, \frac{\partial}{\partial \zeta})$ as a pseudo-differential operator. Now, since $P(z, \frac{\partial}{\partial \zeta})$ is a local operator, it is enough to establish (4.11) for f in the Schwartz space $\mathcal{S}(V)$. But on $\mathcal{S}(V)$

$$P\left(z, \frac{\partial}{\partial \zeta}\right) f(\zeta) = \frac{1}{(2\pi)^{\frac{1}{2}n}} \int_V e^{i(\zeta, s)} P(z, is) \hat{f}(s) \, ds \qquad (n = \dim V) ,$$

extending P to a polynomial on $V \times V_{\mathbb{C}}$ in an obvious way. Thus, by (4.9),

$$(\rho(k)f)\hat{}(s) = \int_V e^{-i(s, \zeta)} f(\zeta k) \, d\zeta$$

$$= J(k) \int_V e^{-i(s, \zeta k^{-1})} f(\zeta) \, d\zeta = J(k) \hat{f}\big(s(k^{-1})'\big)$$

where $J(k)$ is the Jacobian of the transformation $\zeta \to \zeta k^{-1}$. Consequently,

$$P\left(z, \frac{\partial}{\partial \zeta}\right) (\rho(k)f) = \frac{1}{(2\pi)^{\frac{1}{2}n}} \int_V e^{i(z, s)} P(z, is) \hat{f}\big(s(k^{-1})'\big) J(k) \, ds$$

$$= \frac{1}{(2\pi)^{\frac{1}{2}n}} \int_V e^{i(\zeta k, s)} P(zk, is) \hat{f}(s) \, ds = F(zk, \zeta k)$$

using (4.9) and (4.9)$'$ after a change of variable $s \to sk'$. This proves (4.11). ∎

If V is a left H-module and P has the invariance property

$$(4.12) \qquad P\big(hz, (h')^{-1}s\big) = P(z, s) \qquad (z, s \in V)$$

for all h in H, then an obvious modification to the proof of (4.11) establishes the corresponding result

$$(4.13) \qquad P\left(z, \frac{\partial}{\partial \zeta}\right) : f(h\zeta) \longrightarrow F(hz, h\zeta) \qquad (h \in H)$$

in the left module case. Properties (4.11) and (4.13) can also be combined when P satisfies both of (4.9)', (4.12).

Various examples help bring out these ideas.

(A) Consider the symmetric group S_n acting on $\mathbb{R}^n$ by permutation of coordinates. The S_n-invariants in $\mathcal{P}(\mathbb{R}^n)$ are then the *symmetric polynomials*

$$(4.14) \qquad P(t_1, \ldots, t_n) = P(t_{\sigma(1)}, \ldots, t_{\sigma(n)}), \qquad \sigma \in S_n .$$

For example, if

$$(4.15)(\mathrm{i}) \qquad \prod_{j=1}^{n}(\lambda + t_j) = \lambda^n + \sum_{j=1}^{n} e_j(t)\lambda^{n-j} \qquad \big(t = (t_1, \ldots, t_n)\big) ,$$

then

$$(4.15)(\mathrm{ii}) \quad e_1(t) = \sum_j t_j , \quad e_2(t) = \sum_{j<k} t_j t_k , \quad \ldots , \quad e_n(t) = t_1 \cdots t_n$$

must be invariant under all permutations of $t_1, \ldots, t_n$. These are the so-called *elementary* symmetric polynomials which were encountered already in chapter 1 (see (8.15)). A basic result is that they generate freely all S_n-invariants, i.e., P is a symmetric polynomial if and only if

$$(4.16) \qquad P(t) = Q\big(e_1(t), \ldots, e_n(t)\big) , \qquad t \in \mathbb{R}^n ,$$

for some Q in $\mathcal{P}(\mathbb{R}^n)$, and $Q \to Q(e_1(t), \ldots, e_n(t))$ is an algebra isomorphism from $\mathcal{P}(\mathbb{R}^n)$ onto $\mathcal{P}(\mathbb{R}^n)^{S_n}$.

(B) Let V be $\mathbb{R}^{r \times n}$ regarded both as a Hilbert space with inner product

$$(4.17) \qquad (z, \zeta) = \mathrm{tr}(z\zeta') \qquad (z, \zeta \in \mathbb{R}^{r \times n})$$

as well as a $GL(n)$- and $GL(r)$-module under matrix multiplication, where the notation A' for a matrix A will always denote the transpose of A. Now the adjoint with respect to (4.17) of any k in $GL(n)$ or h in $GL(r)$ is just matrix multiplication by the transpose k' and h' respectively, so the two uses of the notation ( )' coincide for matrices; in particular, therefore, the polynomial

$$(4.18) \qquad P(z, \zeta) = Q(z\zeta') \qquad (z, \zeta \in \mathbb{R}^{r \times n})$$

on $\mathbb{R}^{r \times n} \times \mathbb{R}^{r \times n}$ determined by any Q in $\mathcal{P}(\mathbb{R}^{r \times r})$ satisfies

$$(4.19) \qquad P\big(zk, \zeta(k^{-1})'\big) = P(z, \zeta) \qquad (k \in GL(n)) ,$$

which is the form that $(4.9)'$ takes in this example. The invariant inner product (4.17) arises from the trace function $A \rightarrow \text{tr}\, A$ on $\mathbb{R}^{r \times r}$, for instance, and the First Fundamental Theorem of invariant theory characterizes all such invariants in this way.

(4.20) Theorem.

A polynomial P on $\mathbb{R}^{r \times n} \times \mathbb{R}^{r \times n}$ satisfies (4.19) if and only if $P(z, \zeta) = Q(z\zeta')$ for some Q in $\mathcal{P}(\mathbb{R}^{r \times r})$; furthermore, $Q \rightarrow Q(z\zeta')$ is an algebra isomorphism.

One proof of (4.20) uses the Capelli operators to be introduced in a moment; an alternative proof uses Schur's double commutant theorem (see Notes and remarks). Applying (4.11) we see that

$$(4.21) \qquad Q\left(z\frac{\partial'}{\partial\zeta}\right) : f(\zeta) \longrightarrow F(z,\zeta) = Q\left(z\frac{\partial'}{\partial\zeta}\right)f(\zeta)$$

associates to each Q in $\mathcal{P}(\mathbb{R}^{r \times r})$ a differential operator from $\mathcal{C}^\infty(\mathbb{R}^{r \times n})$ into $\mathcal{C}^\infty(\mathbb{R}^{r \times n} \times \mathbb{R}^{r \times n})$ such that

$$(4.21)(\text{i}) \qquad Q\left(z\frac{\partial'}{\partial\zeta}\right) : f(\zeta k) \longrightarrow F(zk, \zeta k) \qquad (k \in GL(n)) ;$$

if in addition $Q(h\zeta h^{-1}) = Q(h)$ for all h in $GL(r)$, then simple modifications to the proof of (4.11) show that

$$(4.21)(\text{ii}) \qquad Q\left(z\frac{\partial'}{\partial\zeta}\right) : f(h\zeta) \longrightarrow F(hz, h\zeta) \qquad (h \in GL(r)) .$$

In particular, for any Q in $\mathcal{P}(\mathbb{R}^{r \times r})$,

$$(4.21)' \qquad\qquad f \longrightarrow Q\left(z\frac{\partial'}{\partial\zeta}\right)f\bigg|_{\zeta=0}$$

defines a differential operator from $\mathcal{C}^\infty(\mathbb{R}^{r \times n})$ into $\mathcal{P}(\mathbb{R}^{r \times n})$ which is $GL(n)$-invariant in the sense that

$$(4.21)''$$
$$\left(Q\left(z\frac{\partial'}{\partial\zeta}\right)\rho(k)f\right)\bigg|_{\zeta=0} = \rho(k)\left(Q\left(z\frac{\partial'}{\partial\zeta}\right)f\bigg|_{\zeta=0}\right) \qquad (k \in GL(n)) .$$

It is no accident of course that $(4.21)'$ bears a strong similarity to the degree m homogeneous Taylor polynomial $(z \cdot \frac{\partial}{\partial\zeta})^m f\big|_{\zeta=0}$ occurring in (3.7). Indeed, theorem (4.20) says intuitively that $GL(n)$-invariants associated with $\mathcal{P}(\mathbb{R}^{r \times n})$, be they operators or polynomials, must arise from polynomials on $\mathbb{R}^{r \times r}$, and as the previous section confirmed, this is exactly what happened in the case $r = 1$. In fact, the linear basis of monomials $x \rightarrow x^m$, $m \geq 0$, for $\mathcal{P}(\mathbb{R})$ was used in two significant ways:

(4.22)(a) *to define the rotation-invariant $\phi_m^{(\lambda)}$ in $\mathcal{P}_m(\mathbb{R}^n)$ by first considering Gegenbauer polynomials C_m^λ on $\mathbb{R}$, and*

(4.22)(b) *to define $GL(n)$-invariant differential operators $P \to (z \cdot \frac{\partial}{\partial \zeta})^m$ $P|_{\zeta=0}$ on $\mathcal{P}(\mathbb{R}^n)$.*

The Taylor series expansion in (3.7) can thus be interpreted as arising via (b) from the function

$$(4.22)(c) \qquad e^x = \sum_{m=0}^{\infty} \frac{1}{m!} x^m \qquad (x \in \mathbb{R}) \; ;$$

the exponential function thus acts as a reproducing kernel for $\mathcal{P}(\mathbb{R}^n)$ in terms of the Fischer inner product. Exactly the same principles apply to polynomials of matrix argument.

Consider the group $SL(r)$ acting on $\mathbb{R}^{r \times n}$ by left multiplication. Then, by (4.4)′, any polynomial in the $\Delta_{k_1 \ldots k_r}$ will be an $SL(r)$-invariant, and this characterizes all such invariants as we shall see.

(4.23) Theorem.

A polynomial on $\mathbb{R}^{r \times n}$ is an $SL(r)$-invariant if and only if it is a polynomial in the minors $\Delta_{k_1 \ldots k_r}$, $1 \leq k_1 < \cdots < k_r \leq n$.

Thus the functions $z \to \Delta_{k_1 \ldots k_r}(z)$ generate the algebra $\mathcal{P}(\mathbb{R}^{r \times n})^{SL}$, just as the coordinate functions $z \to z_k$ generate $\mathcal{P}(\mathbb{R}^n)$, although they do not generate $\mathcal{P}(\mathbb{R}^{r \times n})^{SL}$ freely since in the case of $\mathbb{R}^{2 \times 4}$, for instance,

$$\Delta_{12}\Delta_{34} = \Delta_{13}\Delta_{24} - \Delta_{23}\Delta_{14}$$

(Sylvester's theorem). This causes no problems however, and will never need to be mentioned again. To prove (4.23) and then exploit it we continue to develop concepts for $\mathcal{P}(\mathbb{R}^{r \times n})^{SL}$ parallel to those for $\mathcal{P}(\mathbb{R}^n)$. By the Cauchy–Binet theorem,

(4.24)

$$\sum_{k_1 < \cdots < k_r} \Delta_{k_1 \ldots k_r}(z)\Delta_{k_1 \ldots k_r}(\zeta) = \det(z\zeta') = \det \begin{bmatrix} z_1 \cdot \zeta_1 & \cdots & z_1 \cdot \zeta_r \\ \vdots & & \vdots \\ z_r \cdot \zeta_1 & \cdots & z_r \cdot \zeta_r \end{bmatrix}$$

where $z_1, \ldots, z_r$ and $\zeta_1, \ldots, \zeta_r$ are the rows of z and ζ respectively; this is intimately connected with the inner product (7.12) in Chapter 1 on $\Lambda^r(\mathbb{R}^n)$, and reduces to the usual inner product on $\mathbb{R}^n$ when $r = 1$. As the analogue of $\frac{\partial}{\partial z_k}$ we use the Cayley operator $\Delta_{k_1 \ldots k_r}(\frac{\partial}{\partial \zeta})$ obtained from the $k_1, \ldots, k_r$ columns of $\frac{\partial}{\partial \zeta}$, $\zeta \in \mathbb{R}^{r \times n}$. Thus the $GL(n)$-invariant differential operator

$$(4.24)' \qquad \det\left(z\frac{\partial'}{\partial \zeta}\right) = \sum_{k_1 < \cdots < k_r} \Delta_{k_1 \ldots k_r}(z)\Delta_{k_1 \ldots k_r}\left(\frac{\partial}{\partial \zeta}\right)$$

determined by the function $t \to \det t$ on $\mathbf{R}^{r \times r}$ will play the role that $z \cdot \frac{\partial}{\partial \zeta}$ did in the previous section. Cayley's name is always attached to his operator because of the 'power rule'

(4.25)
$$\Delta_{k_1 \cdots k_r} \left(\frac{\partial}{\partial \zeta} \right) : \left(\Delta_{k_1 \cdots k_r}(\zeta) \right)^m \to \left(\prod_{j=1}^{r} (m + r - j) \right) \left(\Delta_{k_1 \cdots k_r}(\zeta) \right)^{m-1}$$

he established. Finally, we shall need the analogue of the Euler operator $E = \zeta \cdot \frac{\partial}{\partial \zeta} = \sum_k \zeta_k \frac{\partial}{\partial \zeta_k}$. This is the famous Capelli operator

(4.26)
$$H_r = \sum_{k_1 < \cdots < k_r} \Delta_{k_1 \cdots k_r}(\zeta) \Delta_{k_1 \cdots k_r} \left(\frac{\partial}{\partial \zeta} \right).$$

Because the operators $\zeta_j \cdot \frac{\partial}{\partial \zeta_k}$ do not necessarily commute, the Cauchy–Binet identity (4.24) cannot be used to rewrite (4.26) as a single determinant as we did in (4.24)′: extra terms were added to $\det(\zeta \frac{\partial'}{\partial \zeta})$ by Capelli to compensate for this non-commutativity, foreshadowing much later work by Chevalley and Harish-Chandra. With the convention that the determinant $\det[A_{pq}]$ of an $(r \times r)$-matrix of (possibly) non-commuting variables is given by

$$\det[A_{pq}] = \sum_{\sigma \in S_r} \operatorname{sgn} \sigma A_{\sigma(1)1} \cdots A_{\sigma(r)r} ,$$

Capelli showed that

(4.26)′
$$H_r = \det \begin{bmatrix} D_{11} + (r-1) & D_{12} & \cdots & D_{1r} \\ D_{21} & D_{22} + (r-2) & \cdots & D_{2r} \\ \vdots & \vdots & & \\ D_{r1} & D_{r2} & \cdots & D_{rr} \end{bmatrix}$$

where
$$D_{jk} = \zeta_j \cdot \frac{\partial}{\partial \zeta_k} = \sum_\ell \zeta_{j\ell} \frac{\partial}{\partial \zeta_{k\ell}} .$$

That (4.26) and (4.26)′ define the same operator H_r is usually known as Capelli's identity, and is a cornerstone of classical polynomial invariant theory. We shall take this identity for granted, but prove almost everything else.

To illustrate the role of Euler type operator that H_r plays we prove first the following.

(4.27) Theorem.

If P is a determinantally homogeneous polynomial of degree m in $\mathcal{P}(\mathbb{R}^{r \times n})$, then

$$H_r P = \left(\prod_{j=1}^{r} (m + r - j) \right) P \ .$$

The case $r = 1$ reduces exactly to Euler's theorem for homogeneous polynomials. A preliminary lemma is needed.

(4.28) Lemma.

When $P = P(\zeta)$ is $SL(r)$-invariant, then $D_{jk}P = 0$, $j \neq k$; when it is also determinantally homogeneous of degree m, then, in addition, $D_{jj}P = mP$.

Proof. If E_{jk} is the element of the Lie algebra of $GL(r)$ whose entries are all zero except for the (j,k)-entry which has the value 1, then

$$\frac{d}{dt} P \left(\exp(t E'_{jk}) \zeta \right) \Big|_{t=0} = \left(\zeta_j \cdot \frac{\partial}{\partial \zeta_k} \right) P(\zeta) = D_{jk} P \ .$$

Hence $E_{jk}P = 0$, $j \neq k$, when P is $SL(r)$-invariant, and $E_{jj}P = mP$ when it is also determinantally homogeneous of degree m. ∎

Proof of (4.27). Lemma (4.28) ensures that

$$H_r P = \left(D_{11} + (r-1) \right) \left(D_{22} + (r-2) \right) \cdots D_{rr} P$$

whenever P is an $SL(r)$-invariant. Hence

$$H_r P = \left((m+r-1)(m+r-2) \cdots m \right) P$$

when $P(h\zeta) = (\det h)^m P(\zeta)$, $h \in GL(r)$. ∎

In view of (4.26),

$$\Delta_{k_1 \cdots k_r}(\zeta) \Delta_{k_1 \cdots k_r} \left(\frac{\partial}{\partial \zeta} \right) \left(\Delta_{k_1 \cdots k_r}(\zeta) \right)^m$$

$$= H_r \left(\Delta_{k_1 \cdots k_r}(\zeta) \right)^m = \left(\prod_{j=1}^{r} (m+r-j) \right) \left(\Delta_{k_1 \cdots k_r}(\zeta) \right)^m \ ;$$

consequently, Cayley's result (4.25) for $m = 1, 2, \ldots$ is a special case of (4.27).

Proof of (4.23). It is enough to show that any $SL(r)$-invariant is a polynomial in the minors $\Delta_{k_1 \cdots k_r}$. Now every P in $\mathcal{P}(\mathbb{R}^{r \times n})$ can be written uniquely as

(4.29) $$P = P_0 + P_1 + \cdots + P_m$$

for some m where

(4.29)′ $\quad P_j(h\zeta) = \alpha^j P(\zeta) \qquad \left(h = \begin{bmatrix} \alpha & & 0 \\ & \ddots & \\ 0 & & \alpha \end{bmatrix} = \alpha I \right)$.

Indeed, if polynomials on $\mathbb{R}^{r \times n}$ are regarded simply as polynomials on $\mathbb{R}^{rn}$, then a polynomial satisfies (4.29)′ precisely when it is homogeneous of degree j in the usual sense as an element of $\mathcal{P}(\mathbb{R}^{rn})$. Thus (4.29) follows immediately from (3.4). On the other hand, if P is an $SL(r)$-invariant, then each P_j in (4.29) will be $SL(r)$-invariant also. Hence from the outset we can assume that P is an $SL(r)$-invariant such that

(4.29)″ $\qquad P(h\zeta) = \alpha^m P(\zeta) \qquad (h = \alpha I)$

for some m; in particular, $D_{jk}P = 0$ when $j \neq k$. But, for $\alpha > 0$,

$$\begin{bmatrix} \alpha & & & \\ & 1 & & 0 \\ & & \ddots & \\ & 0 & & 1 \end{bmatrix}$$

$$= \begin{bmatrix} \alpha^{1/r} & & & \\ & \alpha^{1/r} & & 0 \\ & & \ddots & \\ & 0 & & \alpha^{1/r} \end{bmatrix} \begin{bmatrix} \alpha^{1-1/r} & & & \\ & \alpha^{-1/r} & & 0 \\ & & \ddots & \\ & 0 & & \alpha^{-1/r} \end{bmatrix},$$

which expresses the left-hand matrix as a product of $(\alpha^{1/r}I)$ and a matrix in $SL(r)$; consequently,

$$P\left(\begin{bmatrix} \alpha\zeta_1 \\ \vdots \\ \zeta_r \end{bmatrix} \right) = \alpha^{m/r} P\left(\begin{bmatrix} \zeta_1 \\ \vdots \\ \zeta_r \end{bmatrix} \right) \qquad (\alpha > 0) .$$

Differentiating with respect to α, we deduce that $D_{11}P = (m/r)P$; similarly, $D_{jj}P = (m/r)P$ for all j. Hence $SL(r)$-invariance together with (4.19)″ ensures that

$$D_{jk}P = \delta_{jk}(m/r)P \qquad (1 \leq j, k \leq r) ,$$

in which case

(4.30) $\qquad H_r P = \left(\prod_{j=1}^{r} \left(\frac{m}{r} + r - j \right) \right) P ,$

modifying trivially the proof of (4.28).

Now define $Q = Q(z, \zeta)$ by

$$Q(z,\zeta) = \det\left(z\frac{\partial'}{\partial\zeta}\right)P(\zeta) = \sum_{k_1 < \cdots < k_r} \Delta_{k_1\cdots k_r}(z)\Delta_{k_1\cdots k_r}\left(\frac{\partial}{\partial\zeta}\right)P$$

$$= \sum_{k_1 < \cdots < k_r} \Delta_{k_1\cdots k_r}(z)P_{k_1\cdots k_r}(\zeta)$$

for some $P_{k_1\cdots k_r}$ in $\mathcal{P}(\mathbb{R}^{r\times n})$. In view of (4.26) and (4.30),

$$P(\zeta) = \text{const.} \sum_{k_1 < \cdots < k_r} \Delta_{k_1\cdots k_r}(\zeta)P_{k_1\cdots k_r}(\zeta)$$

from which the theorem will follow by a simple induction argument provided the $P_{k_1\cdots k_r}$ also are $SL(r)$-invariants satisfying an identity of the form of (4.29)″ for a smaller value of m. But, by (4.21)(ii),

$$\det\left(z\frac{\partial'}{\partial\zeta}\right) : P(h\zeta) \longrightarrow Q(hz, h\zeta)$$

for any h in $GL(r)$. Consequently,

$$Q(hz, h\zeta) = Q(z, \zeta) \qquad \left(h \in SL(r)\right),$$

since P is an $SL(r)$-invariant, while

$$Q(hz, h\zeta) = \alpha^m Q(z, \zeta) \qquad (h = \alpha I)$$

because of (4.29)″. Hence $P_{k_1\cdots k_r}$ is an $SL(r)$-invariant such that

$$P_{k_1\cdots k_r}(h\zeta) = \alpha^{m-r}P_{k_1\cdots k_r}(\zeta) \qquad (h = \alpha I) .$$

This allows the induction argument to proceed, completing the proof.   ∎

Theorem (4.23) prompts the following definition.

(4.31) Definition.
Denote by $\mathcal{P}_m(\mathbb{R}^{r\times n})^{SL}$ the polynomials in $\mathcal{P}(\mathbb{R}^{r\times n})$ which are determinantally homogeneous of degree m.

Since

$$P(hzk) = (\det h)^m P(zk) \qquad \left(P \in \mathcal{P}_m(\mathbb{R}^{r\times n})^{SL}\right),$$

$\mathcal{P}_m(\mathbb{R}^{r\times n})^{SL}$ is a $GL(n)$-submodule of $\mathcal{P}(\mathbb{R}^{r\times n})$; in fact, more is true.

(4.32) Theorem.
The decomposition of $\mathcal{P}(\mathbb{R}^{r\times n})^{SL}$ into its irreducible $GL(n)$-submodules is given by

$$\mathcal{P}(\mathbb{R}^{r\times n})^{SL} = \bigoplus \sum_{m=0}^{\infty} \mathcal{P}_m(\mathbb{R}^{r\times n})^{SL} ;$$

furthermore, $(\mathcal{P}_m(\mathbb{R}^{r \times n})^{SL}, \rho)$ *is a realization of the irreducible representation of* $GL(n)$ *having signature* $\tau_m = (\underbrace{m, \ldots, m}_{r}, 0, \ldots)$.

An unusual Taylor series expansion exhibits (4.32) in a completely explicit way, just as the classical expansion did in (3.7). For each partition $\sigma = (\sigma_1, \ldots, \sigma_s, 0, \ldots, 0)$ with $\sigma_s > 0$ set

$$(4.33) \qquad \sigma! = \prod_{j=1}^{s}(\sigma_j + s - j)! \bigg/ \prod_{i<j}(\sigma_i - \sigma_j + j - i) \; ;$$

technically this is the so-called 'hook number' associated with the partition σ. It reduces to

$$(4.33)' \qquad \tau_m! = \prod_{j=1}^{r}\frac{(m + r - j)!}{(r - j)!} \; ,$$

in the special case of the partition $\tau_m = (\underbrace{m, \ldots, m}_{r}, 0, \ldots)$.

(4.34) Theorem.
Each $SL(r)$*-invariant polynomial* P *admits a unique Taylor series expansion*

$$P(z) = \sum_{m=0}^{\infty}\frac{1}{\tau_m!}\det\left(z\frac{\partial'}{\partial\varsigma}\right)^m P\bigg|_{\varsigma=0} \qquad (z \in \mathbb{R}^{r \times n})$$

as a sum of determinantally homogeneous polynomials.

Thus

$$x \longrightarrow \sum_{m=0}^{\infty}\frac{1}{\tau_m!}(\det x)^m \qquad (x \in \mathbb{R}^{r \times r})$$

acts via (4.11) as a reproducing kernel for $\mathcal{P}(\mathbb{R}^{r \times n})^{SL}$ with respect to the Fischer inner product. Clearly (4.34) reduces to the standard Taylor series on $\mathbb{R}^n$ when $r = 1$.

Proof of (4.34). In view of (4.22), it is enough to show that

$$(4.35) \qquad \det\left(z\frac{\partial'}{\partial\varsigma}\right)^m P(\varsigma) = \delta_{\ell m}\left(\prod_{j=1}^{r}\frac{(m + r - j)!}{(r - j)!}\right)P(z)$$

for P in $\mathcal{P}_\ell(\mathbb{R}^{r \times n})^{SL}$. Set

$$Q(z, \varsigma) = \left(\det\left(z\frac{\partial'}{\partial\varsigma}\right)\right)^m P(\varsigma) \qquad (P \in \mathcal{P}_\ell(\mathbb{R}^{r \times n})^{SL}) \; ;$$

consequently, $Q(hz, \varsigma) = (\det h)^m Q(z, \varsigma)$, $h \in GL(n)$. Now, if $\ell < m$, then $Q \equiv 0$; so we can assume $\ell \geq m$. But, by (4.21)(ii),

$$\det\left(z\frac{\partial'}{\partial\zeta}\right)^m : P(h\zeta) \longrightarrow Q(hz, h\zeta) \qquad (h \in GL(r)) ;$$

thus

$$Q(hz, h\zeta) = (\det h)^\ell Q(z,\zeta) ,$$
$$Q(z, h\zeta) = (\det h)^{\ell-m} Q(z,\zeta) \quad (h \in GL(r)) ;$$

in particular, $Q(z,0) \equiv 0$ when $\ell > m$. On the other hand, when $\ell = m$, $Q = Q(z,\zeta)$ will be a polynomial solely in the first variable. To complete the proof of (4.35) therefore, it is enough to prove that

(4.35)' $$Q(\zeta,\zeta) = \left(\prod_{j=1}^r \frac{(m+r-j)!}{(r-j)!}\right) P(\zeta)$$

when Q is given by (4.25) with $\ell = m$. Define $Q_1, \ldots, Q_m$ successively by

$$Q_1(z,\zeta) = \det\left(z\frac{\partial'}{\partial\zeta}\right) P(\zeta),$$

$$Q_k(z,\zeta) = \det\left(z\frac{\partial'}{\partial\zeta}\right) Q_{k-1}(z,\zeta)(2 \leq k \leq m) ,$$

By (4.21)(ii) yet again, each Q_k is determinantally homogeneous of degree $m - k$ in ζ, and so, by (4.24)', (4.26) and (4.27),

$$Q_k(\zeta,\zeta) = \left(\prod_{j=1}^r (m-k+r-j)\right) Q_{k-1}(\zeta,\zeta)$$

from which (4.35)' follows easily since

$$Q_1(\zeta,\zeta) = \left(\prod_{j=1}^r (m+r-j)\right) P(\zeta) .$$

This completes the proof.   ∎

   In terms of the parallelism between $\mathcal{P}(\mathbf{R}^n)$ and $\mathcal{P}(\mathbf{R}^{r\times n})^{SL}$ we have reached theorem (3.4) as made explicit by (3.7). To continue the parallelism further therefore, Gegenbauer polynomials of matrix argument, hence hypergeometric functions of matrix argument, have to be introduced so that rotation-invariant polynomials $\phi_m^{(\lambda)}$ can be singled out in $\mathcal{P}_m(\mathbf{R}^{r\times n})^{SL}$. Such hypergeometric functions are not as yet fully understood, however, but enough is known for us to indicate the basic ideas.

   In studying the classical Gegenbauer polynomials we carefully distinguished between even and odd values of m, because C_m^λ is even or odd according as m is even or odd (see (3.20)). But even polynomial functions on $\mathbb{R}$ are precisely the $O(1)$-invariants in $\mathcal{P}(\mathbb{R})$, for which the monomials $x \to x^{2k}$, $k \geq 0$, form a basis. Thus, in studying Gegenbauer polynomials of matrix argument, one begins by fixing a basis for the polynomials P on $\mathbb{R}^{r \times r}$ satisfying

$$(4.36) \qquad P(hak) = P(a) \qquad (a \in \mathbb{R}^{r \times r})$$

for all h, k in $O(r)$, i.e., P in $\mathcal{P}(\mathbb{R}^{r \times r})^{O(r) \times O(r)}$. This two-sided invariance compensates for the failure of commutativity of multiplication in $\mathbb{R}^{r \times r}$. Several consequences of (4.36) follow immediately. For instance, the group $GL(r)$ is Zariski-dense in $\mathbb{R}^{r \times r}$, and so any P in $\mathcal{P}(\mathbb{R}^{r \times r})$ is determined by its restriction to $GL(r)$. But every a in $GL(r)$ can be written as

$$a = h \operatorname{Diag}(\alpha_1, \ldots, \alpha_r) k \qquad (h, k \in O(r))$$

for some diagonal matrix $\operatorname{Diag}(\alpha_1, \ldots \alpha_r)$. Thus, each P in $\mathcal{P}(\mathbb{R}^{r \times r})^{O(r) \times O(r)}$ is completely determined by its restriction to the group of diagonal matrices in $GL(r)$; in particular, the two-sided $O(r)$-invariance of such a polynomial ensures that

$$(4.36)' \qquad P(a) = P(a') \qquad (a \in \mathbb{R}^{r \times r}) \ .$$

To make these ideas more precise we need one more example from invariant theory.

   (C) Each Q in $\mathcal{P}(\mathbb{R}^{r \times r})$ determines a polynomial $P(a) = Q(aa')$ on $\mathbb{R}^{r \times r}$ which is (right) $O(r)$-invariant in the sense that $P(ak) = P(a)$ for all k. Furthermore, the First Fundamental Theorem for orthogonal invariants characterizes all such invariants in this way.

(4.37) Theorem.
   A polynomial P on $\mathbb{R}^{r \times r}$ satisfies $P(ak) = P(a)$ for all k in $O(r)$ if and only if $P(a) = Q(aa')$ for some Q in $\mathcal{P}(\mathbb{R}^{r \times r})$.

The Capelli operators can again be used to prove this result, though a direct proof using induction serves equally well (see Notes and remarks). Now aa' is a symmetric matrix, so it is sufficient in (4.37) to consider Q as a polynomial on the space $\mathbb{R}^{r \times r}_{\mathrm{sym}}$ of symmetric $(r \times r)$-matrices. The polynomials on $\mathbb{R}^{r \times r}$ satisfying (4.36) are thus determined by those Q in $\mathcal{P}(\mathbb{R}^{r \times r}_{\mathrm{sym}})$ satisfying

$$Q(h^{-1}ah) = Q(a) \qquad (a \in \mathbb{R}^{r \times r}_{\mathrm{sym}})$$

for all h in $O(r)$. But every symmetric matrix can be diagonalized by an orthogonal matrix, say

$$a = O \operatorname{Diag}\left(\lambda_1(a), \ldots, \lambda_r(a)\right) O^{-1} \qquad (\lambda_j(a) \in \mathbb{R})$$

for some O in $O(r)$; in addition, the eigenvalues of $h^{-1}ah$ coincide with those of a up to a permutation. From this the following theorem completes the characterization of two-sided $O(r)$-invariants on $\mathbb{R}^{r \times r}$ and brings us back to example **(A)**.

(4.38) Theorem.

The mapping

$$Q \longrightarrow Q\big(\lambda_1(aa'), \ldots, \lambda_r(aa')\big)$$

is an algebra isomorphism from $\mathcal{P}(\mathbb{R}^r)^{S_r}$ *onto* $\mathcal{P}(\mathbb{R}^{r \times r})^{O(r) \times O(r)}$.

Fixing a basis for the two-sided $O(r)$-invariants in $\mathcal{P}(\mathbb{R}^{r \times r})$ thus amounts to fixing a basis for the symmetric polynomials on $\mathbb{R}^r$. There are many such bases, the importance of any particular one depending on context. The basis we shall adopt has also been defined by A.M. James as eigenfunctions of various invariant operators on symmetric polynomials; equivalent definitions have been given by McDonald, based on work of Henry Jack, specifying the basis as just one of a one-parameter family of such bases, and yet other definitions relate the basis to root-systems. But for clarity the definition we shall use emphasizes its group-theoretic origin. Let $\mathcal{V}_\Sigma$ be the subspace of $\mathcal{P}(\mathbb{R}^{r \times r}_{\text{sym}})$ of polynomials which are homogeneous of degree Σ thinking of them simply as polynomials in the $\frac{1}{2}r(r+1)$ coordinate functions $x \to x_{jk}$ ($1 \le j \le k \le r$). Since a congruence transformation $x \to kxk'$ preserves $\mathbb{R}^{r \times r}_{\text{sym}}$, $\mathcal{V}_\Sigma$ is a $GL(r)$-module. Now a well-known result of Thrall shows that $\mathcal{V}_\Sigma$ decomposes into a direct sum of irreducible $GL(r)$-submodules $\mathcal{V}_\sigma$, one for each partition $\sigma = (\sigma_1, \ldots, \sigma_s, 0, \ldots,)$, $\sigma_s > 0$, such that $1 \le s \le r$ and $\sigma_1 + \cdots + \sigma_s = \Sigma$; in fact, $\mathcal{V}_\sigma$ is a realization of the irreducible representation of $GL(r)$ having signature $2\sigma = (2\sigma_1, \ldots, 2\sigma_s, 0, \ldots)$; furthermore, by the Cartan theory of zonal polynomials applied to the pair $(GL(r), O(r))$, there is a unique polynomial C_σ in each $\mathcal{V}_\sigma$ satisfying

(4.39) $C_\sigma(hah^{-1}) = C_\sigma(a) \quad \big(h \in O(r)\big) \, , \; C_\sigma(I) = 1 \, .$

As σ varies over all partitions

(4.39)' $\sigma = (\sigma_1, \ldots, \sigma_s, 0, \ldots) \, , \quad \sigma_s > 0 \, , \quad s \le r \, ,$

the functions $a \to C_\sigma(aa')$ thus form a linear basis for $\mathcal{P}(\mathbb{R}^{r \times r})^{O(r) \times O(r)}$. Since $\mathcal{V}_\Sigma = \mathbb{R}x^\Sigma$ when $r = 1$, this basis reduces to the usual basis of monomials x^m, $m \ge 0$, in the special case $r = 1$. Many properties of the C_σ have been established. We list just three of them.

(4.40)(a)   On $\mathbb{R}^{r \times r}$

$$\int_{O(r)} e^{\operatorname{tr}(ak)}\, dk = \sum_{\sigma} \frac{1}{(2\sigma)!}\, C_\sigma(aa')\,,$$

the sum being taken over all partitions σ in $(4.39)'$; this reduces to

$$\cosh x = \tfrac{1}{2}(e^x + e^{-x}) = \sum_{m=0}^{\infty} \frac{1}{(2m)!}\, x^{2m}$$

when $r = 1$.

(4.40)(b)   On $\mathbb{R}^{r \times r}_{\mathrm{sym}}$

$$e^{\operatorname{tr}(a)} = \sum_{\sigma} \frac{1}{\langle C_\sigma, C_\sigma \rangle}\, C_\sigma(a)\,,$$

where the numerical value of

$$\langle C_\sigma, C_\sigma \rangle = C_\sigma\left(\frac{\partial}{\partial \zeta}\right) C_\sigma\Big|_{\zeta=0}$$

can be calculated in terms of $\sigma!$ and r; since $(\frac{d}{dx})^m x^m = m!$, this reduces to

$$e^x = \sum_{m=0}^{\infty} \frac{1}{m!}\, x^m \qquad (x \in \mathbb{R})$$

when $r = 1$.

(4.40)(c)   For each σ in $(4.39)'$ define Φ_σ in $\mathcal{P}(\mathbb{R}^{r \times r}_{\mathrm{sym}})$ by

$$\Phi_\sigma(a) = \left(\prod_{j=1}^{s-1} (\Delta_{1 \cdots j}^{1 \cdots j}(a))^{\sigma_j - \sigma_{j+1}}\right)(\Delta_{1 \cdots s}^{1 \cdots s}(a))^{\sigma_s}$$

where

$$\Delta_{1 \cdots j}^{1 \cdots j}(a) = \det \begin{bmatrix} a_{11} & \cdots & a_{1j} \\ \vdots & & \vdots \\ a_{j1} & \cdots & a_{jj} \end{bmatrix} \qquad (1 \le j \le s)$$

is the determinantal minor formed from the first j rows and columns of a. Then

$$C_\sigma(a) = \int_{O(r)} \Phi_\sigma(kak^{-1})\, dk \qquad (a \in \mathbb{R}^{r \times r}_{\mathrm{sym}})\,;$$

in particular, $C_\sigma(a) = (\det a)^m$ when $\sigma = \tau_m$.

The introduction of hypergeometric functions on $\mathbb{R}^{r \times r}_{\mathrm{sym}}$ is now a straightforward matter.

(4.41) Definition.

The $_2F_1$-hypergeometric function of matrix argument is defined by

$$F(\alpha, \beta; \gamma; t) = \sum_{\sigma} \frac{(\alpha)_\sigma (\beta)_\sigma}{(\gamma)_\sigma} \frac{1}{\langle C_\sigma, C_\sigma \rangle} C_\sigma(t) \qquad (t \in \mathbb{R}^{r \times r}_{\text{sym}}) ,$$

summing over all σ satisfying (4.39)', where

$$(\delta)_\sigma = \prod_{j=1}^{s} \left(\delta - \tfrac{1}{2}(j-1)\right)_{\sigma_j} \qquad (\delta \in \mathbb{C})$$

is the Pochhammer symbol for a partition.

Just as in the classical case, the series terminates and becomes a polynomial when α or β is a negative integer, the case of interest to us, but convergence results have been established for more general values of α, β. In addition, since $(\delta)_\sigma$ is analytic in δ, $F(\alpha, \beta; \gamma; t)$ certainly is analytic in β when α is a negative integer, but again analyticity results have been established in general by Gross, Herz *et al.* (see Notes and remarks). As our aims are modest, however, we shall give the properties of this hypergeometric function (4.41) in the polynomial case $\alpha = -m$ only. An Euler integral representation

$$(4.42) \qquad \frac{\Gamma_r(\gamma)}{\Gamma_r(\beta)\Gamma_r(\gamma - \beta)} \int_0^I \det(I - Rt)^m \det(R)^{\beta - p} \det(I - R)^{\gamma - \beta - p} \, dR$$

for $F(-m, \beta; \gamma; t)$ can be established where $p = \tfrac{1}{2}(r + 1)$,

$$(4.43) \qquad \Gamma_r(\delta) = \pi^{\frac{1}{4}r(r-1)} \prod_{j=1}^{r} \Gamma\left(\delta - \tfrac{1}{2}(j-1)\right)$$

and the integral is taken over the interval from 0 to I in the cone of positive-definite matrices in $\mathbb{R}^{r \times r}_{\text{sym}}$; the integral is absolutely convergent for $0 < t \le I$ provided $\Re\beta, \Re(\gamma - \beta) > (p - 1)$. The matrix analogue of the classical beta-integral is the identity

$$(4.43)' \qquad \frac{\Gamma_r(a)\Gamma_r(b)}{\Gamma_r(a + b)} = \int_0^I \det(R)^{a - p} \det(I - R)^{b - p} \, dR .$$

Consequently,

$$(4.44) \qquad F(-m, \beta; \gamma; I) = (\gamma - \beta)_{\tau_m} / (\gamma)_{\tau_m}$$

provided $\Re\beta, \Re(\gamma - \beta) > p - 1$. Since both sides extend analytically in β, however, this numerical value for $F(-m, \beta; \gamma; I)$ continues to hold for all β. It is the matrix analogue of the special case due to Vandermonde of Gauss' famous identity; in particular, both of

$$F(-m, m + \lambda; \tfrac{1}{2}r; I) \quad , \quad F(-m, m + \lambda + 1; \tfrac{1}{2}r + 1; I)$$

are non-zero provided $\Re\lambda > \tfrac{1}{2}(r - 2)$. Using (3.19), Herz was thus led to the following definition.

(4.45) Definition.

The (normalized) Gegenbauer polynomials of matrix argument are defined on $\mathbb{R}^{r \times r}$ when $\Re\lambda > \frac{1}{2}(r-2)$ by

$$R_{2m}^{\lambda}(\eta) = \left(1/\rho_{2m}^{\lambda}\right)F(-m, m+\lambda; \tfrac{1}{2}r; \eta\eta')$$

for even integers and

$$R_{2m+1}^{\lambda}(\eta) = \left(1/\rho_{2m+1}^{\lambda}\right)\det\eta \; F(-m, m+\lambda+1; \tfrac{1}{2}r+1; \eta\eta')$$

for odd integers where the constants ρ_m^{λ} are chosen so that $R_m^{\lambda}(I) = 1$.

The actual value of the normalizing constants ρ_m^{λ} can be determined from (4.44). Since $R_m^{\lambda}(I) = 1$, it is clear that R_m^{λ} reduces to $C_m^{\lambda}/C_m^{\lambda}(1)$ when $r = 1$. Some properties of Gegenbauer polynomials are easily established.

(4.46) Theorem.

The normalized Gegenbauer polynomials satisfy

$$(i) \qquad R_m^{\lambda}(\eta) = R_m^{\lambda}(\eta') \qquad\qquad (\eta \in \mathbb{R}^{r \times r}),$$

$$(ii) \qquad R_m^{\lambda}(h\eta) = (\det h)^m R_m^{\lambda}(\eta) \qquad (h \in O(r)).$$

Property (4.46)(ii) is the analogue of (3.20) of course.

Proof. As functions of η, $F(-m, \beta; \gamma; \eta\eta')$ and $F(-m, \beta; \gamma; \eta'\eta)$ are two-sided $O(r)$-invariants in $\mathcal{P}(\mathbb{R}^{r \times r})$ which coincide on diagonal matrices; consequently,

$$F(-m, \beta; \gamma; \eta\eta') = F(-m, \beta; \gamma; \eta'\eta) \qquad (\eta \in \mathbb{R}^{r \times r})$$

(see the discussion preceding (4.36)$'$). Property (4.46)(i) now follows immediately from (4.45). Similarly, the left $O(r)$-invariance of $\eta \rightarrow F(-m, \beta; \gamma; \eta\eta')$ together with (4.45) ensures that (4.46)(ii) holds also. ∎

To extend the R_m^{λ} to determinantally homogeneous polynomials on $\mathbb{R}^{r \times n}$, a spherical polar decomposition of z in $\mathbb{R}^{r \times n}$ is needed. Now zz' is a non-negative matrix in $\mathbb{R}^{r \times r}$ which is positive-definite if rank $z = r$, so in this last case z can be written uniquely as $z = |z|\omega_z$ with $|z| = (zz')^{1/2}$ and $\omega_z = (zz')^{-1/2}z$ an $(r \times n)$-matrix such that $\omega_z\omega_z' = I$, i.e., ω_z is an element of the Stiefel manifold $\Sigma_{r,n}$ of all r-tuples of orthonormal vectors in $\mathbb{R}^n$. The manifold $\Sigma_{r,n}$ will play the same role as the unit sphere Σ_{n-1} ($= \Sigma_{1,n}$) did in the previous section. Fix

$$\mathbf{1} =_r \begin{array}{cc} \scriptstyle n-r & \scriptstyle r \\ \left[\begin{array}{cc} 0 & \vdots & I_r \end{array} \right] \end{array} \qquad (r < n)$$

in $\Sigma_{r,n}$ and let M ($\sim O(n-r)$) be the subgroup of $O(n)$ fixing $\mathbf{1}$; then $O(n)$ acts transitively on $\Sigma_{r,n}$ and $\Sigma_{r,n} \sim O(n)/O(n-r)$. Hence each z in $\mathbb{R}^{r \times n}$ can be written uniquely in 'Cartesian coordinates' as

$(4.46)(i)$ $\qquad z = x + y\mathbf{1} \qquad (x \in \mathbb{R}^{r \times (n-r)} ,\ y \in \mathbb{R}^{r \times r})$,

while, if rank $z = r$, it can be written uniquely in 'spherical polar coordinates' as

$(4.46)(ii)$ $\qquad z = |z|\omega_z = |z|(\xi + \eta\mathbf{1}) \qquad (\xi \in \mathbb{R}^{r \times (n-r)} ,\ \eta \in \mathbb{R}^{r \times r})$

with $0 \leq \eta\eta' \leq I$. After all these preliminaries we finally come to the second of the fundamental results of this section, a direct generalization of (3.21).

(4.47) Theorem.

For each $\lambda \in \mathbb{C}$, $\Re\lambda > \frac{1}{2}(r - 2)$, the function

$$\phi_m^{(\lambda)}(z) = (\det |z|)^m R_m^\lambda(\omega_z \mathbf{1}') \qquad (z = |z|\omega_z)$$

is an M-invariant polynomial in $\mathcal{P}_m(\mathbb{R}^{r \times n})^{SL}$ such that $\phi_m^{(\lambda)}(\mathbf{1}) = 1$, $r < n$.

Proof. As defined, $\phi_m^{(\lambda)}$ is specified only on, say, the set Ω of z in $\mathbb{R}^{r \times n}$ having maximal rank. But Ω is Zariski-dense in $\mathbb{R}^{r \times n}$; on the other hand, Ω contains hzk for all h in $GL(r)$ and k in $O(n)$ when rank $z = r$, since rank$(hzk) = $ rank z. Thus $\phi_m^{(\lambda)}$ will extend uniquely to a polynomial in $\mathcal{P}_m(\mathbb{R}^{r \times n})^{SL}$ once it has been shown to be a polynomial on Ω which is determinantally homogeneous of degree m. But this is less obvious than it was in the case $r = 1$ because in general the spherical polar decomposition of hz, $h \in GL(r)$, cannot be described so simply in terms of the corresponding decomposition of z. Indeed, for z in Ω, $z = |z|\omega_z$ where $|z| = (zz')^{1/2}$ and $\omega_z = (zz')^{-1/2}z$, while

$$|hz| = (hzz'h')^{1/2} \quad , \quad \omega_{hz} = (hzz'h')^{-1/2}hz .$$

But then

$$\omega'_{zh}\omega_{zh} = z'h'(hzz'h')^{-1}hz = z'(zz')^{-1}z = \omega'_z\omega_z$$

since $(hzz'h')^{-1/2}$ is symmetric; in addition,

$$(\det |hz|)^{2m} = \det(hzz'h')^m = (\det h)^{2m}(\det |z|)^{2m} .$$

Consequently, by $(4.46)(i)$

$$\begin{aligned}
\phi_{2m}^\lambda(hz) &= (\det h)^{2m}(\det |z|)^{2m} R_{2m}^\lambda(\mathbf{1}\omega'_{hz}) \\
&= (\det h)^{2m}(\det |z|)^{2m} F(-m, m+\lambda; \tfrac{1}{2}r; \mathbf{1}\omega'_{hz}\omega_{hz}\mathbf{1}') \\
&= (\det h)^{2m}(\det |z|)^{2m} F(-m, m+\lambda; \tfrac{1}{2}r; \mathbf{1}\omega'_z\omega_z\mathbf{1}') .
\end{aligned}$$

Consequently,

$$(4.48) \qquad \phi_{2m}^{\lambda}(hz) = (\det h)^{2m}\phi_{2m}^{\lambda}(z) \qquad \left(h \in GL(r)\right) .$$

The corresponding result for ϕ_{2m+1}^{λ} is proved in exactly the same way.

To prove that ϕ_{2m}^{λ} is polynomial on Ω, observe first that it is a linear combination of terms $(\det|z|)^{2m}C_{\sigma}(\mathbf{1}\omega_z'\omega_z\mathbf{1}')$, and hence of terms

$$(\det zz')^m C_{\sigma}\left(\mathbf{1}z'(zz')^{-1}z\mathbf{1}'\right) \qquad (z \in \Omega)$$

where $\sigma = (\sigma_1, \ldots, \sigma_s, 0, \ldots)$ satisfies $(4.39)'$ with $\sigma_1 \leq m$. But on Ω

$$(zz')^{-1} = (\det zz')^{-1}Z$$

for some $Z = [Z_{jk}]$ in $\mathbb{R}^{r \times r}$, where each Z_{jk} is a polynomial in the entries of z. Hence

$$(\det|z|)^{2m}C_{\sigma}(\mathbf{1}\omega_z'\omega_z\mathbf{1}') = (\det zz')^{m-\sigma_1}C_{\sigma}(\mathbf{1}z'Zz\mathbf{1}') ,$$

and so ϕ_{2m}^{λ} is a polynomial on Ω, in fact a determinantally homogeneous polynomial of degree $2m$ on Ω because of (4.48). Once again the corresponding result for ϕ_{2m+1}^{λ} is proved in the same way. The M-invariance of $\phi_m^{(\lambda)}$ is clear because

$$|zk| = |z| \quad , \quad \omega_{zk}\mathbf{1}' = \omega_z k\mathbf{1}' = \omega_z\mathbf{1}' \qquad (k \in M) .$$

This completes the proof. ∎

Analogously to $(3.31)'$, set

$$(4.49) \qquad Z_m^{(\lambda)}(z, \zeta) = (\det|z|)^m (\det|\zeta|)^m R_m^{\lambda}(\omega_z\omega_{\zeta}') .$$

It is an $O(n)$-invariant polynomial on $\mathbb{R}^{r \times n} \times \mathbb{R}^{r \times n}$ which is determinantally homogeneous of degree m separately in z and ζ. Furthermore,

$$(4.50) \qquad \phi_m^{(\lambda)} = Z_m^{(\lambda)}(z, \mathbf{1}) \quad , \quad Z_m^{(\lambda)}(z, \zeta) = Z_m^{(\lambda)}(\zeta, z) ,$$

the second property being an immediate consequence of $(4.46)(i)$.

5 Harmonic polynomials of matrix argument

In this section we shall give an explicit realization of the irreducible unitary representation of $O(n)$, as well as $SO(n)$ and $\mathrm{Spin}(n)$, having signature $(\underbrace{m, \ldots, m}_{r}, 0, \ldots)$ on harmonic polynomials of matrix argument.

The representations having signature $(m_1, \ldots, m_r, 0, \ldots)$ can also be realized in this way, but a discussion of the more general case would take us too far afield.

Unlike the case of $\mathbb{R}^n$, there are several possible definitions of harmonicity on $\mathbb{R}^{r \times n}$, $r > 1$. Let

$$(5.1) \qquad \frac{\partial}{\partial\zeta}\frac{\partial'}{\partial\zeta} = [\Delta_{jk}]_{j,k=1}^r \qquad \left(\Delta_{jk} = \frac{\partial}{\partial\zeta_j} \cdot \frac{\partial}{\partial\zeta_k}\right) ,$$

be the $r \times r$ matrix of second-order Laplacians

$$(5.1)' \qquad \Delta_{jk} = \sum_\ell \frac{\partial^2}{\partial \zeta_{j\ell} \partial \zeta_{k\ell}} \qquad (1 \le j, k \le r)$$

obtained from the rows of $\frac{\partial}{\partial \zeta}$.

(5.2) Definition.
*A function f on $\mathbb{R}^{r \times n}$ is said to be $O(n)$-harmonic when $\Delta_{jk} f = 0$
for all j and k, i.e., when*

$$\left(\frac{\partial}{\partial \zeta} \frac{\partial'}{\partial \zeta} \right) f = 0 .$$

If f is $O(n)$-harmonic, then obviously

$$P(\Delta_{11}, \Delta_{12}, \ldots, \Delta_{rr}) f = 0$$

whenever P is a polynomial without constant term in the operators Δ_{jk}.
It is a well-known result in polynomial invariant theory (use (4.37) and
the proof of (4.11)) that such operators constitute all the $O(n)$-invariant
differential operators on $C^\infty(\mathbb{R}^{r \times n})$ annihilating constants, i.e., those
finite-order differential operators ∂ on $C^\infty(\mathbb{R}^{r \times n})$ for which $\partial 1 = 0$ and

$$(5.3) \qquad \partial(\rho(k)f) = \rho(k)(\partial f) \qquad (k \in O(n)) .$$

Hence the space $\mathcal{H}(\mathbb{R}^{r \times n})$ of all $O(n)$-harmonic polynomials is an $O(n)$-
submodule of $\mathcal{P}(\mathbb{R}^{r \times n})$ exactly as in the case $r = 1$ studied in section
3. This explains the significance of the term $O(n)$-harmonic. We shall
denote by $\mathcal{H}(\mathbb{R}^{r \times n})^{SL}$ the harmonic polynomials in $\mathcal{P}(\mathbb{R}^{r \times n})^{SL}$ and by
$\mathcal{H}_m(\mathbb{R}^{r \times n})^{SL}$ those which are determinantally homogeneous of degree
m. Clearly $\mathcal{H}(\mathbb{R}^{r \times n})^{SL}$ and $\mathcal{H}_m(\mathbb{R}^{r \times n})^{SL}$ are $O(n)$-modules, but it is
not immediately obvious that they contain many non-trivial elements.
Indeed, when $r = n$, the polynomials $f_0(z) \equiv 1$ and $f_1(z) = \det z$ are a
basis for $\mathcal{H}(\mathbb{R}^{r \times r})^{SL}$. In the case $2r \le n$, however, every $O(n)$-module
$\mathcal{H}_m(\mathbb{R}^{r \times n})^{SL}$ is non-trivial.

Note first that $z \to \det(zZ')^m$ defines a complex-valued polynomial
in $\mathcal{P}_m(\mathbb{R}^{r \times n})^{SL}$ for each fixed Z in $\mathbb{C}^{r \times n}$.

(5.4) Theorem.
The polynomial $z \to \det(zZ')^m$ is $O(n)$-harmonic provided $ZZ' = 0$.

To construct specific examples of such Z in $\mathbb{C}^{r \times n}$ when $2r \le n$, set

$$(5.5) \qquad Z_0 = {}_r \begin{matrix} \overset{r}{} \quad \overset{n-2r}{} \quad \overset{r}{} \\ \left[I_r \; \vdots \; 0 \; \vdots \; iJ_r \right] \end{matrix}$$

where

$$(5.5)' \qquad I_r = \begin{bmatrix} 1 & & 0 \\ & \ddots & \\ 0 & & 1 \end{bmatrix} \quad , \quad J_r = \begin{bmatrix} 0 & & 1 \\ & \ddots & \\ 1 & & 0 \end{bmatrix} .$$

Then $Z_0 Z_0' = 0$; in fact, $ZZ' = 0$ whenever $Z = Z_0 k$, $k \in O(n)$.

Proof of (5.4). Let $P = P(x)$ be any polynomial on $\mathbb{R}^{r \times r}$. Then a routine calculation using the chain rule shows that

$$\Delta_{jk} : P(zZ') \longrightarrow \sum_{\ell, m=1}^{r} (ZZ')_{\ell m} \frac{\partial^2 P}{\partial x_{j\ell} \partial x_{km}} (zZ')$$

where $[(ZZ')_{\ell m}]_{\ell, m=1}^{r}$ is the $(r \times r)$-matrix ZZ'. Thus $z \to P(zZ')$ is $O(n)$-harmonic whenever $ZZ' = 0$, and the theorem follows immediately, taking $P(x) = (\det x)^m$. ∎

Since the real and imaginary parts of $z \to \det(zZ')^m$ are separately $O(n)$-harmonic for each fixed Z in $\mathbb{C}^{r \times n}$ satisfying $ZZ' = 0$, it is clear that $\mathcal{H}_m(\mathbb{R}^{r \times n})^{SL}$ contains many non-trivial functions whether it consists of real- or complex-valued polynomials, provided of course that $2r \leq n$. This *restriction will be imposed on r, n throughout the remainder of this section.*

An alternative definition of harmonicity on $\mathbb{R}^{r \times n}$ comes from regarding a function on $\mathbb{R}^{r \times n}$ simply as a function on rn-dimensional Euclidean space. We might say (the terminology is not standard) that a function f on $\mathbb{R}^{r \times n}$ is *trace-harmonic* when

$$(5.6) \qquad \mathrm{tr} \left(\frac{\partial}{\partial \zeta} \frac{\partial'}{\partial \zeta} \right) f = 0 \qquad (f \in C^\infty(\mathbb{R}^{r \times n})) \ .$$

Since

$$\mathrm{tr} \left(\frac{\partial}{\partial \zeta} \frac{\partial'}{\partial \zeta} \right) = \sum_j \Delta_{jj} = \sum_{j,k} \left(\frac{\partial}{\partial \zeta_{jk}} \right)^2 ,$$

a function f will be trace-harmonic on $\mathbb{R}^{r \times n}$ if and only if it is harmonic in the usual sense on $\mathbb{R}^{rn}$. Lemma (3.12) thus applies to trace-harmonic polynomials. But every $O(n)$-harmonic function is certainly trace-harmonic, and so the analogue

$$(5.7) \qquad (P, Q) = \int_{\mathbb{R}^{r \times n}} P(z) \overline{Q(z)} \, e^{-\frac{1}{2} \mathrm{tr}(zz')} \, dz$$

of (3.11) defines a positive-definite inner product on $\mathcal{H}(\mathbb{R}^{r \times n})$ related to the Fischer inner product by

$$(5.7)' \qquad \int_{\mathbb{R}^{r \times n}} P(z) \overline{Q(z)} \, e^{-\frac{1}{2} \mathrm{tr}(zz')} \, dz = 2\pi^{\frac{1}{2}rn} P\left(\frac{\partial}{\partial z} \right) \overline{Q(z)} \Big|_{z=0}$$

The unitarity

(5.8) $\left(\rho(k)P, \rho(k)Q\right) = (P,Q) \qquad (k \in O(n))$

of the representation ρ of $O(n)$ with respect to this inner product (5.7) on $\mathcal{H}(\mathbb{R}^{r \times n})$ is clear. Thus the scene is set for a development of the theory of $O(n)$-harmonic polynomials of matrix argument parallel to the theory for harmonic polynomials on $\mathbb{R}^n$.

On $\mathcal{H}_m(\mathbb{R}^{r \times n})^{SL}$ the distinction between $O(n)$-harmonicity and trace-harmonicity vanishes, surprisingly enough. In fact, even more is true.

(5.8) Theorem.

   An $SL(r)$-invariant function is $O(n)$-harmonic if and only if it is trace-harmonic.

Proof. It is enough to show that $\Delta_{jk}f = 0$ for all j, k provided $(\sum_j \Delta_{jj})f = 0$. Now, for any f in $C^\infty(\mathbb{R}^{r \times n})$ and any h in $SL(r)$,

$$\mathrm{tr}\left(h' \frac{\partial}{\partial \zeta} \frac{\partial'}{\partial \zeta} h\right) : f(h^{-1}\zeta) \longrightarrow \left(\mathrm{tr}\left(\frac{\partial}{\partial \zeta} \frac{\partial'}{\partial \zeta}\right) f\right)(h^{-1}\zeta) ,$$

as a proof analogous to that for (4.11) shows. But, if f is $SL(r)$-invariant, then $f(h^{-1}\zeta) = f(\zeta)$. Thus

(5.9) $\left(\mathrm{tr}\left(h' \frac{\partial}{\partial \zeta} \frac{\partial'}{\partial \zeta} h\right)\right) f(\zeta) = 0 \qquad \left(h \in SL(r)\right)$

when f in addition is trace-harmonic. Set

$$h = \exp(tE_{jk}) = I + tE_{jk} \qquad (j \neq k)$$

For such h, (5.9) becomes

$$(t^2 \Delta_{jj} + 2t\Delta_{jk} + \cdots)f = 0 ,$$

the omitted terms being independent of t. Thus $\Delta_{jj}f = 0 = \Delta_{jk}f$ for all j, k whenever f is an $SL(r)$-invariant which is trace-harmonic. This ensures that f is also $O(n)$-harmonic. ∎

We establish first the analogue of (3.13).

(5.10) Theorem.

   The orthogonal decomposition of $\mathcal{H}(\mathbb{R}^{r \times n})^{SL}$ into its irreducible $O(n)$-submodules is given by

(5.10)′ $\mathcal{H}(\mathbb{R}^{r \times n})^{SL} = \bigoplus \sum_{m=0}^{\infty} \mathcal{H}_m(\mathbb{R}^{r \times n})^{SL} ;$

furthermore, $(\mathcal{H}_m(\mathbb{R}^{r \times n})^{SL}, \rho)$ *is a realization of the irreducible unitary representation of* $SO(n)$ *having signature* $\tau_m = (\underbrace{m, \ldots, m}_{r}, 0, \ldots)$.

By much the same proof as used for (5.4), it is easy to see that $z \to Q(z \frac{\partial'}{\partial \zeta}) f \big|_{\zeta=0}$ is an $O(n)$-harmonic polynomial for any Q in $\mathcal{P}(\mathbb{R}^{r \times r})$ whenever f is $O(n)$-harmonic. Thus the Taylor series expansion

$$(5.10)'' \qquad P(z) = \sum_{m=0}^{\infty} \frac{1}{\tau_m!} \det \left(z \frac{\partial'}{\partial \zeta} \right)^m P \big|_{\zeta=0} \qquad (z \in \mathbb{R}^{r \times n})$$

(see (4.34)) exhibits the decomposition (5.10)' in a completely explicit way.

Proof of (5.10). The orthogonality of the different spaces $\mathcal{H}_m(\mathbb{R}^{r \times n})^{SL}$ follows just as in the case $r = 1$ (see (3.13)); proof of their irreducibility will be omitted, as it was in the case (3.13) too. To identify the signature of $\mathcal{H}_m(\mathbb{R}^{r \times n})^{SL}$ consider the function

$$\Phi_m(z) = (\det z Z_0')^m = \left(\det(s + it) \right)^m \qquad \left(z = \begin{bmatrix} \overset{r}{s} & \overset{n-2r}{\vdots} & \cdots & \overset{r}{\vdots} & t \end{bmatrix} \right)$$

defined using (5.5). Then

$$\rho(d)\Phi_m = e^{im(\theta_1 + \theta_2 + \cdots)} \Phi_m$$

for each

$$d = \begin{bmatrix} \cos\theta_1 & 0 & \cdots & 0 & \sin\theta_1 \\ 0 & \cos\theta_2 & \cdots & \sin\theta_2 & 0 \\ \vdots & \vdots & & \vdots & \vdots \\ 0 & -\sin\theta_2 & \cdots & \cos\theta_2 & 0 \\ -\sin\theta_1 & 0 & \cdots & 0 & \cos\theta_1 \end{bmatrix}$$

in $SO(n)$. Hence Φ_m is a highest weight vector in $\mathcal{H}_m(\mathbb{R}^{r \times n})^{SL}$ having weight $(\underbrace{m, \ldots, m}_{r}, 0, \ldots)$. This completes the proof because $\rho(k)f$ is real-valued whenever f is real-valued. ∎

We come now to one of the principal results of this section characterizing the M-invariant elements of $\mathcal{H}_m(\mathbb{R}^{r \times n})^{SL}$ where as before M is the subgroup of $O(n)$ fixing the 'North Pole' $\mathbf{1}$ in the Stiefel manifold $\Sigma_{r,n}$ in $\mathbb{R}^{r \times n}$. Recall the definition

$$(5.11) \qquad \phi_m^{(\lambda)}(z) = (\det |z|)^m R_m^{\lambda}(\omega_z \mathbf{1}') \qquad (z = |z|\omega_z)$$

of the polynomial in $\mathcal{P}_m(\mathbb{R}^{r \times n})^{SL}$ associated with the normalized Gegen-bauer polynomial R_m^{λ} of matrix argument (see (4.45)).

(5.12) Theorem.
  (Class 1 property) *When* $\lambda = \frac{1}{2}(n - r - 1)$, $\phi_m^{(\lambda)}$ *is the unique M-invariant polynomial in* $\mathcal{H}_m(\mathbb{R}^{r \times n})^{SL}$ *such that* $\phi_m^{(\lambda)}(\mathbf{1}) = 1$; *further-more,* $\phi_m^{(\lambda)}$ *can be represented*

  (i) *as the average*

$$\phi_m^{(\lambda)}(z) = \int_M P(xk + y\mathbf{1})\, dk \qquad (z = x + y\mathbf{1})$$

of any P in $\mathcal{H}_m(\mathbb{R}^{r \times n})^{SL}$ *for which* $P(\mathbf{1}) = 1$,

  (ii) *as the determinantal derivative*

$$\phi_m^{(\lambda)}(z) = \gamma_m^{\lambda} \det \left(z \frac{\partial'}{\partial \zeta} \right)^m \left(\frac{1}{\det |\zeta|^{2\lambda}} \right) \Big|_{\zeta = \mathbf{1}}$$

of the $O(n)$*-harmonic function* $\zeta \to \det |\zeta|^{-2\lambda}$, *where*

$$\gamma_{mr}^{\lambda} = (-1)^{rm} \prod_{j=1}^{r} \frac{\Gamma(2\lambda + m - (r - j))}{\Gamma(2\lambda - (r - j))} .$$

Now almost all $\omega = \xi + \eta\mathbf{1}$ in the Stiefel manifold $S_{r,n}$ in $\mathbb{R}^{r \times n}$ can be written as

$$\omega = (1 - |\eta|^2)^{1/2}\omega_0 + \eta\mathbf{1} \qquad (\omega_0 \in \mathbb{R}^{r \times (n-r)})$$

with ω_0 in the Stiefel manifold

$$S_{r,n-r} = \left\{ \omega_0 \in \mathbb{R}^{r \times (n-r)} : \omega_0 \omega_0' = I_r \right\} .$$

In view of (5.11) and (5.12)(i) therefore, the Gegenbauer polynomial R_m^{λ}, $\lambda = \frac{1}{2}(n - r - 1)$, admits the integral representation

$$(5.13) \qquad R_m^{\lambda}(\eta) = \int_{S_{r,n-r}} f\left((1 - |\eta|^2)^{1/2}\omega_0 + \eta\mathbf{1}\right) d\omega_0 \qquad (|\eta| < 1)$$

where f is any polynomial in $\mathcal{H}_m(\mathbb{R}^{r \times n})^{SL}$ such that $f(\mathbf{1}) = 1$ and $d\omega_0$ is normalized rotation-invariant measure on $S_{r,n-r}$. Property (5.13), of course, reduces to the Laplace type representation (3.33) for the usual Gegenbauer polynomial C_m^{λ} when $r = 1$.

Proof of (5.12). Set

$$\Phi_P(z) = \int_M P(xk + y\mathbf{1})\, dk \qquad (z = x + y\mathbf{1})$$

for any P in $\mathcal{H}_m(\mathbb{R}^{r \times n})^{SL}$ with $P(\mathbf{1}) = 1$. Clearly Φ_P is an M-invariant polynomial in $\mathcal{H}_m(\mathbb{R}^{r \times n})^{SL}$ such that $\Phi_P(\mathbf{1}) = 1$. Hence $O(n)$-harmonic normalized M-invariants exist. On the other hand, $\phi_m^{(\lambda)}$ is a normalized M-invariant in $\mathcal{P}_m(\mathbb{R}^{r \times n})^{SL}$ virtually by construction. Hence, if we can show that $\phi_m^{(\lambda)}$ is $O(n)$-harmonic and that M-invariants in $\mathcal{H}_m(\mathbb{R}^{r \times n})^{SL}$ are unique up to multiplicative constants (the class 1 property), then the first part of the theorem, as well as (5.12)(i), will have been established. Unfortunately, neither result seems to be easy to prove, and both require considerable technical machinery. For these reasons we shall merely sketch their proof (see Notes and remarks for references to detailed proofs).

(a) *Uniqueness.* By Frobenius reciprocity the dimension of the space of M-invariants in $\mathcal{H}_m(\mathbb{R}^{r \times n})^{SL}$ is equal to the number of times the representation of $O(n)$ having signature $(\underbrace{m, \ldots, m}_{r}, 0, \ldots)$ occurs in the decomposition of $\mathcal{H}(\mathbb{R}^{r \times n})$ into its irreducible M-submodules. This is a general result which in the case of $\mathcal{H}(\mathbb{R}^n)$ is made explicit by (3.31). But, for more general r, a result of Gelbart, often referred to as an 'Act of Providence,' shows that the number of occurrences coincides with the dimension of the irreducible representation of $GL(r)$ having signature $(\underbrace{m, \ldots, m}_{r})$. However, this is just the representation $x \to (\det x)^m$ of $GL(r)$, which is one-dimensional. Hence the space of M-invariants in $\mathcal{H}_m(\mathbb{R}^{r \times n})^{SL}$ is one-dimensional, establishing uniqueness.

(b) *Harmonicity.* Herz establishes (5.13) by a very circuitous route independently of theorem (5.12). His proof uses the theory of Bessel functions of matrix argument, i.e., on $\mathbb{R}^{r \times r}$, together with

(i) an extension of Bochner's theorem expressing Fourier transforms of radial functions as Hankel Transforms,

(ii) an extension of Gegenbauer's own generalization of the Poisson integral representation for ordinary Bessel functions, i.e., Bessel functions on $\mathbb{R}^{1 \times 1}$.

Granted (5.13), the representation (5.12) for $\phi_m^{(\lambda)}$ follows immediately from definition (5.11).

Finally, using the uniqueness of M-invariants we can establish (5.12)(ii). Now a straightforward calculation shows that the function $\zeta \to (\det |\zeta|)^{-(n-r-1)}$ is $O(n)$-harmonic wherever it is defined. But then, as in the proof of (5.10)″, the function

$$z \longrightarrow Q\left(z \frac{\partial'}{\partial \zeta}\right) (\det |\zeta|)^{-(n-r-1)}$$

is $O(n)$-harmonic in z for every Q in $\mathcal{P}(\mathbb{R}^{r \times r})$; in particular, as a function of z,

$$F(z, \zeta) = \det \left(z \frac{\partial'}{\partial \zeta} \right)^m (\det |\zeta|)^{-2\lambda} \qquad (\lambda = \tfrac{1}{2}(n - r - 1))$$

is an $O(n)$-harmonic polynomial in $\mathcal{H}_m(\mathbb{R}^{r \times n})^{SL}$ for each ζ for which $(\det |\zeta|)^{-2\lambda}$ is defined. On the other hand, since $|\zeta|$ is $O(n)$-invariant, property (4.21)(ii) ensures that

$$F(zk, \zeta k) = F(z, \zeta) \qquad (k \in O(n)) \,,$$

i.e., F is an $O(n)$-invariant. Hence $F(z, \mathbf{1})$ is an M-invariant in $\mathcal{H}_m(\mathbb{R}^{r \times n})^{SL}$, and so by uniqueness

$$\det \left(z \frac{\partial'}{\partial \zeta} \right)^m (\det |\zeta|)^{-2\lambda} \bigg|_{\zeta = \mathbf{1}} = \text{const.} \; \phi_m^{(\lambda)}(z)$$

when $\lambda = \tfrac{1}{2}(n - r - 1)$. As $\phi_m^{(\lambda)}(\mathbf{1}) = 1$, this constant is just the numerical value of

$$\det \left(z \frac{\partial'}{\partial \zeta} \right)^m (\det |\zeta|)^{-2\lambda} \qquad (z = \zeta = \mathbf{1}) \,,$$

and by the Cauchy–Binet identities in (4.24), (4.24)′, this last numerical value is just the value of

$$(5.14) \qquad \Delta_{1 \cdots r} \left(\frac{\partial}{\partial \zeta} \right)^m \left(\frac{1}{\Delta_{1 \cdots r}(\zeta)} \right)^{n - r - 1} \qquad (\zeta = \mathbf{1}) \,.$$

But Cayley's identity (4.25) is valid for all $m \in \mathbb{R}$. A simple calculation now shows that (5.14) is just

$$(-1)^{rm} \prod_{j=1}^{r} \frac{\Gamma(2\lambda + m - (r - j))}{\Gamma(2\lambda - (r - j))} \,,$$

which is the indicated value of γ_{mr}^{λ}. This proves theorem (5.12) (modulo some non-trivial technical details). ∎

Finally, we can use $Z_m^{(\lambda)}$ as a reproducing kernel for $\mathcal{H}_m(\mathbb{R}^{r \times n})^{SL}$ just as in (3.32). Indeed, exactly the same proof as for (3.32) shows that

$$(5.15) \qquad \int_{\mathbb{R}^{r \times n}} Z_m^{(\lambda)}(z, \zeta) \overline{P(\zeta)} e^{-\frac{1}{2} \text{tr}(\zeta \zeta')} \, d\zeta = (\phi_m^{(\lambda)}, \phi_m^{(\lambda)}) P(z)$$

for all P in $\mathcal{H}_m(\mathbb{R}^{r \times n})^{SL}$, where now

$$(\phi_m^{(\lambda)}, \phi_m^{(\lambda)}) = \int_{\mathbb{R}^{r \times n}} |\phi_m^{(\lambda)}(z)|^2 e^{-\frac{1}{2} \text{tr}(zz')} \, dz \,.$$

Notes and remarks for chapter 3

1. The book by Brocker and tom Dieck ([19]) gives a very readable account of the general results needed for this chapter. The lecture notes by Coifman and Weiss ([25]) should also be consulted.

2. Brief accounts of the material in this section can be found in [19] and [61].

3. This material is completely standard. Coifman and Weiss ([25]) give a particularly good illustration of the analytic consequences of the class 1 property. Stein and Weiss ([96]) establish the relation between spherical harmonics and the Euclidean Fourier transform, and Stein ([94]) considered also the connections to singular integrals. All three accounts are extremely readable.

4. Algebras of polynomial functions have been studied by many authors from equally many points of view. The papers of Gelbart ([42]), Herz ([58]), Maas ([77]), and Ton-That ([99]) are particularly relevant. Kostant's paper ([72]) and Helgason's book ([57]) should also be consulted. Invariant theory has been widely discussed. Weyl's book ([103]) is a standard work, if difficult to penetrate; Procesi's lecture notes ([89]) are more accessible, and Turnbull's book [100] gives a classical viewpoint, but Howe's widely circulated and recently published paper ([59]) makes the most fascinating reading of all. For hypergeometric functions of matrix argument the papers of Gross–Richards ([53]), Herz ([58]), James ([62]), Koornwinder ([69]), Macdonald ([78]), are important; they also contain other relevant references. The presentation of this section has some novelty; many of the ideas are drawn from [46].

5. The papers of Gelbart, Herz, Kostant, Maas and Ton-That cited earlier are basic. It would be especially informative to have a more direct proof of theorem (5.12).

4

Constant coefficient operators of Dirac type

In this chapter we study a special class of first-order linear elliptic systems of partial differential equations having constant coefficients. These are the systems obtained by considering the kernel of an *operator of Dirac type*. An operator of Dirac type is essentially the restriction of the standard Euclidean Dirac operator to a fixed subspace of the Clifford module on which it acts. There are two main methods of selecting the subspace, and thereby specifying the operator. The first of these is the method of involutions, giving rise to a important class of operators, the *graded Dirac operators*. The second method produces rotation-invariant differential operators by selecting subspaces which are spin submodules of the original Clifford module.

Because of the restriction imposed on the range space, functions in the kernel of a Dirac type operator are better behaved than Clifford analytic functions as a class. In particular, they often satisfy a stronger subharmonicity property, and consequently exhibit better boundary regularity than do Clifford analytic functions in general. Thus there are clear implications for the theory of Hardy spaces and boundary value problems for operators of Dirac type.

1 First-order systems: some general results

Let $\mathcal{A}, \mathcal{B}$ be finite-dimensional Hilbert spaces, both over the same field $\mathbb{F} = \mathbb{R}$ or $\mathbb{C}$, and let $\mathcal{A} \otimes \mathbb{F}^n$ be the usual Hilbert space tensor product.

If $\{e_j\}_{j=1}^n$ is the standard basis for $\mathbf{R}^n$, this tensor product consists of the elements

$$(1.1) \qquad \xi = \sum_{j=1}^{n} \xi_j \otimes e_j , \qquad \xi_j \in \mathcal{A}$$

having Hilbert–Schmidt inner product and norm

$$(1.2) \qquad (\xi, \eta) = \sum_{j=1}^{n} (\xi_j, \eta_j)_{\mathcal{A}} , \qquad \|\xi\| = \left(\sum_{j=1}^{n} |\xi_j|_{\mathcal{A}}^2 \right)^{1/2}$$

respectively. To each linear operator $A : \mathcal{A} \otimes \mathbf{F}^n \to \mathcal{B}$ there corresponds an n-tuple $(A_1, \ldots, A_n)$ of linear operators

$$(1.3) \qquad A_j : \mathcal{A} \to \mathcal{B} , \qquad A_j(a) = A(a \otimes e_j) \qquad (1 \le j \le n) .$$

With this notation, we make the following definition.

(1.4) Definition.
  If Ω is a domain in $\mathbf{R}^n$ and $F \in \mathcal{C}^\infty(\Omega, \mathcal{A})$, let

$$\eth_A F = (A \circ \nabla) F = \sum_{j=1}^{n} A_j \frac{\partial F}{\partial x_j}$$

be the differential operator obtained by composing the usual derivative

$$\nabla : \mathcal{C}^\infty(\Omega, \mathcal{A}) \to \mathcal{C}^\infty(\Omega, \mathcal{A} \otimes \mathbf{F}^n) , \qquad \nabla F = \sum_{j=1}^{n} \frac{\partial F}{\partial x_j} \otimes e_j$$

of $\mathcal{A}$-valued functions with a linear operator $A : \mathcal{A} \otimes \mathbf{F}^n \to \mathcal{B}$.

Each first-order linear homogeneous system of constant-coefficient differential operators defined on $\mathcal{C}^\infty(\Omega, \mathcal{A})$ may be expressed in the form $\eth_A F = 0$ for some choice of $\mathcal{B}$ and linear operator $A : \mathcal{A} \otimes \mathbf{F}^n \to \mathcal{B}$. For example, the Dirac operator D (defined in chapter 2, (3.1)) may be written as $\eth_A$ where $\mathcal{A}$ and $\mathcal{B}$ are both taken to be the same $\mathfrak{A}_n$-module $\mathfrak{H}$, and A_j is multiplication by e_j as an imaginary unit in $\mathcal{L}(\mathfrak{H})$, $1 \le j \le n$. More generally, let $\mathfrak{B}$ be a subspace of $\mathfrak{H}$ and set $\mathcal{A} = \mathfrak{B}$, but retain the same $\mathcal{B} = \mathfrak{H}$ and $A_j = e_j$ as before. In this case $\eth_A$ is just the restriction $D : \mathcal{C}^\infty(\Omega, \mathfrak{B}) \to \mathcal{C}^\infty(\Omega, \mathfrak{H})$ of the standard Dirac operator to $\mathfrak{B}$-valued functions.

To understand the algebraic structure of $\eth_A$, we need to know how the properties of $A : \mathcal{A} \otimes \mathbf{F}^n \to \mathcal{B}$ determine those of $\eth_A : \mathcal{C}^\infty(\Omega, \mathcal{A}) \to \mathcal{C}^\infty(\Omega, \mathcal{B})$. Let $\eth_B : \mathcal{C}^\infty(\Omega, \mathcal{A}) \to \mathcal{C}^\infty(\Omega, \mathcal{B}')$ be the differential operator determined by a linear operator B mapping $\mathcal{A} \otimes \mathbf{F}^n$ into a (possibly different) range space $\mathcal{B}'$. We shall say that $\eth_A$ is *equivalent* to $\eth_B$ if they have the same kernel, i.e., $\eth_A F = 0$ if and only if $\eth_B F = 0$.

Now the condition $\delta_A F(\omega) = 0$ is essentially a condition on the first-order Taylor coefficients of the function F at $\omega \in \Omega$. From this it is not difficult to see that δ_A and δ_B are equivalent if and only if the linear operators A and B have the same kernel. Indeed,

$$(1.5) \qquad \xi = \sum_{j=1}^{n} \xi_j \otimes e_j \longrightarrow P_\xi(x) = \sum_{j=1}^{n} x_j \xi_j \qquad (x \in \mathbb{R}^n)$$

identifies the tensor product $\mathcal{A} \otimes \mathbb{F}^n$ with the space of degree 1 homogeneous, $\mathcal{A}$-valued polynomials in $\mathcal{C}^\infty(\mathbb{R}^n, \mathcal{A})$. The subspace

$$(1.6) \qquad \mathcal{V}_A = \{\xi \in \mathcal{A} \otimes \mathbb{F}^n : \delta_A(P_\xi) = 0\}$$

is just the kernel of A; hence if δ_A is equivalent to δ_B then $\mathcal{V}_A = \mathcal{V}_B$, i.e., $\ker A = \ker B$. Conversely, suppose that $\ker A = \ker B$, i.e., $\mathcal{V}_A = \mathcal{V}_B$. Any F in $\mathcal{C}^\infty(\Omega, \mathcal{A})$ has a Taylor expansion of the form

$$(1.7) \quad F(x) = F(\omega) + \sum_{j=1}^{n} (x_j - \omega_j)\xi_j + o(|x - \omega|) \qquad (\xi_j \in \mathcal{A}, \, 1 \le j \le n)$$

near $\omega \in \Omega$. From this it easily follows that $\delta_A F(\omega) = 0$ if and only if $\partial_A(P_\xi) = 0$, and similarly $\delta_B F(\omega) = 0$ if and only if $\delta_B(P_\xi) = 0$. Hence $\mathcal{V}_A = \mathcal{V}_B$ implies the equivalence of δ_A and δ_B.

Furthermore, if $\mathcal{V}$ is any subspace of $\mathcal{A} \otimes \mathbb{F}^n$, and if $A = A_\mathcal{V}$ is the orthogonal projection of $\mathcal{A} \otimes \mathbb{F}^n$ onto the orthogonal complement $\mathcal{V}^\perp$ of $\mathcal{V}$ in $\mathcal{A} \otimes \mathbb{F}^n$, then $\mathcal{V} = \mathcal{V}_A$. To summarize, we have the following.

(1.8) Theorem.
The first-order differential operators δ_A and δ_B on $\mathcal{C}^\infty(\Omega, \mathcal{A})$ are equivalent if and only if A, B have the same kernel in $\mathcal{A} \otimes \mathbb{F}^n$. Moreover, there is a one-to-one correspondence between the subspaces of $\mathcal{A} \otimes \mathbb{F}^n$ and the equivalence classes of all first-order linear homogeneous systems of constant coefficient differential operators on $\mathcal{C}^\infty(\mathbb{R}^n, \mathcal{A})$, namely, that induced by the map $\mathcal{V} \to \partial_{A_\mathcal{V}}$.

The mapping

$$(1.9) \qquad A(\lambda) : a \longrightarrow A(a \otimes \lambda) = \sum_{j=1}^{n} \lambda_j A_j a \qquad (a \in \mathcal{A})$$

defined from $\mathcal{A}$ to $\mathcal{B}$ for each $\lambda = (\lambda_1, \ldots, \lambda_n) \in \mathbb{R}^n$ is just the *symbol mapping* of δ_A. We say that the operator δ_A is *injectively elliptic* when $A(\lambda)$ is injective for all non-zero λ in $\mathbb{R}^n$; but if $A(\lambda)$ is both injective and surjective for all non-zero λ, it is said to be *elliptic*. Ellipticity clearly requires $\dim_\mathbb{F} \mathcal{A} = \dim_\mathbb{F} \mathcal{B}$, whereas injective ellipticity requires only that $\dim_\mathbb{F} \mathcal{A} \le \dim_\mathbb{F} \mathcal{B}$; if $\dim_\mathbb{F} \mathcal{A} < \dim_\mathbb{F} \mathcal{B}$, the operator δ_A is said to be *algebraically over-determined*. Now ellipticity imposes a lower bound on the allowable values of $k = \dim_{\mathbf{F}} \mathcal{A}$. This lower bound is a corollary of the celebrated theorem of J.F. Adams on the maximum number of linearly independent unit tangent vector fields on the sphere in $\mathbf{R}^k$, and is expressed in terms of the so-called *Radon–Hurwitz numbers*. Set $k = (2a + 1)16^d 2^c$, where a, c, d are nonnegative integers, $c \leq 3$; this is a unique decomposition of k. The real and complex Radon–Hurwitz numbers are then given by

$$\rho_{\mathbf{R}}(k) = 2^c + 8d \quad , \quad \rho_{\mathbf{C}}(k) = 8d + 2c + 2 ,$$

and from Adams' theorem it follows easily that if $A(\lambda)$ is bijective for all non-zero λ in $\mathbf{R}^n$, then $\rho_{\mathbf{F}}(k) \geq n$. Note that $\rho_{\mathbf{F}}(k) \leq 2$, if k is odd, so ellipticity forces $\dim_{\mathbf{F}} \mathcal{A}$ to be even when $n \geq 3$. If we write $n = 8m + r + 1$, $0 \leq r \leq 7$, then straightforward calculation yields the following minimum values of k:

(1.10)(i) $\dim_{\mathbf{C}} \mathcal{A} \geq 16^m 2^{[r/2]}$;

(1.10)(ii) $\dim_{\mathbf{R}} \mathcal{A} \geq \begin{cases} 16^m 2^{[r/2]} , & r = 0, 6, 7 , \\ 16^m 2^{[r/2]+1} , & r = 1, 2, 3, 4, 5 . \end{cases}$

The minimum value given in (1.10)(i) coincides with the dimension of the complex spinor space $\mathfrak{S}_{n-1}$, while the minimum value in (1.10)(ii) is precisely the dimension of the real spinor space $\mathfrak{R}_{n-1}$ (see chapter 1, section 7). The significance of this will be brought out in a moment. But first, to conclude the discussion of general results and to prepare for later sections, let us consider some algebraic ideas associated with the Cauchy–Riemann $(\partial, \bar{\partial})$-operators on $\mathbf{R}^2$. Every solution of $\partial F = 0$ or of $\bar{\partial} F = 0$ is automatically harmonic, a property shared by the standard Dirac operator on $\mathbf{R}^n$ too. This prompts the following definition.

(1.11) Definition.

The differential operator $\eth_A$ is said to be a Generalized Cauchy–Riemann (GCR) operator when all components of solutions of $\eth_A F = 0$ are harmonic.

Every GCR operator is injectively elliptic. For if $\eth_A$ is not injectively elliptic, there exist non-zero λ in $\mathbf{R}^n$ and ξ in $\mathcal{A}$ such that $A(\lambda)\xi = 0$. In turn, this ensures that

$$F(x) = e^{x \cdot \lambda} \xi \qquad (x \in \mathbf{R}^n)$$

defines a function in $C^\infty(\mathbf{R}^n, \mathcal{A})$ satisfying

$$\eth_A F(x) = e^{x \cdot \lambda} A(\lambda)\xi = 0 \quad , \quad \Delta F = |\lambda|^2 F \neq 0 ,$$

in violation of the GCR property. The GCR operators also have subharmonicity properties just as $\partial, \bar{\partial}$ did (see chapter 2, (1.16)): there exists $p = p_A < 1$ such that for all $p > p_A$ and for all $F \in C^\infty(\Omega, \mathcal{A})$ with $\eth_A F = 0$, the function $|F|_A^p$ is subharmonic in Ω. This subharmonicity property is at the heart of Hardy space theory, as we have seen already in Chapter 2; we shall discuss it further in section 5 of the present chapter.

In chapter 2 we considered the standard Euclidean Dirac operator D and its adjoint $\overline{D} = -D$, acting on $\mathfrak{H}$-valued functions on Ω, where $\mathfrak{H}$ is an $\mathfrak{A}_n$-module and Ω is any domain in $\mathbb{R}^n$. Now in the case $n = 2$, the prototypical $\mathfrak{A}_2$-modules are $\mathbb{C}^2$, as a realization of $\mathfrak{S}_2$, and $\mathbb{H}$, as a realization of $\mathfrak{A}_2$ or $\mathfrak{R}_2$; and, as we saw in section 2 of that chapter, ∂ and $\bar{\partial}$ were obtained by restricting D to functions taking values in subspaces of $\mathbb{H}$ isomorphic to $\mathbb{C}$. The same result would have been obtained, taking the subspaces

$$\left\{ \begin{bmatrix} z \\ 0 \end{bmatrix} : z \in \mathbb{C} \right\} \quad , \quad \left\{ \begin{bmatrix} 0 \\ z \end{bmatrix} : z \in \mathbb{C} \right\}$$

of $\mathbb{C}^2$. But $\mathbb{C}$ is the prototypical $\mathfrak{A}_1$-module. This suggests a somewhat more direct generalization of the Cauchy–Riemann operators

$$\bar{\partial} = \frac{\partial}{\partial x_0} + i \frac{\partial}{\partial x_1} \quad , \quad \partial = \frac{\partial}{\partial x_0} - i \frac{\partial}{\partial x_1} \quad ((x_0, x_1) \in \mathbb{R}^2) .$$

Let $\mathfrak{H}$ be an $\mathfrak{A}_{n-1}$-module and let $e_1, \ldots, e_{n-1}$ be skew-adjoint operators on $\mathfrak{H}$ having the Clifford property; the identity operator on $\mathfrak{H}$ will be denoted variously by I and e_0.

(1.12) Definition.
The Dirac $\mathcal{D}$-operator and its adjoint $\overline{\mathcal{D}}$ are the first-order operators on $C^\infty(\Omega, \mathfrak{H})$ defined by

$$\mathcal{D}F := \sum_{j=0}^{n-1} e_j \frac{\partial F}{\partial x_j} \quad , \quad \overline{\mathcal{D}}F = \sum_{j=0}^{n-1} \bar{e}_j \frac{\partial F}{\partial x_j} \quad \left(x = (x_0, \ldots, x_{n-1}) \right)$$

for any open set Ω in $\mathbb{R}^n$.

When $\mathfrak{H} = \mathbb{C}$ and e_1 is scalar multiplication by $i = \sqrt{-1}$, then $\mathcal{D}$ and $\overline{\mathcal{D}}$ are just the classical Cauchy–Riemann operators on complex-valued functions on $\mathbb{R}^2 \sim \mathbb{C}$. When $\mathfrak{H} = \mathbb{H}$ and e_1, e_2, e_3 denote multiplication by the imaginary units i, j, k of $\mathbb{H}$, $\mathcal{D}$ is just the Fueter operator

$$\mathcal{D}F = \frac{\partial F}{\partial x_0} + i \frac{\partial F}{\partial x_1} + j \frac{\partial F}{\partial x_2} + k \frac{\partial F}{\partial x_3}$$

on $\mathbb{H}$-valued functions on $\mathbb{R}^4 \sim \mathbb{H}$ (see chapter 2, (1.2)). In the general case, $\mathcal{D}\overline{\mathcal{D}} = \overline{\mathcal{D}}\mathcal{D} = \Delta_n$, so $\mathcal{D}$ and $\overline{\mathcal{D}}$ are elliptic. But every real $\mathfrak{A}_{n-1}$-module has the form $\mathfrak{R}_{n-1} \otimes \mathcal{V}$ for some real finite-dimensional Hilbert space $\mathcal{V}$, while every complex one has the form $\mathfrak{S}_{n-1} \otimes \mathcal{V}$ with $\mathcal{V}$ complex. Thus the space of real spinors, $\mathfrak{R}_{n-1}$, is the smallest $\mathfrak{A}_{n-1}$-module; and by (1.10)(ii) its dimension is equal to the minimal real dimension for ellipticity. Thus in the real case the operators $\mathcal{D}$ and $\overline{\mathcal{D}}$ acting on $\mathcal{C}^{\infty}(\Omega, \mathfrak{R}_{n-1})$, $\Omega \subseteq \mathbb{R}^n$, are the 'smallest' linear elliptic first-order systems of differential operators. Likewise, $\mathcal{D}$ and $\overline{\mathcal{D}}$ acting on $\mathcal{C}^{\infty}(\Omega, \mathfrak{S}_{n-1})$ give rise to the 'smallest' linear elliptic first-order systems of differential operators, in the complex case, since $\mathfrak{S}_{n-1}$ has minimal dimension (in the sense of (1.10)(i)). In the next section we shall see that $\mathcal{D}$ and $\overline{\mathcal{D}}$ are the prototypes of so-called *graded Dirac operators* which are basic to the Atiyah–Singer index theorem to be studied in chapter 4.

In summary then, there are two important ways of constructing first-order systems: the first by specifying a subspace $\mathcal{V}$ of $\mathcal{A} \otimes \mathbb{F}^n$, and the second by specifying a subspace $\mathfrak{B}$ of an $\mathfrak{A}_n$-module $\mathfrak{H}$. The $\mathcal{D}$ and $\overline{\mathcal{D}}$ operators, for example, come from the second construction, because, as we shall soon see, an $\mathfrak{A}_{n-1}$-module $\mathfrak{H}$ can be identified canonically with a subspace of an $\mathfrak{A}_n$-module. Our task in the rest of this chapter will be to see how to choose $\mathcal{V}$ in $\mathcal{A} \otimes \mathbb{F}^n$ or $\mathfrak{B}$ in $\mathfrak{H}$ and to relate the properties of the corresponding first-order systems to each other and to the Dirac operator. The algebraic problems involved are formidable and are not as yet completely resolved. For instance, how does the choice of $\mathcal{V}$ in $\mathcal{A} \otimes \mathbb{R}^n$ ensure that $\eth_{A_\mathcal{V}}$ has the GCR property? Choosing $\mathcal{V}$ is an algebraic problem, whereas the GCR property is an analytic property.

2 Operators of Dirac type

In this section, we focus our attention on an important class of injectively elliptic first-order operators, *the operators of Dirac type*, which arise by restricting the standard Dirac operator D to subspaces of the module $\mathfrak{H}$ on which its coefficients act.

(2.1) Definition.
   The operator $\eth_A : \mathcal{C}^{\infty}(\Omega, \mathcal{A}) \to \mathcal{C}^{\infty}(\Omega, \mathcal{B})$, $\Omega \subseteq \mathbb{R}^n$, *is said to be of Dirac type when there is an $\mathfrak{A}_n$-module $\mathfrak{H}$ containing $\mathcal{A}$ as a subspace such that $\eth_A$ is equivalent to the restriction*

$$(2.2) \qquad D : \mathcal{C}^{\infty}(\Omega, \mathcal{A}) \longrightarrow \mathcal{C}^{\infty}(\Omega, \mathfrak{H}) \ .$$

The first-order operator defined by (2.2) is usually algebraically overdetermined, but if $\mathcal{A}$ is also an $\mathfrak{A}(\mathfrak{H})$-submodule of $\mathfrak{H}$, then D actually maps $\mathcal{C}^{\infty}(\Omega, \mathcal{A})$ to itself, and the corresponding symbol mapping is invertible for each non-zero $\lambda \in \mathbb{R}^n$; in this case D is elliptic in the usual sense of the term, i.e., the symbol mapping is bijective. An operator of Dirac type is automatically a GCR operator, because D is; moreover, a wide variety of GCR operators of importance in mathematics and physics are operators of Dirac type, as we shall see. It is natural to ask, for a given Clifford module $\mathfrak{H}$, which subspaces $\mathfrak{B}$ of $\mathfrak{H}$ give rise to interesting and important operators of Dirac type. How are these subspaces chosen? In practice, there are two main methods for specifying $\mathfrak{B}$. One method, originating essentially with Stein and Weiss, assumes that $\mathfrak{H}$ has a Spin(n)-module structure compatible in a certain sense with the $\mathfrak{A}_n$-module structure and takes for $\mathfrak{B}$ a Spin(n)-submodule of $\mathfrak{H}$. The corresponding Dirac type operator then has a rotational invariance. We shall study such operators of Dirac type in considerable detail in the next section, but here we shall concentrate on the second method of constructing operators of Dirac type: those which arise when there is a $\mathbb{Z}(2)$-grading on $\mathfrak{H}$. Let $\eta : \mathfrak{H} \to \mathfrak{H}$ be an involutive linear mapping of an $\mathfrak{A}_n$-module $\mathfrak{H}$ (i.e., $\eta^2 = I$). Then $\mathfrak{H}$ admits the decomposition $\mathfrak{H} = \mathfrak{H}_+ \oplus \mathfrak{H}_-$, where

$$(2.3) \qquad \mathfrak{H}_+ = \{\xi \in \mathfrak{H} : \eta(\xi) = \xi\} \quad , \quad \mathfrak{H}_- = \{\xi \in \mathfrak{H} : \eta(\xi) = -\xi\}$$

are the $+1$ and -1 eigenspaces, respectively. The involution η is said to be *compatible* with the $\mathfrak{A}_n$-module structure when

$$(2.4) \qquad \eta(e_j \xi) = -e_j \eta(\xi) \qquad (\xi \in \mathfrak{H} , \ 1 \le j \le n) ;$$

in this case each e_j permutes the eigenspaces $\mathfrak{H}_+, \mathfrak{H}_-$, and the eigenspaces are themselves $\mathfrak{A}_{n-1}$-modules relative to the skew-adjoint operators $\varepsilon_j = e_j e_n$, $1 \le j \le n-1$. The restriction of the standard Dirac operator D to $\mathcal{C}^{\infty}(\Omega, \mathfrak{H}_{\pm})$ then defines operators $D_{\pm}$ of Dirac type such that

$$(2.5)$$
$$D_+ : \mathcal{C}^{\infty}(\Omega, \mathfrak{H}_+) \longrightarrow \mathcal{C}^{\infty}(\Omega, \mathfrak{H}_-) \quad , \quad D_- : \mathcal{C}^{\infty}(\Omega, \mathfrak{H}_-) \longrightarrow \mathcal{C}^{\infty}(\Omega, \mathfrak{H}_+)$$

and

$$(2.6) \qquad D_+ D_- = \Delta \quad , \quad D_- D_+ = \Delta .$$

Furthermore, D_+ and D_- are elliptic operators, each of which is the adjoint of the other in the sense that

$$(2.7) \qquad \int_{\Omega} (D_+ f(x), g(x)) \, dx = \int_{\Omega} (f(x), D_- g(x)) \, dx$$

holds for all compactly supported f in $\mathcal{C}^{\infty}(\Omega, \mathfrak{H}_+)$ and g in $\mathcal{C}^{\infty}(\Omega, \mathfrak{H}_-)$.

Because η introduces a $\mathbf{Z}(2)$-grading on $\mathfrak{H}$, these operators D_+ and D_- are called *graded Dirac operators*.

There are several important examples of such involutions.

(2.8) Examples.

(i) Let $\mathfrak{H} = \mathfrak{A}_n$ and let $\eta : \xi \to \xi'$ be the principal automorphism on $\mathfrak{A}_n$. Since

$$\eta(e_j \xi) = e_j' \xi' = -e_j \eta(\xi) \qquad (\xi \in \mathfrak{A}_n , \ 1 \le j \le n) ,$$

η is an involution on $\mathfrak{A}_n$ satisfying (2.4), its ± 1-eigenspaces coincide with the respective spaces $\mathfrak{A}_n^+$, $\mathfrak{A}_n^-$ of even and odd elements in $\mathfrak{A}_n$. Clearly the extension of η to the complex space $\mathfrak{C}_n$ has the same properties.

(ii) By identifying $\mathfrak{A}_n$ with $\Lambda^*(\mathbf{R}^n)$ and $\mathfrak{C}_n$ with $\Lambda^*(\mathbf{C}^n)$ via the linear isomorphism given in (7.10) in chapter 1, we obtain from (i) an involution $\eta : \Lambda^*(\mathbf{F}^n) \to \Lambda^*(\mathbf{F}^n)$, $\mathbf{F} = \mathbf{R}$ or $\mathbf{C}$, such that $\eta = (-1)^k$ on $\Lambda^k(\mathbf{F}^n)$. Its eigenspaces coincide with

$$\Lambda_+^*(\mathbf{F}^n) = \sum_{k=0}^{[\frac{1}{2}n]} \Lambda^{2k}(\mathbf{F}^n) \quad , \quad \Lambda_-^*(\mathbf{F}^n) = \sum_{k=0}^{[\frac{1}{2}n]} \Lambda^{2k+1}(\mathbf{F}^n) .$$

Since $e_j = \mu_j - \mu_j^*$ on $\Lambda^*(\mathbf{F}^n)$ (see chapter 1, (7.54)(**B**)), this involution clearly satisfies (2.4). But, in view of (3.10) in chapter 2, D_+ is just the restriction

$$d - d^* : C^\infty(\Omega, \Lambda_+^*) \longrightarrow C^\infty(\Omega, \Lambda_-^*)$$

mapping even to odd forms, while D_- reverses the parity. In practice it is often useful to have an explicit description of this involution. Now $\mathcal{L}(\Lambda^*(\mathbf{F}^n))$ has dimension 2^{2n}, whereas the subalgebra generated by the imaginary units

$$e_j = \mu_j - \mu_j^* \qquad (1 \le j \le n)$$

has dimension 2^n. Additional generators are needed therefore. Set

$$\tilde{e}_j = \mu_j + \mu_j^* \qquad (1 \le j \le n) .$$

By contrast with the e_j, these $\tilde{e}_j$ are *self-adjoint*; in addition, they satisfy a Clifford condition

$$\tilde{e}_j \tilde{e}_k + \tilde{e}_k \tilde{e}_j = 2\delta_{jk} I \qquad (1 \le j, k \le n) .$$

In particular, $(x_1, \ldots, x_n) \to \sum_j x_j \tilde{e}_j$ is an embedding of $\mathbf{R}^{n,0}$ into $\mathcal{L}(\Lambda^*(\mathbf{R}^n))$, which extends to a isomorphism from $\mathfrak{A}_{n,0}$ onto the sub-algebra generated in $\mathcal{L}(\Lambda^*(\mathbf{R}^n))$ by the $\tilde{e}_j$, whereas the e_j generate an algebra isomorphic to $\mathfrak{A}_{0,n}$ ($= \mathfrak{A}_n$). But $e_j \tilde{e}_k + \tilde{e}_k e_j = 0$ for all j, k. Consequently, the 2^{2n} reduced products

$$\tilde{e}_\alpha e_\beta = \tilde{e}_{\alpha_1} \cdots \tilde{e}_{\alpha_j} e_{\beta_1} \cdots e_{\beta_k}$$

are a basis for $\mathcal{L}(\Lambda^*(\mathbf{F}^n))$. A simple calculation now shows that

$$\eta(u) = (\tilde{e}_j \cdots \tilde{e}_n e_1 \cdots e_n)u = \left(\prod_{j=1}^{n} \mu_j^* \mu_j - \mu_j \mu_j^*\right)u$$

defines an involution on $\Lambda^*(\mathbf{F}^n)$ such that $\eta = (-1)^k$ on $\Lambda^k(\mathbf{F}^n)$ (see chapter 3, (2.27), (2.30) for earlier uses of these ideas).

(iii) When $n = 2m$, let $\mathfrak{H} = \Lambda^*(\mathbf{C}^m)$ as a realization of $\mathfrak{S}_{2m}$, and, as in (2.23)' in chapter 3, define η by

$$\eta(\xi) = i^m(e_1 \cdots e_{2m})\xi = \left(\prod_{j=1}^{m}(\mu_j^* \mu_j - \mu_j \mu_j^*)\right)\xi \qquad (\xi \in \Lambda^*(\mathbf{C}^m)) .$$

Then η is an involution on $\Lambda^*(\mathbf{C}^m)$ compatible with its $\mathfrak{A}_n$-module structure; furthermore,

$$\mathfrak{H}_+ = \Lambda_+^*(\mathbf{C}^m) \quad , \quad \mathfrak{H}_- = \Lambda_-^*(\mathbf{C}^m)$$

(see chapter 3, (2.24)'). In view of (3.9) in chapter 2, the graded Dirac operators associated with this involution are just the restriction of $\sum_{j=1}^{m}(\mu_j \bar{\partial}_j - \mu_j^* \partial_j)$, mapping even forms to odd forms and vice versa.

(iv) When $n = 2m$, let $\mathfrak{H} = \Lambda^*(\mathbf{C}^n)$ and set

$$\eta(\xi) = i^m(e_1 \cdots e_{2m})\xi = i^m\left(\prod_{j=1}^{n}(\mu_j - \mu_j^*)\right)\xi \qquad (\xi \in \Lambda^*(\mathbf{C}^n))$$

(see chapter 3, (2.23)). By (2.21)(ii) in chapter 3, η is an involution on $\Lambda^*(\mathbf{C}^n)$ compatible with its $\mathfrak{A}_n$-module structure. This involution is just the Hodge *-operator (up to a constant). Just as in example (ii), D_+ and D_- are the restriction of $d - d^*$.

(v) Suppose η_0 is an involution on $\mathfrak{H}_0$ compatible with an $\mathfrak{A}_n$-module structure on $\mathfrak{H}_0$. Then, for any finite-dimensional Hilbert space $\mathfrak{B}$, the tensor product $\mathfrak{H} = \mathfrak{H}_0 \otimes \mathfrak{B}$ is an $\mathfrak{A}_n$-module relative to skew-adjoint operators $E_j = e_j \otimes I$; and $\eta = \eta_0 \otimes I$ is an involution on $\mathfrak{H}$ compatible with this $\mathfrak{A}_n$-module structure. The corresponding D_+, D_- associated with $\mathfrak{H}, \eta$ are just the *twisted* versions of the D_+, D_- determined by $\mathfrak{H}_0$ and η_0, the twisting being accomplished by tensoring with $\mathfrak{B}$.

As we shall see in the next chapter, involutions (ii)–(iv) are intimately connected with the geometry of a manifold. Indeed, to each pair D_+, D_- of graded Dirac operators determined by an involution η on $\mathfrak{H}$, there correspond elliptic complexes

$$(2.9)_+ \qquad 0 \longrightarrow C^\infty(\Omega, \mathfrak{H}_+) \xrightarrow{D_+} C^\infty(\Omega, \mathfrak{H}_-) \longrightarrow 0$$

and

$$(2.9)_- \qquad 0 \longrightarrow C^\infty(\Omega, \mathfrak{H}_-) \xrightarrow{\ D_-\ } C^\infty(\Omega, \mathfrak{H}_+) \longrightarrow 0$$

having length 2. In this interpretation, example (ii) becomes the 'rolled-up' deRham complex, while (iii) and (iv) become the Spin and Hirzebruch signature complexes respectively. Example (v) gives the 'twisted' versions of these complexes. In this context the notion of *supersymmetry* associated with η has come to play an important role. Let $\mathfrak{H} = \mathfrak{H}_+ \oplus \mathfrak{H}_-$ be the *super-* or $\mathbf{Z}(2)$-*grading* on $\mathfrak{H}$ determined by an involution η compatible with the $\mathfrak{A}_n$-module structure on $\mathfrak{H}$. The algebra $\mathcal{L}(\mathfrak{H})$ is a *superalgebra* with the *even* (resp. *odd*) operators being those commuting (resp. anti-commuting) with η. Thus $T : \mathfrak{H}_\pm \to \mathfrak{H}_\pm$ when T is even, while $T : \mathfrak{H}_\pm \to \mathfrak{H}_\mp$ when T is odd. Multiplication by any imaginary unit is odd, for example. The *supertrace* on $\mathcal{L}(\mathfrak{H})$ is the scalar-valued function tr_s defined by

$$(2.10) \qquad \mathrm{tr}_s : \mathcal{L}(\mathfrak{H}) \longrightarrow \mathbf{F} \quad , \qquad \mathrm{tr}_s(T) = \mathrm{tr}(\eta T) \ .$$

Clearly $\mathrm{tr}_s(T) = 0$ when T is odd; but, crucially, if T is even,

$$(2.11) \qquad \mathrm{tr}_s(T) = \mathrm{tr}\big(T\big|_{\mathfrak{H}_+}\big) - \mathrm{tr}\big(T\big|_{\mathfrak{H}_-}\big) \ ,$$

i.e., $\mathrm{tr}_s(T)$ is the *difference* of the traces obtained by restricting T to $\mathfrak{H}_+$ and $\mathfrak{H}_-$. For the examples in (2.8) the value of $\mathrm{tr}_s(T)$ will be computed explicitly during the course of the proof of the Atiyah–Singer index theorem (see section 6 in chapter 5).

Finally, let us interpret the Dirac $(\mathcal{D}, \overline{\mathcal{D}})$-operators as graded operators.

(2.12) Theorem.

To each $\mathfrak{A}_{n-1}$-*module* $\mathfrak{H}_0$ *there correspond an* $\mathfrak{A}_n$-*module* $\mathfrak{H}$ *and involution* η *on* $\mathfrak{H}$ *compatible with its* $\mathfrak{A}_n$-*module structure such that*

(i) *the eigenspaces* $\mathfrak{H}_+, \mathfrak{H}_-$ *may be identified with* $\mathfrak{H}_0$,

(ii) *the operators* D_+, D_- *defined relative to* $(\mathfrak{H}, \eta)$ *are equivalent to the operators* $\mathcal{D}, \overline{\mathcal{D}}$ *defined relative to* $\mathfrak{H}_0$ *once the identifications in* (i) *are made.*

Proof. Set

$$\mathfrak{H} = \left\{ \begin{bmatrix} u \\ v \end{bmatrix} : u, v \in \mathfrak{H}_0 \right\} \ ,$$

and define $\eta : \mathfrak{H} \to \mathfrak{H}$ by

$$\eta = \begin{bmatrix} u \\ v \end{bmatrix} \longrightarrow \begin{bmatrix} -u \\ v \end{bmatrix} \qquad (u, v \in \mathfrak{H}_0) \ .$$

Then

$$\mathfrak{H}_+ = \left\{ \begin{bmatrix} 0 \\ v \end{bmatrix} : v \in \mathfrak{H}_0 \right\} \quad , \quad \mathfrak{H}_- = \left\{ \begin{bmatrix} u \\ 0 \end{bmatrix} : u \in \mathfrak{H}_0 \right\} ,$$

each of which may be identified with $\mathfrak{H}_0$ in the obvious way, proving (2.12)(i). Now $\mathcal{L}(\mathfrak{H}) = M(2, \mathcal{L}(\mathfrak{H}_0))$; in particular, if $\varepsilon_1, \ldots, \varepsilon_n$ are skew-adjoint operators on $\mathfrak{H}_0$ having the Clifford property, then

$$e_j = \begin{bmatrix} 0 & \varepsilon_j \\ \varepsilon_j & 0 \end{bmatrix} \quad (1 \leq j \leq n-1) , \qquad e_n = \begin{bmatrix} 0 & I \\ -I & 0 \end{bmatrix}$$

are skew-adjoint operators on $\mathfrak{H}$ such that

(2.13)(i) $e_j e_k + e_k e_j = -2\delta_{jk} I$ $(1 \leq j, k \leq n)$,

(2.13)(ii) $\eta(e_j \xi) = -e_j \eta(\xi)$ $(\xi \in \mathfrak{H} , \ 1 \leq j \leq n)$.

Thus $\mathfrak{H}$ is an $\mathfrak{A}_n$-module, and (2.13)(ii) shows that η is an involution which is compatible with this $\mathfrak{A}_n$-module structure on $\mathfrak{H}$. Now, the operator D corresponding to $\mathfrak{H}$ is plainly given by

$$(2.14) \qquad D = \sum_{j=1}^n e_j \frac{\partial}{\partial x_j} = \begin{bmatrix} 0 & \mathcal{D} \\ -\overline{\mathcal{D}} & 0 \end{bmatrix} ,$$

where we have set

$$\mathcal{D} = \frac{\partial}{\partial x_0} + \sum_{j=1}^{n-1} \varepsilon_j \frac{\partial}{\partial x_j} \quad , \quad \overline{\mathcal{D}} = \frac{\partial}{\partial x_0} - \sum_{j=1}^{n-1} \varepsilon_j \frac{\partial}{\partial x_j} .$$

Consequently,

$$D_+ = \begin{bmatrix} 0 & \mathcal{D} \\ 0 & 0 \end{bmatrix} \quad , \quad D_- = \begin{bmatrix} 0 & 0 \\ -\overline{\mathcal{D}} & 0 \end{bmatrix}$$

from which (2.12)(ii) follows.   ∎

When $n = 2$, equation (2.14) reduces to the realization of the Dirac operator D for $\mathfrak{A}_2 \cong \mathbb{H}$ obtained in chapter 2, section 2, since $\overline{\partial} = \mathcal{D}$ and $\partial = \overline{\mathcal{D}}$.

3 Rotation-invariant systems

The importance of the differential operator introduced originally by Dirac owed as much to its invariance under $\mathrm{Spin}(1,3) \ \textcircled{S} \ \mathbb{R}^{1,3}$ – the two-fold covering group of the rigid motions of Minkowski space $\mathbb{R}^{1,3}$ – as to its factoring of the wave-operator. For Euclidean space, it is invariance under the two-fold covering group of the rigid motions of $\mathbb{R}^n$ generated by translations and proper orthogonal transformations that is important. Thus in this section we begin the study of Dirac type operators which are invariant under the group $\mathrm{Spin}(n) \ \textcircled{S} \ \mathbb{R}^n$. Since any

constant-coefficient differential operator commutes with translation, the
main emphasis will be on invariance under Spin(n). As throughout all
the previous chapters, Spin(n) will act on $\mathbb{R}^n$ by the rotation

$$(3.1) \qquad \sigma(a) : x \longrightarrow axa^* , \qquad (x \in \mathbb{R}^n \subseteq \mathfrak{A}_n) .$$

Let $\mathcal{A}, \mathcal{B}$ be finite-dimensional Hilbert spaces, both real or both complex, which are Spin(n)-modules on which Spin(n) acts by isometries,
i.e., $|au|_\mathcal{A} = |u|_\mathcal{A}$, $|av|_\mathcal{B} = |v|_\mathcal{B}$ hold for all a in Spin(n). The associated
representations of Spin(n) will be denoted by $(\mathcal{A}, \tau)$ and $(\mathcal{B}, \omega)$; for instance, these might be reducible or irreducible unitary representations
of Spin(n). Then it is easy to check that

$$(3.2) \qquad \big(\pi_\tau(a)f\big)(x) = \tau(a)f\big(\sigma(a^{-1})x\big) = \tau(a)f(a^*xa) , \qquad (x \in \mathbb{R}^n)$$

defines a representation of Spin(n) on $\mathcal{C}^\infty(\mathbb{R}^n, \mathcal{A})$, 'rotating' both the
domain and the range of f. There is a corresponding representation π_ω
on $\mathcal{C}^\infty(\mathbb{R}^n, \mathcal{B})$.

(3.3) Definition.
   A first-order system

$$\eth_A = A \circ \nabla = \sum_{j=1}^{n} A_j \frac{\partial}{\partial x_j} : \mathcal{C}^\infty(\mathbb{R}^n, \mathcal{A}) \longrightarrow \mathcal{C}^\infty(\mathbb{R}^n, \mathcal{B})$$

*is said to be rotation-invariant when it commutes with π_τ and π_ω in the
sense that*

$$\eth_A\big(\pi_\tau(a)f\big) = \pi_\omega(a)(\eth_A f) \qquad \big(a \in \text{Spin}(n)\big) ,$$

for all f in $\mathcal{C}^\infty(\mathbb{R}^n, \mathcal{A})$.

This rotation-invariance property can be expressed solely as a condi-
tion on the A_j, i.e., as a condition on the symbol of $\eth_A$. For $a \in \text{Spin}(n)$,
let $[a_{jk}]$ denote the matrix image of a in $SO(n)$ under the covering homomorphism σ from Spin(n) onto $SO(n)$. Then
(3.4)

$$\sigma(a^{-1})x = \sigma(a)^{-1}x = y = (y_1, \ldots, y_n) , \quad y_j = \sum_{k=1}^{n} a_{jk}x_k \ (1 \le j \le n) ,$$

where we have made use of the fact that $[a_{jk}]^{-1} = [a_{kj}]$.

(3.5) Theorem.
   The first-order system $\eth_A$ in (3.3) is rotation-invariant if and only if

$$\omega(a)A_j\tau(a)^{-1} = \sum_{k=1}^{n} a_{jk}A_k \qquad (1 \le j \le n) ,$$

holds for each a in Spin(n).

Proof. The proof is a routine application of the chain rule. By (3.2),
$$\big(\pi_\tau(a)f\big)(x) = \tau(a)f(y)$$
where the coordinates of $y = a^*xa$ are related to those of x by (3.4). Thus,

$$\eth_A\big(\pi_\tau(a)f\big)(x) = \sum_{k=1}^n A_k \frac{\partial}{\partial x_k}\big(\tau(a)f(y)\big) = \sum_{j,k=1}^n A_k \tau(a)\, \frac{\partial f}{\partial y_j}\, \frac{\partial y_j}{\partial x_j}$$

$$= \sum_{j,k=1}^n a_{jk} A_k \tau(a)\, \frac{\partial f}{\partial y_j} \ .$$

On the other hand,

$$\big(\pi_\omega(a)\eth_A f\big)(x) = \sum_{j=1}^n \omega(a) A_j\, \frac{\partial f}{\partial y_j} \ ,$$

so that $\eth_A$ is rotation-invariant if and only if

$$\omega(a)A_j = \sum_{k=0}^n a_{jk} A_k \tau(a)$$

for all $a \in \mathrm{Spin}(n)$, $1 \le j \le n$. This completes the proof. ∎

Theorem (3.5) gives a very useful criterion for establishing the rotation-invariance of the standard Dirac operator D on $\mathbb{R}^n$. Let $\mathfrak{H}$ be an $\mathfrak{A}_n$-module relative to skew-adjoint operators $e_1, \ldots, e_n$. Then, as we saw in theorem (7.47) in chapter 1, the embedding $x \to \sum_{j=1}^n x_j e_j$ of $\mathbb{R}^n$ into $\mathcal{L}(\mathfrak{H})$ extends to a homomorphism from $\mathfrak{A}_n$ into $\mathcal{L}(\mathfrak{H})$. We shall denote by $S_{\mathfrak{H}} : \mathrm{Spin}(n) \to \mathcal{L}(\mathfrak{H})$ the representation of $\mathrm{Spin}(n)$ that this homomorphism defines when restricted to $\mathrm{Spin}(n)$ as a group in $\mathfrak{A}_n$. In view of (7.25) in chapter 1, each $S_{\mathfrak{H}}(a)$, $a \in \mathrm{Spin}(n)$, is an isometry on $\mathfrak{H}$. The choice of notation $(\mathfrak{H}, S_{\mathfrak{H}})$ is consistent with (2.29) in chapter 3, because it reduces to the spin representation S on $\mathfrak{S}_n$ when $\mathfrak{H} = \mathfrak{S}_n$. Thus we might (and shall) call $(\mathfrak{H}, S_{\mathfrak{H}})$ the *spin representation* of $\mathrm{Spin}(n)$ on $\mathfrak{H}$. The next result, a simple basic result, is the analogue for Euclidean space of Dirac's invariance result for Minkowski space.

(3.6) Theorem.
The standard Dirac operator D on $C^\infty(\mathbb{R}^n, \mathfrak{H})$ is rotation-invariant with respect to the spin representation $(\mathfrak{H}, S_{\mathfrak{H}})$ of $\mathrm{Spin}(n)$ on an $\mathfrak{A}_n$-module $\mathfrak{H}$, i.e.,

$$D\big(\pi(a)f\big) = \pi(a)(Df) \qquad \big(a \in \mathrm{Spin}(n)\big) \ ,$$

for all f in $C^\infty(\mathbb{R}^n, \mathfrak{H})$, $\pi_{\mathfrak{H}} = \pi_{S_{\mathfrak{H}}}$.

Proof. In view of (3.5), it is enough to show that

$$S_{\mathfrak{H}}(a)e_j S_{\mathfrak{H}}(a)^{-1} = \sum_{k=1}^{n} a_{jk}e_k \qquad (1 \le j \le n)$$

for all a in Spin(n). But

$$S_{\mathfrak{H}}(a)e_j S_{\mathfrak{H}}(a)^{-1} = ae_j a^* = \sigma(a)e_j = \sum_{j=1}^{n} a_{jk}e_k \ ,$$

since $\sigma(a) = [a_{jk}]$. This completes the proof. ∎

Now let η be an involution on $\mathfrak{H}$ which is compatible with its $\mathfrak{A}_n$-module structure. Then, by (2.4),

(3.7) $\eta(S_{\mathfrak{H}}(a)\xi) = S_{\mathfrak{H}}(a)\eta(\xi) \qquad (a \in \mathrm{Spin}(n))$

since every a in Spin(n) is an *even* product of elements in $\mathbb{R}^n$. Consequently, the eigenspaces $\mathfrak{H}_+, \mathfrak{H}_-$ of η are Spin(n)-submodules of $\mathfrak{H}$ with respect to $S_{\mathfrak{H}}$. From (3.6) we thus obtain the following.

(3.8) Corollary.

If η is an involution on $\mathfrak{H}$ compatible with its $\mathfrak{A}_n$-module structure, then the associated graded Dirac operators

$$D_+ : C^\infty(\mathbb{R}^n, \mathfrak{H}_+) \longrightarrow C^\infty(\mathbb{R}^n, \mathfrak{H}_-), \ D_- : C^\infty(\mathbb{R}^n, \mathfrak{H}_-) \longrightarrow C^\infty(\mathbb{R}^n, \mathfrak{H}_+)$$

are rotation-invariant operators with respect to the spin representation $S_{\mathfrak{H}}$ of Spin(n) on $\mathfrak{H}$.

Extensions of (3.6) and (3.8) to representations other than $S_{\mathfrak{H}}$ are very important. Let $(\mathfrak{H}, \tau)$ be a representation of Spin(n) on $\mathfrak{H}$. Criterion (3.5) together with the proof of (3.6) gives the following.

(3.9) Theorem.

The standard Dirac operator

$$D : C^\infty(\mathbb{R}^n, \mathfrak{H}) \longrightarrow C^\infty(\mathbb{R}^n, \mathfrak{H})$$

is rotation-invariant with respect to a representation $(\mathfrak{H}, \tau)$ of Spin(n), i.e., $D \circ \pi_\tau(a) = \pi_\tau(a) \circ D$, if and only if

(3.9)′ $\tau(a)e_j \tau(a)^{-1} = \sigma(a)e_j \qquad (1 \le j \le n)$

for all a in Spin(n).

To extend (3.8) to the representation $(\mathfrak{H}, \tau)$, we have to ensure that the eigenspaces of η are invariant under τ.

(3.10) Corollary.

Let η be an involution on $\mathfrak{H}$ compatible with its $\mathfrak{A}_n$-module structure. Then

$$D_+ : C^\infty(\mathbb{R}^n, \mathfrak{H}_+) \longrightarrow C^\infty(\mathbb{R}^n, \mathfrak{H}_-), \ D_- : C^\infty(\mathbb{R}^n, \mathfrak{H}_-) \longrightarrow C^\infty(\mathbb{R}^n, \mathfrak{H}_+)$$

are rotation-invariant operators with respect to $(\mathfrak{H}, \tau)$ provided

$$(3.10)' \qquad \eta\big(\tau(a)\xi\big) = \tau(a)\eta(\xi) \qquad \big(a \in \mathrm{Spin}(n)\big)$$

for all ξ in $\mathfrak{H}$.

Conditions $(3.9)'$ and $(3.10)'$ apply to the important examples studied in (2.8).

(3.11) Examples.

(i) Let $(\mathfrak{C}_n, \sigma)$, $\sigma(a) : u \to aua^*$, be the extension to $\mathfrak{C}_n$ of the familiar representation σ of $\mathrm{Spin}(n)$ (see chapter 3, (2.19), (2.27)). Then, as an operator on $\mathfrak{C}_n$,

$$\big(\sigma(a)e_j\sigma(a)^{-1}\big)u = \sigma(a)(e_j a^* ua) = (ae_j a^*)u = \big(\sigma(a)e_j\big)u$$

for all u in $\mathfrak{C}_n$ since $aa^* = I$. Then $(3.9)'$ holds, and so

$$D : C^\infty(\mathbb{R}^n, \mathfrak{C}_n) \longrightarrow C^\infty(\mathbb{R}^n, \mathfrak{C}_n)$$

is rotation-invariant with respect to σ. The same result obviously applies to $\mathfrak{A}_n$ as well. To check $(3.10)'$ for the involution $\eta : u \to u'$, observe that

$$\eta\big(\sigma(a)u\big) = (aua^*)' = au'a^* = \sigma(a)\eta(u) \qquad (u \in \mathfrak{C}_n)$$

since the principal automorphism reduces to the identity on $\mathfrak{A}_n^+$. Hence

$$D : C^\infty(\mathbb{R}^n, \mathfrak{A}_n^+) \longrightarrow C^\infty(\mathbb{R}^n, \mathfrak{A}_n^-), \ D : C^\infty(\mathbb{R}^n, \mathfrak{A}_n^-) \longrightarrow C^\infty(\mathbb{R}^n, \mathfrak{A}_n^+)$$

are rotation-invariant with respect to σ.

(ii) By identifying $\mathfrak{A}_n$ with $\Lambda^*(\mathbb{R}^n)$ and the principal automorphism with the involution η which is $(-1)^k$ on $\Lambda^k(\mathbb{R}^n)$, we see that

$$d - d^* : C^\infty(\mathbb{R}^n, \Lambda_+^*) \longrightarrow C^\infty(\mathbb{R}^n, \Lambda_-^*), \ C^\infty(\mathbb{R}^n, \Lambda_-^*) \longrightarrow C^\infty(\mathbb{R}^n, \Lambda_+^*)$$

are rotation-invariant relative to the representation $(2.17)'$ in chapter 3.

(iii) The rotation invariance of

$$D : C^\infty(\mathbb{R}^n, \mathfrak{S}_n) \longrightarrow C^\infty(\mathbb{R}^n, \mathfrak{S}_n)$$

with respect to the spin representation is covered already by (3.6); similarly, condition $(3.10)'$ for the involution in $(2.8)(iii)$ is established in $(2.24)(i)$ in chapter 3. Thus

$$D : C^\infty(\mathbb{R}^n, \mathfrak{S}_n^+) \longrightarrow C^\infty(\mathbb{R}^n, \mathfrak{S}_n^-) \ , \ \ C^\infty(\mathbb{R}^n, \mathfrak{S}_n^-) \longrightarrow C^\infty(\mathbb{R}^n, \mathfrak{S}_n^+)$$

are rotation-invariant relative to the spin representation when $n = 2m$.

(iv) In view of (2.21)(i) in chapter 3, the involution

$$\eta(\xi) = i^m (e_1 \cdots e_{2m}) \xi = i^m \left(\prod_{j=1}^{n} (\mu_j - \mu_j^*) \right) \xi \qquad (\xi \in \Lambda^*(\mathbb{C}^{2m}))$$

on $\Lambda^*(\mathbb{C}^{2m})$ satisfies $(3.10)'$ with respect to $(2.17)'$ in chapter 3. Hence the corresponding graded Dirac operators defined as restrictions of $d - d^*$ are rotation-invariant.

(v) If $(\mathfrak{H}_0 \otimes \mathfrak{B}, \tau_0 \otimes \tau_1)$ is a tensor product of representations of $\mathrm{Spin}(n)$ and η_0 satisfies $(3.10)'$ with respect to τ_0, then clearly $\eta = \eta_0 \otimes I$ satisfies $(3.10)'$ relative to $\tau_0 \otimes \tau_1$. In particular,

$$D : C^\infty(\mathbb{R}^n, \mathfrak{S}_n \otimes \cdots \otimes \mathfrak{S}_n) \longrightarrow C^\infty(\mathbb{R}^n, \mathfrak{S}_n \otimes \cdots \otimes \mathfrak{S}_n)$$

is rotation-invariant with respect to the m-fold tensor product $S \otimes \cdots \otimes S$ of the spin representation S (see chapter 3, (2.32)), as is the graded Dirac operator

$$D : C^\infty(\mathbb{R}^n, \mathfrak{S}_n^+ \otimes \mathfrak{S}_n \cdots \otimes \mathfrak{S}_n) \longrightarrow C^\infty(\mathbb{R}^n, \mathfrak{S}_n^- \otimes \mathfrak{S}_n \cdots \otimes \mathfrak{S}_n) \ .$$

It is instructive to bring out the representation-theoretic significance of these ideas. Let $\mathrm{Spin}(n) \textcircled{s} \mathbb{R}^n$ be the two-fold covering of the (connected) group of rigid motions of $\mathbb{R}^n$. This group is the semidirect product

$$(3.12)(\mathrm{i}) \qquad \{ g = (a, u) : a \in \mathrm{Spin}(n) \ , \ u \in \mathbb{R}^n \}$$

of 'rotations' and translations, with group multiplication

$$(3.12)(\mathrm{ii}) \qquad g_1 g_2 = (a, u)(b, v) = \left(ab, \sigma(b)^{-1} u + v \right) \ .$$

Now translation $f(x) \to f(x + u)$ is well-defined on $C^\infty(\mathbb{R}^n, \mathcal{A})$, whether or not $\mathcal{A}$ is a $\mathrm{Spin}(n)$-module. But if $(\mathcal{A}, \tau)$, say, is a representation of $\mathrm{Spin}(n)$, then it is easily checked that

$$(3.13) \qquad \left(\pi_\tau(g) f \right)(x) = \tau(a) f \left(\sigma(a)^{-1} x + u \right) \qquad (g = (a, u))$$

defines a representation of $\mathrm{Spin}(n) \textcircled{s} \mathbb{R}^n$ on $C^\infty(\mathbb{R}^n, \mathcal{A})$, combining (3.2) with translation, i.e.,

$$(3.14) \qquad f(x) \longrightarrow \tau(a) f \left(\sigma(a)^{-1} x \right) \longrightarrow \tau(a) f \left(\sigma(a)^{-1} x + u \right) \ .$$

Indeed,

$$\pi_\tau(g_1) \left(\pi_\tau(g_2) f \right)(x)$$
$$= \tau(a) \left(\pi_\tau(g_2) f \right) \left(\sigma(a)^{-1} x + u \right)$$
$$= \tau(a) \tau(b) f \left(\sigma(b)^{-1} \left(\sigma(a)^{-1} x + u \right) + v \right)$$
$$= \tau(ab) f \left(\sigma(ab)^{-1} x + \left(\sigma(b)^{-1} u + v \right) \right) = \left(\pi_\tau(g_1 g_2) f \right)(x) \ .$$

This representation π_τ of $\mathrm{Spin}(n) \textcircled{s} \mathbb{R}^n$ on $C^\infty(\mathbb{R}^n, \mathcal{A})$ is said to be the representation of $\mathrm{Spin}(n) \textcircled{s} \mathbb{R}^n$ *induced* from $(\mathcal{A}, \tau)$. Notice that the representation π_τ of $\mathrm{Spin}(n)\circledS\mathbb{R}^n$ becomes the representation (3.2) of $\mathrm{Spin}(n)$ by regarding $\mathrm{Spin}(n)$ as a subgroup of the semidirect product.

Now let π_ω be the corresponding representation of $\mathrm{Spin}(n)\circledS\mathbb{R}^n$ on $\mathcal{C}^\infty(\mathbb{R}^n,\mathcal{B})$ induced from a representation $(\mathcal{B},\omega)$ as in the beginning of the section. Since any constant-coefficient differential operator commutes with translation, we see, using the composition (3.14), that a first-order $\eth_A : \mathcal{C}^\infty(\mathbb{R}^n,\mathcal{A}) \to \mathcal{C}^\infty(\mathbb{R}^n,\mathcal{B})$ satisfies

$$(3.15) \qquad \eth_A\big(\pi_\tau(g)f\big) = \pi_\omega(g)(\eth_A f) \qquad \big(g \in \mathrm{Spin}(n)\circledS\mathbb{R}^n\big) \;,$$

i.e., commutes with the induced representations π_τ and π_ω of $\mathrm{Spin}(n) \otimes \mathbb{R}^n$, if and only if it is rotation-invariant in the sense of definition (3.3). Using criterion (3.5) we thus deduce the following.

(3.16) Theorem.

The first-order operator

$$\eth_A = \sum_{j=1}^n A_j \frac{\partial}{\partial x_j} : \mathcal{C}^\infty(\mathbb{R}^n,\mathcal{A}) \longrightarrow \mathcal{C}^\infty(\mathbb{R}^n,\mathcal{B})$$

commutes with $\pi_\tau(g)$ *and* $\pi_\omega(g)$, $g \in \mathrm{Spin}(n)\circledS\mathbb{R}^n$, *if and only if*

$$\omega(a)A_j\tau(a)^{-1} = \sum_{k=1}^m a_{jk}A_k \qquad (1 \le j \le n) \;.$$

Theorem (3.16) certainly applies to all the examples in this section. But if $\eth_A$ does commute with $\pi_\tau(g)$ and $\pi_\omega(g)$, then its kernel

$$(3.17) \qquad \ker\eth_A = \big\{f \in \mathcal{C}^\infty(\mathbb{R}^n,\mathcal{A}) : \eth_A f = 0\big\}$$

will be a $\mathrm{Spin}(n)\circledS\mathbb{R}^n$-submodule of $\mathcal{C}^\infty(\mathbb{R}^n,\mathcal{A})$ because

$$\eth_A\big(\pi_\tau(g)f\big) = \pi_\omega(g)(\eth_A f) = 0 \qquad \big(g \in \mathrm{Spin}(n)\circledS\mathbb{R}^n\big)$$

whenever $\eth_A f = 0$. Thus the representation-theoretic role of such a $\eth_A$ is to pick out a $\mathrm{Spin}(n)\circledS\mathbb{R}^n$-submodule from the reducible module $\mathcal{C}^\infty(\mathbb{R}^n,\mathcal{A})$. More generally, suppose $(\mathcal{A},\tau) = (\mathcal{B},\omega)$. Then the eigenspace

$$\mathcal{E}_\lambda(\mathbb{R}^n,\mathcal{A}) = \big\{f \in \mathcal{C}^\infty(\mathbb{R}^n,\mathcal{A}) : \eth_A f = \lambda f\big\} \qquad (\lambda \in \mathbb{C})$$

is well-defined (though possibly trivial), and each $\mathcal{E}_\lambda(\mathbb{R}^n,\mathcal{A})$ is a $\mathrm{Spin}(n)\circledS\mathbb{R}^n$-submodule of $\mathcal{C}^\infty(\mathbb{R}^n,\mathcal{A})$. The associated representation $(\mathcal{E}_\lambda(\mathbb{R}^n,\mathcal{A}), \pi_\tau)$ of $\mathrm{Spin}(n)\circledS\mathbb{R}^n$ is known, not surprisingly, as an *eigenspace representation* and has been extensively studied. In Dirac's original equation for Minkowski space this eigenvalue was related to 'mass'.

4 The operators $\eth_\tau$

Let $\tau = (m_1, \ldots, m_p)$ be the signature of an equivalence class in $\mathcal{M}_{\mathbb{C}}(\mathrm{Spin}(n))$, $p = [\frac{1}{2}n]$, and let $\mathcal{V}$ be any irreducible $\mathrm{Spin}(n)$-module in this equivalence class. We shall attempt to define a constant-coefficient, rotation-invariant operator $\eth_\tau$ of Dirac type having domain $\mathcal{C}^\infty(\mathbb{R}^n, \mathcal{V})$. Granted such a definition, $\eth_\tau$ becomes an operator also on $\mathcal{C}^\infty(\Omega, \mathcal{V})$, $\Omega \subseteq \mathbb{R}^n$; furthermore, if $\mathbf{V} = \mathcal{V}_1 \oplus \cdots \oplus \mathcal{V}_k$ is a decomposition of a $\mathrm{Spin}(n)$-module into its irreducible $\mathrm{Spin}(n)$-submodules having respective signatures $\tau_1, \ldots, \tau_k$, then $\partial_{\tau_1} \oplus \cdots \oplus \partial_{\tau_k}$ is an operator of Dirac type on $\mathcal{C}^\infty(\Omega, \mathbf{V})$, thereby recovering the standard Dirac operator on the various examples in the previous section. In this way, every $\mathrm{Spin}(n)$-module would have associated with it a constant-coefficient operator of Dirac type defined on open subsets of $\mathbb{R}^n$. The case of non-constant-coefficient operators on Riemannian manifolds will be studied in the next chapter.

In principle the construction is simple. For, by theorem (2.32) in chapter 3, $\mathcal{V}$ can be identified with a subspace $\mathfrak{B}$ of a tensor product $\mathfrak{H} = \Lambda^*(\mathbb{C}^p) \otimes \cdots \otimes \Lambda^*(\mathbb{C}^p)$ so that the representation, say τ, of $\mathrm{Spin}(n)$ on $\mathcal{V}$ coincides with the restriction to $\mathfrak{B}$ of the representation $S \otimes \cdots \otimes S$ of $\mathrm{Spin}(n)$ on this tensor product. On the other hand, by (3.6) and (3.11)(i), the standard Dirac operator D is rotation-invariant on $\mathcal{C}^\infty(\mathbb{R}^n, \mathfrak{H})$ with respect to $S \otimes \cdots \otimes S$. This suggests that we should identify $\eth_\tau$ on $\mathcal{C}^\infty(\mathbb{R}^n, \mathcal{V})$ with the restriction of D to $\mathcal{C}^\infty(\mathbb{R}^n, \mathfrak{B})$. The difficulty with this approach is that there may be many different choices of $\mathfrak{B}$ in $\mathfrak{H}$, for each of which the representation $(\mathfrak{B}, S \otimes \cdots \otimes S)$ has signature τ; but it is not obvious that one obtains the same $\eth_\tau$ on $\mathcal{C}^\infty(\mathbb{R}^n, \mathcal{V})$ by restricting D to the different spaces $\mathcal{C}^\infty(\mathbb{R}^n, \mathfrak{B})$. What is needed is an intrinsic definition of $\eth_\tau$ solely in terms of the equivalence class in $\mathcal{M}_{\mathbb{C}}(\mathrm{Spin}(n))$ having signature τ. It is in attempts to formulate this definition that the work of Stein and Weiss played such a basic role through use of theorems (1.8) and (3.5). As in the previous sections we begin with a general $\eth_A$, before specializing to $\eth_\tau$.

Let $\mathcal{V}$ be a finite-dimensional complex $\mathrm{Spin}(n)$-module. Then, by (1.8), the subspaces V of $\mathcal{V} \otimes \mathbb{C}^n$ determine the equivalence classes of first-order linear homogeneous systems of constant-coefficient differential operators on $\mathcal{C}^\infty(\mathbb{R}^n, \mathcal{V})$. Let $\eth_V$ be the differential operator $P_V \circ \nabla$ where P_V is the orthogonal projection of $\mathcal{V} \otimes \mathbb{C}^n$ onto the orthogonal complement $V^\perp$ of V in $\mathcal{V} \otimes \mathbb{C}^n$. Now the covering homomorphism $\sigma : \mathrm{Spin}(n) \to SO(n)$ defines a representation $(\mathbb{C}^n, \sigma)$ of $\mathrm{Spin}(n)$; hence the tensor product $\mathcal{V} \otimes \mathbb{C}^n$ is a Spin(n)-module (see chapter 3, (1.9)). Let $(\mathcal{V} \otimes \mathbb{C}^n, \tau \otimes \sigma)$ be the corresponding representation of Spin(n). From (3.5) we obtain the following.

(4.1) Theorem.
The operator
$$\eth_V = P_V \circ \nabla : \mathcal{C}^\infty(\mathbb{R}^n, \mathcal{V}) \longrightarrow \mathcal{C}^\infty(\mathbb{R}^n, V \otimes \mathbb{R}^n)$$
is rotation-invariant, i.e.,
$$\eth_V\big(\tau(a)f\big) = (\tau \otimes \sigma)(a)(\eth_V f) \qquad \big(a \in \mathrm{Spin}(n)\big),$$
if and only if V is a Spin(n)-submodule of $\mathcal{V} \otimes \mathbb{C}^n$.

Constructing rotation-invariant operators $\eth_V$ thus amounts to the *algebraic problem of selecting* Spin(n)-*submodules of* $\mathcal{V} \otimes \mathbb{C}^n$.

Proof of (4.1). Let $\eth_V = \sum_j A_j \frac{\partial}{\partial x_j}$, where $A_j : \mathcal{V} \to V$ is defined by $A_j(\xi) = P_V(\xi \otimes e_j)$, $\xi \in \mathcal{V}$. Then, by (3.5), $\eth_V$ is rotation-invariant if and only if
$$\big((\tau \otimes \sigma)(a)A_j\tau(a)^{-1}\big)\xi = \sum_k a_{jk}A_k\xi \qquad (\xi \in \mathcal{V}).$$
By definition, therefore,
$$(\tau \otimes \sigma)(a)P_V\big(\tau(a)^{-1}\xi \otimes e_j\big) = P_V\big(\xi \otimes \sigma(a)e_j\big) \qquad (\xi \in \mathcal{V})$$
since $\sigma(a)e_j = \sum_k a_{jk}e_k$; consequently, $\eth_V$ is rotation-invariant if and only if
$$(\tau \otimes \sigma)(a)P_V(\xi \otimes e_j) = P_V\big(\tau(a)\xi \otimes \sigma(a)e_j\big) \qquad (\xi \in \mathcal{V})$$
holds for each j, $1 \le j \le n$. Hence the projection P_V from $\mathcal{V} \otimes \mathbb{R}^n$ onto $\mathcal{V}^\perp$ commutes with $(\tau \otimes \sigma)(a)$, $a \in \mathrm{Spin}(n)$, and so $\eth_V$ is rotation-invariant if and only if P_V is equivariant, i.e., precisely when V is a Spin(n)-submodule of $\mathcal{V} \otimes \mathbb{C}^n$. ∎

Of course, all that (4.1) has done is convert the problem into two difficult algebraic problems:

(4.2)(a) *describe all irreducible* Spin(n)-*submodules of the tensor product* $\mathcal{V} \otimes \mathbb{C}^n$,

(4.2)(b) *choose for V some combination of these irreducible submodules, but which ones, and why?*

Problem (4.2)(a) has been completely resolved when $\mathcal{V}$ is irreducible; its solution identifies the irreducible submodules of $\mathcal{V} \otimes \mathbb{C}^n$ by their signature. The first attempt to solve (4.2)(b) was given by Stein and Weiss (see Notes and remarks). Since their answer does not depend on the precise answer of (4.2)(a) in complete generality, we shall concentrate first on this solution, though in general it does not give an operator $\eth_V$ of the required Dirac type. Note that the Cartan composition $\mathcal{V} \boxtimes \mathbf{C}^n$ is a canonically defined irreducible submodule of $\mathcal{V} \otimes \mathbf{C}^n$.

(4.3) Definition.

(Stein-Weiss) *If* $(\mathcal{V}, \tau)$ *is an irreducible unitary representation of* Spin(n) *and* $V = \mathcal{V} \boxtimes \mathbf{C}^n$ *is the Cartan composition of* $(\mathcal{V}, \tau)$ *and* $(\mathbf{C}^n, \sigma)$ *in* $\mathcal{V} \otimes \mathbf{C}^n$, *then* $\overline{\partial}_\tau$ *denotes the operator* $P_V \circ \nabla$ *on* $C^\infty(\Omega, \mathcal{V})$ *with* P_V *the orthogonal projection from* $\mathcal{V} \otimes \mathbf{C}^n$ *onto the complement of* $\mathcal{V} \boxtimes \mathbf{C}^n$.

The basic result of Stein–Weiss identifies $\overline{\partial}_\tau$ for the fundamental representations of Spin(n). Remarkably enough, $\overline{\partial}_\tau$ coincides in these cases with the basic geometric first-order operators in Euclidean analysis: the Hodge–deRham (d, d^*)-system on k-forms and the Dirac operator on spinor-valued functions. But, of course, we have seen that these are just operators of Dirac type obtained from $\mathfrak{C}_n$ or $\mathfrak{S}_n$. More precisely, Stein–Weiss proved the following.

(4.4) Theorem.

The $\overline{\partial}_\tau$-*operator is equivalent*

(i) *to the* (d, d^*)-*system on* k-*forms when* $\tau = d_k$, $1 \leq 2k < n$, *and to the restriction of* (d, d^*) *to* $\Lambda_\pm^p(\mathbf{C}^n)$-*valued functions when* $\tau = d_p^\pm$, $n = 2p$,

(ii) *to the standard Dirac operator* D *on spinor-valued functions when* $\tau = s$, $n = 2p + 1$, *and to the restriction of* D *to* $\mathfrak{S}_n^\pm$-*valued functions when* $\tau = s_\pm$, $n = 2p$.

They then showed by Cartan composition arguments that $\overline{\partial}_\tau$ is always a GCR operator. Rather than prove any of these results immediately, however, it will be instructive to investigate first the underlying ideas. The emphasis on *rotation*-invariance comes from the corresponding properties

$$(4.5) \qquad \overline{\partial} : f(e^{i\theta}z) \longrightarrow e^{i\theta}\overline{\partial}f(z) \quad , \quad \partial : f(e^{i\theta}z) \longrightarrow e^{-i\theta}\partial f(z)$$

of the Cauchy–Riemann operators. For these properties are the basis of the link between Fourier series expansions and Taylor series, making possible joint use of real- and complex-variable techniques. Furthermore, since $\overline{\partial}$ commutes with rotations, $F(z) \to F(e^{i\theta}z)$ defines a representation of $SO(2)$ on the kernel

$$(4.6) \qquad \ker \overline{\partial}(\Omega) = \{F \in C^\infty(\Omega) : \overline{\partial}F = 0\}$$

of $\overline{\partial}$ whenever Ω is an open disk centered at the origin, and the Taylor series expansion $F(z) = \sum_{n=0}^{\infty} a_n z^n$ can be interpreted as the decomposition of $\ker \overline{\partial}$ into its $SO(2)$-irreducible subspaces indexed by the 'cone' $\{ n \in \mathbb{Z} : n \geq 0 \}$ of non-negative integers. From a Hardy space point of view, this is the basis for the classical F. and M. Riesz theorem for H^1. Invariance under *translation*, a consequence of the constant coefficient property, allows Taylor series expansions to be developed also in any open disk not necessarily centered at the origin. Such analytic properties as the Cauchy integral theory and associated H^p theory for functions in $\ker \overline{\partial}$ were studied in detail in chapter 2, of course; the notion of *over-determinedness* was central to H^p theory (see, for instance, (4.33) in chapter 2). Analytically, this over-determinedness can be expressed by characterizing the solutions of $\overline{\partial} F = 0$ as being locally of the form $F = \partial \Phi$ with Φ a real-valued harmonic function. But the Laplacian on $\mathbb{R}^2$ is $SO(2)$-invariant, and so $\Phi(z) \to \Phi(e^{i\theta} z)$ defines a representation of $SO(2)$ on the space

$$(4.6)(i) \qquad \mathcal{H}(\Omega) = \left\{ \Phi \in \mathcal{C}^{\infty}(\Omega, \mathbb{R}) : \Delta \Phi = 0 \right\}$$

of real-valued harmonic functions on any open disk centered at the origin. Consequently, the rotation-invariance of ∂ ensures that

$$(4.6)(ii) \qquad \partial : \mathcal{H}(\Omega) \longrightarrow \ker \overline{\partial}(\Omega) \quad , \quad \partial : \Phi \longrightarrow \partial \Phi$$

is an $SO(2)$-equivariant mapping from $\mathcal{H}(\Omega)$ *onto* $\ker \overline{\partial}(\Omega)$ for all such Ω. This is a representation-theoretic interpretation of over-determinedness. Thus all these considerations go into choosing the subspace V in (4.2)(b). For simplicity of exposition, however, only those $\tau = (m_1, \ldots, m_r, 0, \ldots)$ with $r + 1 \leq [\frac{1}{2} n]$ will be studied, i.e., at least one 0 occurs in τ (see Notes and remarks). By (2.34) in chapter 3, Spin(n)-modules having such signature will also be irreducible $O(n)$-modules on which the representation of $O(n)$ is single-valued.

Proof of (4.4). By restricting attention to d_k, $1 \leq k \leq p - 1$, $p = [\frac{1}{2} n]$, we can take $V = \Lambda^k(\mathbb{C}^n)$ and

$$(4.7) \qquad \Lambda^k(\mathbb{C}^n) \otimes \mathbb{C}^n \cong \left(\Lambda^k(\mathbb{C}^n) \,[\times]\, \mathbb{C}^n \right) \oplus \Lambda^{k+1}(\mathbb{C}^n) \oplus \Lambda^{k-1}(\mathbb{C}^n) \;,$$

so that with obvious notation

$$(4.7)(i) \qquad d_k \otimes d_1 = \underbrace{(2, 1, \ldots, 1, 0, \ldots)}_{k} + d_{k+1} + d_{k-1} \;,$$

since the signature of $(\mathbb{C}^n, \sigma)$ is d_1. In the simplest case, $k = 1$, (4.7) corresponds to the decomposition

$$A = \left(\tfrac{1}{2}(A + A') - \tfrac{1}{n}\operatorname{tr}(A)I\right) + \tfrac{1}{2}(A - A') + \tfrac{1}{n}\operatorname{tr}(A)I$$

of any $A \in \mathbb{C}^{n \times n}$ into the sum respectively of a symmetric traceless matrix, a skew-symmetric matrix, and a constant multiple of the identity matrix. The operator $\bar{\partial}_\tau$, $\tau = d_1$, on 1-forms $F = \sum_{j=1}^n F_j e_j$ is thus given by

$$\bar{\partial}_\tau F = \tfrac{1}{2}\left[\frac{\partial F_j}{\partial x_k} - \frac{\partial F_k}{\partial x_j}\right] + \tfrac{1}{n}\left(\sum_j \frac{\partial F_j}{\partial x_j}\right)I \ .$$

Hence $\bar{\partial}_\tau$, $\tau = d_1$, is equivalent to the 'div–curl' system

$$(4.8) \qquad \sum_j \frac{\partial F_j}{\partial x_j} = 0 \quad , \qquad \frac{\partial F_j}{\partial x_k} - \frac{\partial F_k}{\partial x_j} = 0 \qquad (1 \le j, k \le n) \ .$$

Because of the 'curl-free' condition every solution of $\bar{\partial}F = 0$ is locally the gradient $\frac{\partial \Phi}{\partial x} = (\frac{\partial \Phi}{\partial x_1}, \ldots, \frac{\partial \Phi}{\partial x_n})$ of a scalar-valued function Φ, while the 'divergence-free' condition ensures that Φ is harmonic. In fact, it is easy to see that $F = \frac{\partial \Phi}{\partial x}$, Φ harmonic, characterizes locally the solutions of $\bar{\partial}_\tau F = 0$, $\tau = d_1$.

The case of $\tau = d_k$ is much the same except for one simple but important complication. For any $\xi = \sum_j \xi_j \otimes e_j$ in $\Lambda^k(\mathbb{C}^n) \otimes \mathbb{C}^n$, the second and third terms in (4.7) are given respectively by $\sum_j \mu_j(\xi_j)$ and $\sum_j \mu_j^*(\xi_j)$. Consequently, on k-forms

$$\bar{\partial}_\tau \omega = \sum_j \mu_j \frac{\partial \omega}{\partial x_j} + \sum_j \mu_j^* \frac{\partial \omega}{\partial x_j} = d\omega + d^*\omega \ ;$$

and so $\bar{\partial}_\tau$, $\tau = d_k$, is equivalent to the Hodge–deRham (d, d^*)-system on k-forms. Now, classically (Poincaré), any k-form solution of $d\omega = 0$ can be written locally as $\omega = d\phi$, with ϕ a $(k-1)$-form; specifically, in an open ball centered at the origin,

$$(4.9) \qquad \phi(x) = \int_0^1 \left(\sum_{j=1}^n x_j \mu_j^*\right) \omega(sx) \, ds \ .$$

But, if $\omega = d\phi$,

$$d^*\omega = d^* d\phi = (d^*d + dd^*)\phi + d(d^*\phi) = \Delta\phi + d(d^*\phi) \ .$$

Thus $\omega = d\phi$ will be a k-form solution of the (d, d^*)-system if ϕ is a harmonic $(k-1)$-form solution of $d^*\phi = 0$. It is easy to see that this characterizes locally the k-form solutions of the (d, d^*)-system. Hence for $\tau = d_k$, the over-determinedness property can be expressed by saying that the exterior derivative d is an $O(n)$-equivariant operator from the space

$$\left\{\phi \in C^\infty\left(\Omega, \Lambda^{k-1}(\mathbb{C})\right) : \Delta\phi = 0 \ , \ d^*\phi = 0\right\}$$

of harmonic $(k-1)$-form solutions of $d^*\phi = 0$ onto the kernel of $\overline{\partial}_\tau$, $\tau = d_k$, in $\mathcal{C}^\infty(\Omega, \Lambda^k(\mathbf{C}^n))$, Ω an open ball in $\mathbf{R}^n$ centered at the origin. The proof in all the remaining cases of (4.4) is much the same. ∎

The next step on the road to studying $\overline{\partial}_\tau$ (and hence finally δ_τ) is the case $\tau = (m, \ldots, m, 0, \ldots)$ of multiples of the fundamental representations d_r. As we saw in the previous chapter, the space $\mathcal{H}_m(\mathbf{R}^{r \times n})^{SL}$ of degree m determinantally homogeneous $O(n)$-harmonic polynomials on $\mathbf{R}^{r \times n}$ is a very convenient realization of an irreducible $\mathrm{Spin}(n)$-module having such signature. Now, just as in (4.7)(i), it can be shown that

$$(4.10) \qquad \underbrace{(m, \ldots, m, 0, \ldots)}_{r} \otimes d_1 = \underbrace{(m+1, m, \ldots, m, 0, \ldots)}_{r}$$

$$+ \underbrace{(m, \ldots, m, 1, 0, \ldots)}_{r} + \underbrace{(m, \ldots, m, m-1, 0, \ldots)}_{r},$$

with the first term on the right being the signature of the Cartan composition $\mathcal{H}_m(\mathbf{R}^{r \times n})^{SL} \boxtimes \mathbf{C}^n$. But an explicit description of the $\mathrm{Spin}(n)$-submodules corresponding to this description cannot be derived as easily as it was in (4.7) for the case $m = 1$. It is at times such as this that the polynomial invariant ideas in sections 4 and 5 of chapter 3 become decidedly useful. We consider first the case $\tau = (m, 0, \ldots, 0)$ corresponding to the familiar representation on the space $\mathcal{H}_m(\mathbf{R}^n)$ (see chapter 3, (3.13)). The operator $\overline{\partial}_\tau$, $\tau = (m, 0, \ldots)$, is then usually known as the 'higher gradients operator' because of the Stein–Weiss local characterization of solutions of $\overline{\partial}_\tau F = 0$ as mth-order gradients $F = (\frac{\partial}{\partial x})^m \Phi$ of scalar-valued harmonic functions Φ. In such a characterization, however, F would take values in a space of symmetric traceless tensors rather than in the more analytically appealing spaces of harmonic polynomials. Thus we shall formulate the results for functions $F : \Omega \to \mathcal{P}(\mathbf{R}^n)$, often regarding them as scalar-valued functions $F = F(x, \xi)$ on $\Omega \times \mathbf{R}^n$; restrictions on the range of F then become restrictions on $F = F(x, \xi)$ as a function of ξ. By identifying any (x, ξ) in $\Omega \times \mathbf{R}^n$ with the matrix $z = [\begin{smallmatrix} x \\ \xi \end{smallmatrix}]$ in $\mathbf{R}^{2 \times n}$, we can also regard F as a function of matrix argument. In all three cases the action of $O(n)$ is natural and familiar: when Ω is an open ball in $\mathbf{R}^n$ centered at the origin each a in $O(n)$ acts by

(4.11)

$$\begin{cases} \text{(i)} & F(x) \longrightarrow \rho(a)F(xa) & \big(F : \Omega \to \mathcal{P}(\mathbf{R}^n)\big) \,, \\ \text{(ii)} & F(x, \xi) \longrightarrow F(xa, \xi a) & (F : \Omega \times \mathbf{R}^n \to \mathbf{C}) \,, \\ \text{(iii)} & F(z) \longrightarrow F(za) & \big(z = [\begin{smallmatrix} x \\ \xi \end{smallmatrix}] \in \mathbf{R}^{2 \times n}\big) \,, \end{cases}$$

where in (i) ρ is the representation $(\rho(a)\phi)(\xi) = \phi(\xi a)$ on $\mathcal{P}(\mathbf{R}^n)$ defined earlier in III (3.2); clearly these all coincide under the various identifications of F. Analogously to (4.9) we prove first the following.

(4.12) Theorem.

Let $F = F(x,\xi)$ be a smooth scalar-valued function on $\Omega \times \mathbf{R}^n$ such that $F(x,\lambda\xi) = \lambda^m F(x,\xi)$, and set

$$\Phi(x) = \frac{1}{(m-1)!} \int_0^1 \left(x \cdot \frac{\partial}{\partial\xi}\right)^m F(sx,\xi)(1-s)^{m-1}\,ds \qquad (x \in \Omega)$$

where Ω is an open ball in $\mathbf{R}^n$ centered at the origin. Then

$$F(x,\xi) = \frac{1}{m!}\left(\xi \cdot \frac{\partial}{\partial x}\right)^m \Phi$$

on $\Omega \times \mathbf{R}^n$ provided $\frac{\partial^2 F}{\partial x_j \partial \xi_k} = \frac{\partial^2 f}{\partial x_k \partial \xi_j}$, $1 \le j, k \le n$.

Both in the statement and proof of (4.12), the notation

(4.13) $\qquad \dfrac{\partial}{\partial x} = \left(\dfrac{\partial}{\partial x_1}, \dots, \dfrac{\partial}{\partial x_n}\right) \quad, \quad \dfrac{\partial}{\partial \xi} = \left(\dfrac{\partial}{\partial \xi_1}, \dots, \dfrac{\partial}{\partial \xi_n}\right)$

is being used for the usual *gradient* relative to x and ξ respectively. Consequently, (4.12) expresses F explicitly as the mth-order gradient of a scalar-valued function provided

(4.14) $\qquad \dfrac{\partial^2 F}{\partial x_j \partial \xi_k} = \dfrac{\partial^2 F}{\partial x_k \partial \xi_j} \qquad (1 \le j, k \le n)\,.$

But if F is known to be of the form

(4.14)′ $\qquad\qquad F(x,\xi) = \dfrac{1}{m!}\left(\xi \cdot \dfrac{\partial}{\partial x}\right)^m \Phi\,,$

then

$$\frac{\partial^2 F}{\partial x_j \partial \xi_k} = \frac{1}{(m-1)!}\left(\xi \cdot \frac{\partial}{\partial x}\right)^{m-1} \frac{\partial^2 \Phi}{\partial x_j \partial x_k} = \frac{\partial^2 F}{\partial x_k \partial \xi_j}\,.$$

Hence (4.14) is both necessary and sufficient for (4.14)′ to hold.

Proof of (4.12). The proof is little more than repeated integration by parts. Indeed,

$$\frac{\partial \Phi}{\partial x} = \frac{1}{(m-1)!}$$

$$\times \int_0^1 (1-s)^{m-1}\left(x \cdot \frac{\partial}{\partial\xi}\right)^{m-1}\left\{m\frac{\partial F}{\partial\xi}(sx,\xi) + \sum_j sx_j \frac{\partial^2 F}{\partial\xi_j \partial x}(sx,\xi)\right\}ds\,.$$

But condition (4.14) on F ensures that

$$\sum_j sx_j \frac{\partial^2 F}{\partial\xi_j \partial x} = \sum_j sx_j \frac{\partial^2 F}{\partial\xi \partial x_j} = \frac{\partial}{\partial\xi}\left(\frac{dF}{ds}(sx,\xi)\right)\,.$$

For $m > 1$, integration by parts thus gives

$$\frac{\partial \Phi}{\partial x} = \frac{1}{(m-2)!} \frac{\partial}{\partial \xi} \int_0^1 \left(x \cdot \frac{\partial}{\partial \xi}\right)^{m-1} F(sx, \xi)(1-s)^{m-2}\, ds \ ,$$

and repeated use of this argument shows that

$$\left(\xi \cdot \frac{\partial}{\partial x}\right)^m \Phi(x) = \left(\xi \cdot \frac{\partial}{\partial \xi}\right)^m F(x, \xi) \ ,$$

from which the result follows applying Euler's theorem, since F is homogeneous of degree m in ξ. ∎

(4.15) Corollary.

Suppose further that $F = F(x, \xi)$ is harmonic in ξ and $\sum_j \frac{\partial^2 F}{\partial x_j \partial \xi_j} = 0$. Then $F = F(x, \xi)$ is $O(n)$-harmonic as a function of $\begin{bmatrix} x \\ \xi \end{bmatrix}$ on $\mathbb{R}^{2 \times n}$, while $\Phi = \Phi(x)$ is harmonic on $\mathbb{R}^n$.

Proof. The $O(n)$-harmonicity of F requires that

$$\sum_j \left(\frac{\partial}{\partial x_j}\right)^2 F = \sum_j \frac{\partial^2 F}{\partial x_j \partial \xi_j} = \sum_j \left(\frac{\partial}{\partial \xi_j}\right)^2 F = 0 \ .$$

Now the last equalities hold by hypothesis. But, by homogeneity,

$$m \left(\frac{\partial}{\partial x_j}\right)^2 F = \sum_k \xi_k \left(\frac{\partial}{\partial x_j}\right)^2 \frac{\partial F}{\partial \xi_k} = \sum_k \xi_k \frac{\partial^2}{\partial x_j \partial \xi_k}\left(\frac{\partial F}{\partial x_j}\right)$$

$$= \sum_k \xi_k \frac{\partial}{\partial x_k}\left(\frac{\partial^2 F}{\partial x_j \partial \xi_j}\right)$$

using (4.14). With this the remaining equality follows from the condition $\sum_j \frac{\partial^2 F}{\partial x_j \partial \xi_j} = 0$. On the other hand,

$$\sum_j \left(\frac{\partial}{\partial x_j}\right)^2 \Phi = \frac{1}{(m-1)!} \int_0^1 \left(x \cdot \frac{\partial}{\partial \xi}\right)^{m-2} G(sx, \xi)(1-s)^{m-1}\, ds$$

where $G(sx, \xi)$ is given by

$$m(m-1) \sum_j \left(\frac{\partial}{\partial \xi_j}\right)^2 F + 2ms \left(x \cdot \frac{\partial}{\partial \xi}\right) \sum_j \frac{\partial^2 F}{\partial x_j \partial \xi_j}$$

$$+ s^2 \left(x \cdot \frac{\partial}{\partial \xi}\right)^2 \sum_j \left(\frac{\partial}{\partial x_j}\right)^2 F \ .$$

Hence Φ is harmonic as function of x on $\mathbb{R}^n$. ∎

Combining (4.12) and (4.15) we see that a function $F = F(x, \xi)$ which is harmonic and homogeneous of degree m in ξ can be represented locally as the mth-order gradient

(4.16) $$F(x,\xi) = \frac{1}{m!}\left(\xi \cdot \frac{\partial}{\partial x}\right)^m \Phi$$

of a scalar-valued harmonic function Φ when

(4.17) $\left(\dfrac{\partial}{\partial x} \cdot \dfrac{\partial}{\partial \xi}\right)F = 0$, $\dfrac{\partial^2 F}{\partial x_j \partial \xi_k} = \dfrac{\partial^2 F}{\partial x_k \partial \xi_j}$ $(1 \le j, k \le n)$.

On the other hand, (4.17) certainly holds whenever F admits such a local characterization. Hence (4.17) defines a first-order system (in x) which must be equivalent to the higher gradients operator. This can be proved directly.

(4.18) Theorem.
    The operator $\overline{\partial}_\tau$, $\tau = (m, 0, \dots)$ is equivalent to the first-order system (4.17) on functions $F = F(x, \xi)$ which are harmonic and homogeneous of degree m in ξ.

Proof. On $F : \Omega \to \mathcal{H}_m(\mathbf{R}^n)$,

$$\overline{\partial}_\tau F = A_\tau \left(\sum_{j=1}^n \frac{\partial F}{\partial x_j} \otimes e_j\right) \qquad (\tau = (m, 0, \dots))$$

where A_τ is the orthogonal projection onto the complement $(\mathcal{H}_m(\mathbf{R}^n)$ $\boxtimes \mathbf{C}^n)^\perp$ of the Cartan composition in $\mathcal{H}_m(\mathbf{R}^n) \otimes \mathbf{C}^n$. But, by (4.10), this Cartan composition has signature $(m+1, 0, \dots)$, the same as that of $\mathcal{H}_{m+1}(\mathbf{R}^n)$. Thus we need to recognize an isomorphic copy of $\mathcal{H}_{m+1}(\mathbf{R}^n)$ inside $\mathcal{H}_m(\mathbf{R}^n) \otimes \mathbf{C}^n$. Set

$$\nu : \mathcal{H}_{m+1}(\mathbf{R}^n) \longrightarrow \mathcal{H}_m(\mathbf{R}^n) \otimes \mathbf{C}^n \ , \ \nu : \phi \longrightarrow \sum_j \phi_j \otimes e_j \ \left(\phi_j = \frac{\partial \phi}{\partial \xi_j}\right) .$$

Since $\sum_j \xi_j \frac{\partial \phi}{\partial \xi_j} = m\phi(\xi)$, ν actually is an $O(n)$-equivariant embedding; furthermore, the components ϕ_j in $\nu(\phi)$ satisfy

(4.19) $\displaystyle\sum_j \frac{\partial \phi_j}{\partial \xi_j} = 0$, $\dfrac{\partial \phi_j}{\partial \xi_k} = \dfrac{\partial \phi_k}{\partial \xi_j}$ $(1 \le j, k \le n)$.

Conversely, set $\phi(\xi) = \sum_j \xi_j \phi_j$ corresponding to any $\sum_j \phi_j \otimes e_j$ in $\mathcal{H}_m(\mathbf{R}^n) \otimes \mathbf{C}^n$. Then, under condition (4.19), ϕ is a polynomial in $\mathcal{H}_{m+1}(\mathbf{R}^n)$ and

$$\frac{\partial \phi}{\partial \xi_k} = \phi_k + \sum_j \xi_j \frac{\partial \phi_j}{\partial \xi_k} = \phi_k + \sum_j \xi_j \frac{\partial \phi_k}{\partial \xi_j} = (m+1)\phi_k \ ,$$

as simple calculations show. Hence $\mathcal{H}_m(\mathbf{R}^n) \boxtimes \mathbf{C}^n$ consists precisely of those $\sum_j \phi_j \otimes e_j$ satisfying (4.19); and then $\phi_j = \frac{\partial \phi}{\partial \xi_j}$ for some ϕ in

$\mathcal{H}_{m+1}(\mathbb{R}^n)$. Consequently, $\overline{\partial}_\tau F = 0$ if and only if

$$\sum_j \frac{\partial^2 F}{\partial x_j \partial \xi_j} = 0 \quad , \quad \frac{\partial^2 F}{\partial x_j \partial \xi_k} = \frac{\partial^2 F}{\partial x_k \partial \xi_j} \quad (1 \leq j, k \leq n) ,$$

completing the proof.   ∎

Other properties of $\overline{\partial}_\tau$, such as its relation to the Dirac operator and to Taylor series expansions, now follow easily.

(4.20) Theorem.
The operator $\overline{\partial}_\tau$, $\tau = (m, 0, \ldots)$, *is equivalent to*

$$(D_x D_\xi) F = \left(\sum_j e_j \frac{\partial}{\partial x_j} \right) \left(\sum_k e_k \frac{\partial}{\partial \xi_k} \right) F$$

on functions $F = F(x, \xi)$ *which are harmonic and homogeneous of degree* m *in* ξ.

Proof. By direct calculation

$$D_x D_\xi = -\left(\sum_j \frac{\partial^2}{\partial x_j \partial \xi_j} \right) + \sum_{j \neq k} \left(\frac{\partial^2}{\partial x_j \partial \xi_k} - \frac{\partial^2}{\partial x_k \partial \xi_j} \right) e_j e_k .$$

Hence $\overline{\partial}_\tau F = 0$ if and only if $(D_x D_\xi) F = 0$.   ∎

By the proof of (4.12) and (4.18), the mapping $\phi \to (\xi \cdot \frac{\partial}{\partial x})^m \phi$ is an $O(n)$-equivariant embedding of $\mathcal{H}_{m+\ell}(\mathbb{R}^n)$ into the kernel of $\overline{\partial}_\tau$, $\ell \geq 0$. On the other hand, the (formal) Taylor series expansion

$$\Phi(x) = \sum_{k=0}^{\infty} \frac{1}{k!} \left(x \cdot \frac{\partial}{\partial \zeta} \right)^k \phi \Big|_{\zeta=0} = \sum_{k=0}^{\infty} \Phi_k$$

of (3.7) in chapter 3 represents each harmonic function Φ as a sum of polynomials Φ_k in $\mathcal{H}_k(\mathbb{R}^n)$. Since every solution of $\overline{\partial}_\tau F = 0$ can be represented locally as $(\xi \cdot \frac{\partial}{\partial x})^m \Phi$ for such a Φ, we thus deduce the following.

(4.21) Theorem.
Locally every solution of $\overline{\partial}_\tau F = 0$, $\tau = (m, 0, \ldots)$, *has a (formal) Taylor series expansion* $F = \sum_{k=0}^{\infty} F_k$ *where each* F_k *lies in an irreducible* $SO(n)$-*submodule of* $\ker \overline{\partial}_\tau$ *having signature* $(m + k, 0, \ldots)$.

Intuitively, therefore, the kernel of $\overline{\partial}_\tau$ has a local decomposition

$$\ker \overline{\partial}_\tau \sim \sum_{k=0}^{\infty} \mathcal{H}_{m+k}(\mathbb{R}^n)$$

indexed by a 'cone' of integers exactly as in the case of the Cauchy–Riemann $\bar{\partial}$-operator.

By writing the second of the conditions in (4.17) as

$$\frac{\partial^2 F}{\partial x_j \partial \xi_k} - \frac{\partial^2 F}{\partial x_k \partial \xi_j} = \Delta_{jk}\left(\frac{\partial}{\partial z}\right) F = 0 \qquad \left(z = \begin{bmatrix} x \\ \xi \end{bmatrix}\right)$$

where $\Delta_{jk}\left(\frac{\partial}{\partial z}\right)$ is the Cayley operator on $\mathbb{R}^{2 \times n}$ (see section 4 of chapter 3), we see that $\bar{\partial}_\tau$ is equivalent to the restriction of either of
(4.22)

$$\begin{cases} \text{(a)} & \left(\frac{\partial}{\partial x} \cdot \frac{\partial}{\partial \xi}\right) F = 0 \quad , \quad \Delta_{jk}\left(\frac{\partial}{\partial z}\right) F = 0 \qquad (1 \le j < k \le n) \,, \\[2mm] \text{(b)} & D_x D_\xi F = 0 \end{cases}$$

to functions $F = F(x, \xi)$ harmonic and homogeneous in ξ. But both of these last systems generalize readily to functions $F = F(x, \xi)$ with $\xi \in \mathbb{R}^{r \times n}$; herein lies a clue to characterizations of more general $\bar{\partial}_\tau$ and τ_τ, with the Cauchy–Binet identity

$$\sum_{1 \le k_1 < \cdots < k_r \le n} \Delta_{k_1 \cdots k_r}(z) \Delta_{k_1 \cdots k_r}(\zeta) = \det(z\zeta')$$

for z, ζ in $\mathbb{R}^{r \times n}$ replacing the usual inner product

$$\sum_{1 \le k \le n} z_k \zeta_k = z \cdot \zeta \qquad (z, \zeta \in \mathbb{R}^n) \,,$$

in the proofs of results corresponding to (4.12),...,(4.21), and with
(4.23)

$$\begin{cases} \text{(i)} & F(x) \to \rho(a) F(a) & \left(F : \Omega \to \mathcal{P}(\mathbb{R}^{r \times n})\right) \,, \\[2mm] \text{(ii)} & F(x, \xi) \to F(xa, \xi a) & \left(F : \Omega \times \mathbb{R}^{r \times n} \to \mathbb{C}\right) \,, \\[2mm] \text{(iii)} & F(z) \to F(za) & \left(z = \begin{bmatrix} x \\ \xi \end{bmatrix} \in \mathbb{R}^{(r+1) \times n}\right) \end{cases}$$

replacing (4.11). The rows of each ξ in $\mathbb{R}^{r \times n}$ will be denoted by $\xi_1, \ldots, \xi_r$, where $\xi_j \in \mathbb{R}^n$.

(4.24) Theorem.
The operator $\bar{\partial}_\tau$, $\tau = (m, \ldots, m, 0, \ldots)$, is equivalent to the restriction of

$$\text{(i)} \quad \left(\frac{\partial}{\partial x} \cdot \frac{\partial}{\partial \xi_j}\right) F = 0 \quad (1 \le j \le r) \,,$$

$$\text{(ii)} \quad \Delta_{k_1 \cdots k_{r+1}}\left(\frac{\partial}{\partial z}\right) F = 0 \quad (1 \le k_1 < \cdots < k_{r+1} \le n)$$

to functions $F = F(x, \xi)$ which are $O(n)$-harmonic and determinan-

tally homogeneous of degree m *in* ξ*; in addition,* $\overline{\partial}_\tau$ *is equivalent to the restriction of*

$$(iii) \quad (D_x D_{\xi_1} \ldots D_{\xi_r}) F = 0$$

to these same functions where

$$D_x = \sum_{k=1}^{n} e_k \frac{\partial}{\partial x_k} \quad , \quad D_{\xi_j} = \sum_{k=1}^{n} e_k \frac{\partial}{\partial \xi_{jk}} \quad (1 \le j \le r) \ .$$

Lack of space prevents the inclusion of the proof of (4.24) as well as of the local characterization of solutions of $\overline{\partial}_\tau F = 0$ and the 'cone' type description the Taylor series of these solutions (see Notes and remarks). We shall concentrate instead on the definition of δ_τ. The simple case of $\tau = (2, 1, 0, \ldots)$ indicates already the extra complications that arise when τ is not a multiple of a fundamental representation. For

$$(4.25) \quad (2, 1, 0, \ldots) \otimes d_1$$

$$= (3, 1, 0, \ldots) + (2, 2, 0, \ldots) + (2, 1, 1, 0, \ldots) + (2, 0, \ldots) \ ,$$

with $(3, 1, 0, \ldots)$ being the signature of the Cartan composition; in particular, the tensor product contains four irreducible submodules rather than the three in (4.10). But in each of (4.7)(i) and (4.10) the second and third terms on the right-hand side are the signature of irreducible submodules which account separately for the two operators d, d^* characterizing $\overline{\partial}_\tau$, $\tau = d_k$, and for the two operators in (4.22)(a) characterizing $\overline{\partial}_\tau$, $\tau = (m, \ldots, m, 0, \ldots)$. In (4.25) the corresponding signatures are $(2, 1, 1, 0, \ldots)$ and $(2, 0, \ldots)$, leaving $(2, 2, 0, \ldots)$ in addition to the highest weight $(3, 1, 0, \ldots)$. Simply for expediency, therefore, one might take as subspace V used to define $\delta_\tau = P_V \circ \nabla$, $\tau = (2, 1, 0, \ldots)$, the *sum* of the irreducible submodules having signatures $(3, 1, 0, \ldots)$ and $(2, 2, 0, \ldots)$. Such arbitrariness of choice, however, ignores the intended 'universal' nature of the δ_τ: they should be associated with more general n-dimensional Riemannian manifolds, hyperbolic or spherical space for instance, not just with $\mathbb{R}^n$, and they should be intimately related to the underlying geometric structure and any symmetry structure the manifold has. This is the significance of the Stein–Weiss result (4.4), for it would show that the basic geometric differential operators on Riemannian manifolds (with spin structure if need be) arise from the *fundamental* representations of Spin(n). The δ_τ for more general signatures would thus form a much larger family of such operators on manifolds, with the earlier criteria used to 'motivate' the choice of V indicating the properties this family would have as operators on one particular manifold: Euclidean space. The choice of V, which in the case $\tau = (2, 1, 0, \ldots)$ is

the one suggested, emerged from attempts to realize representations of
the group of isometries of hyperbolic space in just the same way as (3.17)
realizes representations of the group of isometries, the rigid motions, of
Euclidean space. For a general $\tau = (m_1, m_2, \ldots, m_r, 0, \ldots)$ it is known
that

$$(4.26) \qquad \tau \otimes d_1 = \left(\sum_{j=1}^{r} (m_1, \ldots, m_j + 1, \ldots, m_r, 0, \ldots) \right) + \cdots$$

where a term $(m_1, \ldots, m_j + 1, \ldots, m_r, 0, \ldots)$ is to be omitted if $m_{j-1} = m_j$ because it fails condition (2.6) in chapter 3. For $\tau = (m, \ldots, m, 0, \ldots)$
therefore, there is only one such term, the highest weight, but in (4.25)
there were more terms, as there will be in general.

(4.27) Definition.
   Let V be an irreducible Spin(n)-module having signature $\tau = (m_1, \ldots, m_r, 0, \ldots)$. Then $\eth_\tau$ is the operator $P_V \circ \nabla$ on $C^\infty(\mathbb{R}^n, V)$ with V the sum of the irreducible Spin(n)-submodules having signature $(m_1, \ldots, m_j + 1, \ldots, m_r, 0, \ldots)$, $1 \leq j \leq r$.

   The relationship of $\eth_\tau$ to the (d, d^*)-system is known, and hence its
relationship to the Dirac operator is known also. For lack of space we
omit the details (see Notes and remarks).

5 Critical indices of subharmonicity

The subharmonicity property

$$(5.1) \qquad \Delta |F|^p \geq 0 \quad , \quad p \geq \frac{n-2}{n-1} ,$$

for Clifford analytic functions F in $C^\infty(\Omega, \mathfrak{H})$ – where Ω is an open set in
$\mathbb{R}^n$ and $\mathfrak{H}$ is an $\mathfrak{A}_n$-module – plays an important role in boundary regularity problems, as was seen in chapter 2. Specifically, Clifford analytic
functions exhibit better boundary regularity than harmonic functions
do in general, precisely because (5.1) holds for values of p less than 1.
In view of the fact that the kernel of an operator of Dirac type consists
precisely of those functions in the kernel of D having range in some fixed
subspace $\mathfrak{B}$ of the $\mathfrak{A}_n$-module $\mathfrak{H}$, it is natural to ask whether this restriction improves the boundary regularity still further by reducing the lower
bound on the allowed range of p in (5.1). As it happens, this turns out
to be an algebraic question, depending only on the symbol of the operator. With this in mind, we shall develop some basic ideas for general

first-order systems, considering in turn the refinements that are possible for GCR operators, rotation-invariant GCR operators, and (finally) the rotation-invariant operators $\eth_\tau$. We begin with a definition.

(5.2) Definition.

Let $\mathcal{A}, \mathcal{B}$ be finite-dimensional Hilbert spaces over $\mathbb{F} = \mathbb{R}$ or $\mathbb{C}$, and let $A : \mathcal{A} \otimes \mathbb{F}^n \to \mathcal{B}$ be a linear operator. The differential operator $\eth_A$ is said to have the subharmonicity property when there exists p_0 such that, for every open subset Ω of $\mathbb{R}^n$ and for all $f \in C^\infty(\Omega, \mathcal{A})$ satisfying $\eth_A F \equiv 0$,

(5.3) $\Delta |F(x)|^p \geq 0$ *for all $x \in \Omega$ and $p > p_0$.*

Moreover, the smallest choice of p_0 for which (5.3) is true will be called the critical index of subharmonicity for $\eth_A$.

When F is itself harmonic and $p \geq 1$, the mean-value property of harmonic functions ensures that $x \to |F(x)|^p$ is automatically subharmonic. For GCR systems, therefore, the main concern is in seeing whether, and how far, the critical index p_0 falls below 1. Since p_0 will depend only upon the symbol of $\eth_A$, the critical index of subharmonicity as defined in (5.2) will be independent of the open set Ω.

There is a very useful criterion for subharmonicity of solutions of $\eth_A F = 0$ relating the norm in $\mathcal{A}$ of directional derivatives

(5.4) $(\alpha \cdot \nabla)F = \displaystyle\sum_{j=1}^n \alpha_j \frac{\partial F}{\partial x_j}$, $\alpha = (\alpha_1, \ldots, \alpha_n) \in \Sigma_{n-1}$,

of F to the norm in $\mathcal{A} \otimes \mathbb{F}^n$ of its gradient $\nabla F = \overline{\Sigma}_j \frac{\partial F}{\partial x_j} \otimes e_j$.

(5.5) Theorem.

Let $\eth_A = A \circ \nabla$ be a GCR operator and let $p < 2$ be fixed. Then $\Delta |F|^p \geq 0$ on Ω for all solutions of $\eth_A F = 0$ in $C^\infty(\Omega, \mathcal{A})$ if and only if

(5.6) $|(\alpha \cdot \nabla)F(x)|_\mathcal{A} \leq \left(\dfrac{1}{2-p} \right)^{1/2} \|\nabla F(x)\|_{\mathcal{A} \otimes \mathbb{R}^n}$, $\alpha \in \Sigma_{n-1}$,

holds everywhere on Ω for all such solutions.

As a condition on the first-order Taylor coefficients of solutions, (5.6) can be used to express the critical index in terms of the symbol of $\eth_A$.

(5.7) Corollary.

When $\eth_A = A \circ \nabla$ is a GCR operator its critical index of subharmonicity p_0 is given by

$$\left(\frac{1}{2 - p_0} \right)^{1/2} = \max \left| \sum_{j=1}^{n} \alpha_j \xi_j \right|_A,$$

the maximum being taken over all $\alpha = (\alpha_1, \ldots, \alpha_n)$ in Σ_{n-1} and $\xi = \sum_{j=1}^{n} \xi_j \otimes e_j$ in $\mathcal{V}_A$ (= ker A) of norm 1. Furthermore, p_0 is strictly smaller than 1.

Proof of theorem (5.5). When f is harmonic, $x \to |f(x)|_A^p$ is subharmonic on Ω if and only if

$$\sum_{j=0}^{n} \left\{ \Re \left(\frac{\partial f}{\partial x_j}(x), f(x) \right)_A \right\}^2 \le \left(\frac{1}{2 - p} \right) |f(x)|_A^2 \|\nabla f(x)\|_{A \otimes \mathbb{R}^n}^2$$

holds for all x in Ω (see chapter 2, (3.34)). But

$$\left(\sum_{j=1}^{n} \left\{ \Re \left(\frac{\partial f}{\partial x_j}(x), f(x) \right)_A \right\}^2 \right)^{1/2} = \sup \left| \sum_{j=1}^{n} \alpha_j \Re \left(\frac{\partial f}{\partial x_j}(x), f(x) \right)_A \right|$$

$$= \sup \left| \Re ((\alpha \cdot \nabla) f(x), f(x))_A \right|,$$

taking the supremum over all α in Σ_{n-1}. Consequently $x \to |f(x)|_A^p$ is subharmonic on Ω if and only if
(5.8)

$$\sup_{\alpha \in \Sigma_{n-1}} \left| \Re ((\alpha \cdot \nabla) f(x), f(x))_A \right| \le \left(\frac{1}{2 - p} \right)^{1/2} |f(x)|_A \|\nabla f(x)\|_{A \otimes \mathbb{R}^n}$$

holds for all x in Ω.

Now let $\eth_A$ be a GCR operator. Then, in view of (5.8), the Cauchy–Schwarz inequality

$$\left| ((\alpha \cdot \nabla) F(x), F(x))_A \right| \le |(\alpha \cdot \nabla) F(x)|_A |F(x)|_A$$

ensures that $x \to |F(x)|_A^p$ is subharmonic on Ω for every solution of $\eth_A F = 0$ whenever (5.6) holds for all such solutions. To establish the converse, choose any solution of $\eth_A F = 0$, and fix $x \in \Omega$. We have to show that (5.6) holds for this choice of F and x whenever (5.8) holds for all f in the kernel of $\eth_A$. But near x there exist $\xi_0, \xi_1, \ldots, \xi_n$ in $\mathcal{A}$ so that the Taylor expansion of F of order 1 is

$$F(y) = F(x) + \sum_{j=1}^{n} (y_j - x_j) \xi_j + o(|x - y|)$$

with $\xi = \sum_{j=1}^{n} \xi_j \otimes e_j$ in $\mathcal{V}_A$ since $\eth_A F(x) = 0$ (see the proof of theorem

(1.8) in the present chapter). In this case, (5.6) reads

$$(5.9) \qquad \sup_{\alpha} \left| \sum_{j=1}^{n} \alpha_j \xi_j \right|_{\mathcal{A}} \le \left(\frac{1}{2-p} \right)^{1/2} \|\xi\|_{\mathcal{A} \otimes \mathbb{R}^n} .$$

Now fix α in Σ_{n-1}. Then there exists $\eta \in \mathcal{A}$, $|\eta|_{\mathcal{A}} = 1$, so that

$$\left| \sum_{j=1}^{n} \alpha_j \xi_j \right|_{\mathcal{A}} = \sum_{j=1}^{n} \alpha_j (\xi_j, \eta)_{\mathcal{A}} .$$

On the other hand, as a member of the kernel of $\delta_{\mathcal{A}}$, the first-order polynomial

$$f(y) = \eta + \sum_{j=1}^{n} (y_j - x_j) \xi_j$$

must satisfy (5.8), i.e.,

$$\sum_{j=1}^{n} \alpha_j (\xi_j, \eta)_{\mathcal{A}} \le \left(\frac{1}{2-p} \right)^{1/2} \|\xi\|_{\mathcal{A} \otimes \mathbb{R}^n} ,$$

establishing (5.9) and completing the proof of the theorem. ∎

Proof of corollary (5.7). By (5.9),

$$\left(\frac{1}{2-p_0} \right)^{1/2} = \sup \left| \sum_{j=1}^{n} \alpha_j \xi_j \right|_{\mathcal{A}} ,$$

the supremum being taken over all α in Σ_{n-1} and ξ in $\mathcal{V}_{\mathcal{A}}$ of norm 1. But compactness of the unit sphere in a finite-dimensional Hilbert space ensures that this supremum is attained, i.e.,

$$(5.10) \qquad \left(\frac{1}{2-p_0} \right)^{1/2} = \left| \sum_{j=1}^{n} \lambda_j \xi_j \right|_{\mathcal{A}} \le 1$$

for a particular choice of $\lambda \in \Sigma_{n-1}$ and $\xi \in \mathcal{V}_{\mathcal{A}}$ of norm 1; in particular, $p_0 \le 1$. Suppose now that $p_0 = 1$, so that

$$(5.11) \qquad \left| \sum_{j=1}^{n} \lambda_j \xi_j \right|_{\mathcal{A}} = \left(\sum_{j=1}^{n} |\xi_j|_{\mathcal{A}}^2 \right)^{1/2} = 1 .$$

Now choose $\sigma(a) = [a_{jk}]$ so that $\sigma(a^{-1})\lambda = \mathbf{1} = (1, 0, \dots, 0)$, and define $\eta = \sum_{j=1}^{n} \eta_j \otimes e_j$ in $\mathcal{A} \otimes \mathbb{R}^n$ by

$$\eta_j = \sum_{j=1}^{n} a_{jk} \xi_k , \qquad 1 \le j \le n .$$

Then

$$\sum_{j=1}^{n} |\eta_j|_{\mathcal{A}}^2 = \sum_{j=1}^{n} |\xi_j|_{\mathcal{A}}^2 = 1$$

since $[a_{jk}]$ is orthogonal, while

$$\sum_{j=1}^{n} \lambda_j \xi_j = \sum_{j,k} \lambda_k a_{jk} \eta_j = \eta_1$$

by the choice of $[a_{jk}]$ in $SO(n)$. In view of (5.11), therefore, $\eta_j = 0$ when $1 < j \le n$. Hence

$$\left(\sum_{k=1}^{n} \lambda_k A_k\right) \eta_1 = \sum_{k=1}^{n} a_{1k} A_k \eta_1 = \sum_{k=1}^{n} A_k \left(\sum_{j=1}^{n} a_{jk} \eta_j\right) = \sum_{k=1}^{n} A_k \xi_k = 0 \ ,$$

contradicting the ellipticity of the GCR operator $\eth_A$. Thus $p_0 < 1$, and the proof is complete. ∎

When $\eth_A$ is rotation-invariant, the criterion of corollary (5.7) simplifies further because $\mathcal{V}_A$ is then a $\mathrm{Spin}(n)$-module.

(5.12) Theorem.

Fix $\nu = (\nu_1, \ldots, \nu_n)$ in Σ_{n-1}. Then the critical index of subharmonicity p_0 of any rotation-invariant GCR operator $\eth_A = A \circ \nabla$ on $C^\infty(\mathbb{R}^n, \mathcal{A})$ satisfies

$$(5.13) \qquad \left(\frac{1}{2 - p_0}\right)^{1/2} = \max \left| \sum_{j=1}^{n} \nu_j \eta_j \right|_{\mathcal{A}} \ ,$$

the maximum being taken over all $\eta = \sum_{j=1}^{n} \eta_j \otimes e_j$ in $\mathcal{V}_A$ of norm 1.

Proof. By (5.10) the left-hand side of (5.13) satisfies

$$(5.14) \qquad \left(\frac{1}{2 - p_0}\right)^{1/2} = \left| \sum_{j=1}^{n} \lambda_j \xi_j \right|_{\mathcal{A}} \ge \max \left| \sum_{j=1}^{n} \nu_j \eta_j \right|_{\mathcal{A}}$$

for some $\lambda \in \Sigma_{n-1}$ and $\xi \in \mathcal{V}_A$ of norm 1. Now choose a in $\mathrm{Spin}(n)$ so that

$$\sigma(a^{-1}) \lambda = \nu \ , \qquad \nu_j = \sum_{k=1}^{n} \lambda_k a_{jk} \qquad (1 \le j \le n) \ ,$$

where $[a_{jk}]$ is the image of $\sigma(a)$ in $SO(n)$ under the covering homomorphism from $\mathrm{Spin}(n)$ onto $SO(n)$ (see (3.10)). Define $\eta \in \mathcal{A} \otimes \mathbb{R}^n$ of norm 1 by

$$\eta_j = \sum_{k=1}^{n} a_{jk} \tau(a^{-1}) \xi_k \qquad (0 \le j \le n) \ .$$

By the rotation-invariance of δ_A, together with theorem (3.9), we have

$$A\eta = \sum_{j=1}^{n} A_j\eta_j = \sum_{j,k} A_j a_{jk}\tau(a^{-1})\xi_k = \sum_k \left(\sum_j A_j a_{jk} \right)\tau(a^{-1})\xi_k$$

$$= \sum_k \omega(a^{-1})A_k\xi_k = \omega(a^{-1})A\xi = 0 \ .$$

On the other hand,

$$\sum_{j=1}^{n} \nu_j\eta_j = \sum_{j=1}^{n} \left(\sum_{r,s} \lambda_r a_{jr} a_{js}\tau(a^{-1})\xi_s \right)$$

$$= \tau(a^{-1}) \sum_{s=1}^{n} \lambda_s\xi_s \ .$$

Consequently, for this choice of $\eta \in \mathcal{V}_A$ with norm 1, we have

$$\left| \sum_{j=1}^{n} \lambda_j\xi_j \right|_{\mathcal{A}} = \left| \sum_{j=1}^{n} \nu_j\eta_j \right|_{\mathcal{A}}$$

which establishes (5.13). ∎

In practice, one natural choice of ν is $\mathbf{1} = (0,0,\ldots,1)$, in which case (5.13) becomes

$$(5.15) \qquad \left(\frac{1}{2-p_0} \right)^{1/2} = \max |\eta_n|_{\mathcal{A}} \ ,$$

the maximum being taken over all $\eta \in \mathcal{A} \otimes \mathbb{R}^n$ satisfying

$$(5.15)' \qquad A\eta = \sum_{j=1}^{n} A_j\eta_j = 0 \ , \qquad \sum_{j=1}^{n} |\eta_j|_{\mathcal{A}}^2 = 1 \ .$$

In sufficiently simple cases, (5.15) can be evaluated by direct algebraic calculation. For instance, in the case of the Spin(n)-module $\mathcal{A} = \Lambda^r(\mathbb{R}^n)$, Stein–Weiss have proved

(5.16) Theorem.

The critical index of subharmonicity for the restriction of the Hodge–deRham (d, d^*)-system to r-forms on $\mathbb{R}^n$ satisfies

$$p_0 \le \frac{n-r-1}{n-r} \ , \qquad 1 \le r < \left[\frac{n}{2} \right] \ .$$

Proof. As in the proof of the corresponding result for the Dirac operator D on Clifford-algebra-valued functions, the proof hinges on the fact that the inequality

$$(5.17) \qquad |c_1|^2 \le \frac{m-1}{m}(|c_1|^2 + \cdots + |c_m|^2) \ , \qquad c_j \in \mathbb{C} \ ,$$

always holds whenever $c_1 + c_2 + \cdots + c_m = 0$. There is a natural identification of $\Lambda^r(\mathbb{R}^n) \otimes \mathbb{R}^n$ with the set of sequences of the form $\{\frac{\partial \phi}{\partial x_j}\}_{j=1}^n$, where ϕ is allowed to range over a special class of r-forms. Specifically, ϕ is given by

$$\phi(x) = \sum_\alpha (x \cdot a_\alpha) e_{\alpha_1} \wedge \cdots \wedge e_{\alpha_r}$$

where the sum is taken over all ordered r-element subsets $\alpha = \{\alpha_1 < \alpha_2 < \cdots < \alpha_r\}$ of $\{1, 2, \ldots, n\}$, and, for each such subset, a_α is an element of $\mathbb{R}^n$. The corresponding subspace $\mathcal{V}$ of $\Lambda^r(\mathbb{R}^n) \otimes \mathbb{R}^n$ determined by (d, d^*) consists of all ϕ such that $d\phi = 0$, $d^*\phi = 0$. Thus the critical index p_0 is given by

$$\frac{1}{2 - p_0} = \max \left| \frac{\partial \phi}{\partial x_n} \right|^2 = \max \sum_\alpha (e_n \cdot a_\alpha)^2 \, ,$$

the maximum being taken over all choices $\{a_\alpha\} \subseteq \mathbb{R}^n$ satisfying
(5.18)

$$\begin{cases} \text{(i)} \quad \sum_{j,\alpha} (e_j \cdot a_\alpha) e_j \wedge e_{\alpha_1} \wedge \cdots \wedge e_{\alpha_r} = 0 \, , \\[2em] \text{(ii)} \quad \sum_{j,\alpha} \sum_{k=1}^r (-1)^{k-1} (e_j \cdot a_\alpha) \delta_{j\alpha_k} e_{\alpha_1} \wedge \cdots \wedge \hat{e}_{\alpha_k} \wedge \cdots \wedge e_{\alpha_r} = 0 \end{cases}$$

as well as

$$\sum_\alpha |a_\alpha|^2 = \sum_{j,\alpha} (e_j \cdot a_\alpha)^2 = 1 \, .$$

Now, let $\alpha = (\alpha_1, \ldots, \alpha_r)$ be fixed, and assume first that $n \notin \alpha$. For each k, $1 \leq k \leq r$, let $\beta(k)$ denote the r-element set obtained from α by deleting α_k and adding n. Then the coefficient of $e_{\alpha_1} \wedge \cdots \wedge e_{\alpha_r} \wedge e_n$ in (5.18)(i) is given by

$$(-1)^r (e_n \cdot a_\alpha) + \sum_{k=1}^r (-1)^{k-1} (e_{\alpha_k} \cdot a_{\beta(k)}) = 0 \, .$$

Thus, by (5.17),

$$(e_n \cdot a_\alpha)^2 \leq \frac{r}{r+1} \left\{ (e_n \cdot a_\alpha)^2 + \sum_{k=1}^r (e_{\alpha_k} \cdot a_{\beta(k)})^2 \right\} \, .$$

Next, we consider the case in which $n \in \alpha$. Let $\mu = \{\mu_1 < \mu_2 < \cdots < \mu_{n-r}\}$ denote the complement of α in $\{1, 2, \ldots, n\}$, and, for $1 \leq k \leq n-r$, let $\gamma(k)$ denote the ordered r-element set obtained from α by adding μ_k and deleting n. Then the coefficient of $e_{\alpha_1} \wedge e_{\alpha_2} \wedge \cdots \wedge e_{\alpha_{r-1}}$

in (5.18)(ii) is given by

$$(-1)^{r-1}(e_n \cdot a_\alpha) + \sum_{k=1}^{n-r} \pm(e_{\mu_k} \cdot a_{\gamma(k)}) = 0 \,,$$

where the sign associated to $(e_{\mu_k} \cdot a_{\gamma(k)})$ depends upon the placement of μ_k in $\gamma(k)$. In this case, (5.17) implies

$$(e_n \cdot a_\alpha)^2 \leq \frac{n-r}{n-r+1}\left\{(e_n \cdot a_\alpha)^2 + \sum_{k=1}^{n-r}(e_{\mu_k} \cdot a_{\gamma(k)})^2\right\} \,.$$

Consequently,

$$\sum_\alpha (e_n \cdot a_\alpha)^2 \leq \max\left\{\frac{r}{r+1} \,,\, \frac{n-r}{n-r+1}\right\} \sum_\alpha |a_\alpha|^2 \,.$$

Since

$$\frac{n-r}{n-r+1} > \frac{r}{r+1}$$

when $r < [\frac{n}{2}]$, the inequality on p_0 follows immediately. ∎

Next, we compute the critical index of subharmonicity for the higher gradients operator of degree m, a result originally due to Calderón and Zygmund:

(5.19) Theorem.
  For the representation $(\mathcal{H}_\tau, \tau)$, $\tau = (m, 0, \ldots, 0)$ of $SO(n)$ on the space $\mathcal{H}_m(\mathbb{R}^n)$ of spherical harmonics homogeneous of degree m, the critical index of subharmonicity for the higher gradients operator $\eth_\tau$ is given by

$$p_\tau = \frac{n-2}{n+m-2} \,.$$

Proof. In section 4 we wrote $\eth_\tau$ as $A_\tau \circ \nabla$, and saw that the kernel of A_τ in $\mathcal{H}_m(\mathbb{R}^n) \otimes \mathbb{R}^n$ is given by

$$\mathcal{V}_\tau = \left\{\sum_{j=1}^n \frac{\partial f}{\partial \xi_j} \otimes e_j : f \in \mathcal{H}_{m+1}(\mathbb{R}^n)\right\} \,.$$

Thus, by (5.13),
(5.20)

$$M_\tau \equiv (2 - p_\tau)^{-1} = \max\left\{\left\|\frac{\partial f}{\partial \xi_n}\right\|^2 \Big/ \sum_{j=1}^n \left\|\frac{\partial f}{\partial \xi_n}\right\|^2 : f \in \mathcal{H}_{m+1}(\mathbb{R}^n)\right\} \,,$$

the norm $\|\cdot\|$ denoting the L^2-norm with respect to normalized surface measure on the unit sphere Σ_{n-1} of $\mathbf{R}^n$. We shall prove first that

$$(5.21) \qquad \sum_{j=1}^{n}\left\|\frac{\partial f}{\partial \xi_j}\right\|^2 = (m+1)(n+2m)\|f\|^2 , \qquad f \in \mathcal{H}_{m+1}(\mathbf{R}^n) .$$

To see this, note that by Green's theorem and homogeneity,

$$\int_{|\xi|\leq 1}\left(\sum_j\left|\frac{\partial f}{\partial \xi_j}\right|^2\right)d\xi = \int_{\Sigma_{n-1}} f(\xi')\frac{\partial f}{\partial r}\,d\omega(\xi')$$

$$= (m+1)\|f\|^2$$

(where $d\omega$ denotes normalized surface measure on Σ_{n-1}). On the other hand, since $\frac{\partial f}{\partial \xi_j}$ is homogeneous of degree m,

$$\int_{|\xi|\leq 1}\left(\sum_j\left|\frac{\partial f}{\partial \xi_j}\right|^2\right)d\xi = \int_0^1 r^{n+2m-1}\left\{\int_{\Sigma_{n-1}}\left(\sum_j\left|\frac{\partial f}{\partial \xi_j}\right|^2\right)d\omega(\xi')\right\}dr$$

$$= \frac{1}{n+2m}\sum_{j=1}^{n}\left\|\frac{\partial f}{\partial \xi_j}\right\|^2 ,$$

whence (5.21) follows.

In view of (5.21), the calculation of M_τ is reduced to that of
(5.22)
$$M_\tau' \equiv \max\left\{\left\|\frac{\partial f}{\partial \xi_n}\right\|^2 : f \in \mathcal{H}_{m+1}(\mathbf{R}^n) , \ \|f\| = 1\right\} = (m+1)(n+2m)M_\tau ,$$

which may be interpreted as the norm of the operator $\frac{\partial}{\partial \xi_n}$ as a mapping from $\mathcal{H}_{m+1}(\mathbf{R}^n)$ into (in fact, onto) $\mathcal{H}_m(\mathbf{R}^n)$. The special role assigned to the **1**-direction will be critical in evaluating M_τ'.

By the class 1 property (chapter 3, (3.27)) the function $\phi_m^{(\lambda)}$, $\lambda = \frac{1}{2}(n-2)$, is the unique element of $\mathcal{H}_m(\mathbf{R}^n)$ satisfying

$$(5.23) \qquad \phi_m^{(\lambda)}(\mathbf{1}) = \phi_m^{(\lambda)}(0,0,\ldots,0,1) = 1$$

which is invariant under the subgroup $SO(n-1)$ of $SO(n)$ of rotations leaving **1** fixed. Moreover,

$$(5.24) \qquad \|\phi_m^{(\lambda)}\|^2 = \int_{\Sigma_{n-1}} |\phi_m(\xi')|^2\,d\omega(\xi') = d_{m,n}^{-1}$$

where

$$(5.25) \qquad d_{m,n} = \dim \mathcal{H}_m(\mathbf{R}^n) = \left(\frac{n+2m-2}{n+m-2}\right)\frac{(n+m-2)!}{m!(n-2)!} .$$

Now, by SO_{n-1}-invariance, $\partial\phi_m^{(\lambda)}/\partial\xi_n$ is a constant multiple of $\phi_{m-1}^{(\lambda)}$; but, by homogeneity,

$$\sum_{j=1}^{n}\xi_j\frac{\partial\phi_m}{\partial\xi_j} = m\phi_m(\xi) .$$

Evaluating at $\xi = 1$, we obtain

(5.26)
$$\frac{\partial \phi_m^{(\lambda)}}{\partial \xi_n} = m \phi_{m-1}^{(\lambda)} \ .$$

Thus, (5.24) and (5.26) yield

(5.27)
$$\left\| \frac{\partial \phi_{m+1}^{(\lambda)}}{\partial \xi_n} \right\| \Big/ \|\phi_{m+1}^{(\lambda)}\| = (m+1) \left(\frac{d_{m+1,n}}{d_{m,n}} \right)^{1/2}$$

Next, we claim that

(5.28)
$$\left\| \frac{\partial f}{\partial \xi_n} \right\| \Big/ \|f\| \le \left\| \frac{\partial \phi_{m+1}^{(\lambda)}}{\partial \xi_n} \right\| \Big/ \|\phi_{m+1}^{(\lambda)}\| \ , \qquad f \in \mathcal{H}_{m+1}(\mathbb{R}^n) \ .$$

Once we have proved this, the theorem follows: for (5.22), (5.27), and (5.28) together imply that

$$M_\tau = \frac{m+1}{n+2m} \frac{d_{m+1,n}}{d_{m,n}} \ ;$$

a simple calculation using (5.25) yields

(5.29)
$$M_\tau = \frac{n+m-2}{n+2m-2} \ , \quad p_\tau = 2 - M_\tau^{-1} = \frac{n-2}{n+m-2} \ .$$

Thus it remains to prove (5.28), which will be accomplished by means of the explicit decomposition of $\mathcal{H}_{m+1}(\mathbb{R}^n)$ into its SO_{n-1}-invariant subspaces (see chapter 3, (3.30)). Each f in $\mathcal{H}_{m+1}(\mathbb{R}^n)$ may be written uniquely as

$$f(\xi) = \sum_{j=0}^{m+1} \phi_{m+1-j}^{(\lambda+j)}(\xi) P_j(x) \ , \qquad \xi = (x, \xi_n) \ , \ P_j \in \mathcal{H}_j(\mathbb{R}^{n-1}) \ .$$

In particular, if $\|f\| = 1$, we may write

(5.30)
$$f(\xi) = \sum_{j=0}^{m+1} \alpha_j d_{m+1-j, n+2j}^{1/2} \ \phi_{m+1-j}^{(\lambda+j)}(\xi) u_j(x)$$

where $\sum_{j=0}^{m+1} \alpha_j^2 = 1$ and each u_j is an element of an orthonormal basis for $\mathcal{H}_j(\mathbb{R}^{n-1})$. Since the u_j are independent of ξ_n, (5.30) and (5.26) imply

(5.31)
$$\frac{\partial f}{\partial \xi_n} = \sum_{j=0}^{m} \alpha_j d_{m+1-j, n+2j}^{1/2} (m+1-j) \phi_{m-j}^{(\lambda+j)}(\xi) u_j(x)$$

$$= \sum_{j=0}^{m} \alpha_j \beta_j d_{m-j, n+2j}^{1/2} \ \phi_{m-j}^{(\lambda+j)}(\xi) u_j(x)$$

where

(5.32) $$\beta_j = (m+1-j)\left(\frac{d_{m+1-j,\,n+2j}}{d_{m-j,\,n+2j}}\right)^{1/2}$$

$$= \left[\frac{(m+1-j)(n+2m)(n+m+j-2)}{n+2m-2}\right]^{1/2}$$

by (5.25). Now

$$d_{m-j,\,n+2j}^{1/2}\,\phi_{m-j}^{(\lambda+j)}(\xi)u_j(x)\,,\qquad j=0,1,\ldots,m\,,$$

are elements of an orthonormal basis for $\mathcal{H}_m(\mathbb{R}^n)$, so that (5.31) gives

$$\left\|\frac{\partial f}{\partial\xi_n}\right\|^2 = \sum_{j=0}^m \alpha_j^2\beta_j^2 \le \beta_0^2\sum_{j=0}^m \alpha_j^2 \le \beta_0^2\,,$$

since (5.32) shows that β_j^2 is a strictly decreasing function of j. But, by (5.27) and (5.32),

$$\beta_0^2 = \left\|\frac{\partial\phi_{m+1}^{(\lambda)}}{\partial\xi_n}\right\|^2 \Big/ \|\phi_{m+1}^{(\lambda)}\|^2\,,$$

from which (5.28) and the theorem follow.   ∎

Estimation of the critical index of subharmonicity for more complicated rotationally invariant operators of Dirac type is the subject of ongoing research and little is known. It involves the subtle interplay between the representation theory and the geometry of how the subspace $\mathfrak{B}$ 'lies' inside the Clifford module $\mathfrak{H}$.

We conclude this section with a discussion of the effect of restricting the range space of the Dirac operator on the boundary regularity of the Clifford analytic functions in its kernel. Suppose that $\mathfrak{H}$ is an $\mathfrak{A}_n$-module and $\mathfrak{B}$ is a subspace of $\mathfrak{H}$. Let M be a domain in $\mathbb{R}^n$ which is either a bounded Lipschitz domain, or the unbounded special Lipschitz domain above the graph of a Lipschitz function. As in chapter 2, section 7, we consider the Hardy spaces $H^p(M,\mathfrak{B})$ and $H^p(\partial M,\mathfrak{B})$. Let $p_{\mathfrak{B}}$ denote the critical index of subharmonicity for the restriction of D to $\mathfrak{B}$-valued functions; by (5.1) clearly $p_{\mathfrak{B}} \le \frac{n-2}{n-1}$. Now let q_0 denote the larger of 1, $p_{\mathfrak{B}}(2-\varepsilon(\beta))$. Straightforward modification of the proof of theorem (7.9) of chapter 2 yields the following.

(5.33) Theorem.
Let $p > q_0$ and suppose $F \in H^p(M,\mathfrak{B})$. Then F^+ exists almost everywhere and belongs to $L^p(\partial M,\mathfrak{B})$, and $F = C_M F^+$ in M. Moreover, when $q_0 = 1$ the result is also true for $p = 1$.

We also obtain the following analogues of corollaries (7.21) and (7.22) of chapter 2.

(5.34) Corollary.
Let $p > q_0$ and suppose $f \in L^p(\partial M, \mathfrak{B})$. Then there exists $F \in H^p(M, \mathfrak{B})$ such that $f = F^+$ almost everywhere on ∂M if and only if $\mathcal{H}_M f = f$. If $p = q_0 = 1$, then given $f \in L^1(\partial M, \mathfrak{B})$ there exists $F \in H^1(M, \mathfrak{B})$ with $F^+ = f$ almost everywhere if and only if $\mathcal{H}_M f \in L^1(\partial M, \mathfrak{B})$ and $\mathcal{H} f = f$.

(5.35) Corollary.
For $p > q_0$, $H^p(\partial M, \mathfrak{B})$ is the subspace of $L^p(\partial M, \mathfrak{B})$ consisting of those functions f for which $\mathcal{H}_M f = f$. If $q_0 = 1$, then $H^1(\partial M, \mathfrak{B})$ is the subspace of $L^1(\partial M, \mathfrak{B})$ consisting of those functions f for which $\mathcal{H}_M f \in L^1(\partial M, \mathfrak{B})$ and $\mathcal{H}_M f = f$.

For a given β and $\mathfrak{B}$, q_0 is always strictly less than $2 - \varepsilon(\beta)$. In particular, this shows that the $L^p(\partial M, \mathfrak{B})$-Dirichlet problem for the operator $\mathcal{D}$ in M is uniquely solvable for a larger range of ps than the corresponding problem for the Laplacian. Moreover, $H^p(\partial M, \mathfrak{B})$ is a space of functions for $q_0 < p < 2 - \varepsilon(\beta)$; recent results of Carlos Kenig and Jill Pipher show that for $p < 2 - \varepsilon(\beta)$, the space of boundary values of harmonic $H^p(M)$-functions is a space of distributions. The trade-off for this improved boundary regularity is expressed in the condition $\mathcal{H}_M f = f$: the Hilbert transform of a $\mathfrak{B}$-valued function is constrained to live in $\mathfrak{B}$. This forces f to be in the kernel of a (possibly very large) system of singular integral operators on ∂M. Indeed, if P is the projection onto the orthogonal complement $\mathfrak{B}^\perp$ in $\mathfrak{H}$, then the system of singular integral equations is given by $P(\mathcal{H}_M f) = 0$. This system may be thought of as imposing 'gauge conditions' or 'moment conditions' on the boundary.

Note that if $\varepsilon(\beta)$ is very small, $p_{\mathfrak{B}}$ must be less than $\frac{1}{2}$ in order to satisfy $q_0 = 1$. The critical index is substantially less than $\frac{1}{2}$ for higher gradients operators of sufficiently high degree (i.e., in $\mathbb{R}^n$, consider $m > n - 2$). Of course, if M is already C^1, then $q_0 = 1$ and we obtain an appropriate characterization of $H^p(\partial M, \mathfrak{B})$ as a space of functions for $1 \le p < \infty$.

We note, also, that the condition $P(\mathcal{H}_M f) = 0$ often imposes a simplified structure on the operator $\mathcal{H}_M$. For example, consider the (d, d^*) system on smooth 1-forms, which is equivalent to

$$D : C^\infty(M, \mathfrak{A}_n^{(1)}) \longrightarrow C^\infty(M, \mathfrak{A}_n^{(0)} \oplus \mathfrak{A}_n^{(2)}) .$$

Now suppose that $f \in H^p(\partial M, \mathfrak{A}_n^{(1)})$ for some $p > (2 - \varepsilon)\frac{n-2}{n-1}$. For ease of notation, we write

(5.36)

$$W(x - u) = \frac{u - x}{|u - x|^n} = \sum_{j=1}^n W_j(x - u)e_j , \quad f = \sum_{j=1}^n f_j e_j , \quad \eta = \sum_{j=1}^n \eta_j e_j .$$

Then

$$(5.37) \qquad \mathcal{H}_M f(x) = \frac{2}{\omega_n} \int_{\partial M} W(x - u)\eta(u)f(u)\, dS(u)$$

and the projection onto $\mathfrak{A}_n^{(1)}$ of the integrand $W(x-u)\eta(u)f(u)$ of (5.37) is easily computed to be

$$(5.38) \qquad - \sum_{\ell,m=1}^{n} W_\ell(x - u)\eta_\ell(u)f_m(u)e_m$$

$$+ \sum_{\ell \neq m} \left[W_\ell(x - u)\eta_m(u) - W_m(x - u)\eta_\ell(u) \right] f_\ell(u)e_m \ .$$

For $1 \leq m \leq n$, the e_m-component of (5.38) is equal to

$$(5.39) \qquad \sum_{\nu=1}^{n} T_{m\nu}\big(W(x - u), f(u) \big)\eta_\nu(u)$$

where $T_{m\nu} : \mathbb{R}^n \times \mathbb{R}^n \to \mathbb{R}$ is a bilinear map defined for $x, y \in \mathbb{R}^n$ by

$$(5.40) \qquad T_{m\nu}(x, y) = \delta_{m\nu}(x \cdot y) - (x_\nu y_m + x_m y_\nu) \ ,$$

which is the bilinear form associated to Maxwell's stress tensor. Thus the condition $\mathcal{H}_M f = f$ becomes, in this case,

$$(5.41) \qquad f(x) = \sum_{m,\nu=1}^{n} \frac{2}{\omega_n} \int_{\partial M} T_{m\nu}\big(W(x - u), f(u) \big)\eta_\nu(u)\, dS(u)e_m \ .$$

Notes and remarks for chapter 4

1. For a thorough discussion of generalized Cauchy–Riemann systems, see [96]. The results of Adams are discussed in [1] and [61]. An extended analysis of the Dirac $\mathcal{D}$ and $\overline{\mathcal{D}}$-operators has been made in [17].

2. The notion of involution compatible with a Clifford module structure has become very important in the developments surrounding the Atiyah–Singer index theorem (see [52], [90], [104]).

3. The paper of Stein and Weiss [95] is an important historical reference for the balance of the chapter.

4. The principal ideas are taken from [29], based on work summarized in [30], [31]. The explicit decomposition of tensor products has generated a large industry demanding great skill and insight (see for instance [32], [60], [76], [103]).

5. The subharmonicity property is discussed in [96]. and several explicit estimates are obtained in [95]. For the proof of theorem (5.16), the

reader may wish to consult [21] and [74]. The work of Kenig and Pipher is found in [65]. The representation formula (5.41) was originally proved by Paul Weiss in 1946 by means of Maxwell's reciprocal theorem (see [102]).

5

Dirac operators on manifolds

In this chapter we turn to the analysis of Clifford algebras and modules, together with an analysis of the associated Dirac operators, on Riemannian manifolds more general than the open subsets of Euclidean space studied in the previous chapter. Concepts from differential geometry are needed from the outset, but in keeping with the spirit of making the material available to more classically trained analysts we have attempted to minimize the use of differential geometric machinery, possibly at the expense of clarity and elegance. In the first section, therefore, Dirac operators are introduced explicitly on a single coordinate patch of a manifold. This serves several useful purposes. It helps bring out quite simply the role that the curvature of the manifold plays in the expression for D^2. For the 'flat' case studied earlier, $(-D^2)$ was the Euclidean Laplacian, and solutions of $DF = 0$ were automatically harmonic: the forerunner of the GCR property of operators. But the fundamental Bochner–Weitzenböck theorem expresses $(-D^2)$ in general as a second-order Laplacian together with a zero-order curvature operator. This idea is a basic one throughout the chapter. In the specific examples of the spinor Laplacian and the Hodge Laplacian, the curvature operator is explicitly calculated. By assuming that the coordinates form a normal coordinate system, we also express $(-D^2)$ asymptotically as a sum of an operator, which will play a fundamental role in the proof of the Atiyah–Singer index theorem, and a remainder operator.

Section 2 deals with the problem of passing from a local setting to a global setting. It is accomplished at the cost of technical complexity through the use of principal bundles, but this brings out clearly the need for a spin structure on the manifold if Clifford modules are to be defined on the manifold. Implicit in the section is the importance of the same linear algebra associated with $SO(n)$- and $\text{Spin}(n)$-modules as in previous chapters. Technically speaking, the analysis exploits the $SO(n)$- or $\text{Spin}(n)$-structure on the manifold, whether it is flat or curved.

Section 3 shows how the passage from $\mathbb{R}$ and $\mathbb{C}$ to more general Clifford algebras allows the notion of fractional linear transformations on $\mathbb{R}$ and $\mathbb{C}$ to be defined naturally on higher-dimensional Euclidean space. Then in turn this is used in section 4 to give a brief, introductory account of representation theory for $\text{Spin}_0(n,1)$ in parallel with the well-known description of the unitary representations of $SL(2,\mathbb{R})$, thinking of $SL(2,\mathbb{R})$ as a transformation group on $\mathbb{C}$. We derive in the process explicit realizations of the Dirac operator on hyperbolic and spherical space. No mention is made of the Bochner–Weitzenböck formulas associated with these operators, simply because we do not develop the representation theory sufficiently far to get eigenspace representations, but their role is still a basic one.

Finally in sections 5 and 6 we give the analytic proof of the Atiyah–Singer index theorem for Dirac operators, using Getzler's simplification of the heat kernel approach for the heat kernel associated with $-D^2$. Of course, the curvature of the manifold enters in a fundamental way once again.

Except for section 4, $\mathfrak{H}$ throughout the chapter will denote an $\mathfrak{A}_n$-module on which there is a representation τ of $\text{Spin}(n)$ such that

$$\tau(a)e_j\tau(a)^{-1} = \sigma(a)e_j \qquad (1 \leq j \leq n)$$

for all a in $\text{Spin}(n)$, where $\sigma : \text{Spin}(n) \to SO(n)$ is the two-fold covering homomorphism.

1 Local theory

Let

(1.1) $g : U \to \mathbb{R}^{n \times n}$, $g(x) = \big[g_{ij}(x)\big]$

be a smooth, matrix-valued function defined on an open set U in $\mathbb{R}^n$; $g(x)$ will always be taken to be positive-definite and symmetric. Then

(1.2)(i) $g_x(\xi,\eta) = \sum_{i,j} g_{ij}(x)\xi_i\eta_j \qquad (\xi,\eta \in \mathbb{R}^n)$

is a positive-definite inner product on $\mathbb{R}^n$ varying smoothly with x, while

$$(1.2)(ii) \quad g_x(X,Y) = \sum_{i,j} g_{ij}(x)a_i b_j \quad \left(X = \sum_i a_i \frac{\partial}{\partial x_i} \,,\, Y = \sum_j b_j \frac{\partial}{\partial x_j}\right)$$

defines a positive-definite inner product on the tangent space $T_x(U)$ to U at x. Thus U can be regarded as a coordinate neighborhood for a Riemannian manifold M taking $x = (x_1, \ldots, x_n)$ as coordinates and (1.2)(ii) as inner product, making it possible to study the Dirac operator on M by introducing it as a non-constant-coefficient non-homogeneous first-order system of differential operators on $C^\infty(U, \mathfrak{H})$. Its properties can then be developed in a completely explicit way without excessive use of differential geometry. The global definition will be given in section 2, before discussing hyperbolic and spherical space in section 3. There are two basic problems to be dealt with first:

(A) *the definition of D, proof of ellipticity and essential self-adjointness*;

(B) *the identification of D^2 as the sum of a second-order Laplacian and a curvature operator* (Bochner–Weitzenböck theorem).

Let

$$(1.3)(i) \qquad g^{-1} : U \to \mathbb{R}^{n \times n} \quad , \quad g^{-1}(x) = \left[g^{ij}(x)\right]$$

be the matrix-valued function inverse to g in the sense that

$$(1.3)(ii) \qquad g(x)g^{-1}(x) = I = [\delta_{ij}] \qquad (x \in U) ,$$

and let $\gamma, \gamma^{-1} : U \to \mathbb{R}^{n \times n}$,

$$(1.4) \qquad \gamma(x) = \left[\gamma_{ij}(x)\right] \quad , \quad \gamma^{-1}(x) = \left[\gamma^{ij}(x)\right] \qquad (x \in U) ,$$

be the unique square roots of g and g^{-1} respectively. Every one of g^{-1}, γ and γ^{-1} is positive-definite and symmetric because g is.

(1.5) Definition.
    When $e_1, \ldots, e_n$ are the skew-adjoint operators on $\mathfrak{H}$ satisfying the Clifford condition $e_j e_k + e_k e_j = -2\delta_{jk}$, set

$$e_i(x) = \sum_{j=1}^n \gamma^{ij}(x)e_j \qquad (x \in U)$$

for each i, $1 \le i \le n$.

By construction

$$(1.6) \qquad e_j(x)e_k(x) + e_k(x)e_j(x) = -2g^{jk}(x) \qquad (x \in U) .$$

Now let $d\tau : \mathrm{spin}(n) \to \mathfrak{A}(\mathfrak{H})$ be the representation of the Lie algebra of

Spin(n) derived from the representation $\tau : \mathrm{Spin}(n) \to \mathfrak{A}(\mathfrak{H})$ of Spin(n). Our first task is to make precise and explicit the following definition.

(1.7) Definition.

As a differential operator on $C^\infty(U, \mathfrak{H})$ the standard Dirac operator D on (U, g) is defined by

$$D = \sum_{i=1}^{n} e_i(x) \left(\frac{\partial}{\partial x_i} + d\tau\big(\omega_i(x)\big) \right) \qquad x \in U$$

where $\omega_1, \ldots, \omega_n : U \to \mathrm{spin}(n)$ are smooth functions uniquely determined by g.

In view of (1.6),

$$D^2 = -\sum_{j,k} g^{jk}(x) \frac{\partial^2}{\partial x_j \partial x_k} + \text{ lower-order terms.}$$

These terms have a profound algebraic and geometric significance, as we shall see; but an excursion into Riemannian differential geometry is needed first to define $\omega_1, \ldots, \omega_n$ and to introduce the notion of curvature. The group-theoretic ideas needed were dealt with in chapter 3. Now differentiation of 'vector-valued functions' on a manifold requires a *connection*. Let $C^\infty(U, T(U))$ be the space of smooth vector fields $X = \sum_i a_i(x) \frac{\partial}{\partial x_i}$ on U; it is a $C^\infty(U)$-module under pointwise multiplication.

(1.8) Definition.

A connection on (U, g) is any bilinear mapping $\nabla : (X, Y) \to \nabla_X(Y)$ from $C^\infty(U, T(U)) \times C^\infty(U, T(U))$ into $C^\infty(U, T(U))$ which satisfies

(i) $\nabla_{\phi X + \psi Y}(Z) = \phi \nabla_X(Z) + \psi \nabla_Y(Z),$

(ii) $\nabla_X(\phi Y) = (X\phi)Y + \phi \nabla_X(Y),$

for all X, Y, Z in $C^\infty(U, T(U))$ and ϕ, ψ in $C^\infty(U)$.

For example,

$$(1.9) \qquad \nabla_X(Y) = \sum_{i,k} \left(a_i(x) \frac{\partial b_k}{\partial x_i} \right) \frac{\partial}{\partial x_k} \qquad \big(X, Y \in C^\infty(U, T(U)) \big)$$

is the usual connection on Euclidean space, and it is not hard to show that connections always exist on (U, g) whatever the choice of g. But the Fundamental Theorem of Riemannian differential geometry as applied to (U, g) asserts the following.

(1.10) Theorem.

There is a unique connection on (U, g) having the further properties

(i) $\nabla_X(Y) - \nabla_Y(X) = [X, Y]$,

(ii) $X g_x(Y, Z) = g_x(\nabla_X(Y), Z) + g_x(Y, \nabla_X(Z))$ $(x \in U)$

for all X, Y, Z in $C^\infty(U, T(U))$.

This unique connection on (U, g), the so-called Riemannian connection, can be made completely explicit. For, if

(1.11) $$\nabla_{\frac{\partial}{\partial x_i}} \left(\frac{\partial}{\partial x_j} \right) = \sum_k \Gamma_{ij}^k(x) \frac{\partial}{\partial x_k} \qquad (x \in U) ,$$

then (1.10)(i) ensures that

(1.12)(i) $$\Gamma_{ij}^k(x) = \Gamma_{ji}^k(x) \qquad (1 \le i, j \le n)$$

since $\left[\frac{\partial}{\partial x_i}, \frac{\partial}{\partial x_j} \right] = 0$, while a calculation using (1.8) and (1.10) shows that

(1.12)(ii) $$\Gamma_{ij}^k(x) = \frac{1}{2} \sum_\ell g^{k\ell} \left\{ \frac{\partial g_{j\ell}}{\partial x_i} + \frac{\partial g_{i\ell}}{\partial x_j} - \frac{\partial g_{ij}}{\partial x_\ell} \right\} .$$

Hence the Riemannian connection on (U, g) is given by

$$\nabla_X(Y) = \sum_k \left(\sum_i a_i(x) \frac{\partial b_k}{\partial x_i} + \sum_{i,j} \Gamma_{ij}^k a_i(x) b_j(x) \right) \frac{\partial}{\partial x_k}$$

with $\{ \Gamma_{ij}^k \}$ defined by (1.12). Clearly this reduces to the Euclidean connection (1.9) when $g(x) = I = [\delta_{ij}]$. But the vector fields $\partial/\partial x_i$ on U arising from its coordinates are not always the most convenient ones to use, since in general they are not orthonormal. Set

(1.13)(i) $$E_i(x) = \sum_j \gamma^{ij}(x) \frac{\partial}{\partial x_j} \qquad (x \in U , \ 1 \le i \le n) ,$$

and define $\omega_{ij}^k \in C^\infty(U)$ by

(1.13)(ii) $$\nabla_{\frac{\partial}{\partial x_i}} (E_j(x)) = \sum_k \omega_{ij}^k(x) E_k(x) \qquad (x \in U) .$$

Since by construction $g_x(E_i, E_j) = \delta_{ij}$, (1.10)(ii) ensures that

$$\omega_{ij}^k(x) + \omega_{ik}^j(x) = g_x(\nabla_{\frac{\partial}{\partial x_i}}(E_j), E_k) + g_x(E_j, \nabla_{\frac{\partial}{\partial x_i}}(E_k)) = 0 ,$$

and so $[\omega_{ij}^k]_{j,k}$ is skew-symmetric for each i $(1 \le i \le n)$ and x in U. The spin(n)-valued functions

(1.13)(iii) $$\omega_i(x) = \frac{1}{2} \sum_{j<k} \omega_{ij}^k(x) e_j e_k = \frac{1}{4} \sum_{j,k} \omega_{ij}^k(x) e_j e_k$$

(see chapter 1, (8.10)) will be the ones used in the definition of the Dirac operator. More precisely,

(1.14) $$D = \sum_i e_i(x) \left(\frac{\partial}{\partial x_i} + d\tau\big(\omega_i(x)\big) \right)$$

$$= \sum_i e_i(x) \left(\frac{\partial}{\partial x_i} + \frac{1}{4} \sum_{j,k} \omega_{ij}^k(x)\, d\tau(e_j e_k) \right)$$

where ω_i and ω_{ij}^k are determined by (1.13) and $e_i(x)$ is defined by (1.5). The reasons why the definition of D takes this particular form will emerge as its properties are developed.

Now define $\nu_{ij}^k \in C^\infty(U)$ by

(1.15) $$\nabla_{E_i}\big(E_j(x)\big) = \sum_k \nu_{ij}^k(x) E_k(x) \; .$$

The relation between ν_{ij}^k and $\gamma^{ij}, \omega_{ij}^k$ becomes important.

(1.16) Theorem.

  The function

$$\nu_i(x) = \frac{1}{2} \sum_{j<k} \nu_{ij}^k(x) e_j e_k = \frac{1}{4} \sum_{j,k} \nu_{ij}^k(x) e_j e_k$$

is a smooth spin(n)*-valued function on* U *such that*

$$\nu_{ij}^k(x) = \sum_\ell \gamma^{i\ell}(x) \omega_{\ell j}^k(x) \qquad (1 \le i, j, k \le n) \; .$$

Proof. As with ω_{ij}^k, the orthogonality of the $E_i(x)$ ensures that $\nu_{ij}^k(x) + \nu_{ik}^j(x) = 0$. Thus $\nu_i : U \to$ spin(n). On the other hand, by (1.8)(i) and (1.13),

$$\nabla_{E_i}\big(E_j(x)\big) = \sum_\ell \gamma^{i\ell}(x) \nabla_{\frac{\partial}{\partial x_\ell}}\big(E_j(x)\big) = \sum_{k,\ell} \gamma^{i\ell}(x) \omega_{\ell j}^k(x) E_k(x)$$

on U. The expression for ν_{ij}^k now follows.   ∎

The Riemannian connection on (U, g) extends to a linear connection $(X, F) \to \nabla_X(F)$ from $C^\infty(U, T(U)) \times C^\infty(U, \mathfrak{H})$ into $C^\infty(U, \mathfrak{H})$ setting

(1.17)(i) $$\nabla^\tau_{\frac{\partial}{\partial x_i}}(F) = \left(\frac{\partial}{\partial x_i} + d\tau\big(\omega_i(x)\big) \right) F \qquad (x \in U)$$

and extending by linearity so that

(1.17)(ii) $$\nabla^\tau_X = \sum_i a_i(x) \nabla^\tau_{\frac{\partial}{\partial x_i}} \qquad \left(X = \sum_i a_i(x) \frac{\partial}{\partial x_i} \right) \; .$$

In analogy with (1.8), it has the properties

$$(1.18) \quad \begin{cases} \text{(i)} \quad \nabla^\tau_{\phi X + \psi Y}(F) = \phi \nabla^\tau_X(F) + \psi \nabla^\tau_X(F) \,, \\[2mm] \text{(ii)} \quad \nabla^\tau_X(\phi F) = (X\phi)F + \phi \nabla^\tau_X(F) \,, \end{cases}$$

for all F, G in $\mathcal{C}^\infty(U, \mathfrak{H})$, and X in $\mathcal{C}^\infty(U, T(U))$, since $\mathcal{C}^\infty(U, \mathfrak{H})$ is a $\mathcal{C}^\infty(U)$-module; in this context (1.10)(i) does not make sense, but (1.10)(ii) becomes

$$(1.18)\text{(iii)} \quad X\big(F(x), G(x)\big) = \big(\nabla^\tau_X F(x), G(x)\big) + \big(F(x), \nabla^\tau_X G(x)\big)$$

writing $(\cdot, \cdot)$ for the inner product on $\mathfrak{H}$. It is customary to refer to these 'directional derivatives' ∇^τ_X as *covariant derivatives*.

(1.19) Theorem.
As a differential operator on $\mathcal{C}^\infty(U, \mathfrak{H})$ the standard Dirac operator on (U, g) becomes

$$D = \sum_{i,j} \gamma^{ij}(x) e_j \nabla^\tau_{\frac{\partial}{\partial x_j}} = \sum_i e_i \nabla^\tau_{E_i} \,;$$

more explicitly,

$$D = \sum_{i,j} \gamma^{ij}(x) e_j \left(\frac{\partial}{\partial x_j} + d\tau\big(\omega_j(x)\big) \right) = \sum_i e_i \Big(E_i(x) + d\tau\big(\nu_i(x)\big) \Big)$$

with $\omega_i(x)$, $\nu_i(x)$ given by (1.13) and (1.16) respectively.

Proof. In view of (1.16) and (1.17),

$$\nabla^\tau_{E_i(x)} = \sum_j \gamma^{ij}(x) \nabla^\tau_{\frac{\partial}{\partial x_j}} = \sum_j \gamma^{ij} \left(\frac{\partial}{\partial x_j} + d\tau\big(\omega_j(x)\big) \right)$$

$$= E_i(x) + d\tau\big(\nu_i(x)\big)$$

from which the various expressions for D follow at once. ∎

Since the covariant derivative $\nabla^\tau_{E_i}$ becomes simply the vector field $\frac{\partial}{\partial x_i}$ on $\mathcal{C}^\infty(U, \mathfrak{H})$ when (U, g) is an open subset of $\mathbb{R}^n$ with its usual Euclidean structure, it is clear that the operator D in (1.19) reduces to the standard Dirac operator on Euclidean space (see chapter 2, (3.1)). To complete the answer to problem (A) and to begin answering (B) we derive from (1.19) the following corollaries.

(1.20) Corollary.

As a differential operator on $C^\infty(U, \mathfrak{H})$ the Dirac operator is elliptic and essentially self-adjoint in the sense that

$$\int_U \big(DF(x), G(x)\big)\, d\,\mathrm{vol} = \int_U \big(F(x), DG(x)\big)\, d\,\mathrm{vol}$$

for all compactly supported F, G in $C^\infty(U, \mathfrak{H})$.

(1.21) Corollary.

As an operator on $C^\infty(U, \mathfrak{H})$

$$D^2 = -\sum_j \left(\nabla^\tau_{E_j} \nabla^\tau_{E_j} - \left(\sum_k \nu^k_{jj}(x) \nabla^\tau_{E_k} \right) \right)$$

$$+ \sum_{i<j} e_i e_j \big(\nabla^\tau_{E_i} \nabla^\tau_{E_j} - \nabla^\tau_{E_j} \nabla^\tau_{E_i} - \nabla^\tau_{[E_i, E_j]} \big)\ .$$

In (1.20) $d\,\mathrm{vol}$ denotes the volume element

$$d\,\mathrm{vol} = \big(\det g(x)\big)^{1/2}\, dx_1 \cdots dx_n$$

on (U, g). On the other hand, despite its non-intuitive appearance, the formula for D^2 in (1.21) lies at the heart of the properties of the Dirac operator on manifolds; for the second term on the right is a '*curvature*' operator, while the other is a second order *Laplacian* on $C^\infty(U, \mathfrak{H})$, as we shall see.

Proof of (1.20). Since

$$D^2 = -\sum_{j,k} g^{jk}(x) \frac{\partial^2}{\partial x_j \partial x_k} + \quad \text{lower-order terms,}$$

the principal symbol of $(-D^2)$ is the same as that of the elliptic operator $\sum_{jk} g^{jk} \frac{\partial^2}{\partial x_j \partial x_k}$; hence D also must be elliptic. The self-adjointness property is a simple consequence of Stokes' theorem. ∎

Proof of (1.21). By (1.19)

$$D^2 = \sum_{i,j} e_i \Big(E_i(x) + d\tau\big(\nu_i(x)\big) \Big) e_j \Big(E_j(x) + d\tau\big(\nu_j(x)\big) \Big)$$

$$= \sum_{i,j} e_i e_j \nabla^\tau_{E_i} \nabla^\tau_{E_j} + \sum_{i,j} e_i \Big(d\tau\big(\nu_i(x)\big) e_j - e_j\, d\tau\big(\nu_i(x)\big) \Big) \nabla^\tau_{E_j}\ .$$

But

$$d\tau\big(\nu_i(x)\big) e_j - e_j\, d\tau\big(\nu_i(x)\big) = d\sigma\big(\nu_i(x)\big) e_j = [\nu_i(x), e_j] = \sum_k \nu^k_{ij}(x) e_k$$

because of the invariance property $\tau(a)e_j\tau(a)^{-1} = \sigma(a)e_j$. Thus

$$(1.22)(i) \quad D^2 = \sum_{i,j} e_i e_j \nabla^\tau_{E_i} \nabla^\tau_{E_j} + \sum_{i,k} e_i e_k \left(\sum_j \nu^k_{ij}(x) \nabla^\tau_{E_j} \right)$$

$$= \sum_{i,j} e_i e_j \left(\nabla^\tau_{E_i} \nabla^\tau_{E_j} - \left(\sum_k \nu^k_{ij}(x) \nabla^\tau_{E_k} \right) \right)$$

since $\nu^k_{ij} = -\nu^j_{ik}$. Now

$$(1.22)(ii) \quad \sum_{i \neq j} e_i e_j \left(\nabla^\tau_{E_i} \nabla^\tau_{E_j} - \left(\sum_k \nu^k_{ij}(x) \nabla^\tau_{E_k} \right) \right)$$

$$= \sum_{i<j} e_i e_j \left\{ \left(\nabla^\tau_{E_i} \nabla^\tau_{E_j} - \nabla^\tau_{E_j} \nabla^\tau_{E_i} \right) - \sum_k (\nu^k_{ij}(x) - \nu^k_{ji}(x)) \nabla^\tau_{E_k} \right\} .$$

On the other hand, by (1.10)(i),

$$\sum_k (\nu^k_{ij}(x) - \nu^k_{ji}(x)) E_k = \nabla_{E_i}(E_j) - \nabla_{E_j}(E_i) = [E_i, E_j] ;$$

consequently,

$$(1.22)(iii) \quad \sum_k (\nu^k_{ij}(x) - \nu^k_{ji}(x)) \nabla^\tau_{E_k} = \nabla^\tau_{[E_i, E_j]}$$

(see (1.18)(i)). Formula (1.21) for D^2 now follows immediately, substituting (1.22)(ii) and (iii) into (1.22)(i). ∎

The definition of the second-order Laplacian Δ_τ follows the $\Delta = \mathrm{div}(\mathrm{grad})$ form of the classical Laplacian on Euclidean space where in the present context 'grad' is the total derivative

$$(1.23) \quad \nabla_\tau : F \longrightarrow \sum_j \nabla^\tau_{E_j} F(x) \otimes e_j \quad (F \in \mathcal{C}^\infty(U, \mathfrak{H}))$$

mapping $\mathcal{C}^\infty(U, \mathfrak{H})$ into $\mathcal{C}^\infty(U, \mathfrak{H} \otimes \mathbb{R}^n)$. As $\mathfrak{H} \otimes \mathbb{R}^n$ is an $\mathfrak{A}_n$-module on which $\tau \otimes \sigma$ defines a representation of $\mathrm{Spin}(n)$, a linear connection $(X, G) \to \nabla^{\tau \otimes \sigma}_X(G)$ on $\mathfrak{H} \otimes \mathbb{R}^n$-valued functions is obtained by replacing τ with $\tau \otimes \sigma$ throughout (1.17). To identify the form of the divergence operator we first prove the following.

(1.24) Lemma.
For each $G = \sum_j G_j(x) \otimes e_j$ with $G_j \in \mathcal{C}^\infty(U, \mathfrak{H})$,

$$\nabla^{\tau \otimes \sigma}_{E_i} : G \longrightarrow \sum_j \left(\nabla^\tau_{E_i}(G_j) - \sum_k \nu^k_{ij}(x) G_k \right) \otimes e_j .$$

Proof. By definition

$$\nabla_{E_i}^{\tau \otimes \sigma} : G \longrightarrow \left(\frac{\partial}{\partial x_i} + d(\tau \otimes \sigma)\big(\nu_i(x)\big) \right) G$$

$$= \left(\frac{\partial}{\partial x_i} + d\tau\big(\nu_i(x)\big) \otimes I + I \otimes d\sigma\big(\nu_i(x)\big) \right) G$$

$$= \sum_j \left(\left(\frac{\partial}{\partial x_i} + d\tau\big(\nu_i(x)\big) \right) G_j \right) \otimes e_j + \sum_j G_j \otimes d\sigma\big(\nu_i(x)\big)e_j \ .$$

But

$$d\sigma\big(\nu_i(x)\big)e_j = \sum_k \nu_{ij}^k(x)e_k = -\sum_k \nu_{ik}^j(x)e_k \ .$$

Consequently,

$$\nabla_{E_i}^{\tau \otimes \sigma} : G \longrightarrow \sum_j (\nabla_{E_i}^\tau G_j) \otimes e_j - \sum_{j,k} \nu_{ik}^j(x)G_j \otimes e_k \ ,$$

proving the lemma. ∎

Thus the 'divergence' operator from $\mathcal{C}^\infty(U, \mathfrak{H} \otimes \mathbb{R}^n)$ into $\mathcal{C}^\infty(U, \mathfrak{H})$ is given by

$$G \longrightarrow \sum_j \left(\nabla_{E_j}^\tau(G_j) - \sum_k \nu_{jj}^k(x)G_k \right) \qquad \left(G = \sum_j G_j \otimes e_j \right) ;$$

in a very precise sense it is the adjoint operator ∇_τ^* to (1.23).

(1.25) Definition.
   The Laplacian Δ_τ is the second-order differential operator

$$\Delta_\tau F = (\nabla_\tau^* \nabla_\tau)F = \sum_j \left(\nabla_{E_j}^\tau \nabla_{E_j}^\tau(F) - \left(\sum_k \nu_{jj}^k(x)\nabla_{E_k}^\tau(F) \right) \right)$$

on $\mathcal{C}^\infty(U, \mathfrak{H})$.

Clearly

$$\Delta_\tau = \sum_{j,k} g^{jk}(x)\frac{\partial^2}{\partial x_j \partial x_k} + \quad \text{lower-order terms,}$$

so every Δ_τ has the same principal symbol; in particular, Δ_τ is elliptic. This identifies one set of terms in (1.21). To complete the solution of problem (B) we shall relate the remaining terms in D^2 to curvature. Now, intuitively, the curvature of (U, g) is a measure of the failure of ∇_X, ∇_Y to commute as operators on $\mathcal{C}^\infty(U, T(U))$, while the curvature of $\mathfrak{H}$ in relation to (U, g) measures the failure of ∇_X^τ, ∇_Y^τ to commute as operators on $C^\infty(U, \mathfrak{H})$. More precisely, for each X, Y in $C^\infty(U, T(U))$ let

(1.26) $$R(X, Y) = \nabla_X \nabla_Y - \nabla_Y \nabla_X - \nabla_{[X,Y]}$$

and

(1.27) $$R^\tau(X, Y) = \nabla_X^\tau \nabla_Y^\tau - \nabla_Y^\tau \nabla_X^\tau - \nabla_{[X,Y]}^\tau$$

be *curvature operators* on $C^\infty(U, T(U))$ and $C^\infty(U, \mathfrak{H})$ respectively. Clearly

(1.28)(i) $$R(X, Y) + R(Y, X) = 0 \qquad (X, Y \in C^\infty(U, T(U))),$$

but such additional properties as

(1.28)(ii) $$\begin{cases} R(X,Y)Z + R(Y,Z)X + R(Z,X)Y = 0, \\ g(R(X,Y)Z, W) + g(R(X,Y)W, Z) = 0, \\ g(R(X,Y)Z, W) = g(R(Z,W)X, Y) \end{cases}$$

are well-known and follow more or less easily from the definition of $R(X, Y)$.

Now define functions $R_{ijk\ell}$ in $C^\infty(U)$ by

(1.29)(i) $$R(E_i, E_j) : E_k(x) \longrightarrow \sum_\ell R_{ijk\ell}(x) E_\ell(x) \qquad (x \in U);$$

the function

(1.29)(ii) $$x \longrightarrow \{R_{ijk\ell}(x)\}_{ijk\ell} \qquad (x \in U)$$

on U is called the *Riemannian curvature tensor*. A routine calculation using (1.15) shows that

(1.30) $$\begin{aligned} R_{ijk\ell}(x) = {}& E_i(\nu_{jk}^\ell) - E_j(\nu_{ik}^\ell) \\ & + \sum_r (\nu_{jk}^\tau \nu_{ir}^\ell - \nu_{ik}^\tau \nu_{jr}^\ell + \nu_{ij}^\tau \nu_{rk}^\ell - \nu_{ji}^\tau \nu_{rk}^\ell), \end{aligned}$$

from which $R_{ijk\ell}$ can be calculated; symmetry properties of the $R_{ijk\ell}$ follow from (1.28):

(1.30)(i) $$\begin{cases} R_{ijk\ell} + R_{ij\ell k} = 0 = R_{ijk\ell} + R_{jik\ell}, \\ R_{ijk\ell} = R_{k\ell ij}, \\ R_{ijk\ell} + R_{jki\ell} + R_{kij\ell} = 0 \quad \text{(Bianchi)}. \end{cases}$$

In particular,

(1.31) $$R_{ij}(x) = \frac{1}{2} \sum_{k<\ell} R_{ijk\ell}(x) e_k e_\ell \qquad (x \in U)$$

defines a spin(n)-valued function on U.

(1.32) Lemma.

When $x \to \{R_{ijk\ell}(x)\}$ is the Riemannian curvature tensor,

$$R^\tau(E_i, E_j) = d\tau\big(R_{ij}(x)\big) = \frac{1}{2}\sum_{k<\ell} R_{ijk\ell}(x)\, d\tau(e_k e_\ell) \qquad (x \in U) .$$

Proof. By definition

$$R^\tau(E_i, E_j) = \nabla^\tau_{E_i}\nabla^\tau_{E_j} - \nabla^\tau_{E_j}\nabla^\tau_{E_i} - \nabla^\tau_{[E_i, E_j]}$$

$$= \big(E_i\, d\tau(\nu_j) - d\tau(\nu_j)E_i\big) - \big(E_j\, d\tau(\nu_i) - d\tau(\nu_i)E_j\big)$$

$$+ \,[d\tau(\nu_i), d\tau(\nu_j)] + \sum_r (\nu^\tau_{ij} - \nu^\tau_{ji})\, d\tau(\nu_r)$$

$$= d\tau\Big(E_i(\nu_j) - E_j(\nu_i) + [\nu_i, \nu_j] + \sum_r (\nu^\tau_{ij} - \nu^\tau_{ji})\nu_r \Big) .$$

But

$$E_i(\nu_j) - E_j(\nu_i) = \frac{1}{2}\sum_{k<\ell}\big(E_i(\nu^\ell_{jk}) - E_j(\nu^\ell_{ik})\big)e_k e_\ell$$

while

$$[\nu_i, \nu_j] + \sum_r (\nu^\tau_{ij} - \nu^\tau_{ji})\nu_r = \frac{1}{2}\sum_{k<\ell}\sum_r \big(\nu^\tau_{jk}\nu^\ell_{ir} - \nu^\tau_{ik}\nu^\ell_{jr} + \nu^\tau_{ij}\nu^\ell_{rk} - \nu^\tau_{ji}\nu^\ell_{rk}\big)e_k e_\ell .$$

Hence, by (1.30),

$$R^\tau(E_i, E_j) = d\tau\Big(\frac{1}{2}\sum_{k<\ell} R_{ijk\ell}(x)e_k e_\ell\Big) = d\tau\big(R_{ij}(x)\big) ,$$

completing the proof. ∎

Combining this last lemma with (1.25) we thus obtain the fundamental Bochner–Weitzenböck type formula for the Dirac operator, completing the solution of problem (**B**).

(1.33) Theorem.

On $C^\infty(U, \mathfrak{H})$

$$D^2 = -\Delta_\tau + \sum_{i<j} e_i e_j R^\tau(E_i, E_j)$$

$$= -\Delta_\tau + \frac{1}{2}\sum_{i<j} e_i e_j\Big(\sum_{k<\ell} R_{ijk\ell}(x)\, d\tau(e_k e_\ell)\Big)$$

where $\Delta_\tau = \nabla^*_\tau\nabla_\tau$ is the Laplacian on $C^\infty(U, \mathfrak{H})$ and $\{R_{ijk\ell}\}$ is the curvature tensor of (U, g).

In some important special cases the curvature term in D^2 further simplifies. Recall first that the *scalar curvature* of (U, g) is the scalar curvature operator

$$(1.34) \qquad \kappa(x) = -\sum_{i,j} R_{ijij}(x) \qquad (x \in U) ,$$

in terms of the $R_{ijk\ell}$ as we have defined them.

(1.35) **(A) Spinor Laplacian.** When $\mathfrak{H} = \mathfrak{S}_n$ and τ is the Spin representation S of $\mathrm{Spin}(n)$ on $\mathfrak{S}_n$, $n = 2m$, the second-order operator D^2 is often called the *spinor Laplacian*, and the particular form

$$D^2 = -\Delta_S + \tfrac{1}{4}\kappa$$

that (1.33) then takes is frequently referred to as Lichnerowicz' formula. To establish this formula, observe that as an operator on $\mathfrak{S}_n$,

$$S\!\left(\exp\!\left(\tfrac{1}{2} t e_k e_\ell\right)\right) u = \exp\!\left(\tfrac{1}{2} t e_k e_\ell\right) u = \left(\cos\!\left(\tfrac{1}{2}t\right) + e_k e_\ell \sin\!\left(\tfrac{1}{2}t\right)\right) u$$

for $u \in \mathfrak{S}_n$ (see chapter 3, section 2); thus

$$\tfrac{1}{2} dS(e_k e_\ell) = \tfrac{1}{2} e_k e_\ell$$

and so

$$\tfrac{1}{2} \sum_{i<j} e_i e_j \left(\sum_{k<\ell} R_{ijk\ell}(x)\, dS(e_k e_\ell) \right) = \tfrac{1}{2} \sum_{i<j} \sum_{k<\ell} R_{ijk\ell}(x) e_i e_j e_k e_\ell$$

$$= \tfrac{1}{8} \sum_{i,j,k,\ell} R_{ijk\ell}(x) e_i e_j e_k e_\ell = \tfrac{1}{8}\left(\Sigma_1 + \Sigma_2\right)$$

where Σ_2 is the sum over all i, j, k, ℓ such that i, j, k are distinct, while Σ_1 is the sum over those i, j, k, ℓ in which at least two of i, j, k are equal. But by the Bianchi identities

$$R_{ijk\ell}(x) + R_{jki\ell}(x) + R_{kij\ell}(x) = 0$$

for all i, j, k, ℓ, whereas

$$e_i e_j e_k e_\ell = e_j e_k e_i e_\ell = e_k e_i e_j e_\ell$$

when i, j, k are all distinct. Thus $\Sigma_2 = 0$. But

$$\Sigma_1 = \sum_{i,j,\ell} \left(R_{iji\ell}(x) e_i e_j e_i e_\ell + R_{ijj\ell}(x) e_i e_j e_j e_\ell \right) .$$

If $j \neq \ell$, then

$$R_{iji\ell}(x) = R_{i\ell ij}(x) , \qquad e_j e_\ell + e_\ell e_j = 0 ;$$

similarly, if $i \neq \ell$, then

$$R_{ijj\ell}(x) + R_{\ell jji}(x) = 0 , \qquad e_i e_j e_j e_\ell = e_\ell e_j e_j e_i .$$

Consequently,

$$\Sigma_1 = 2 \sum_{i,j} R_{ijij}(x) e_i e_j e_i e_j = -2 \sum_{i,j} R_{ijij}(x)$$

from which Lichnerowicz' formula follows using (1.34).

(1.35) **(B) Hodge Laplacian.** When $\mathfrak{H} = \Lambda^*(\mathbb{F}^n)$, $\mathbb{F} = \mathbb{R}$ or $\mathbb{C}$, and τ is the ubiquitous representation σ as in $(2.17)'$ in chapter 3, then $D = d - d^*$; thus

$$D^2 = (d - d^*)^2 = -(dd^* + d^*d) .$$

Consequently, up to sign, D^2 is the Hodge Laplacian, and (1.33) takes the form

$$D^2 = -\Delta_\sigma - \tfrac{1}{4}\kappa + \tfrac{1}{8} \sum_{i,j,k,\ell} R_{ijk\ell}(x) e_i e_j \tilde{e}_k \tilde{e}_\ell$$

where κ is the scalar curvature operator and $e_i, e_j, \tilde{e}_k, \tilde{e}_\ell$ generate $\mathcal{L}(\Lambda^*(\mathbb{F}^n))$ (see chapter 4, (2.8)(ii)). To establish this formula, observe first that on $\mathbb{F}^n$

$$\sigma\Big(\exp(\tfrac{1}{2} t e_k e_\ell)\Big) : e_m \longrightarrow \begin{cases} \cos t e_k + \sin t e_\ell & (m = k) , \\ -\sin t e_k + \cos t e_\ell & (m = \ell) , \end{cases}$$

while at the same time it fixes all other e_m (see chapter 3, (2.4)). Thus

$$\tfrac{1}{2} d\sigma(e_k e_\ell) = \mu_\ell \mu_k^* - \mu_k \mu_\ell^* = \mu_\ell \mu_k^* + \mu_\ell^* \mu_k = -\tfrac{1}{2}(e_k e_\ell - \tilde{e}_k \tilde{e}_\ell)$$

since $\mu_k \mu_\ell^* + \mu_\ell^* \mu_k = 0$ when $k \neq \ell$. Consequently,

$$\tfrac{1}{2} \sum_{i<j} e_i e_j \Big(\sum_{k<\ell} R_{ijk\ell}(x) \, d\sigma(e_k e_\ell) \Big)$$

$$= (-\tfrac{1}{2}) \sum_{i<j} \sum_{k<\ell} R_{ijk\ell}(x)(e_i e_j e_k e_\ell - e_i e_j \tilde{e}_k \tilde{e}_\ell)$$

$$= -\tfrac{1}{4}\kappa + \tfrac{1}{8} \sum_{i,j,k,\ell} R_{ijk\ell}(x) e_i e_j \tilde{e}_k \tilde{e}_\ell ,$$

using the symmetries of $R_{ijk\ell}$ as in the previous examples. This completes the proof. Strictly in terms of the Hodge Laplacian,

$$dd^* + d^*d = \Delta_\sigma + \tfrac{1}{4}\kappa - \tfrac{1}{8} \sum_{i,j,k,\ell} R_{ijk\ell}(x) e_i e_j \tilde{e}_k \tilde{e}_\ell .$$

(1.35) **(C) Twisted spinor Laplacian.** Let $\mathfrak{H} = \mathfrak{S}_n \otimes \mathfrak{B}$ where $(\mathfrak{B}, \tau_0)$ is a unitary representation of $\mathrm{Spin}(n)$. Then, with $\tau = S \otimes \tau_0$, D^2 is now a 'twisted' version of the spinor Laplacian, and (1.33) becomes

$$D^2 = -\Delta_\tau + \tfrac{1}{4}\kappa + \tfrac{1}{2} \sum_{i<j} e_i e_j \Big(\sum_{k<\ell} d\tau_0(e_k e_\ell) \Big) ,$$

showing the simple effect of the twisting on the curvature term. This formula follows immediately from (1.33) and example (**A**) because

$$d\tau = d(S \otimes \tau_0) = dS \otimes I + I \otimes d\tau_0 \ .$$

It applies, for instance, to the m-fold tensor product $\mathfrak{H} = \mathfrak{S}_n \otimes \cdots \otimes \mathfrak{S}_n$ (see chapter 3, (2.32)). Example (**B**) also has this form.

There are two very useful variations on this theme. Let $(\mathcal{A}, \tau)$, $(\mathcal{B}, \omega)$ be unitary representations of $\mathrm{Spin}(n)$ and let $A : \mathcal{A} \otimes \mathbb{C}^n \to \mathcal{B}$ be a linear operator as in the previous chapter. Now set

$$(1.36) \qquad \eth_A = \sum_i A_i \nabla^\tau_{E_i} = \sum_j a_j(x) \Big(\frac{\partial}{\partial x_j} + d\tau \big(\omega_j(x) \big) \Big) \ ,$$

so that $\eth_A : \mathcal{C}^\infty(U, \mathcal{A}) \to \mathcal{C}^\infty(U, \mathcal{B})$, where

$$A_i \xi = A(\xi \otimes e_i) \quad , \quad a_j(x) = \sum_i \gamma^{ij}(x) A_i \ .$$

These are just the 'local, curved' versions of the operators constructed in the previous chapter. After assuming that A is $\mathrm{Spin}(n)$-equivariant, i.e.

$$A\big((\tau \otimes \sigma)(a) u \big) = \omega(a) A(u) \qquad \big(a \in \mathrm{Spin}(n) \big)$$

on $\mathcal{A} \otimes \mathbb{C}^n$ (see chapter 4, (3.5)), we shall derive the corresponding 'global, curved' version of $\eth_A$ in the next section. In practice, of course, this process is usually reversed: one starts with a globally defined differential operator, and in studying its local properties chooses a local realization of it in the coordinate system best suited to the problem at hand. In the context of this section, this means choosing the matrix $g = [g_{jk}(x)]$ so that it and all the functions derived from it have a particularly convenient 'form'.

One frequent choice of coordinate system is a so-called normal system. Let U be an open set in $\mathbb{R}^n$ which is star-shaped about the origin. Then the usual coordinates $x = (x_1, \ldots, x_n)$ for U are a *normal coordinate system* for (U, g) if the rays through the origin are geodesics starting from the origin. A necessary and sufficient condition for this is that

$$(1.37)(\mathrm{i}) \qquad \sum_j g_{ij}(x) x_j = x_i \qquad (x \in U) \ ,$$

or, equivalently, that

$$(1.37)(\mathrm{ii}) \qquad g_{ij}(0) = \delta_{ij} \quad , \quad \frac{\partial g_{ij}}{\partial x_k}(0) = 0$$

for all i, j, k; in particular, the Christoffel symbols Γ^k_{ij} all vanish at the origin, and

$$(1.38) \quad \begin{cases} \text{(i)} \quad g_{ij}(x) = \delta_{ij} + \tfrac{1}{3}\sum_{p,q} R_{ipjq}(0)x_p x_q + O(|x|^3) , \\ \text{(ii)} \quad \gamma^{ij}(x) = \delta_{ij} + O(|x|^2) , \end{cases}$$

near the origin. It can then be shown that

$$\omega_i(x) = -\tfrac{1}{2}\sum_j x_j R_{ij}(0) + \sum_{i,j} x_i x_j \phi_{ij}(x)$$

for some $\phi_{ij} : U \to \mathrm{spin}(n)$. Hence ν^k_{jj} is a real-valued function such that $\nu^k_{jj}(x) = O(|x|)$ and

$$(1.38)\text{(iii)} \qquad \nabla_{E_i} = \nabla_{\frac{\partial}{\partial x_i}} + \sum_j \psi_{ij}(x)\nabla_{\frac{\partial}{\partial x_j}}$$

where ψ_{ij} is a real-valued function such that $\psi_{ij}(x) = O(|x|^2)$. Lengthy but straightforward calculations using (1.33) and (1.38) enable us now to determine the essential behavior of D^2 near the origin. We consider the cases of the spinor Laplacian and Hodge Laplacian separately, using the notation of (1.35).

(1.39) **(A) Spinor Laplacian.** As $\mathcal{L}(\mathfrak{S})$ is a realization of $\mathfrak{C}_n$, $n = 2m$, let

$$\mathcal{L}(\mathfrak{S}) = \mathcal{A}_0 \oplus \mathcal{A}_1 \oplus \cdots \oplus \mathcal{A}_n \qquad (\mathcal{A}_k = \mathfrak{C}_n^{(k)}) ,$$

be the decomposition of $\mathcal{L}(\mathfrak{S})$ into its subspaces of k-multivectors. By (1.35),

$$\nabla_{\frac{\partial}{\partial x_i}} = \frac{\partial}{\partial x_i} + dS(\omega_i(x))$$

$$= \frac{\partial}{\partial x_i} - \tfrac{1}{2}\sum_j x_j R_{ij}(0) + \sum_{i,j} x_i x_j \phi_{ij}(x)$$

for some $\phi_{ij} : U \to \mathcal{A}_2$, since $\mathrm{spin}(n) \subseteq \mathfrak{C}_n^{(2)}$. But then by the Bochner–Weitzenböck formula and a lengthy calculation,

$$D^2 = -\sum_{i=1}^n \left(\frac{\partial}{\partial x_i} - \tfrac{1}{2}\sum_{j=1}^n x_i R_{ij}(0)\right)^2 + \mathcal{R}_S ,$$

where $\mathcal{R}_S$ is the second-order operator

$$\mathcal{R}_S = \sum_{i,j}\left(a_{ij}(x)\frac{\partial^2}{\partial x_i \partial x_j} + x_i b_{ij}(x)\frac{\partial}{\partial x_j} + x_i x_j c_{ij}(x)\right) + c(x)$$

whose coefficients satisfy

(i) $a_{ij} : U \to \mathbb{R}$, $b_{ij} : U \to \mathcal{A}_0 \oplus \mathcal{A}_2$, $c_{ij} : U \to \mathcal{A}_0 \oplus \mathcal{A}_2 \oplus \mathcal{A}_4$ *and* $a_{ij}(x) = O(|x|^2)$, $b_{ij} = O(|x|)$, $c_{ij} = O(|x|)$,

(ii) $c : U \to \mathcal{A}_0 \oplus \mathcal{A}_2$ *with*

$$c(x) = \xi_0 + \sum_{i=1}^{n} x_i \xi_i + O(|x|^2)$$

where $\xi_0 \in \mathcal{A}_0$ and $\xi_i \in \mathcal{A}_2$, $1 \le i \le n$.

The term $\mathcal{R}_S$ is just the 'remainder' differential operator with a careful description of the subspaces of $\mathfrak{C}_n$ in which its coefficients take their values. A similar comment applies to the next example.

(1.39) **(B) Hodge Laplacian.** In chapter 4, (2.8)(ii) we saw that the reduced products

$$\tilde{e}_\alpha e_\beta = \tilde{e}_{\alpha_1} \cdots \tilde{e}_{\alpha_j} e_{\beta_1} \cdots e_{\beta_k}$$

form a basis for $\mathcal{L}(\mathfrak{H})$, $\mathfrak{H} = \Lambda^*(\mathbf{F}^n)$. Now set

$$\mathcal{L}(\mathfrak{H}) = \mathcal{A}_0 \oplus \mathcal{A}_1 \oplus \cdots \oplus \mathcal{A}_{2n}$$

where

$$\mathcal{A}_\ell = \operatorname{span}_{\mathbf{F}}\{\tilde{e}_\alpha e_\beta : |\alpha| + |\beta| = \ell\} \ .$$

By (1.35),

$$\nabla_{\frac{\partial}{\partial x_i}} = \frac{\partial}{\partial x_i} + d\sigma\big(\omega_i(x)\big) = \frac{\partial}{\partial x_i} + \sum_{j=1}^{n} x_i \phi_{ij}(x)$$

for some $\phi_{ij} : U \to \mathcal{A}_2$. But then, by the Bochner–Weitzenböck formula and another lengthy calculation,

$$dd^* + d^*d = \sum_{i=1}^{n} \left(\frac{\partial}{\partial x_i}\right)^2 - \tfrac{1}{8} \sum_{i,j,k,\ell} R_{ijk\ell}(0) e_i e_j \tilde{e}_k \tilde{e}_\ell + \mathcal{R}_\sigma$$

with $\mathcal{R}_\sigma$ the second-order operator

$$\mathcal{R}_\sigma = \sum_{i,j=1}^{n} \left(a_{ij}(x)\frac{\partial^2}{\partial x_i \partial x_j} + x_i b_{ij} \frac{\partial}{\partial x_j}\right) + \sum_{i=1}^{n} x_i c_i(x) + c_0(x)$$

whose coefficients satisfy

$$a_{ij} : U \to \mathbf{R} \quad ; \quad b_{ij}, c_0 : U \to \mathcal{A}_0 \oplus \mathcal{A}_1 \oplus \mathcal{A}_2 \quad ; \quad c_i : U \to \mathcal{A}_2 \oplus \mathcal{A}_4$$

with $a_{ij}(x) = O(|x|^2)$.

2 Global theory

A Clifford algebra bundle can be constructed over any Riemannian manifold, but topological obstructions may prevent the construction of Clifford modules. Let (M, g) be an n-dimensional Riemannian manifold; then, by definition, there is a positive-definite inner product g_x on the tangent space $T_x M$ to M at x, and so the universal Clifford algebra $\mathfrak{A}(T_x M, g_x)$ exists for every x in M. Since $(V, Q) \to \mathfrak{A}(V, Q)$ is functorial in the category of finite-dimensional inner product spaces over $\mathbb{R}$ (or $\mathbb{C}$), general results ensure that the individual $\mathfrak{A}(T_x M, g_x)$ can be glued together to form a bundle over M, proving the following.

(2.1) Theorem.
For any n-dimensional Riemannian manifold (M, g) the set

$$\mathfrak{A}_n(M, g) = \bigcup_{x \in M} \mathfrak{A}(T_x M, g_x)$$

is an algebra bundle over M, the so-called Clifford algebra bundle.

This bundle inherits a canonical C^*-algebra structure from the C^*-algebra structure on each of its fibers, but we shall omit details of it and of (2.1), favoring instead a less direct construction which involves the structure of (M, g) in a much more fundamental way. For this, however, the notion of *principal fiber bundle* is needed.

(2.2) Definition.
Let M be a C^∞-manifold and K a Lie group. Then a principal K-bundle over M consists of a triple $(P, \pi, \cdot)$ where

  (i) the total space P is a C^∞-manifold on which the structure group K acts (on the left) as a Lie transformation group $k : p \to k \cdot p$,
  (ii) the projection map $P \xrightarrow{\pi} M$ is a smooth mapping from P onto M such that $\pi(k \cdot p) = \pi(p)$,
  (iii) for each x in M there is a neighborhood U of x and diffeomorphism $\kappa : \pi^{-1}(U) \to K \times U$ of the form $\kappa(p) = (\phi(p), \pi(p))$ where $\phi : \pi^{-1}(U) \to K$ is K-equivariant, i.e., $\phi(k \cdot p) = k\phi(p)$.

Although the choice of side on which K acts on P is a matter of convention, it is customary to assume that K acts on the right. Our choice was made because of the first of the following two examples of principal K-bundles.

(2.3) Examples.
(a) Let M be a manifold on which a Lie group G acts transitively (on the right) as a Lie transformation group, and let K be the subgroup of G fixing some base point x_0 in M. Then M is diffeomorphic to the homogeneous space $K \backslash G$ of right cosets; the canonical homomorphism $\pi : G \to K \backslash G$ can thus be regarded as a G-equivariant projection $G \xrightarrow{\pi} M$ such that $\pi(kg) = \pi(g)$, $g \in G$. It is also known that there is a neighborhood U of the image $\pi(e)$ in M of the identity e in G and a

diffeomorphism $\nu : U \to G$ from U into G such that $\pi \circ \nu$ is the identity on U. From this it follows easily that $(G, \pi, \cdot)$ is a principal K-bundle over M with '$\cdot$' being group multiplication in G.

(b) Now let (M, g) be an n-dimensional Riemannian manifold, and denote by $O(M, g)$ the *orthonormal frame bundle* of M whose fiber $\pi^{-1}(x)$ at x consists of all orthonormal bases for $T_x M$. Since

$$(2.3)(i) \qquad k = [k_{ij}] : E_i(x) \longrightarrow F_j(x) = \sum_i k_{ij} E_i(x) \qquad (k \in O(n))$$

maps one orthonormal basis of $T_x M$ to a second one, it is easily seen that $O(M, g)$ is a principal $O(n)$-bundle with '$\cdot$' defined by (2.3)(i). When M is orientable, the *oriented orthonormal frame bundle* $SO(M, g)$ of M is a principal $SO(n)$-bundle whose fiber $\pi^{-1}(x)$ consists of all coherently *oriented* orthonormal bases for $T_x M$. More generally, let $(P, \pi', \cdot)$ be a principal K-bundle over M for which there is an isomorphism $\gamma : K \to O(n)$ from K into $O(n)$ and a embedding $\beta : P \to O(M, g)$ of P in $O(M, g)$ such that

$$(2.3)(ii) \qquad \pi' = \pi \circ \beta \quad , \quad \beta(k \cdot p) = \gamma(k) \cdot \beta(p) \qquad (k \in K , \ p \in P) .$$

Then $\beta : P \to O(M, g)$ is said to *reduce* $O(n)$ to K.

These two examples are closely related when the group G in (2.3)(a) is a group of isometries preserving some Riemannian metric g on M. For then each $a \in G$ determines an orthogonal transformation $a_* : T_x M \to T_{x \cdot a} M$ such that $a \to a_*$ is a homomorphism from K into $O(T_{x_0} M)$ whose kernel N is a subgroup of K which is normal in K, though not necessarily normal in G; in particular, $N \backslash K$ is isomorphic to a subgroup of $O(n)$. The group $K_0 = N \backslash K$ acts on $P = N \backslash G$ by

$$(2.4) \qquad k : Na \longrightarrow (Nh)(Na) = Nha \qquad (k = Nh) ,$$

and $\pi : P \to K_0 \backslash P \cong K \backslash G \cong M$. Thus $(N \backslash G, \pi, \cdot)$ is a principal $(N \backslash K)$-bundle over M defining '$\cdot$' by (2.4).

(2.5) Theorem.

If G is a group of isometries on (M, g), then there is an embedding $\beta : N \backslash G \to O(M, g)$ reducing $O(n)$ to $N \backslash K$.

Proof. An orthonormal basis for a n-dimensional inner product V can be thought of as an isometry from $\mathbb{R}^n$ to V. Thus, let $f : \mathbb{R}^n \to T_{x_0} M$ be an isometry determining an orthonormal basis of $T_{x_0} M$. For any $a \in G$, the composition $a_* \circ f$ determines an orthonormal basis of $T_{x_0 \cdot a} M$, and $a_* \circ f = f$ if and only if $a \in N$. Now set

$$\beta(Na) = (a^{-1})_* \circ f \quad , \quad \gamma(Nh) = f^{-1} \circ (h^{-1})_* \circ f .$$

It is easily seen that γ is an isomorphism from $N \setminus K$ into $O(n)$ and that β satisfies (2.3)(ii), completing the proof.   ∎

(2.6) Examples.

(i) When $\mathbb{R}^n$ has the Euclidean metric its orthonormal frame bundle is diffeomorphic to the motion group $O(n)\circledS\mathbb{R}^n$, and its oriented orthonormal frame bundle is diffeomorphic to the identity component $SO(n)\circledS\mathbb{R}^n$ of this motion group.

(ii) Let (Σ_{n-1}, g) be the unit sphere in $\mathbb{R}^n$ with g the restriction to Σ_{n-1} of the Euclidean metric on $\mathbb{R}^n$. Now Σ_{n-1} can be realized both as $M \setminus O(n)$ and as $M_0 \setminus SO(n)$, where M and M_0 are the respective subgroups fixing the North Pole **1**. But the mapping $a \to a_*$ from M into $O(T_1(\Sigma_{n-1}))$ is an isomorphism. Hence $O(\Sigma_{n-1}, g)$ is diffeomorphic to $O(n)$, while $SO(n)$ is diffeomorphic to the oriented orthonormal frame bundle.

(iii) Let $(\mathbb{R}P^{n-1}, g)$ be $(n-1)$-dimensional real projective space realized as the quotient manifold $O(1)\setminus\Sigma_{n-1}$ by identifying antipodal points of Σ_{n-1}. Then $\mathbb{R}P^{2k}$ is not orientable; but $\mathbb{R}P^{2k-1}$ is orientable and its oriented orthonormal frame bundle is diffeomorphic to $\{\pm I\} \setminus SO(2k)$.

(iv) The oriented orthonormal frame bundle of n-dimensional hyperbolic space is diffeomorphic to $SO_0(1, n)$.

Many analytic constructions on a manifold M are conveniently described via a principal K-bundle $(P, \pi, \cdot)$ over that manifold. Let $\mathcal{H}$ be a K-module and let K act on $P \times \mathcal{H}$ by $k : (p, \xi) \to (k \cdot p, k\xi)$. The quotient space $P \times_K \mathcal{H}$ then consists of all equivalence classes

$$(2.7) \qquad [p, \xi] = \{(q, \eta) : q = k \cdot p \ , \ \eta = k\xi \ , \ k \in K\} \ ,$$

and there is a natural differentiable structure on $P \times_K \mathcal{H}$ with respect to which it becomes a vector bundle over M with projection $\pi_{\mathcal{H}} : [p, \xi] \to \pi(p)$ and fiber

$$\pi_{\mathcal{H}}^{-1}(x) = \{[p, \xi] : \pi(p) = x \ , \ \xi \in \mathcal{H}\}$$

isomorphic to $\mathcal{H}$. For convenience, this vector bundle, said to be *associated to* $\mathcal{H}$, will be denoted by $E_{\mathcal{H}}$. Any additional structure on $\mathcal{H}$ is inherited by $E_{\mathcal{H}}$ provided that it is appropriately K-invariant. For instance, if $\mathcal{H}$ is a Hilbert space on which K acts by isometries, then $E_{\mathcal{H}}$ becomes a Hilbert bundle, defining the inner product on $\pi_{\mathcal{H}}^{-1}(x)$ in terms of the inner product $(\cdot, \cdot)$ on $\mathcal{H}$ by

$$\big([p, \xi] \, , \, [p, \eta]\big) = (\xi, \eta) \qquad \big(\pi(p) = x\big) \ .$$

The isometric action of K ensures that this is well-defined; for if $[p, \xi] = [q, \xi']$ and $[p, \eta] = [q, \eta']$, then $q = k \cdot p$, $\xi' = k\xi$ and $\eta' = k\eta$ for some $k \in K$, in which case

$$\left([q, \xi'], [q, \eta']\right) = (k\xi, \, k\eta) = (\xi, \eta) \, .$$

Both the tangent bundle TM and the cotangent bundle T^*M are isomorphic to the Hilbert bundle $E_{\mathbb{R}^n}$ associated via $O(M, g)$ with the canonical action of $O(n)$ on $\mathbb{R}^n$; but from now on, for technical simplicity, we shall consider only orientable Riemannian manifolds M and their oriented orthonormal frame bundle $SO(M)$ $(= SO(M, g))$.

Now the representation (2.17) in chapter 3 of $\mathrm{Spin}(n)$ on $\mathfrak{A}_n$ descends to a single-valued representation of $SO(n)$, denoted again by σ, which on $\mathbb{R}^n$ is just the canonical action of $SO(n)$ by rotation on $\mathbb{R}^n$; furthermore, each $\sigma(k)$, $k \in SO(n)$, is an isometry on $\mathfrak{A}_n$ with respect to both the Clifford norm and Clifford algebra norm. The associated bundle $E_{\mathfrak{A}_n} = SO(M) \times_K \mathfrak{A}_n$, $K = SO(n)$, can thus be regarded as a Hilbert bundle over M with fiber norm

(2.8)(i) $\qquad\qquad |[p, \xi]| = |\xi|_{\mathfrak{A}} \qquad (\xi \in \mathfrak{A}_n)$

coming from the Clifford norm on $\mathfrak{A}_n$. On the other hand, the norm

(2.8)(ii) $\qquad\qquad \|[p, a]\| = \|a\|_{\mathfrak{A}} \qquad (a \in \mathfrak{A}_n)$

coming from the Clifford algebra norm on $\mathfrak{A}_n$ also is well-defined on $E_{\mathfrak{A}_n}$. Since

$$\sigma(k)(ab) = \left(\sigma(k)a\right)\left(\sigma(k)b\right) \qquad (a, b \in \mathfrak{A}_n) \, ,$$

(see chapter 3, (2.17)), it is clear that

$$[p, a][p, b] = [p, ab] \qquad (a, b \in \mathfrak{A}_n)$$

defines a multiplicative structure on $E_{\mathfrak{A}_n}$ such that (2.8)(ii) is just the operator norm of $E_{\mathfrak{A}_n}$ as an algebra of operators on $E_{\mathfrak{A}_n}$ regarding the latter as a Hilbert bundle. In addition, both the principal automorphism and principal anti-automorphism on $\mathfrak{A}_n$ pass to $E_{\mathfrak{A}_n}$ setting

$$[p, u]' = [p, u'] \quad , \quad [p, u]^* = [p, u^*]$$

since $(\sigma(k)u)' = \sigma(k)a'$ and $(\sigma(k)u)^* = \sigma(k)u^*$, $k \in O(n)$. As $E_{\mathbb{R}^n}$ is isomorphic to the tangent bundle TM, the following theorem is now virtually obvious on a functorial level.

(2.9) Theorem.

Let (M, g) be an arbitrary n-dimensional Riemannian manifold. Then the associated bundle $E_{\mathfrak{A}_n}$ is isomorphic to the Clifford algebra bundle $\mathfrak{A}_n(M, g)$ over M.

Thus we shall identify $E_{\mathfrak{A}_n}$ with $\mathfrak{A}_n(M)$ ($= \mathfrak{A}_n(M,g)$), and retain the names Clifford norm and Clifford algebra norm for the respective norms (2.8)(i),(ii). The complex Clifford algebra bundle $\mathfrak{C}_n(M)$ is constructed analogously.

The associated bundle construction suggests that there should be a $\mathfrak{A}_n(M)$-module bundle $E_{\mathfrak{H}}$ associated with each $\mathfrak{A}_n$-module $\mathfrak{H}$. Difficulties arise, however, if $\mathfrak{H}$ is a Spin(n)-module, not an $SO(n)$-module, as the construction will not go through unless $(P, \pi, \cdot)$ has Spin(n) for structure group. Existence of such principal bundles imposes an important topological restriction on a manifold.

(2.10) Definition.

An n-dimensional orientable Riemannian manifold M is said to have a Spin structure when there is a principal Spin(n)-bundle $(P, \rho, \cdot)$ over M extending the principal $SO(n)$-bundle $(SO(M), \pi, \cdot)$ in the sense that there exists a smooth mapping $\mu : P \to SO(M)$ with $\rho = \pi \circ \mu$ and

$$\mu(a \cdot p) = \sigma(a) \cdot \mu(p) \qquad \big(p \in P \, , \, a \in \text{Spin}(n)\big) \, .$$

A manifold may have exactly one, several, or no spin structures at all.

(2.11) Examples.

(i) The unit sphere in $\mathbb{C}$ has two inequivalent spin structures. But the unit sphere in $\mathbb{R}^n$, $n \geq 3$, has exactly one spin structure; its principal Spin(n)-bundle is diffeomorphic to Spin(n).

(ii) For each $\ell \geq 1$, real projective space $\mathbb{R}P^{4\ell-1}$ has two inequivalent spin structures, whereas $\mathbb{R}P^{4\ell+1}$ does not have any.

Throughout the rest of this section (M,g) will be an *n-dimensional oriented Riemannian manifold having spin structure* specified by a principal Spin(n)-bundle $(P, \rho, \cdot)$ satisfying the conditions of (2.10). Such a manifold is said to be a Spin-manifold. Let $(\mathcal{A}, \tau)$ be, say, a unitary representation of Spin(n) and let $C^\infty(M, E_{\mathcal{A}})$ be the space of smooth sections $F : M \to E_{\mathcal{A}}$ of $E_{\mathcal{A}}$. On the other hand, let $C^\infty_\tau(P, \mathcal{A})$ be the space of smooth functions $f : P \to \mathcal{A}$ satisfying

(2.12) $$f(a \cdot p) = \tau(a)f(p) \qquad \big(a \in \text{Spin}(n)\big)$$

on P. Then

(2.13) $$F(x) = [p, f(p)] \qquad \big(\rho(p) = x\big)$$

defines a 1–1 correspondence between these two spaces. Furthermore, if $\phi : U \to P$ is a local cross-section of P, $U \subseteq M$, then $F \to f \circ \phi$ identifies F on U with a smooth $\mathcal{A}$-valued function on U, bringing us back to the local case studied previously. The unique Riemannian connection on M will be used to associate to each vector field X on M differential operators on sections F and ones on functions f, which are then used to define δ_A (see Notes and remarks for more detailed accounts).

Intuitively, the Riemannian connection singles out a subspace of *horizontal* vectors in the tangent space $T_p P$ to each p in P; furthermore, this subspace is invariant under Spin(n) and varies smoothly with p. Thus to each vector field X on M there is assigned uniquely a *horizontal lift* X^* of X so that X^* is a vector field on P for which X_p^* is horizontal and $\rho(X_p^*) = X_{\rho(p)}$. In particular, $f \to X^* f$ is well-defined on smooth A-valued *functions* on P. To define derivatives of sections of E_A, let $\gamma : (-\varepsilon, \varepsilon) \to M$ be a smooth curve in M with $\gamma(0) = x_0$ for some fixed x_0. Then to each ξ in the fiber $\pi_A^{-1}(x_0)$ of E_A over x_0, there corresponds a unique *horizontal lift* of γ to a smooth curve $\gamma^* : (-\varepsilon, \varepsilon) \to E_A$ such that

$$\gamma^*(0) = \xi \quad , \quad \pi_A\big(\gamma^*(t)\big) = \gamma(t) \qquad (-\varepsilon < t < \varepsilon) \ ;$$

here horizontal means that the tangent vector to γ^* is horizontal at each point. In particular, $\xi \to \gamma^*(t)$ defines an isomorphism $\gamma_0^t : \pi_A^{-1}(\gamma(0)) \to \pi_A^{-1}(\gamma(t))$. Let γ_t^0 be its inverse. Now choose γ as above so that in addition $\dot{\gamma}(0) = X(x_0)$. Then

$$(2.14) \qquad \nabla_X F(x_0) = \lim_{t \to 0} \frac{1}{t} \left\{ \gamma_t^0 \Big(F\big(\gamma(t)\big) \Big) - F\big(\gamma(0)\big) \right\}$$

makes good sense because $\gamma_t^0(F(\gamma(t)))$ and $F(\gamma(0))$ belong to the same fiber $\pi_A^{-1}(x_0)$. Clearly, $\nabla_X F$ is again a section of E_A and $(X, F) \to \nabla_X F$ has all the properties of (1.18) replacing $\mathfrak{H}$-valued functions by sections of E_A. It is a well-known result that

$$(2.15) \qquad \nabla_X F = [p, X^* f(p)] \qquad \big(\rho(p) = x\big)$$

when F and f are related by (2.13); in particular, $X^* f(a \cdot p) = \tau(a) X^* f(p)$.

To define the 'total' derivative, whether on sections or on functions, we exploit the realization of the cotangent bundle $T^* M$ with the bundle $E_{\mathbb{R}^n}$ associated with the representation $(\mathbb{R}^n, \sigma)$ of Spin(n). Thus, by (2.13), 1-forms Θ on M are in 1–1 correspondence with functions $\theta : P \to \mathbb{R}^n$ such that $\theta(a \cdot p) = \sigma(a)\theta(p)$. Now each p in P has the form $p = (x, \{X_j(x)\})$ where $\{X_1(x), \dots, X_n(x)\}$ is an oriented basis for $T_x M$ determined by an orthonormal frame $\{X_1, \dots, X_n\}$. Let $X_1^*, \dots, X_n^*$ be the horizontal lifts of this frame, and let $\Theta_1, \dots, \Theta_n$ be the co-frames dual to $X_1, \dots, X_n$, i.e., $\Theta_j(X_k) = \delta_{jk}$.

(2.16) Definition.

With the notation above define the total derivative ∇ on $C^\infty(M, E_\mathcal{A})$
by

$$\nabla : F \longrightarrow \sum_{j=1}^{n} \nabla_{X_j} F \otimes \Theta_j \ ,$$

and on $C_\tau^\infty(P, \mathcal{A})$ by

$$\nabla : f \longrightarrow \sum_{j=1}^{n} \partial_j f \otimes e_j$$

where $\partial_j f(p) = X_j^* f(p)$ when $p = (x, \{X_j(x)\})$ and $\{e_1, \ldots, e_n\}$ is the
standard basis of $\mathbb{R}^n$.

Clearly ∇F is a section of $E_{\mathcal{A} \otimes \mathbb{C}^n}$, identifying $E_\mathcal{A} \otimes T^*M$ with $E_\mathcal{A} \otimes E_{\mathbb{R}^n}$, and thence with $E_{\mathcal{A} \otimes \mathbb{C}^n}$; while ∇f is an $\mathcal{A} \otimes \mathbb{C}^n$-valued function on P. The following fundamental result brings out the true meaning of these operators; its proof will conclude this section.

(2.17) Theorem.

When F and f are related by $F(x) = [p, f(p)]$, then
$$\nabla F(x) = \big[p, \nabla f(p)\big] \qquad \big(\rho(p) = x\big) \ ;$$

in particular,

$$\nabla f(a \cdot p) = (\tau \otimes \sigma)(a) \nabla f(p) \qquad \big(a \in \mathrm{Spin}(n)\big) \ .$$

Defining first-order differential operators has now been reduced to the same algebraic setting as the rotation-invariant case of the previous chapter. For let $(\mathcal{B}, \omega)$ be a second unitary representation of $\mathrm{Spin}(n)$ and $A : \mathcal{A} \otimes \mathbb{C}^n \to \mathcal{B}$ a linear operator such that

(2.18) $A\big((\tau \otimes \sigma)(a)u\big) = \omega(a)A(u) \qquad \big(a \in \mathrm{Spin}(n)\big)$

on $\mathcal{A} \otimes \mathbb{C}^n$. Then

(2.19) $\eth_A f = (A \circ \nabla) f = \sum_{j=1}^{n} (A_j \circ \partial_j) f$

is a first-order linear differential operator from $C_\tau^\infty(P, \mathcal{A})$ into $C_\omega^\infty(P, \mathcal{B})$, setting $A_j(\xi) = A(\xi \otimes e_j)$ as in (1.3) of chapter 4. Indeed,

$$(\eth_A f)(a \cdot p) = A\big((\tau \otimes \sigma)(a) \nabla f(p)\big) = \omega(a)(\eth f)(p)$$

because of (2.17) and (2.18); consequently, $\eth_A f$ is a function in $C_\omega^\infty(P, \mathcal{B})$. Thus for each of the differential operators in sections 3 and 4 of the previous chapter having a $\mathrm{Spin}(n)$-equivariant symbol there will be a corresponding operator on any Spin-manifold; if the symbol is actually

$SO(n)$-equivariant, then the operator will be defined on the manifold, whether or not it has a spin structure. By using (2.17), these operators can then be realized on sections of bundles over M. For instance, let $\mathfrak{H}$ be an $\mathfrak{A}_n$-module on which there is a unitary representation $(\mathfrak{H}, \tau)$ of $\text{Spin}(n)$ such that

(2.20) $\qquad \tau(a)e_j\tau(a)^{-1} = \sigma(a)e_j \qquad \big(a \in \text{Spin}(n)\big)$.

As a differential operator on $C^\infty_\tau(P, \mathfrak{H})$, the standard Dirac operator is simply

(2.21)(i) $\qquad\qquad Df = \sum_{j=1}^{n} e_j \partial_j f$.

But $E_{\mathfrak{H}}$ and $E_{\mathfrak{A}_n}$ are well-defined, thinking of $(\mathfrak{A}_n, \sigma)$ as a representation of $\text{Spin}(n)$. In addition, (2.19) ensures that $E_{\mathfrak{H}}$ becomes an $E_{\mathfrak{A}_n}$-module setting

$$[p, e_{i_1} \ldots e_{i_k}][p, \xi] = [p, (e_{i_1} \ldots e_{i_k})\xi] .$$

(Although the earlier construction of $E_{\mathfrak{A}_n}$ regarded $(\mathfrak{A}_n, \sigma)$ as an $SO(n)$-module, the compatibility conditions between $(SO(M), \pi, \cdot)$ and $(P, \rho, \cdot)$ guarantee that the two constructions yield isomorphic bundles; consequently, use of the notation $E_{\mathfrak{A}_n}$ is consistent.) But $E_{\mathbb{R}^n}$ is just a subbundle of $E_{\mathfrak{A}_n}$. Hence, by (2.17) and (2.19), the standard Dirac operator on $C^\infty(M, E_{\mathfrak{H}})$ is given by

(2.21)(ii) $\qquad\qquad DF(x) = \sum_{j=1}^{n} \Theta(x)\nabla_{X_j} F(x)$.

For many practical purposes, however, the local coordinate realizations of section 1 are surely preferable.

Proof of (2.17). The definition of ∇F in (2.16) is independent of the choice of orthonormal frame. For, if $X'_j = \sum_i a_{ij}(x)X_k$, $1 \leq j \leq n$, is another orthonormal frame, the corresponding dual frame is $\Theta'_j = \sum_k a_{jk}\Theta_k$, $1 \leq j \leq k$, and the properties in (1.18) ensure that

$$\sum_j \nabla_{X'_j} F \otimes \Theta'_j = \sum_{i,k} \bigg(\sum_j a_{ij}a_{jk} \bigg) \nabla_{X_i} F \otimes \Theta_k = \sum_i \nabla_{X_i} F \otimes \Theta_i .$$

Given x in M, choose any p in the fiber $\rho^{-1}(x)$ in P over x. It has the form $(x, \{X_j(x)\})$ determined by a orthonormal frame $\{X_1, \ldots, X_n\}$; let $\{\Theta_1, \ldots, \Theta_n\}$ be the dual co-frame. We can use these to define $\nabla F(x)$. But, under the identification of TM and T^*M with $E_{\mathbb{R}^n}$, there correspond ψ_j, θ_k in $C^\infty_\sigma(P, \mathbb{R}^n)$ so that

$$X_j(x) = [p, \psi_j(p)] \quad , \quad \Theta_k(x) = [p, \theta_k(p)] ,$$

while

$$\nabla F(x) = \left[p, \sum_j X_j^* f(p) \otimes \theta_j(p) \right].$$

Since $p = (x, \{X_j(x)\})$, however, $\psi_j(p) = e_j$ where $\{e_1, \ldots, e_n\}$ is the standard basis for $\mathbb{R}^n$; and by duality, $\theta_k(p) = e_k$. This completes the proof. ∎

Since the Bochner–Weitzenböck property of the Dirac operator is really an algebraic property, actually a representation-theoretic one, the use of differential operators on vector-valued functions on P to specify differential operators on sections on M has been used to good effect to derive Bochner–Weitzenböck results globally (see Notes and remarks).

3 Dirac operators on hyperbolic and spherical space

There is a well-known realization of real two-dimensional hyperbolic space H_2 as the upper half-plane $\mathbb{C}_+$ with the Poincaré metric, in which the action of $SL(2, \mathbb{R}) \sim \mathrm{Spin}_0(2, 1)$ by fractional linear transformations

$$(3.1) \qquad g = \begin{bmatrix} a & b \\ c & d \end{bmatrix} : z \longrightarrow z.g = \frac{b + zd}{a + zc} \qquad (z \in \mathbb{C}_+)$$

is the (connected) group of transformations of H_2 preserving the hyperbolic metric. Extensive use has been made of this, for instance, in constructing irreducible unitary representations of $SL(2, \mathbb{R})$ on spaces of analytic functions on $\mathbb{C}_+$. In this section we shall study the corresponding realization of real n-dimensional hyperbolic space H_n as the upper half-space $\mathbb{R}_+^n$. The Dirac operator on H_n then takes on a very simple explicit form. More remarkably still perhaps, the multiplicative structure on $\mathfrak{A}_n$ will be used to describe the isometry group of H_n by the action of $\mathrm{Spin}(n, 1)$ as 'fractional linear transformations' on $\mathbb{R}_+^n$. In this way the familiar harmonic analysis on $\mathbb{R}^{n-1}$ and on $\mathbb{R}_+^n$ has an appearance much closer to that on $\mathbb{R}$ and $\mathbb{C}_+$ than would be possible without the use of Clifford algebra theory (see section 4).

There are entirely analogous results for spherical space, realizing the unit sphere Σ_n in $\mathbb{R}^{n+1}$ via stereographic projection as the compactification $\mathbb{R}^n \cup \{\infty\}$ of $\mathbb{R}^n$. The prototypical example is the Riemann sphere thought of as $\mathbb{C} \cup \{\infty\}$ with $SU(2) \sim \mathrm{Spin}(3)$ acting by linear fractional transformations

$$(3.2) \qquad g = \begin{bmatrix} \alpha & -\overline{\beta} \\ \beta & \overline{\alpha} \end{bmatrix} : \longrightarrow \frac{\overline{\alpha} z - \overline{\beta}}{\alpha + \beta z} \qquad (z \in \mathbb{C} \cup \{\infty\}).$$

Once again the Dirac operator has a simple form, but now the action of Spin($n+1$) by 'fractional linear transformations' on $\mathbb{R}^n \cup \{\infty\}$ replaces (3.2).

Under the coordinate system

$$\left\{ z = (x,y) : x = (x_1, \ldots, x_{n-1}) \in \mathbb{R}^{n-1}, y > 0 \right\}$$

on $\mathbb{R}^n_+$ the hyperbolic metric is determined by the function

(3.3) $g : \mathbb{R}^n_+ \longrightarrow \mathbb{R}^{n \times n}$, $g(z) = \dfrac{1}{y^2} I$.

Since $\gamma^{-1}(z) = yI$ and

(3.4)(i) $E_i(z) = y\dfrac{\partial}{\partial x_i}$ $(1 \leq i < n)$, $E_n(z) = y\dfrac{\partial}{\partial y}$,

straightforward calculations show that

(3.4)(ii) $\omega_{ij}^k = \dfrac{1}{y}(\delta_{ij}\delta_{kn} - \delta_{jn}\delta_{ik})$, $\nu_{ij}^k = y\omega_{ij}^k$ $(1 \leq i,j,k \leq n)$.

Hence the functions $\omega_i, \nu_i : \mathbb{R}^n_+ \to \mathrm{spin}(n)$ are given by

(3.4)(iii) $\omega_i(z) = \dfrac{1}{2y}e_ie_n$, $\nu_i(z) = \dfrac{1}{2}e_ie_n$ $(1 \leq i < n)$.

Putting all these together we deduce (see (1.19)) the following.

(3.5) Theorem.
   As a differential operator on $\mathcal{C}^\infty(\mathbb{R}^n_+, \mathfrak{H})$ the standard Dirac operator D on H_n is defined by

$$D = \left(\sum_{i=1}^{n-1} e_i \left(y\dfrac{\partial}{\partial x_i} + \dfrac{1}{2}d\tau(e_ie_n) \right) + e_ny\dfrac{\partial}{\partial y} \right) .$$

To realize n-dimensional spherical space S_n, define $g : \mathbb{R}^n \to \mathbb{R}^{n \times n}$ by

(3.6) $g(x) = \dfrac{1}{(1 + |x|^2)^2} I$ $(x \in \mathbb{R}^n)$;

symbolically, the induced Riemannian metric tensor is

$$g(x) = \left(\dfrac{1}{1 + |x|^2} \right)^2 \sum_{j=1}^n dx_j^2 .$$

Now the stereographic projection

(3.7) $x \longrightarrow \left(\dfrac{2}{|x|^2 + 1} \right)x + \left(\dfrac{|x|^2 - 1}{|x|^2 + 1} \right)\mathbf{1}$ $(x \in \mathbb{R}^n)$

is a diffeomorphism from $\mathbb{R}^n$ into the unit sphere Σ_n in $\mathbb{R}^{n+1}$, omitting only the North Pole $\mathbf{1}$; furthermore, it can be shown that this is an isometric mapping between $(\mathbb{R}^n, g)$ and $(\Sigma_n \setminus \mathbf{1}, g_s)$ where g_s is the usual Riemannian metric on Σ_n. Hence we can think of $\widehat{\mathbb{R}}^n = \mathbb{R}^n \cup \{\infty\}$ as a realization of S_n. Since

$$g^{-1}(x) = (1 + |x|^2)^2 I \quad , \quad \frac{\partial g}{\partial x_i} = -\frac{4x_i}{(1+|x|^2)^3} I \quad (1 \le i \le n) \,,$$

straightforward calculations show that

$$\nabla_{\frac{\partial}{\partial x_j}} \left(\frac{\partial}{\partial x_i} \right) = -\frac{2}{(1+|x|^2)} \sum_k (x_j \delta_{ik} + x_i \delta_{jk} - x_k \delta_{ij}) \frac{\partial}{\partial x_k} \,.$$

But

$$\gamma(x)^{-1} = (1 + |x|^2) I \quad , \quad E_i(x) = (1 + |x|^2) \frac{\partial}{\partial x_i} \quad (1 \le i \le n) \,,$$

and so

$$\nabla_{\frac{\partial}{\partial x_i}} \big(E_j(x) \big) = \frac{\partial}{\partial x_i} (1 + |x|^2) \frac{\partial}{\partial x_j} + (1 + |x|^2) \nabla_{\frac{\partial}{\partial x_i}} \left(\frac{\partial}{\partial x_j} \right)$$

$$= \frac{2}{(1+|x|^2)} \sum_k (x_k \delta_{ij} - x_j \delta_{ik}) E_k(x) = \sum_k \omega_{ij}^k(x) E_k(x) \,.$$

Hence

$$\omega_i(x) = \tfrac{1}{2} \sum_{j<k} \omega_{ij}^k(x) e_j e_k = \frac{1}{2(1+|x|^2)} \sum_{j,k} (x_k \delta_{ij} - x_j \delta_{ik}) e_j e_k$$

$$= \frac{1}{(1+|x|^2)} \sum_k x_k e_i e_k = \frac{1}{(1+|x|^2)} e_i x \,,$$

identifying $x = (x_1, \ldots, x_n)$ with $x = \sum_k x_k e_k$ in $\mathfrak{A}(\mathfrak{H})$. This gives the following.

(3.8) Theorem.
As a differential operator on $C^\infty(\widehat{\mathbb{R}}^n, \mathfrak{H})$ the standard Dirac operator D on S_n is defined by

$$D = (1 + |x|^2) \left(\sum_{i=1}^n e_i \left(\frac{\partial}{\partial x_i} + \frac{1}{(1+|x|^2)} \, d\tau(e_i x) \right) \right) \,.$$

The realization of $\mathrm{Spin}_0(n, 1)$ as a multiplicative group of 2×2 matrices arises from identifying $\mathrm{Spin}_0(n, 1)$ with a group of transformers in $\mathfrak{A}_{n,0}$, thence with a multiplicative group in $M(2, \mathfrak{A}_{n-2})$. During this construction extensive use will be made of (2.29) and sections 5 and 6 in chapter 1. First we realize Minkowski space $\mathbb{R}^{n,1}$ as a subspace of $M(2, \mathfrak{A}_{n-2})$. Let $e_1, \ldots, e_{n-2}$ be imaginary units generating $\mathfrak{A}_{n-2}$ as usual, and let e_0 ($= 1$) be the identity in $\mathfrak{A}_{n-2}$. Then

$$(x_0, \ldots, x_{n-2}) \longrightarrow x = \sum_{j=0}^{n-2} x_j e_j$$

embeds $\mathbb{R}^{n-1}$ into $\mathfrak{A}_{n-2}$ so that $|x|^2 = \bar{x}x = (x_0^2 + \cdots + x_{n-2}^2)^{1/2}$. Now define $\gamma_0, \ldots, \gamma_{n-1}$ in $M(2, \mathfrak{A}_{n-2})$ by
(3.9)

$$\gamma_0 = \begin{bmatrix} 0 & 1 \\ 1 & 0 \end{bmatrix} , \ \gamma_j = \begin{bmatrix} 0 & e_j \\ -e_j & 0 \end{bmatrix} \ (1 \le j < n-1) , \ \gamma_{n-1} = \begin{bmatrix} 1 & 0 \\ 0 & -1 \end{bmatrix} ;$$

clearly

$$(3.9)(\text{i}) \qquad z \longrightarrow \sum_{j=0}^{n-2} x_j \gamma_j + y\gamma_{n-1} = \begin{bmatrix} y & x \\ x' & -y \end{bmatrix} \qquad \left(z = (x, y) \right)$$

embeds $\mathbb{R}^{n,0}$ into $M(2, \mathfrak{A}_{n-2})$ so that

$$\left(\sum_{j=0}^{n-2} x_j \gamma_j + y\gamma_{n-1} \right)^2 = |z|^2 \qquad (z \in \mathbb{R}^n) .$$

Thus, as we saw earlier in I (2.29), the Clifford algebra generated by $\gamma_0, \ldots, \gamma_{n-1}$ identifies $M(2, \mathfrak{A}_{n-2})$ with $\mathfrak{A}_{n,0}$; furthermore,

$$w' = \begin{bmatrix} d' & -c' \\ -b' & a' \end{bmatrix} , \quad w^* = \begin{bmatrix} \bar{a} & \bar{c} \\ \bar{b} & \bar{d} \end{bmatrix} , \quad \bar{w} = \begin{bmatrix} d^* & -b^* \\ -c^* & a^* \end{bmatrix}$$

whenever $w = \begin{bmatrix} a & b \\ c & d \end{bmatrix}$. In particular,

$$(3.10)(\text{i}) \qquad \bar{w}w = \begin{bmatrix} d^*a - b^*c & d^*b - b^*d \\ a^*c - c^*a & a^*d - c^*b \end{bmatrix} ,$$

while

$$(3.10)(\text{ii}) \qquad (z, t) \longrightarrow \zeta = \begin{bmatrix} t+y & x \\ x' & t-y \end{bmatrix}$$

embeds $\mathbb{R}^{n,1}$ in $M(2, \mathfrak{A}_{n-2})$ so that

$$|\zeta|^2 = \bar{\zeta}\zeta = t^2 - |z|^2 .$$

Hence by (6.12) and (6.15) in chapter 1, $\mathrm{Spin}_0(n, 1)$ is the group

$$(3.11) \qquad \left\{ \zeta_1 \cdots \zeta_k : \zeta_j \in \mathbb{R}^{n,1} , \ |\zeta_j|^2 = 1 , \ k \ge 1 \right\}$$

of all finite products in $M(2, \mathfrak{A}_{n-2})$ of elements from the unit sphere of $\mathbb{R}^{n,1}$. To associate a group of linear fractional transformations of $\mathbb{R}^n$ with $\mathrm{Spin}_0(n, 1)$ we shall identify (3.11) with a group of 2×2 matrices whose entries are transformers in $\mathfrak{A}_{n-2}$. Recall that the Clifford semigroup Λ_{n-2} consists of all a in $\mathfrak{A}_{n-2}$ having the property: *for each x in $\mathbb{R}^{n-1}$ there exists ξ in $\mathbb{R}^{n-1}$ such that $ax = \xi a'$* (see chapter 1, (5.29),(7.23)).

(3.12) Definition.

For each $n \geq 2$, let $SL_2(\Lambda_{n-2})$ be the subset of all $g = \begin{bmatrix} a & b \\ c & d \end{bmatrix}$ in $M(2, \mathfrak{A}_{n-2})$ whose matrix entries satisfy

(i) $a, b, c, d \in \Lambda_{n-2}$,   (ii) $a^*c, b^*d \in \mathbb{R}^{n-1}$ (iii) $d^*a - b^*c = 1$.

Since the principal anti-automorphism reduces to the identity on $\mathbb{R}^{n-1}$ as a subspace of $\mathfrak{A}_{n-2}$ (see chapter 1, (5.23)(iii)), condition (3.12)(ii) ensures that $a^*c = c^*a$ and $b^*d = d^*b$. Hence $SL_2(\Lambda_{n-2})$ is the subset of all g in $M(2, \mathfrak{A}_{n-2})$ for which

(3.12)'   (i) $|g|^2 = \bar{g}g = 1$ (ii) $a^*c, b^*d \in \mathbb{R}^{n-1}$.

The latter is a particularly appealing definition of $SL_2(\Lambda_{n-2})$ because of our original motivation of $|\cdot|^2$ as a generalization of the determinant function (see chapter 1, (5.14)). Indeed, $\Lambda_{n-2} = \mathbb{R}$ when $n = 2$, while $\Lambda_{n-2} = \mathbb{C}$ when $n = 3$; consequently, $SL_2(\Lambda_{n-2})$ becomes $SL_2(\mathbb{R}) \sim \mathrm{Spin}_0(2,1)$ when $n = 2$, and it becomes $SL_2(\mathbb{C}) \sim \mathrm{Spin}_0(3,1)$ when $n = 3$. The notation of (3.12) is thus deliberately chosen to suggest the following fundamental result.

(3.13) Theorem.

As a multiplicative group in $M(2, \mathfrak{A}_{n-2})$, $n \geq 2$, the spin group $\mathrm{Spin}_0(n, 1)$ coincides with $SL_2(\Lambda_{n-2})$.

Assuming (3.13) for the moment, let $\mathbf{j}$ be an imaginary unit which together with $e_1, \ldots, e_{n-2}$ generates $\mathfrak{A}_{n-1}$; then $\mathfrak{A}_{n-1} = \mathfrak{A}_{n-2} \oplus \mathbf{j}\mathfrak{A}_{n-2}$ and

$$(3.14) \qquad (x_0, \ldots, x_{n-2}, y) \longrightarrow z = x + y\mathbf{j} = \sum_{i=0}^{n-2} x_i e_i + y\mathbf{j}$$

is an embedding of $\mathbb{R}^n$ into $\mathfrak{A}_{n-1}$ such that $|z|^2 = \bar{z}z = |x|^2 + y^2$. In this notation, $\mathbb{R}_+^n = \{x + y\mathbf{j} : y > 0\}$ and $\mathbb{R}_-^n = \{x + y\mathbf{j} : y < 0\}$. Notice that these two realizations of $\mathbb{R}^n$, one in $M(2, \mathfrak{A}_{n-2})$ via (3.9)(i), and the other in $\mathfrak{A}_{n-1}$ via (3.14) are analogous to the classical identifications of $\mathbb{C}$ as matrices $\begin{bmatrix} a & b \\ -b & a \end{bmatrix}$ and as $a + \mathbf{j}b$. Definition (3.12) now has the right form to ensure the following.

(3.15) Theorem.

The fractional linear transformation

$$g = \begin{bmatrix} a & b \\ c & d \end{bmatrix} : z \longrightarrow z \cdot g = (a + zc)^{-1}(b + zd) , \quad \infty \cdot g = c^{-1}d \quad (z \in \mathbb{R}^n)$$

is well-defined and bijective on $\mathbb{R}^n \cup \{\infty\}$ if g is in $SL_2(\Lambda_{n-2})$, hence in $\mathrm{Spin}_0(n, 1)$; furthermore, under this action,

(i) the orbits of $\mathrm{Spin}_0(n,1)$ consist of the upper and lower half-spaces $\mathbb{R}^n_+$, $\mathbb{R}^n_-$ as well as their compactified mutual boundary $\mathbb{R}^{n-1} \cup \{\infty\}$,

(ii) $\mathrm{Spin}_0(n,1)$ is a two-fold covering of the connected group of transformations preserving the hyperbolic metric on $\mathbb{R}^n_+$, while

(iii) $\mathrm{Spin}_0(n,1)$ is a two-fold covering of the group of sense-preserving conformal transformations of $\mathbb{R}^{n-1} \cup \{\infty\}$.

Hence theorems (3.13) and (3.15) show that $\mathrm{Spin}_0(n,1)$, alias $SL_2(\Lambda_{n-1})$, has exactly the same relationship to $\mathbb{R}^n$ as $SL_2(\mathbb{R})$ does to $\mathbb{C}$. The proofs of these results proceed in several interlocking steps.

(A) $SL_2(\Lambda_{n-2}) \supseteq \mathrm{Spin}_0(n,1)$. Since Euclidean space is a positive-definite quadratic space, every non-zero element of Λ_{n-2} is invertible (see chapter 1, (5.16)). Hence, by chapter 1, (5.20), Λ_{n-2} is the union of $\{0\}$ and the Clifford group

(3.16) $$\Gamma_{n-2} = \{\xi_1 \xi_2 \ldots : \xi_j \in \mathbb{R}^{n-1}, \ \xi_j \neq 0\}$$

of all finite products in $\mathfrak{A}_{n-2}$ of non-zero elements in $\mathbb{R}^{n-1}$; in particular, every ω in the unit sphere of $\mathbb{R}^{n,1}$ is in $SL_2(\Lambda_{n-2})$ since all such ω have the form

$$\omega = \begin{bmatrix} \alpha & x \\ x' & \beta \end{bmatrix}, \quad \alpha\beta + |x|^2 = 1 \quad (\alpha, \beta \in \mathbb{R}, \ x \in \mathbb{R}^{n-1}).$$

Consequently, to prove **(A)** it is enough to show for such ω that ωg is in $SL_2(\Lambda_{n-2})$ whenever g is. Now

$$\omega g = \begin{bmatrix} \alpha a + xc & \alpha b + xd \\ x'a + \beta c & x'b + \beta d \end{bmatrix} \quad \left(g = \begin{bmatrix} a & b \\ c & d \end{bmatrix} \right).$$

If $a = 0$, or $c = 0$, then by (3.16) it is clear that $\alpha a + xc$ and $x'a + \beta c$ are in Λ_{n-2}. So suppose $a, c \neq 0$; in this case,

$$c = \frac{1}{|a|^2} a'a^*c, \quad a = \frac{1}{|c|^2} c'c^*a.$$

But condition (3.12)(ii) on $SL_2(\Lambda_{n-1})$ ensures that $a^*c = \xi$ with ξ in $\mathbb{R}^{n-1}$. Thus

$$c = \frac{1}{|a|^2} a'\xi = \frac{1}{|a|^2}(\sigma(a')\xi)a = va$$

for some v in $\mathbb{R}^{n-1}$, since a' is a transformer; similarly, $a = uc$ with u in $\mathbb{R}^{n-1}$ since $c^*a = a^*c = \xi$. Consequently, both of $(\alpha a + xc)$, $(x'a + \beta c)$ can still be written as finite products

$$\alpha a + xc = (\alpha u + x)c, \quad x'a + \beta c = (x' + \beta v)a$$

of elements from $\mathbb{R}^{n-1}$ even when $a, c \neq 0$. This together with a corresponding argument for the other entries shows that ωg is in $M(2, \Lambda_{n-2})$.

On the other hand,

$$(\alpha a + xc)^*(x'a + \beta c) = (\alpha a^* + c^*x)(x'a + \beta c)$$
$$= (\alpha\beta + |x|^2)a^*c + \alpha a^* x'a + \beta c^* xc$$

(see chapter 1, (5.23)(iii)). Now

$$a^*x'a = \xi_0 \bar{a}a = |a|^2\xi_0 \quad , \quad c^*xc = \xi_1 \bar{c}c = |c|^2\xi_1$$

for some ξ_0, ξ_1 in $\mathbf{R}^{n-1}$ since a, c are transformers. Consequently, the product $(\alpha a + xc)^*(x'a + \beta c)$ is in $\mathbf{R}^{n-1}$, as also will be the product $(\alpha b + xd)^*(x'b + \beta d)$. Hence wg has property (3.12)(ii). Finally, to establish (3.12)(iii) for wg, it is enough to observe that

$$(wg)^-(wg) = \bar{g}(\bar{w}w)g = 1$$

(see (3.12)'(i)). This proves (**A**).   ∎

(**B**) $SL_2(\Lambda_{n-2})$-*decompositions*. There is a basic decomposition of $SL_2(\Lambda_{n-2})$ which is the classical Bruhat decomposition for $\mathrm{Spin}_0(n,1)$ once (3.13) has been proved. Set

(3.17)(i)
$$\begin{cases} M = \left\{ \begin{bmatrix} a & 0 \\ 0 & a' \end{bmatrix} : a \in \Lambda_{n-1} , |a|^2 = 1 \right\} , \\ A = \left\{ \begin{bmatrix} \lambda^{-1/2} & 0 \\ 0 & \lambda^{1/2} \end{bmatrix} : \lambda > 0 \right\} \end{cases}$$

and

(3.17)(ii)
$$V = \left\{ \begin{bmatrix} 1 & v \\ 0 & 1 \end{bmatrix} : v \in \mathbf{R}^{n-1} \right\} .$$

These are multiplicative groups in $M(2, \mathfrak{A}_{n-2})$; in fact

$$M \sim \mathrm{Spin}(n-1) \quad , \quad A \sim \mathbf{R} \quad , \quad V \sim \mathbf{R}^{n-1} .$$

They are also related to the geometric properties of the fractional linear transformation in (3.15). For, when $z = x + y\mathbf{j}$,

(3.18)(i)
$$\begin{bmatrix} a & 0 \\ 0 & a' \end{bmatrix} : z \longrightarrow a^{-1}za' = \sigma(a^{-1})z = \sigma(a^{-1})x + y\mathbf{j}$$

is a *rotation* of $\mathbf{R}^n$ fixing the $\mathbf{j}$-direction, while

(3.18)(ii)
$$\begin{bmatrix} \lambda^{-1/2} & 0 \\ 0 & \lambda^{1/2} \end{bmatrix} : z \longrightarrow \lambda z = \lambda x + \lambda y\mathbf{j}$$

defines *dilation* of $\mathbf{R}^n$ by λ, and

(3.18)(iii)
$$\begin{bmatrix} 1 & v \\ 0 & 1 \end{bmatrix} : z \longrightarrow z + v = (x + v) + y\mathbf{j}$$

defines *translation* by v in any hyperplane parallel to $\mathbf{R}^{n-1}$. These are the basic transformations of harmonic analysis on Euclidean space. On hyperbolic space, however, one has to add the *inversion*

(3.18)(iv) $$\omega = \begin{bmatrix} 0 & -1 \\ 1 & 0 \end{bmatrix} : z \longrightarrow -z^{-1}$$

determined by the so-called *Weyl group element* ω in $SL_2(\Lambda_{n-1})$.

(3.19) Lemma.
Every g in $SL_2(\Lambda_{n-2})$ can be written as a product in $M(2, \mathfrak{A}_{n-2})$ of elements from M, A, V and $\{\omega\}$.

Proof. Suppose first that $g = \begin{bmatrix} a & b \\ c & d \end{bmatrix}$ with $c \neq 0$. Then ac^{-1}, $c^{-1}d$ are elements of Λ_{n-2} such that

$$\begin{bmatrix} 1 & ac^{-1} \\ 0 & 1 \end{bmatrix} \begin{bmatrix} (c^*)^{-1} & 0 \\ 0 & c \end{bmatrix} \begin{bmatrix} 0 & -1 \\ 1 & 0 \end{bmatrix} \begin{bmatrix} 1 & c^{-1}d \\ 0 & 1 \end{bmatrix} = \begin{bmatrix} a & B \\ c & d \end{bmatrix}$$

where $B = ac^{-1}d - (c^*)^{-1}$. But

$$c^*B = (c^*a)c^{-1}d - 1 = a^*(cc^{-1})d - 1 = c^*b ,$$

using (3.12)(ii), (iii), so

(3.20) $$g = \begin{bmatrix} 1 & ac^{-1} \\ 0 & 1 \end{bmatrix} \begin{bmatrix} (c^*)^{-1} & 0 \\ 0 & c \end{bmatrix} \begin{bmatrix} 0 & -1 \\ 1 & 0 \end{bmatrix} \begin{bmatrix} 1 & c^{-1}d \\ 0 & 1 \end{bmatrix} .$$

As we saw already in the proof of step (**A**), $a = uc$ with u in $\mathbb{R}^{n-1}$ when $c \neq 0$; consequently, the first term in (3.20) is in V. To show that the last term also is in V, observe that

$$c^{-1}d = (d^*a - b^*c)c^{-1}d = d^*ud - b^*d = |d|^2\xi - b^*d$$

for some ξ in $\mathbb{R}^{n-1}$ because of (3.12)(iii) and the fact that d is a transformer. Thus $c^{-1}d \in \mathbb{R}^{n-1}$, and so the last term in (3.20) is in V. On the other hand,

$$\begin{bmatrix} (c^*)^{-1} & 0 \\ 0 & c \end{bmatrix} = \begin{bmatrix} \alpha' & 0 \\ 0 & \alpha \end{bmatrix} \begin{bmatrix} \lambda^{-1/2} & 0 \\ 0 & \lambda^{1/2} \end{bmatrix}$$

with $\alpha = (1/|c|^2)^{1/2}c$, $\lambda = |c|^2$. Hence the second term in (3.20) is a product of elements from M to A, showing that g is a product in $M(2, \mathfrak{A}_{n-2})$ of elements from M, A, V and $\{\omega\}$ when $c \neq 0$. But if $c = 0$, then $d^*a = 1$, so d is non-zero and hence invertible. Consequently,

$$g = \begin{bmatrix} a & b \\ 0 & d \end{bmatrix} = \begin{bmatrix} 1 & bd^{-1} \\ 0 & 1 \end{bmatrix} \begin{bmatrix} a & 0 \\ 0 & d \end{bmatrix} .$$

By the arguments used before, the right-hand side is a product of elements from M, A, and V, completing the proof. ∎

To finish the proof of (3.13) it is enough now to show that the group $\mathrm{Spin}_0(n, 1)$ contains M, A, V and ω. In view of the proof of step (**B**), this would in fact prove the following more precise statement (Bruhat decomposition).

(**C**) $\mathrm{Spin}_0(n,1) = VMA\omega V \cup VMA$. Clearly M, A, and ω all lie in (3.11). But for any v in $\mathbb{R}^{n-1}$, $v \neq 0$,

$$\begin{bmatrix} 1 & v \\ 0 & 1 \end{bmatrix} = \begin{bmatrix} 0 & v \\ v' & 0 \end{bmatrix} \begin{bmatrix} 0 & (1/|v|^2)v \\ (1/|v|^2)v' & 1 \end{bmatrix}.$$

Hence V also lies in (3.11).   ∎

Theorem (3.15) follows easily from a further series of brief steps. Recall that

(3.21)

$$g = \begin{bmatrix} a & b \\ c & d \end{bmatrix} : z \longrightarrow z \cdot g = (a+zc)^{-1}(b+zd), \quad \infty \cdot g = c^{-1}d \qquad (z \in \mathbb{R}^n).$$

(**D**) *Well-definedness.* If (3.21) is well-defined for g_1, g_2 in $M(2, \mathfrak{A}_{n-2})$, then the product $g_1 g_2$ determines the fractional linear transformation

(3.22) $g_1 g_2 : z \longrightarrow (z \cdot g_1) \cdot g_2 \qquad (z \in \mathbb{R}^n \cup \{\infty\})$.

But, as we have observed already, the groups M, A, and and V give fractional linear transformations which preserve $\mathbb{R}^n$; each of these groups also fixes $\{\infty\}$. On the other hand, the fractional linear transformation (3.18)(ii) determined by the Weyl group element ω preserves $\mathbb{R}^n \setminus \{0\}$ and interchanges $0, \infty$. Hence by step (**C**), the fractional linear transformation (3.21) is well-defined on $\mathbb{R}^n \cup \{\infty\}$ (and clearly is bijective). ∎

It is not hard to show that if (3.21) is well-defined and bijective on $\mathbb{R}^n \cup \{\infty\}$, then g must be a non-zero real multiple αg_0 of some g_0 in $SL_2(\Lambda_{n-2})$.

(**E**) *Orbits.* Since $\mathbb{R}^n_+$ is preserved by each fractional linear transformation in (3.18), it is preserved by $\mathrm{Spin}_0(n,1)$. But any $z = x + y\mathbf{j}$, $y > 0$, can be obtained from $\mathbf{j}$ as the composition

$$\mathbf{j} \longrightarrow \left(\frac{1}{y}\right)x + \mathbf{j} \longrightarrow y\left(\left(\frac{1}{y}\right)x + \mathbf{j}\right) = x + y\mathbf{j}$$

of a translation and a dilation. Hence $\mathbb{R}^n_+$ is the $\mathrm{Spin}_0(n,1)$-orbit of $\mathbf{j}$. Similarly, $\mathbb{R}^n_-$ is the $\mathrm{Spin}_0(n,1)$-orbit of $-\mathbf{j}$. On the other hand, $\mathbb{R}^{n-1}$ is preserved by each of (3.18)(i),(ii) and (iii), while (3.18)(iv) interchanges 0 and ∞. Clearly $\mathbb{R}^{n-1} \cup \{\infty\}$ is then the $\mathrm{Spin}_0(n,1)$-orbit of 0. ∎

(**F**) *Jacobians.* Several algebraic results related to the transformation $z \to z \cdot g$ are often needed. They will all follow from the identity

$$(3.23) \qquad (a + zc)(z \cdot g - \zeta \cdot g) = (z - \zeta) \frac{(a + \zeta c)'}{|a + \zeta c|^2} \qquad (z, \zeta \in \mathbb{R}^n)$$

for g in $\mathrm{Spin}_0(n, 1)$. As a first application set $z = \zeta + h$, $h \in \mathbb{R}^n_+$ and let $h \to 0$. Thus the derivative of $z \to z \cdot g$ as a function from $\mathbb{R}^n_+$ to $\mathbb{R}^n_+$ is the linear mapping $J_g : \mathbb{R}^n \to \mathbb{R}^n$ defined by

$$J_g : \zeta \longrightarrow \frac{1}{|a + \zeta c|^2} \, \sigma\big((a + \zeta c)^{-1}\big)\zeta \qquad (\zeta \in \mathbb{R}^n) \ .$$

Similarly, as a mapping from $\mathbb{R}^{n-1}$ to $\mathbb{R}^{n-1}$, its derivative is the linear mapping $J_g : \mathbb{R}^{n-1} \to \mathbb{R}^{n-1}$ defined by

$$J_g : x \longrightarrow \frac{1}{|a + xc|^2} \sigma\big((a + xc)^{-1}\big)x \qquad (x \in \mathbb{R}^{n-1})$$

whenever x and $x \cdot g \neq \infty$.

The proof of (3.23) makes essential use of the defining properties for $SL_2(\Lambda_{n-2})$. For the condition $ad^* - cb^* = 1$ ensures that

$$(b + zd)(a + \zeta c)^* - (a + zc)(b + \zeta d)^*$$
$$= (ba^* - ab^*) + (z - \zeta) + z(dc^* - cd^*)\zeta \ .$$

But then, by (3.12)(ii),

$$(a + zc)(z \cdot g - \zeta \cdot g)(a + \zeta c)^* = (z - \zeta)$$

from which (3.23) follows almost immediately. ∎

(G) *Isometries.* No translation or dilation other than the identity can fix $\mathbb{R}^n_+$; and inversion does not fix $\mathbb{R}^n_+$ either. Thus a fractional linear transformation fixing $\mathbb{R}^n_+$ must be a rotation such that

$$z \longrightarrow \sigma(a^{-1})x + y\mathbf{j} = x + y\mathbf{j} \qquad (x \in \mathbb{R}^{n-1} \ , \ y > 0)$$

for some $a \in \mathrm{Spin}(n - 1)$. This occurs if and only if $a = \pm I$, since $\sigma : \mathrm{Spin}(n - 1) \to SO(n - 1)$ is a two-fold covering. Hence by (3.22) the mapping from $\mathrm{Spin}_0(n, 1)$ to the fractional linear transformations (3.21) is an anti-homomorphism with kernel $\pm I$. But any diffeomorphism of $\mathbb{R}^n_+$ in the connected group of isometries of H_n is known to be the composition of a rotation fixing the $\mathbf{j}$-direction, a translation in any hyperplane parallel to $\mathbb{R}^{n-1}$, a dilation and an inversion. Consequently, $\mathrm{Spin}_0(n, 1)$ is a two-fold covering of the connected group of isometries of H_n $(\sim \mathbb{R}^n_+)$. ∎

(H) *Conformal transformations.* As we saw earlier (see chapter 1, (4.16)), the sense-preserving conformal transformations of $\mathbb{R}^{n-1} \cup \{\infty\}$ are compositions of rotations, translations, dilations and inversions. Hence by much the same proof as in step (**G**), we see that $\mathrm{Spin}_0(n, 1)$

is a two-fold covering of the group of sense-preserving conformal transformations of $\mathbb{R}^{n-1} \cup \{\infty\}$. This completes the proof of theorem (3.15).
∎

It is perhaps of interest to relate (3.15)(ii) and (iii) to the well-known
theorem of Liouville which asserts that any diffeomorphism of $\mathbb{R}^n_+$ preserving the hyperbolic metric extends to a conformal transformation of
the boundary $\mathbb{R}^{n-1} \cup \{\infty\}$ of $\mathbb{R}^n_+$; it asserts further that the sense-
preserving conformal transformations are the extensions in the identity
component of the isometry group of H_n. But theorem (3.15) exhibits
this extension very explicitly through the use of fractional linear transformations on $\mathbb{R}^n \cup \{\infty\}$.

We turn now to fractional linear transformations of $\mathbb{R}^n \cup \{\infty\}$ derived
from $\mathrm{Spin}(n + 1)$. As this group and $\mathrm{Spin}(n, 1)$ both contain $\mathrm{Spin}(n)$
as a subgroup, it will be important to arrange the algebra so that the
fractional linear transformations of $\mathbb{R}^n \cup \{\infty\}$ obtained from $\mathrm{Spin}(n+1)$
coincide with those defined by (3.21) on this common subgroup $\mathrm{Spin}(n)$.
Thus, let $e_1, \ldots, e_{n-2}, \mathbf{j}$ be the same imaginary units generating $\mathfrak{A}_{n-1}$
as in the previous construction corresponding to the embedding (3.14)
of $\mathbb{R}^n$ into $\mathfrak{A}_{n-1}$. Then, by (2.29) in chapter 1, the imaginary units in
(3.9) together with $\gamma_n = \begin{bmatrix} 0 & \mathbf{j} \\ -\mathbf{j} & 0 \end{bmatrix}$ generate the Clifford algebra $\mathfrak{A}_{n+1,0}$,
realizing it as $M(2, \mathfrak{A}_{n-1})$ with

$$(z, t) \longrightarrow \zeta = \begin{bmatrix} t & z \\ z' & -t \end{bmatrix} , \qquad |\zeta|^2 = -(|z|^2 + t^2) ,$$

being the embedding of $\mathbb{R}^{n+1,0}$ into $M(2, \mathfrak{A}_{n-1})$. Hence, by (6.2) in
chapter 1, $\mathrm{Spin}(n + 1)$ is the group

$$\{\zeta_1 \cdots \zeta_{2k} : \zeta_j \in \mathbb{R}^{n+1,0} , \ |\zeta_j|^2 = -1 , \ k \geq 1\}$$

of all *even* products in $M(2, \mathfrak{A}_{n-1})$ of elements from the unit sphere of
$\mathbb{R}^{n+1,0}$. But the even part of $\mathfrak{A}_{n+1,0}$ is the subalgebra

$$\left\{ \begin{bmatrix} a & -b' \\ b & a \end{bmatrix} : a, b \in \mathfrak{A}_{n-1} \right\}$$

of $M(2, \mathfrak{A}_{n-1})$, so the problem becomes one of characterizing $\mathrm{Spin}(n+1)$
as a multiplicative subgroup of this subalgebra. Now $\mathrm{Spin}(3)$ can be
identified with

$$SU(2) = \left\{ \begin{bmatrix} z & -\bar{w} \\ w & z \end{bmatrix} : z, w \in \mathbb{C} , \ |z|^2 + |w|^2 = 1 \right\}$$

(see chapter 1, (9.11)(i)). Thus in complete analogy with $SL_2(\Lambda_{n-2})$ we
are led to introduce a subset of the space $M(2, \Lambda_{n-1})$ of 2×2 matrices

having entries from the Clifford semigroup

$$\Lambda_{n-1} = \{a_1 a_2 \ldots : a_j \in \mathbb{R}^n\}$$

in $\mathfrak{A}_{n-1}$.

(3.24) Definition.

For each $n \geq 2$, let $SU_2(\Lambda_{n-1})$ be the subset of all $g = \begin{bmatrix} a & -b' \\ b & a' \end{bmatrix}$ in $M(2, \Lambda_{n-1})$ whose matrix entries satisfy

    (i) $a^* b \in \mathbb{R}^n$ *(ii)* $|a|^2 + |b|^2 = 1$.

Analogously to $(3.12)'$, $SU_2(\Lambda_{n-1})$ is the subset of all $g = \begin{bmatrix} a & -b' \\ b & a' \end{bmatrix}$ in $M(2, \Lambda_{n-1})$ for which

$(3.24)'$ *(i)* $|g|^2 = \bar{g}g = 1$ *(ii)* $a^* b \in \mathbb{R}^n$.

But Λ_{n-1} reduces to $\mathbb{C}$ when $n = 2$, so $SU_2(\Lambda_{n-1})$ becomes the group $SU_2 \sim \mathrm{Spin}(3)$. There are results for spherical space corresponding to (3.13) and (3.15) for hyperbolic space.

(3.25) Theorem.

As a multiplicative group in $M(2, \mathfrak{A}_{n-1})$, $n \geq 2$, the spin group $\mathrm{Spin}(n+1)$ coincides with $SU_2(\Lambda_{n-1})$.

(3.26) Theorem.

The fractional linear transformation

$$g = \begin{bmatrix} a & -b' \\ b & a' \end{bmatrix} : a \longrightarrow z \cdot g = (a + zb)^{-1}(za' - b') , \quad \infty \cdot g = b^{-1}a'$$

is well-defined and bijective on $\mathbb{R}^n \cup \{\infty\}$ if g is in $SU_2(\Lambda_{n-1})$, hence in $\mathrm{Spin}(n+1)$; furthermore,

    (i) the only $\mathrm{Spin}(n+1)$-orbit is $\mathbb{R}^n \cup \{\infty\}$,

    (ii) $\mathrm{Spin}(n+1)$ is a two-fold covering of the group of all transformations of $\mathbb{R}^n \cup \{\infty\}$ preserving the spherical metric.

The proof of these results will be omitted; the techniques used are much the same as before. Instead, let us check that the two fractional linear transformations on $\widehat{\mathbb{R}}^n = \mathbb{R}^n \cup \{\infty\}$ coincide on the subgroup

$(3.27)(i)$ $K = SL_2(\Lambda_{n-2}) \cap SU_2(\Lambda_{n-1})$

common to $SL_2(\Lambda_{n-2})$ and $SU_2(\Lambda_{n-1})$. But, clearly,

$(3.27)(ii)$ $K = \left\{ \begin{bmatrix} a & -b' \\ b & a' \end{bmatrix} : a^* b \in \mathbb{R}^{n-1} , \ |a|^2 + |b|^2 = 1 \right\}$.

In view of the characterizations of $SL_2(\Lambda_{n-2})$ and $SU_2(\Lambda_{n-1})$, however,

K is just a realization of $\mathrm{Spin}(n)$ and

(3.27)(iii) $k : z \longrightarrow z \cdot k = (a + zb)^{-1}(za' - b')$, $\infty \cdot k - b^{-1}a'$

describes its action on $\mathbb{R}^n \cup \{\infty\}$ as a subgroup of either group. Hence all the algebra is consistent.

4 Representation theory for $\mathrm{Spin}_0(n,1)$

The action of $\mathrm{Spin}_0(n,1)$ by fractional linear transformations on $\mathbb{R}_+^n$ and on $\mathbb{R}^{n-1} \cup \{\infty\}$ allows us to construct two quite different looking families of representations on $\mathfrak{H}$-valued functions on these $\mathrm{Spin}_0(n,1)$-orbits. The distinction between the two families is more apparent than real, however, since they turn out to be equivalent, so from an algebraic point of view one family is just as good as the other and one elects to study whichever is more convenient. But, analytically, one is interested also in studying the equivariant operators from one family to the other which establish the equivalence of the families. It is precisely at this point that contact is made with elliptic boundary value problems after interpreting $\mathbb{R}^{n-1} \cup \{\infty\}$ as the boundary of $\mathbb{R}_+^n$. The classical case of $SL(2,\mathbb{R})$ acting by fractional linear transformations on $\mathbb{R}^2 \sim \mathbb{C}$, which the previous section sought to generalize, was used by Bargmann many years ago to construct explicitly the unitary representations of $SL(2,\mathbb{R})$. Later, Kunze and Stein exploited the group-invariance of such familiar operators in harmonic analysis on $\mathbb{R}$ as fractional integrals, Hilbert transforms and the like to analyze intertwining operators between representations of $SL(2,\mathbb{R})$. Both studies laid the groundwork for what is now a highly developed theory valid for general semi-simple Lie groups (see Notes and remarks for references). Although great progress has been made in the more algebraic aspects of these ideas, the analytic aspects are less understood. In this section we give a brief and incomplete introductory account for $\mathrm{Spin}_0(n,1)$, using its action by fractional linear transformations to develop very close parallels between the Bargmann–Kunze–Stein theory for $\mathrm{Spin}_0(2,1) \sim SL(2,\mathbb{R})$ and that for $\mathrm{Spin}_0(n,1)$. Throughout this section we shall use the same notation as in section 3, with imaginary units $e_1, \ldots, e_{n-2}, \mathbf{j}$ as generators for $\mathfrak{A}_{n-1}$ and $\mathrm{Spin}_0(n,1)$ realized as $SL_2(\Lambda_{n-2})$. To each g in $\mathrm{Spin}_0(n,1)$ corresponds the fractional linear transformation

$$g : z \to z \cdot g = (a + zc)^{-1}(b + zd) \qquad \left(g = \begin{bmatrix} a & b \\ c & d \end{bmatrix} \right)$$

of $\mathbb{R}_+^n$ and of $\mathbb{R}^{n-1} \cup \{\infty\}$ (see (3.15)). The group $\mathrm{Spin}(n)$ will be realized as the group

$$\mathrm{Spin}(n) = \{a \in \Lambda_{n-1} : \|a\|_{\mathfrak{A}} = 1\}$$

in $\mathfrak{A}_{n-1}$ (see chapter 1, (7.26)).

Let $\mathfrak{H}$ be a $\mathfrak{A}_{n-1}$-module on which there is a representation τ of $\mathrm{Spin}(n)$. For instance, if $\mathfrak{H}$ is any $\mathfrak{A}_{n-1}$-module, the homomorphism $\mathfrak{A}_{n-1} \to \mathcal{L}(\mathfrak{H})$ defines such a τ on restriction to $\mathrm{Spin}(n)$; in fact, each irreducible unitary representation of $\mathrm{Spin}(n)$ can be realized as a subrepresentation of a representation on an $\mathfrak{A}_{n-1}$-module $\mathfrak{H}$.

(4.1) Theorem.

For each $n \geq 2$,

$$\big(\pi_\tau(g)F\big)(z) = \tau\!\left(\frac{a+zc}{|a+zc|}\right)F(z \cdot g) \qquad (z \in \mathbb{R}_+^n)$$

defines a representation of $\mathrm{Spin}_0(n,1)$ *on* $C^\infty(\mathbb{R}_+^n, \mathfrak{H})$.

Proof. It is necessary first to show that $(a+zc)/|a+zc|$ is in $\mathrm{Spin}(n)$ when $z \in \mathbb{R}_+^n$ and $g \in \mathrm{Spin}_0(n,1)$. For this it is enough to prove that $(a+zc)$ is a non-zero element of Λ_{n-1}. Now $z = x + \mathbf{j}y$, $y > 0$; consequently,

$$\begin{aligned}
|a+zc|^2 &= \big(\bar{a} + \bar{c}(\bar{x} - y\mathbf{j})\big)\big(a + (x + y\mathbf{j})c\big) \\
&= |a|^2 + |z|^2|c|^2 + (\bar{a}xc + \bar{c}\bar{x}a) + y\mathbf{j}(a^*c - c^*a) \\
&= |a + xc|^2 + y^2|c|^2
\end{aligned}$$

using (7.18), (7.22), and (7.25) in chapter 1. But a, c cannot both be zero since $d^*a - b^*c = 1$. Thus $|a + zc|^2 > 0$. On the other hand, if $a = 0$ or $c = 0$, then $(a + zc)$ is clearly in Λ_{n-1}. So suppose $c \neq 0$; by the proof in step (**A**) of section 2, there exists u in $\mathbb{R}^{n-1}$ such that $a = uc$, and so $a + zc = (u + z)c \in \Lambda_{n-1}$. Hence $(a + zc)/|a + zc|$ is always well-defined in $\mathrm{Spin}(n)$ for any z in $\mathbb{R}_+^n$ and g in $\mathrm{Spin}_0(n,1)$; in particular, $z \to (a + zc)/|a + zc|$ will be smooth as a mapping from $\mathbb{R}_+^n$ into $\mathrm{Spin}(n)$. Thus $\pi_\tau(g) : C^\infty(\mathbb{R}_+^n, \mathfrak{H}) \to C^\infty(\mathbb{R}_+^n, \mathfrak{H})$.

Finally, to check that $g \to \pi_\tau(g)$ defines a representation on $C^\infty(\mathbb{R}_+^n, \mathfrak{H})$, choose any

$$g = \begin{bmatrix} a & b \\ c & d \end{bmatrix} \quad , \quad g_1 = \begin{bmatrix} \alpha & \beta \\ \gamma & \delta \end{bmatrix}$$

in $\mathrm{Spin}_0(n,1)$. Then

(4.2) $\quad (a + zc)\big(\alpha + (z \cdot g)\gamma\big) = (a + zc)\big(\alpha + (a + ac)^{-1}(b + ad)\gamma\big)$

$$= (a\alpha + b\gamma) + z(c\alpha + d\gamma) ,$$

and so

$$\pi_\tau(g)\big(\pi_\tau(g_1)F\big)(z)$$

$$= \tau\Big(\frac{a+zc}{|a+zc|}\Big)\tau\Big(\frac{\alpha+(z\cdot g)\gamma}{|\alpha+(z\cdot g)\gamma|}\Big)F(z\cdot gg_1) = \big(\pi_\tau(gg_1)F\big)(z)\ .$$

This completes the proof.   ∎

Since the principal automorphism obviously restricts to an automorphism $k \to k'$ of $\mathrm{Spin}(n)$ as a group in $\mathfrak{A}_{n-1}$, we obtain immediately from (4.1) a second representation of $\mathrm{Spin}_0(n,1)$.

(4.3) Corollary.

For each $n \geq 2$,

$$\big(\pi'_\tau(g)F\big)(z) = \tau\Big(\frac{(a+zc)'}{|a+zc|}\Big)F(z\cdot g) \qquad (z \in \mathbb{R}^n_+)$$

defines a representation of $\mathrm{Spin}_0(n,1)$ *on* $C^\infty(\mathbb{R}^n_+, \mathfrak{H})$.

Theorem (4.1) is exactly the analogue for $\mathrm{Spin}_0(n,1)$ of the representation π_τ of $\mathrm{Spin}(n)\circledS\mathbb{R}^n$ on $C^\infty(\mathbb{R}^n, \mathfrak{H})$ defined by (3.13) in chapter 4, since $\mathrm{Spin}_0(n,1)$ is a two-fold covering of the (connected) isometry group of n-dimensional hyperbolic space just as $\mathrm{Spin}(n)\circledS\mathbb{R}^n$ is the two-fold covering of the (connected) isometry group of n-dimensional Euclidean space. Hence $(C^\infty(\mathbb{R}^n_+, \mathfrak{H}), \pi_\tau)$ is called the representation of *$\mathrm{Spin}_0(n,1)$ induced* from $(\mathfrak{H}, \tau)$. Analogous to the 'flat' case of chapter 4, therefore, we look for differential operators $\eth_A$ on $C^\infty(\mathbb{R}^n_+, \mathfrak{H})$ which are $\mathrm{Spin}_0(n,1)$-invariant in the sense that

$$(4.4) \qquad \eth_A\big(\pi'_\tau(g)F\big) = \pi_\tau(g)(\eth_A F) \qquad (g \in \mathrm{Spin}_0(n,1))\ .$$

For then the kernel

$$\ker \eth_A = \big\{F \in C^\infty(\mathbb{R}^n_+, \mathfrak{H}) : \eth_A F = 0\big\}$$

is a $\mathrm{Spin}_0(n,1)$-submodule of $C^\infty(\mathbb{R}^n_+, \mathfrak{H})$. Such an operator $\eth_A$ is said to *intertwine* the representations π_τ and π'_τ. They can be characterized in the same way as the flat case (see chapter 4,(3.16)), using ideas from earlier sections in this chapter.

In the realization of $\mathrm{Spin}(n)$ as a group in $\mathfrak{A}_{n-1}$, its Lie algebra has a basis

$$\tfrac{1}{2}e_k e_\ell \quad (0 \leq k \leq \ell \leq n-2)\ , \qquad \tfrac{1}{2}e_k\mathbf{j} \quad (0 \leq k \leq n-2)$$

where $e_0 (= I)$ is the identity in $\mathfrak{A}_{n-1}$. Thus, by (1.23) and (3.4), the total derivative on $C^\infty(\mathbb{R}^n_+, \mathfrak{H})$ is given by

$$\nabla_\tau F = \sum_{i=0}^{n-2}\Big(y\frac{\partial}{\partial x_i} + \tfrac{1}{2}d\tau(e_i\mathbf{j})\Big)F\otimes e_i + y\frac{\partial F}{\partial y}\otimes\mathbf{j}\ .$$

To characterize these $\eth_A$, recall yet again the two-fold covering homomorphism

$$\sigma : \mathrm{Spin}(n) \longrightarrow SO(n) \quad , \quad \sigma(a)v = av(a')^{-1}$$

with a in $\mathrm{Spin}(n)$ as a group in $\mathfrak{A}_{n-1}$ (see chapter 1, (7.26)). The analogue of (3.5) in chapter 4 now becomes the following.

(4.5) Theorem.

A differential operator

$$\eth_A = \sum_{i=0}^{n-2} A_i \left(y \frac{\partial}{\partial x_i} + \tfrac{1}{2} d\tau(e_i \mathbf{j}) \right) + A_{n-1} y \frac{\partial F}{\partial y}$$

on $\mathcal{C}^\infty(\mathbb{R}_+^n, \mathfrak{H})$ *intertwines* π_τ *and* π'_τ *in the sense of* (4.4) *if and only if*

$$\tau(a) A_i \tau(a')^{-1} = \sum_j a_{ij} A_j \qquad (0 \le j \le n-1)$$

for all a *in* $\mathrm{Spin}(n)$.

In particular, the Dirac $\mathcal{D}$-operator

$$\mathcal{D} = \sum_{i=0}^{n-2} e_i \left(y \frac{\partial}{\partial x_i} + \tfrac{1}{2} d\tau(e_i \mathbf{j}) \right) + \mathbf{j} y \frac{\partial}{\partial y}$$

on $\mathcal{C}^\infty(\mathbb{R}_+^n, \mathfrak{H})$ is one such operator. We shall omit the proof of this and of (4.5) because a more detailed study will be made elsewhere dealing with two fundamental but difficult problems for the subrepresentation $(\ker \mathcal{D}, \pi'_\tau)$:

(4.6)(i) (non-triviality) *show that there are solutions of* $\mathcal{D}F = 0$ *in* $\mathcal{C}^\infty(\mathbb{R}_+^n, \mathfrak{H})$;

(4.6)(ii) (unitarizability) *exhibit a Hilbert space structure on* $\ker \mathcal{D}$ *with respect to which each* $\pi_\tau(g)$, $g \in \mathrm{Spin}_0(n,1)$ *is a unitary operator.*

A natural candidate for this Hilbert space structure is the Lebesgue L^2-structure defined by the $\mathrm{Spin}_0(n,1)$-invariant measure on $\mathbb{R}_+^n$. Denote by $L^2(\mathbb{R}_+^n, \mathfrak{H})$ the Lebesgue space of all $\mathfrak{H}$-valued functions on $\mathbb{R}_+^n$ for which

$$\|F\|_{L^2} = \left(\int_0^\infty \int_{\mathbb{R}^{n-2}} |F(x + y\mathbf{j})|^2 \frac{dx\,dy}{y^n} \right)^{1/2}$$

is finite; the subspace

(4.7)(i) $B^2(\mathbb{R}_+^n, \mathfrak{H}) = \left\{ F \in L^2(\mathbb{R}_+^n, \mathfrak{H}) : \mathcal{D}F = 0 \right\}$

might well be called a *Bergman* space on n-dimensional hyperbolic space by analogy with the classical Bergman spaces of analytic functions. Similarly, we can introduce *weighted Hardy H^2 spaces* $H^2_\alpha(\mathbf{R}^n_+, \mathfrak{H})$ as those solutions of $\mathcal{D}F = 0$ for which

$$(4.7)(\text{ii}) \qquad \|F\|_{H^2_\alpha} = \sup_{y > 0} \frac{1}{y^\alpha} \left(\int_{\mathbf{R}^{n-1}} |F(x + y\mathbf{j})|^2 \, dx \right)^{1/2}$$

is finite. But what is far from clear is whether these spaces are non-trivial, and in the latter case whether

$$\|\pi_\tau(g)F\|_{H^2_\alpha} = \|F\|_{H^2_\alpha} \qquad (g \in \mathrm{Spin}_0(n, 1))$$

for all F in $H^2_\alpha(\mathbf{R}^n_+, \mathfrak{H})$. Nonetheless, the representations $(B^2(\mathbf{R}^n_+, \mathfrak{H}), \pi_\tau)$ are fully understood; they consitute explicit realizations of the so-called *discrete series* representations of $\mathrm{Spin}_0(n, 1)$ for suitable $\mathfrak{H}$. The Hardy spaces $H^2_\alpha(\mathbf{R}^n_+, \mathfrak{H})$ should be connected for appropriate choices of α with so-called *limits of discrete series* representations.

The second family of representations of $\mathrm{Spin}_0(n, 1)$ is constructed on weighted L^2-spaces on the boundary of $\mathbf{R}^n_+$: given $\lambda \in \mathbf{C}$, denote by $L^2_\lambda(\mathbf{R}^{n-1}, \mathfrak{H})$ the usual Lebesgue space of $\mathfrak{H}$-valued functions on $\mathbf{R}^{n-1}$ for which

$$(4.8) \qquad \|f\|_\lambda = \left(\int_{\mathbf{R}^{n-1}} |f(v)|^2 (1 + |v|^2)^{2\Re \lambda} \, dv \right)^{1/2}$$

is finite.

(4.9) Definition.
   For each $\lambda \in \mathbf{C}$ and g in $\mathrm{Spin}_0(n, 1)$ define $U(\tau, \lambda, g)$ on $L^2_\lambda(\mathbf{R}^{n-1}, \mathfrak{H})$ by

$$\big(U(\tau, \lambda, g)f\big)(v) = \frac{1}{|a + vc|^{2(\lambda + \rho)}} \tau\left(\frac{a + vc}{|a + vc|} \right) f(v \cdot g) \qquad (v \in \mathbf{R}^{n-1})$$

where $\rho = \frac{1}{2}(n - 1)$.

A proof very similar to the second half of the proof of (4.1) shows that

$$\big(U(\tau, \lambda, g_1)U(\tau, \lambda, g_2)\big)f = U(\tau, \lambda, g_1 g_2)f$$

on $L^2_\lambda(\mathbf{R}^{n-1}, \mathfrak{H})$ assuming, of course, that $U(\tau, \lambda, g_1)$ and $U(\tau, \lambda, g_2)$ are bounded operators on $L^2_\lambda(\mathbf{R}^{n-1}, \mathfrak{H})$. But, by step (**C**) in the proof of (3.13), every g in $\mathrm{Spin}_0(n, 1)$ is the product of elements from M, A, V and $\{\omega\}$. Thus in proving the boundedness of $U(\tau, \lambda, g)$ on $L^2_\lambda(\mathbf{R}^{n-1}, \mathfrak{H})$ it is enough to consider rotations, dilations, translations and inversion of $\mathbf{R}^{n-1}$ separately.

(4.10) Theorem.
   For each $n \geq 2$ and $\lambda \in \mathbf{C}$, $g \to U(\tau, \lambda, g)$ defines a representation of $\mathrm{Spin}_0(n, 1)$ by bounded operators on $L^2_\lambda(\mathbf{R}^{n-1}, \mathfrak{H})$.

Proof. We have only to show that $U(\tau, \lambda, g)$ is bounded. Now

$$\|U(\tau,\lambda,g)f\|_\lambda^2 = \int_{\mathbb{R}^{n-1}} \frac{(1+|v|^2)^{2\Re\lambda}}{|a+vc|^{4(\Re\lambda+\rho)}} \, |f(v\cdot g)|^2 \, dv$$

$$= \int_{\mathbb{R}^{n-1}} |f(v\cdot g)|^2 \left(\frac{1+|v|^2}{|a+vc|^2}\right)^{2\Re\lambda} d(v\cdot g)$$

since τ is a unitary representation of $\mathrm{Spin}(n)$ on $\mathfrak{H}$ and

$$d(v\cdot g) = \frac{1}{|a+vc|^{2(n-1)}} \, dv = \frac{1}{|a+vc|^{4\rho}} \, dv$$

(see section 3, (**F**)). But, for any g in M,

$$\frac{1+|v|^2}{|a+vc|^2} = 1 + |v\cdot g|^2 \, ,$$

while corresponding calculations show that

$$\frac{1+|v|^2}{|a+vc|^2} \leq \mathrm{const.}\ (1+|v\cdot g|^2)$$

for some constant (depending on g) whenever g is in A, V, or $\{\omega\}$. Hence $U(\tau,\lambda,g)$ is always bounded on $L_\lambda^2(\mathbb{R}^{n-1},\mathfrak{H})$.   ∎

This result has an obvious corollary for the same reason that (4.1) did.

(4.11) Corollary.
For each $n \geq 2$ and $\lambda \in \mathbb{C}$,

$$\big(U'(\tau,\lambda,g)f\big)(x) = \frac{1}{|a+vc|^{2(\lambda+\rho)}} \tau\Big(\frac{(a+vc)'}{|a+vc|}\Big) f(v\cdot g)$$

defines a representation of $\mathrm{Spin}_0(n,1)$ *by bounded operators on* $L_\lambda^2(\mathbb{R}^{n-1},\mathfrak{H})$.

Since the conformal transformations of $\mathbb{R}^{n-1} \cup \{\infty\}$ include rotations, dilations and translations, it is not surprising that the operators on the family of Hilbert spaces $\{L_\lambda^2 : \lambda \in \mathbb{C}\}$ commuting with $U(\tau,\lambda)$ and $U'(\tau,\lambda)$ are intimately connected with the classical transform operators in Euclidean Fourier analysis, particularly with Calderón–Zygmund type operators of varying degrees of homogeneity. Denote by $\mathcal{S}(\mathbb{R}^{n-1},\mathfrak{H})$ the usual Schwartz space of rapidly decreasing smooth $\mathfrak{H}$-valued functions on $\mathbb{R}^{n-1}$.

(4.12) Definition.

The Calderón–Zygmund type operators $\mathcal{A}_{\tau,\alpha}$, $\alpha \in \mathbb{C}$, are defined formally as the convolution operators

$$\mathcal{A}_{\tau,\alpha} : f \longrightarrow \int_{\mathbb{R}^{n-1}} |x - v|^{2(\alpha-\rho)} \tau\left(\frac{x-v}{|x-v|}\mathbf{j}\right) f(v) \, dv$$

on $\mathcal{S}(\mathbb{R}^{n-1}, \mathfrak{H})$ having kernel

$$(4.13)(i) \qquad A_{\tau,\alpha}(x) = |x|^{2(\alpha-\rho)} \tau\left(\frac{x}{|x|}\mathbf{j}\right) \qquad (x \in \mathbb{R}^{n-1}) \,.$$

There is an entirely analogous operator $\mathcal{A}'_{\tau,\alpha}$ having kernel

$$(4.13)(ii) \qquad A'_{\tau,\alpha}(x) = |x|^{2(\alpha-\rho)} \tau\left(\frac{x'}{|x|}\mathbf{j}\right) \qquad (x \in \mathbb{R}^{n-1}) \,.$$

As $A_{\tau,\alpha}$ and $A'_{\tau,\alpha}$ are just the extension to homogeneous functions of smooth $\mathcal{L}(\mathfrak{H})$-valued functions on the unit sphere of $\mathbb{R}^{n-1}$, $\mathcal{A}_{\tau,\alpha}$ and $\mathcal{A}'_{\tau,\alpha}$ are operators to which the standard results of Fourier analysis apply. For instance, the case $\Re\,\alpha > 0$ is particularly simple because the kernels are then locally integrable, though the corresponding operators are basic to the study of $U(\tau,\lambda)$ and $U'(\tau,\lambda)$ nonetheless.

(4.14) Theorem.

When $\Re\,\lambda > 0$, the operator $\mathcal{A}_{\tau,\lambda}$ is a bounded operator from $L^2_\lambda(\mathbb{R}^{n-1}, \mathfrak{H})$ into $L^2_{-\lambda}(\mathbb{R}^{n-1}, \mathfrak{H})$ intertwining $U(\tau,\lambda)$ and $U(\tau,-\lambda)$ in the sense that

$$\mathcal{A}_{\tau,\lambda} \circ U(\tau,\lambda,g) = U(\tau,-\lambda,g) \circ \mathcal{A}_{\tau,\lambda} \qquad \big(g \in \mathrm{Spin}_0(n,1)\big)$$

always holds.

Proof. (Merryfield) In establishing the boundedness of $\mathcal{A}_{\tau,\lambda}$ it is enough to show that

$$\int_{\mathbb{R}^{n-1}} \left(\int_{\mathbb{R}^{n-1}} |x-v|^{2(\lambda-\rho)} |f(v)| \, dv\right)^2 (1+|x|^2)^{-2\lambda} \, dx$$

$$\leq \text{const.} \int_{\mathbb{R}^{n-1}} |f(v)|^2 (1+|v|^2)^{2\lambda} \, dv$$

with λ real and positive. But, if

$$E_0 = \{v \in \mathbb{R}^{n-1} : |v| \leq 1\} \quad , \quad E_k = \{v \in \mathbb{R}^{n-1} : 2^{k-1} < |v| \leq 2^k\} \,,$$

the last inequality is equivalent to

$$\sum_{j,k\geq 0} 2^{-4\lambda j} \int_{E_j} \left(\int_{E_k} |x-v|^{2(\alpha-\rho)} |f(v)|\, dv \right)^2 dx$$

$$\leq \text{const.} \sum_{k\geq 0} 2^{4\lambda k} \int_{E_k} |f(v)|^2\, dv \; ;$$

this is the inequality we shall establish. Now

$$\text{const. } 2^{\max(j,k)} \leq |x-v| \leq \text{const. } 2^{\max(j,k)}$$

whenever $x \in E_j$, $v \in E_k$ and $|j-k| > 1$. Consequently,

$$(4.15) \qquad 2^{-4\lambda j} \int_{E_j} \left(\int_{E_k} |x-v|^{2(\lambda-\rho)} |f(v)|\, dv \right)^2 dv$$

$$\leq \text{const. } 2^{-2\rho|j-k|} \left(2^{4\lambda k} \int_{E_k} |f(v)|^2\, dv \right)$$

provided $|j-k| > 1$. On the other hand, if χ_j and χ_k denote the characteristic functions of E_j and E_k respectively, Minkowski's inequality gives

$$\left(\int_{E_j} \left(\int_{E_k} |x-v|^{2(\lambda-\rho)} |f(v)|\, dv \right)^2 dx \right)^{1/2}$$

$$\leq \int_{\mathbf{R}^{n-1}} |y|^{2(\lambda-\rho)} \left(\int_{\mathbf{R}^{n-1}} |f(x-y)|^2 \chi_k(x-y)\chi_j(x)\, dx \right)^{1/2} dy$$

$$\leq \int_{\Delta_{jk}} |y|^{2(\lambda-\rho)} \left(\int_{E_k} |f(v)|^2\, dv \right)^{1/2} dy \;,$$

where $\Delta_{jk} = \{u+v : u \in E_j\ ,\ v \in E_k\}$. When $|j-k| \leq 1$, therefore,

$$2^{-4\lambda j} \int_{E_j} \left(\int_{E_k} |x-v|^{2(\lambda-\rho)} |f(v)|\, dv \right)^2 dx$$

$$\leq \text{const. } 2^{4\lambda k} \int_{E_k} |f(v)|^2\, dv \;.$$

Hence (4.15) holds without restriction on j, k, completing proof of the boundedness of $\mathcal{A}_{\tau,\lambda}$ from $L^2_\lambda(\mathbf{R}^{n-1}, \mathfrak{H})$ to $L^2_{-\lambda}(\mathbf{R}^{n-1}, \mathfrak{H})$.

To establish the intertwining property of $\mathcal{A}_{\tau,\lambda}$, note first that

$$\mathcal{A}_{\tau,\lambda}\big(U(\tau,\lambda,g)f\big)(x)$$

$$= \int_{\mathbf{R}^{n-1}} \frac{|x-v|^{2(\lambda-\rho)}}{|a+vc|^{2(\lambda+\rho)}} \tau\left(\frac{(x-v)\mathbf{j}(a+vc)}{|x-v|\,|a+vc|} \right) f(v \cdot g)\, dv \;,$$

which can be written as

(4.16)
$$\int_{\mathbb{R}^{n-1}} \left(\frac{|x-u|}{|a+uc|}\right)^{2(\lambda-\rho)} \tau\left(\frac{(x-u)\mathbf{j}(a+uc)}{|x-u|\,|a+uc|}\right) f(u)\,du \qquad (u = v \cdot g^{-1})$$

after a change of variable $v \to v \cdot g$ (see proof of (4.10)). But, by (3.23),

$$\frac{(x-u)\mathbf{j}(a+uc)}{|x-u|\,|a+uc|} = \frac{(x-u)(a+uc)'}{|x-u|\,|a+uc|}\mathbf{j}$$

$$= \frac{(a+xc)(x\cdot g - u\cdot g)}{|a+xc|\,|x\cdot g - u\cdot g|}\mathbf{j} = \frac{(a+xc)(x\cdot g - v)}{|a+xc|\,|x\cdot g - v|}\mathbf{j},$$

while

$$\frac{|x-u|}{|a+uc|} = |a+xc|\,|x\cdot g - u\cdot g| = |a+xc|\,|x\cdot g - v|.$$

After substitution of these identities into (4.16), the integral thus becomes

$$|a+xc|^{2(\lambda-\rho)}\,\tau\left(\frac{a+xc}{|a+xc|}\right)\int_{\mathbb{R}^{n-1}} |x\cdot g - v|^{2(\lambda-\rho)}\,\tau\left(\frac{x\cdot g - v}{|x\cdot g - v|}\mathbf{j}\right) f(v)\,dv$$

which is just $U(\tau, -\lambda, g)(\mathcal{A}_{\tau,\lambda}f)(x)$. This completes the proof. ∎

Clearly the same proof will show that $\mathcal{A}'_{\tau,\lambda}$ is a bounded operator from $L^2_\lambda(\mathbb{R}^{n-1}, \mathfrak{H})$ into $L^2_{-\lambda}(\mathbb{R}^{n-1}, \mathfrak{H})$ intertwining $U'(\tau, \lambda)$ and $U'(\tau, -\lambda)$, $\Re\lambda > 0$. When $\Re\lambda = 0$, the integrals defining $\mathcal{A}_{\tau,\lambda}$ and $\mathcal{A}'_{\tau,\lambda}$ have to be interpreted in a principal value sense, however. For instance, when $\lambda = 0$,

$$\mathcal{A}_{\tau,0}f(x) = \lim_{\varepsilon \to 0} \int_{|x-v|>\varepsilon} \frac{1}{|x-v|^{2\rho}}\,\tau\left(\frac{x-v}{|x-v|}\mathbf{j}\right) f(v)\,dv.$$

Combining the second half of the last proof with standard singular integral operator theory, we obtain the following.

(4.17) Theorem.
If $x \to \tau(\frac{x}{|x|}\mathbf{j})$ has mean value zero on the unit sphere in $\mathbb{R}^{n-1}$, then $\mathcal{A}_{\tau,0}$ is a bounded operator on $L^2(\mathbb{R}^{n-1}, \mathfrak{H})$ intertwining $U(\tau, 0)$; there is an analogous result for $\mathcal{A}'_{\tau,0}$.

By regarding $\mathbb{R}^{n-1}$ as the boundary of $\mathbb{R}^n_+$, one can expect there will be two operators linking $L^2_\lambda(\mathbb{R}^{n-1}, \mathfrak{H})$ with $C^\infty(\mathbb{R}^n_+, \mathfrak{H})$, the first a potential (or Poisson) type operator extending functions on $\mathbb{R}^{n-1}$ to functions in an eigenspace of the Laplacian Δ_τ on $\mathbb{R}^n_+$, and the second a trace (or boundary) operator taking boundary values of functions in eigenspaces of Δ_τ. For these operators to have a representation-theoretic significance, they must also intertwine the representations $U(\tau, \lambda)$ and π_τ or $U'(\tau, \lambda)$ and π'_τ.

(4.18) Definition.
For each $\lambda \in \mathbb{C}$ the Cauchy–Szegő kernel $S_{\tau,\lambda}$ is defined on $\mathbb{R}^n_+$ by

$$S_{\tau,\lambda}(z) = \left(\frac{y}{|z|^2}\right)^{\rho - \lambda} \tau\left(\frac{z}{|z|}\mathbf{j}\right) \qquad (z = x + \mathbf{j}y) \ ;$$

analogously

$$S'_{\tau,\lambda}(z) = \left(\frac{y}{|z|^2}\right)^{\rho - \lambda} \tau\left(\frac{z'}{|z|}\mathbf{j}\right) \ .$$

The terminology is due to the fact that the associated *Cauchy–Szegő* transform

$$\mathcal{S}'_{\tau,\lambda}f(z) = \int_{\mathbb{R}^{n-1}} \left(\frac{y}{|z - v|^2}\right)^{\rho - \lambda} \tau\left(\frac{(z - v)'}{|z - v|}\mathbf{j}\right) f(v)\, dv$$

becomes

$$\mathcal{S}'_{\tau,0}f(z) = y^\rho \int_{\mathbb{R}^{n-1}} \left(\frac{z' - v'}{|z - v|^n}\right) \mathbf{j}f(v)\, dv$$

when $\lambda = 0$ and τ is the Spin representation $\tau(z) : \xi \to z\xi$ of $\mathrm{Spin}(n)$ on $\mathfrak{H}$; this (up to the factor y^ρ) is essentially the Cauchy integral for functions having range in an $\mathfrak{A}_{n-1}$-module, but in general it might more accurately be called a *Poisson transform*. Such factors $y^{\rho-\lambda}$ account for the appearance of the weight $y^{-\alpha}$ in the definition of Hardy spaces (see (4.7)(ii)).

(4.19) Theorem.
For each $\lambda \in \mathbb{C}$ the Cauchy–Szegő transform

$$f \longrightarrow \mathcal{S}_{\tau,\lambda}f(z) = \int_{\mathbb{R}^{n-1}} \left(\frac{y}{|z - v|^2}\right)^{\rho - \lambda} \tau\left(\frac{z - v}{|z - v|}\mathbf{j}\right) f(v)\, dv$$

is a mapping from $L^2_\lambda(\mathbb{R}^{n-1}, \mathfrak{H})$ into $C^\infty(\mathbb{R}^n_+, \mathfrak{H})$ intertwining $U(\tau, \lambda)$ and π_τ in the sense that

$$\pi_\tau(g) \circ \mathcal{S}_{\tau,\lambda} = \mathcal{S}_{\tau,\lambda} \circ U(\tau, \lambda, g) \qquad \left(g \in \mathrm{Spin}_0(n, 1)\right) \ .$$

Proof. Since

$$\frac{1 + |v|^2}{1 + y^2 + |x|^2} \leq y^2 + |x - v|^2 \leq (1 + y^2 + |x|^2)(1 + |v|^2) \ ,$$

and $|z - v|^2 = y^2 + |x - v|^2$, Hölder's inequality ensures that

$$|\mathcal{S}_{\tau,\lambda}f(z)| \leq \mathrm{const.} \left(\int_{\mathbb{R}^n} \frac{1}{(y^2 + |x - v|^2)^\rho}\, dv\right)^{1/2} \|f\|_\lambda \leq \mathrm{const.} \ \|f\|_\lambda$$

for each z in $\mathbb{R}^n_+$, the constants being allowed to depend on z. Thus

$\mathcal{S}_{\tau,\lambda}f(z)$ is well-defined; in fact, it is easy to see that $z \to \mathcal{S}_{\tau,\lambda}f(z)$ is smooth since the integral kernel of $\mathcal{S}_{\tau,\lambda}$ is smooth. A similar proof to the one in (4.14) for the intertwining property of $\mathcal{A}_{\tau,\lambda}$ establishes also the intertwining property of $\mathcal{S}_{\tau,\lambda}$. ∎

There is an analogous result for $\mathcal{S}'_{\tau,\lambda}$ relating $U'(\tau,\lambda)$ and π'_τ. Discussion of the eigenspace property will be omitted. Instead we concentrate on the boundary operator $\mathcal{B}_{\tau,\lambda}$ that takes traces of functions in $C^\infty(\mathbb{R}^n_+, \mathfrak{H})$. Recall the notion of non-tangential limit

$$(4.20) \qquad \operatorname*{lim-n.t.}_{z \to x} F(z) = \lim_{\substack{z \to x \\ z \in \Gamma_\alpha(x)}} F(z)$$

as $z \in \mathbb{R}^n_+$ approaches x through values in the cone

$$(4.20)(i) \qquad \Gamma_\alpha(x) = \{z = \xi + y\mathbf{j} \in \mathbb{R}^n_+ : |x - \xi| < \alpha y\}$$

(see chapter 2, (5.2), (5.3)). Then

$$(4.21) \qquad (\mathcal{B}_{\tau,\lambda}F)(x) = \operatorname*{lim-n.t.}_{z \to x} \frac{1}{y^{\rho-\lambda}} F(z) \qquad (z = \xi + y\mathbf{j}),$$

where it is assumed that the limit exists in each cone $\Gamma_\alpha(x)$, $0 < \alpha < \infty$, for almost all x. Notice that formally

$$(4.22) \quad \mathcal{B}_{\tau,\lambda}(\mathcal{S}_{\tau,\lambda}f)(x) = \operatorname*{lim-n.t.}_{z \to x} \int_{\mathbb{R}^{n-1}} \tau\left(\frac{z-v}{|z-v|}\mathbf{j}\right) f(v)\,dv$$

$$= \int_{\mathbb{R}^{n-1}} \tau\left(\frac{x-v}{|x-v|}\mathbf{j}\right) f(v)\,dv = \mathcal{A}_{\tau,\lambda}f(x),$$

i.e., $\mathcal{A}_{\tau,\lambda} = \mathcal{B}_{\tau,\lambda} \circ \mathcal{S}_{\tau,\lambda}$, suggesting the intertwining property

$$(4.23) \qquad U(\tau,-\lambda,g)(\mathcal{B}_{\tau,\lambda}F) = \mathcal{B}_{\tau,\lambda}(\pi_\tau(g)F).$$

(4.24) Theorem.
If F is a function in $C^\infty(\mathbb{R}^n_+, \mathfrak{H})$ for which $\mathcal{B}_{\tau,\lambda}F$ is well-defined, then $\mathcal{B}_{\tau,\lambda}(\pi_\tau(g)F)$ exists and

$$\mathcal{B}_{\tau,\lambda}(\pi_\tau(g))F = U(\tau,-\lambda,g)(\mathcal{B}_{\tau,\lambda}F).$$

Proof. Fix $x \in \mathbb{R}^{n-1}$ and $g \in \mathrm{Spin}_0(n,1)$, $a + xc \neq 0$. Then $a + zc \to a + xc$ however $z \to x$. But, for $z = \xi + y\mathbf{j}$,

$$(\pi_\tau(g)F)(z) = \tau\left(\frac{a+zc}{|a+zc|}\right) F(z \cdot g),$$

where

$$z \cdot g = \xi + \eta\mathbf{j} = \xi + \frac{y}{|a+zc|^2}\mathbf{j}$$

for some ζ in $\mathbb{R}^{n-1}$. Consequently, as $z \to x$ in $\Gamma_\alpha(x)$,

$$\lim_{z \to x} \left(\frac{1}{y^{\rho-\lambda}} (\pi_\tau(g)F)(x) \right)$$

$$= \frac{1}{|z + xc|^{2(\rho-\lambda)}} \tau \left(\frac{a + xc}{|a + xc|} \right) \lim_{z \to x} \left(\frac{1}{\eta^{2(\rho-\lambda)}} F(z \cdot g) \right) .$$

Thus the crucial question concerns the image of the cone $\Gamma_\alpha(x)$ under the fractional linear transformation $z \to z \cdot g$, because this will determine whether $F(z \cdot g) \to F(x \cdot g)$ non-tangentially. Now by (3.23),

$$|z \cdot g - x \cdot g| = \frac{\alpha y}{|a + zc| \, |a + xc|} \qquad \left(z \in \Gamma_\alpha(x) \right)$$

$$\leq \alpha \left(\frac{y}{|a + zc|^2} \right) \left(1 + \frac{y^2 |c|^2}{|a + xc|^2} \right)^{1/2}$$

since $|a + zc|^2 = |a + xc|^2 + y^2 |c|^2$. Thus for $0 < y \leq 1$ and $a + xc \neq 0$, $z \cdot g$ lies in a cone $\Gamma_\beta(x \cdot g)$ where β depends only on x and g. Hence, as z approaches x in $\Gamma_\alpha(x)$,

$$\lim_{z \to x} \frac{1}{\eta^{2(\rho-\lambda)}} F(z \cdot g) = (\mathcal{B}_{\tau,\lambda} F)(x \cdot g) .$$

This completes the proof. ∎

In view of the natural parallels between chapter II where $\mathbb{R}^n_+$ had its Euclidean metric and the results so far in the present section where $\mathbb{R}^n_+$ is thought of as a realization of n-dimensional Euclidean space, there should be Hilbert space structures on function spaces on $\mathbb{R}^{n-1}$ with respect to which

(4.25)(i) *each* $U(\tau, \lambda, g)$ *is unitary for fixed* λ *and all* g *in* $\mathrm{Spin}_0(n,1)$,

(4.25)(ii) *the Cauchy–Szegő mapping* $\mathcal{S}_{\tau,\lambda}$ *and trace operator* $\mathcal{B}_{\tau,\lambda}$ *define unitarily equivariant mappings between boundary spaces and the Hilbert subspaces of* $C^\infty(\mathbb{R}^n_+, \mathfrak{H})$ *as prescribed by* (4.6).

One of the important roles of the intertwining operators $\mathcal{A}_{\tau,\lambda}$ is to specify a natural candidate for such a Hilbert space structure. Indeed, with respect to the inner product on $\mathfrak{H}$,

$$(4.26) \qquad (f, \phi)_{\tau,\lambda} = \int_{\mathbb{R}^{n-1}} \left(\mathcal{A}_{\tau,\lambda} f(v), \phi(v) \right) dv \qquad (\lambda > 0)$$

defines an inner product on $L^2_\lambda(\mathbb{R}^{n-1}, \mathfrak{H})$. Theorem (4.10) guarantees both the boundedness

$$(4.27)(i) \qquad |(f, \phi)_{\tau,\lambda}| \leq \mathrm{const.} \, \|f\|_\lambda \|\phi\|_\lambda ,$$

and the unitarity

$$(4.27)(ii) \qquad \left(U(\tau, \lambda, g)f, U(\tau, \lambda, g)\phi \right)_{\tau,\lambda}$$

$$= \int_{\mathbb{R}^{n-1}} \frac{1}{|a+vc|^{4\rho}} \left((\mathcal{A}_{\tau,\lambda} f)(v \cdot g), \phi(v \cdot g) \right) dv$$

$$= (f, \phi)_{\tau,\lambda}$$

since $(\mathfrak{H}, \tau)$ is a unitary representation of $\mathrm{Spin}(n)$ (see also section 3(**F**)). Consequently, the completion $\mathcal{W}^{\tau,\lambda}(\mathbb{R}^{n-1}, \mathfrak{H})$ of $L^2_\lambda(\mathbb{R}^{n-1}, \mathfrak{H})$ with respect to this inner product structure $(\cdot, \cdot)_{\tau,\lambda}$ would become a Hilbert space on which $g \to U(\tau, \lambda, g)$ extends to a unitary representation of $\mathrm{Spin}_0(n, 1)$ once $(\cdot, \cdot)_{\tau,\lambda}$ was known to be *positive-definite*. Herein lies the rub: establishing positive-definiteness is far from easy, and often positivity occurs only on a subspace of $L^2_\lambda(\mathbb{R}^{n-1}, \mathfrak{H})$. The notation $\mathcal{W}^{\tau,\lambda}$ was deliberately chosen to suggest $\mathfrak{H}$-valued Sobolev spaces, because (4.26) when written as a double integral is essentially one of the equivalent norms on a homogeneous L^2-Sobolev space.

Although the irreducible unitary representations of $\mathrm{Spin}_0(n, 1)$ have been characterized algebraically, and some of them realized analytically by the methods of this section, the complete analytic story has yet to be written. The methods of this section will probably suffice.

5 Asymptotics for heat kernels

In preparation for the proof of the Atiyah–Singer index theorem for Dirac operators to be given in the final section we derive asymptotic estimates for the heat kernel of a second-order elliptic operator

$$(5.1) \qquad \Delta = \sum_{i,j} a_{ij}(x) \frac{\partial^2}{\partial x_i \partial x_j} + \sum_i b_i(x) \frac{\partial}{\partial x_i} + c(x) \qquad (x \in \mathbb{R}^n)$$

on $\mathcal{C}^\infty(\mathbb{R}^n, \mathfrak{H})$ where $\mathfrak{H}$ is any $\mathfrak{A}_n$-module. There are various well-known estimates for such kernels when $\mathfrak{H}$ is a Banach space, but the extra $\mathfrak{A}_n$-module structure on $\mathfrak{H}$ will make it possible to associate a one-parameter family $\Delta^{(\varepsilon)}$ of elliptic operators to Δ so that asymptotic estimates for the heat kernel of Δ can be obtained as the limit as $\varepsilon \to 0$ of corresponding estimates for the heat kernel of $\Delta^{(\varepsilon)}$. Quillen has likened the process to the passage from quantum mechanics to classical mechanics by letting Planck's constant approach 0.

Let Δ be a differential operator of the form (5.1) having smooth $\mathcal{L}(\mathfrak{H})$-valued coefficients; in particular, these coefficients could be real-valued, identifying $\mathbb{R}$ with the subspace $\mathbb{R}I$ in $\mathcal{L}(\mathfrak{H})$. By a *heat kernel* for Δ we shall mean a smooth function $\Gamma : \Omega_T \to \mathcal{L}(\mathfrak{H})$, $\Gamma = \Gamma(x, y, t)$, on a cylinder Ω_T such that

(5.2)(i) *as a function of* (x,t),

$$\left(\Delta - \frac{\partial}{\partial t}\right)\Gamma(\cdot, y; \cdot) = 0 \qquad (y \in \mathbb{R}^n) ,$$

(5.2)(ii) *for each compactly supported function* f *in* $C^\infty(\mathbb{R}^n, \mathfrak{H})$

$$\lim_{t \to 0+} \int_{\mathbb{R}^n} \Gamma(x, y; t) f(y) \, dy = f(x) ,$$

i.e., $\Gamma(x, y; t) \to \delta_0(x - y) \otimes I$ as $t \to 0+$. For instance, in the classical case of the Euclidean Laplacian Δ_n, Γ is given in terms of the usual Gauss–Weierstrass kernel W_t by

$$(5.3) \qquad \Gamma(x, y; t) = W_t(x - y) = \frac{1}{(4\pi t)^{n/2}} \exp\left[-\frac{1}{4t}|x - y|^2\right] ,$$

so that

$$(5.3)(i) \qquad \Gamma(0, 0; t) = \frac{1}{(4\pi t)^{n/2}} \qquad (t > 0) .$$

On the other hand, if Δ_H is the harmonic oscillator

$$(5.4) \qquad \Delta_H = \sum_i \frac{\partial^2}{\partial x_i^2} - \sum_{i,j,k} c_{ik} c_{jk} x_i x_j = \Delta_n - |Cx|^2$$

with $C = [c_{ij}]$ a real symmetric matrix, then

$$\Gamma_H(x, y; t) = \frac{1}{(4\pi t)^{n/2}} \hat{A}(2Ct) \exp\left[-\frac{1}{4t}\left\{(L(2Ct)x, x)\right.\right.$$
$$\left.\left. + (L(2Ct)y, y) - 2(B(2Ct)x, y)\right\}\right]$$

where
(5.5)
$$\hat{A}(C) = \det\left(\frac{C}{\sinh C}\right)^{1/2} , \qquad L(C) = \frac{C}{\tanh C} , \qquad B(C) = \frac{C}{\sinh C} .$$

To make sense of $\hat{A}(C)$, $L(C)$ and $B(C)$ observe that the power series expansions

$$(5.5)(i) \qquad \frac{z}{\sinh z} = 1 - \frac{z^2}{3!} + a_2 z^4 + \cdots + a_n z^{2n} + \cdots$$

and

$$(5.5)(ii) \qquad \frac{z}{\tanh z} = 1 + \frac{z^2}{3} + b_2 z^4 + \cdots + b_n z^{2n} + \cdots$$

converge for all sufficiently small z; hence

$$I + \frac{1}{3!}t^2 C^2 + a_2 t^4 C^4 + \cdots , \qquad I + \frac{1}{3}t^2 C^2 + b_2 t^4 C^4 + \cdots$$

define $tC/\sinh(tC)$ and $tC/\tanh(tC)$ as $(n \times n)$-matrices for each C and all small enough t. Thus $\hat{A}(2tC)$, $(L(2tC)x, x)$ and $(B(2Ct)x, y)$ are correspondingly well-defined. On the other hand, as these last expressions are invariant under similarity transformations $C \to OCO^{-1}$, $O \in O(n)$, they could also be defined simply by diagonalizing the real symmetric matrix C, since

$$\hat{A}(D) = \left(\prod_{j=1}^{n} \frac{\lambda_j}{\sinh \lambda_j}\right)^{1/2} \quad , \quad (L(D)x, x) = \sum_j \left(\frac{\lambda_j}{\tanh \lambda_j}\right) x_j \ ,$$

and

$$(B(D)x, y) = \sum_j \left(\frac{\lambda_j}{\sinh \lambda_j}\right) x_j y_j$$

for *any* diagonal matrix $D = \text{Diag}(\lambda_1, \ldots, \lambda_n)$ having real entries.

But in general, let Δ be any elliptic operator having the form (5.1) where $\mathfrak{H}$ is an $\mathfrak{A}_n$-module and

(5.6)(i) $a_{ij} = a_{ij}(x)$ *is a real-valued function such that* $[a_{ij}(x)]$ *is positive-definite and symmetric,*

(5.6)(ii) Δ *coincides with the Euclidean Laplacian* Δ_n *outside some compact set.*

Intuitively therefore, Δ is a compactly supported perturbation of Δ_n; in particular, the coefficients of Δ are uniformly bounded, as is each of their derivatives, and there exist $\lambda_0, \lambda_1 > 0$ so that

$$(5.7) \qquad \lambda_0 |\xi|^2 \leq \sum_{i,j} a_{ij}(x) \xi_i \xi_j \leq \lambda_1 |\xi|^2 \qquad (\xi \in \mathbb{R}^n)$$

uniformly in x. By using an orthogonal change of coordinates if necessary, we can (and shall) assume that $a_{ij}(0) = \delta_{ij}$. Typically, such operators arise geometrically. For instance, if $g = [g_{ij}(x)]$ defines a Riemannian metric on $\mathbb{R}^n$ which is asymptotically Euclidean, i.e., $g_{ij}(x) = \delta_{ij}$ outside some compact set, and D is the Dirac operator on $C^\infty(\mathbb{R}^n, \mathfrak{H})$ determined by the Riemannian manifold $(\mathbb{R}^n, g)$, then

$$(5.8) \qquad -D^2 = \sum_{i,j} g^{ij}(x) \frac{\partial^2}{\partial x_i \partial x_j} + \text{lower-order terms},$$

as we saw earlier in section 1 of this chapter; it will coincide with Δ_n near infinity because the Riemannian metric is Euclidean there. Hence $-D^2$ has properties (i) and (ii) of (5.6). The corresponding Laplace–Beltrami operator on scalar-valued functions and Hodge Laplacian on forms have these properties too, although the scalar field $\mathbb{R}$ or $\mathbb{C}$ is not an $\mathfrak{A}_n$-module, of course.

(5.9) Theorem.

Let Δ be a differential operator satisfying (5.1) and (5.6). Then Δ has a unique heat kernel $\Gamma = \Gamma(x, y; t)$ on Ω_T; furthermore,

(i) *for some $\delta > 0$*

$$\|\Gamma(x,y;t)\| \le \text{const.}\ \frac{1}{(4\pi t)^{n/2}}\ \exp\left[-\frac{\delta}{t}|x-y|^2\right]\quad (\|\cdot\| = \|\cdot\|_{\mathcal{L}(\mathfrak{H})})\ ,$$

(ii) *for each integer $m > n/2$ there exist $a_1,\ldots,a_m$ in $\mathcal{L}(\mathfrak{H})$ such that*

$$\Gamma(0,0;t) = \frac{1}{(4\pi t)^{n/2}}\left(I + a_1 t + \cdots + a_m t^m + O(t^{m+1})\right)\ .$$

Thus (5.9) measures quantitatively how the perturbation from Δ_n to Δ causes the heat kernel for Δ to deviate from the Gauss–Weierstrass kernel. The proof of (5.9) is generally well-known and uses the same strategy, originating with E.E. Levi in 1907, for both parts. Let $G = G(x,y;t)$ be a smooth $\mathcal{L}(\mathfrak{H})$-valued function on Ω_T such that

$$(5.10)\qquad \lim_{t\to 0+}\int_{\mathbf{R}^n} G(x,y;t)f(y)\,dy = f(x)$$

for any compactly supported f. Now set

$$(5.11)\qquad \Gamma(x,y;t) = G(x,y;t) + \int_0^t\int_{\mathbf{R}^n} G(x,\xi;t-s)\Phi(\xi,y;s)\,d\xi\,ds$$

where $\Phi: \Omega_T \to \mathcal{L}(\mathfrak{H})$ is to be determined. Property (5.10) ensures that Γ satisfies (5.2)(i) if Φ is a solution of the integral equation

$$(5.12)\qquad \Phi(x,y;t) = K(x,y;t) + \int_0^t\int_{\mathbf{R}^n} K(x,\xi;t-s)\Phi(\xi,y;s)\,d\xi\,ds$$

where

$$(5.12)(i)\qquad K(x,y;t) = \left(\Delta_x - \frac{\partial}{\partial t}\right)G(x,y;t)\ .$$

Consequently, when $\mathcal{K}$ is the integral operator

$$(5.13)\qquad (\mathcal{K}f)(x,t) = \int_0^t\int_{\mathbf{R}^n} K(x,\xi;t-s)f(\xi,s)\,d\xi\,ds\ ,$$

the integral equation (5.12) becomes

$$K(x,y;t) = (I-\mathcal{K})\big(\Phi(\cdot,y;\cdot)\big)\qquad (y\in\mathbf{R}^n)\ ,$$

so that by formal inversion

$$(5.14)\qquad \Phi(x,y;t) = \sum_{k=0}^{\infty}\mathcal{K}^k\big(K(\cdot,y;\cdot)\big) = \sum_{k=0}^{\infty} K_k(x,y;t)$$

setting

$$(5.14)(i)\qquad K_0(x,y;t) = K(x,y;t)\ ,\quad K_k(x,y;t) = \mathcal{K}\big(K_{k-1}(\cdot,y;\cdot)\big)\ .$$

Hence

$$(5.15)\ \ \Gamma(x,y;t) = G(x,y;t) + \sum_{k=0}^{\infty}\int_0^t\int_{\mathbf{R}^n} G(x,\xi;t-s)K_k(\xi,y;s)\,d\xi\,ds\ .$$

On the other hand, once the requisite convergence has been established, property (5.10) will ensure also that

$$\lim_{t \to 0+} \int_{\mathbf{R}^n} \Gamma(x, y; t) f(y) \, dy$$

$$= f(x) + \lim_{t \to 0+} \int_0^t \int_{\mathbf{R}^n \times \mathbf{R}^n} G(x, \xi; t - s) \Phi(\xi, y; s) f(y) \, dy \, d\xi \, ds$$

$$= f(x)$$

when f has compact support. Hence the function Γ defined by (5.11) will be a heat kernel for Δ. Proof of its uniqueness is standard, and so will be omitted. Thus we are left with constructing G and establishing the convergence of (5.15).

Proof of (5.9)(i). Let $G = G(x, y; t)$ be the heat kernel for the constant-coefficient elliptic operator

$$\Delta_{(y)} = \sum_{i,j} a_{ij}(y) \frac{\partial^2}{\partial x_i \partial x_j}$$

obtained by 'freezing' the coefficients of the principal part of Δ at y. The following properties of G are well-known.

(5.16) Lemma.
  There is a unique heat kernel $G = G(x, y; t)$ for $\Delta_{(y)}$. Furthermore, it is a function in $C^\infty(\Omega_T, \mathcal{L}(\mathfrak{S}))$ such that

$$\left| \left(\frac{\partial}{\partial x} \right)^\alpha G(x, y; t) \right| \le B t^{-\frac{1}{2}|\alpha|} W_{\beta t}(x - y)$$

holds for some choice of B, β and all α, $0 \le |\alpha| \le 2$.

Now, by (5.12)(i) and the definition of $G = G(x, y; t)$,

$$K(x, y; t) = \left(\sum_{i,j} (a_{ij}(x) - a_{ij}(y)) \frac{\partial^2}{\partial x_i \partial x_j} + \sum_i b_i(x) \frac{\partial}{\partial x_i} + c(x) \right) G(x, y; t) ;$$

on the other hand

$$|a_{ij}(x) - a_{ij}(y)| \le \text{const.} \left(\sum_k |x_k - y_k| \right) ,$$

while

$$\frac{|x_k - y_k|}{t^{1/2}} W_{\beta t}(x - y) \le \text{const.} W_{2\beta t}(x - y)$$

since $\alpha e^{-\alpha^2/\beta} \le e^{-\alpha^2/2\beta}$. Using these estimates together with (5.16), we deduce that

(5.17) $$\|K(x, y; t)\| \le C t^{-1/2} W_{2\beta t}(x - y) .$$

Estimates for K_k now follow by induction. Indeed, (5.17) estimates K_0; and so

$$\|K_1(x, y; t)\|$$

$$= \left\| \int_0^t \int_{\mathbf{R}^n} K(x, \xi; t - s) K_0(\xi, y; s) \, d\xi \, ds \right\|$$

$$\leq C^2 \int_0^t s^{-1/2} (t - s)^{-1/2} \left(\int_{\mathbf{R}^n} W_{2\beta(t-s)}(x - \xi) \, W_{2\beta s}(\xi - y) \, d\xi \right) ds$$

$$= C^2 \left(\int_0^1 s^{-1/2} (1 - s)^{-1/2} \, ds \right) W_{2\beta t}(x - y)$$

$$= C^2 \left(\Gamma(\tfrac{1}{2}) \right)^2 W_{2\beta t}(x - y) \ .$$

Suppose then that

$$\|K_{k-1}(x, y; t)\| \leq \left(\frac{C^k \Gamma(\tfrac{1}{2})^k}{\Gamma(\tfrac{1}{2}k)} \right) t^{\frac{1}{2}(k-2)} W_{2\beta t}(x - y) \qquad (k \geq 1) \ .$$

By the argument as for K_1,

$$\|K_k(x, y; t)\| \leq \frac{C^{k+1} \Gamma(\tfrac{1}{2})^k}{\Gamma(\tfrac{1}{2}k)} \left(\int_0^t (t - s)^{-1/2} s^{\frac{1}{2}(k-2)} \, ds \right) W_{2\beta t}(x - y)$$

$$= \left(\frac{C^{k+1} \Gamma(\tfrac{1}{2})^{k+1}}{\Gamma(\tfrac{1}{2}(k + 1))} \right) t^{\frac{1}{2}(k-1)} W_{2\beta t}(x - y) \ .$$

Hence

$$\|\Gamma(x, y; t)\| \leq \text{const.} \ \left(1 + \sum_{k=1}^{\infty} \frac{C^k \Gamma(\tfrac{1}{2})^k}{\Gamma(\tfrac{1}{2}(k + 1))} t^{\frac{1}{2}k} \right) W_{2\beta t}(x - y)$$

using (5.16) again and the estimate on K_k. Since

$$\frac{1}{\Gamma(\tfrac{1}{2}k + \tfrac{1}{2})} \leq \text{const.} \ \frac{(2e)^{\frac{1}{2}k}}{(k + 1)^k}$$

(by Stirling's formula), estimate (5.9)(i) follows easily.

Proof of (5.9)(ii). (Patodi) The function $G = G(x, y; t)$ is now defined via the following known result.

(5.18) Lemma.

For each integer $m > n/2$ there exist unique functions $\phi_1, \ldots, \phi_m$ in $C^{\infty}(\mathbf{R}^n \times \mathbf{R}^n, \mathcal{L}(\mathfrak{S}))$ such that

(i) $\phi_0(x, x) = I$, but $\phi_1, \ldots, \phi_m$ have compact support,

(ii) $\left(\Delta_x - \dfrac{\partial}{\partial t} \right) \colon \left(\displaystyle\sum_{s=0}^m t^s \phi_s(x, y) \right) W_t(x-y) \to t^m W_t(x-y)(\Delta_x \phi_m) \ .$

Now set

$$(5.18)' \qquad G(x, y; t) = \left(\sum_{s=0}^{m} t^s \phi_s(x, y) \right) W_t(x - y)$$

and

$$K(x, y; t) = \left(\Delta_x - \frac{\partial}{\partial t} \right) G = \left(t^m W_t(x - y) \right) \Delta_x \phi_m \; ;$$

in particular,

$$(5.19) \qquad \|K(x, y; t)\| \leq C t^m W_t(x - y) \; .$$

We can now proceed exactly as in the proof of part (i). Suppose that

$$(5.20) \quad \|K_{k-1}(x, y; t)\| \leq C^k \frac{\Gamma(m + 1)^k}{\Gamma(k(m + 1))} t^{km + k - 1} W_t(x - y) \quad (k \geq 1) \; .$$

Inequality (5.19) is the case $k = 1$. But then

$$\|K_k(x, y; t)\| = \left\| \int_0^t \int_{\mathbf{R}^n} K(x, \xi; t - s) K_{k-1}(\xi, y; s) \, d\xi \, ds \right\|$$

$$\leq C^{k+1} \frac{(\Gamma(m + 1))^k}{\Gamma(k(m + 1))} \left(\int_0^t (t - s)^m s^{km + k - 1} \, ds \right) W_t(x - y)$$

$$= C^{k+1} \frac{\Gamma(m + 1)^{k+1}}{\Gamma((k + 1)(m + 1))} t^{(k+1)m + k} W_t(x - y) \; .$$

Thus (5.20) continues to hold with k instead of $k - 1$, and so

$$\|\Gamma(x, y; t) - G(x, y; t)\|$$

$$(5.21) \qquad \leq \sum_{k=0}^{\infty} \left\| \int_0^t \int_{\mathbf{R}^n} G(x, \xi; t - s) K_k(\xi, y; s) \, d\xi \, ds \right\|$$

$$= O(t^{m+1} W_t(x - y))$$

using (5.18)', (5.20), and the inequality

$$\frac{\Gamma(x)^k}{\Gamma(kx)} \leq 2\pi^{\frac{1}{2}(k-1)} \frac{1}{k^{kx - 1/2}} \qquad (x \to \infty)$$

which follows from the Gauss duplication theorem. This is more than enough to prove (5.9)(ii).   ∎

Clearly (5.21) provides another proof of (5.9)(i), but a careful reading of the proof of the principal general result of this section, theorem (5.27), will show that (5.9)(ii) is not really needed. It has been included to bring out more clearly the nature of the asymptotic estimates being derived.

Now we come to the fundamentally new idea: let

$$(5.22)(\text{i}) \qquad \mathcal{A}_0 \oplus \mathcal{A}_1 \oplus \cdots \oplus \mathcal{A}_p \oplus \cdots \oplus \mathcal{A}_{rn} = \mathcal{L}(\mathfrak{H}) \qquad (r \text{ fixed})$$

be a subspace decomposition of $\mathcal{L}(\mathfrak{H})$ such that

(5.22)(ii) $I \in \mathcal{A}_0$, $\mathcal{A}_p \mathcal{A}_q \subseteq \mathcal{A}_0 \oplus \cdots \oplus \mathcal{A}_{p+q}$ $(p + q \leq rn)$.

We shall think of $\mathcal{A}_p$ as being the elements in $\mathcal{L}(\mathfrak{H})$ that are 'homogeneous of degree p'. Define $\{\lambda(\varepsilon) : \varepsilon > 0\}$ on $\mathcal{L}(\mathfrak{H})$ by

$$\lambda(\varepsilon) : \xi \longrightarrow (\varepsilon^{-p})^{1/r} \xi \qquad (\xi \in \mathcal{A}_p) ,$$

and then define $\{M_\varepsilon : \varepsilon > 0\}$ on $C^\infty(\mathbb{R}^n, \mathcal{L}(\mathfrak{H}))$ by

$$(M_\varepsilon F)(x) = \lambda(\varepsilon) F(\varepsilon x) \qquad (x \in \mathbb{R}^n) .$$

In group-theoretic terms, $\varepsilon \to M_\varepsilon$ is a representation of the multiplicative group of positive reals. Thus

$$\Delta^{(\varepsilon)} = \varepsilon^2 M_\varepsilon \Delta M_\varepsilon^{-1} \qquad (\varepsilon > 0)$$

is a differential operator on $C^\infty(\mathbb{R}^n, \mathcal{L}(\mathfrak{H}))$ which because of (5.6)(i) and (5.22)(ii) has the form

$$\Delta^{(\varepsilon)} = \sum_{i,j} a^{(\varepsilon)}(x) \frac{\partial^2}{\partial x_i \partial x_j} + \sum_i b_i^\varepsilon(x) \frac{\partial}{\partial x_i} + c^{(\varepsilon)}(x) ,$$

where $a_{ij}^{(\varepsilon)}(x) = a_{ij}(\varepsilon x)$ and

$$b_i^{(\varepsilon)}(x) = \varepsilon \lambda(\varepsilon) b_i(\varepsilon x) \lambda(\varepsilon)^{-1} \quad , \quad c^{(\varepsilon)}(x) = \varepsilon^2 \lambda(\varepsilon) c(\varepsilon x) \lambda(\varepsilon)^{-1} ;$$

in particular, each $\Delta^{(\varepsilon)}$ is an elliptic operator satisfying (5.6)(i),(ii). By (5.9) it therefore has a unique heat kernel $\Gamma_{(\varepsilon)} = \Gamma_{(\varepsilon)}(x, y; t)$ which is easily expressed in terms of the heat kernel for Δ.

(5.23) Lemma.
   The heat kernel $\Gamma_{(\varepsilon)}$ for $\Delta^{(\varepsilon)}$ is given on Ω_T by

$$\Gamma_{(\varepsilon)}(x, y; t) = \varepsilon^n \lambda(\varepsilon) \Gamma(\varepsilon x, \varepsilon y; \varepsilon^2 t) \lambda(\varepsilon)^{-1}$$

where Γ is the heat kernel for Δ.

Proof. It is enough to check that $\varepsilon^n M_\varepsilon \Gamma(\cdot, \varepsilon y; \varepsilon^2 t) \lambda(\varepsilon)^{-1}$ satisfies (5.2) when Δ is replaced by $\Delta^{(\varepsilon)}$. But

$$\frac{\partial}{\partial t} \left(M_\varepsilon \Gamma(\cdot, \varepsilon y; \varepsilon^2 t) \lambda(\varepsilon)^{-1} \right) = M_\varepsilon \frac{\partial}{\partial t} \Gamma(\cdot, \varepsilon y; \varepsilon^2 t) \lambda(\varepsilon)^{-1}$$

$$= \varepsilon^2 M_\varepsilon \Delta \left(\Gamma(\cdot, \varepsilon y; \varepsilon^2 t) \lambda(\varepsilon)^{-1} \right)$$

$$= \Delta^{(\varepsilon)} \left(M_\varepsilon \Gamma(\cdot, \varepsilon y; \varepsilon^2 t) \lambda(\varepsilon)^{-1} \right) ,$$

while

$$\varepsilon^n \int_{\mathbb{R}^n} M_\varepsilon \Gamma(\cdot, \varepsilon y; \varepsilon^2 t) \lambda(\varepsilon)^{-1} f(y) \, dy = \int_{\mathbb{R}^n} \lambda(\varepsilon) \Gamma(\varepsilon x, y; \varepsilon^2 t) \lambda(\varepsilon)^{-1} f(\tfrac{1}{\varepsilon} y) \, dy$$

$$\longrightarrow \int_{\mathbb{R}^n} \left(\delta_0(\varepsilon x - y) \otimes I \right) f(\tfrac{1}{\varepsilon} y) \, dy = f(x)$$

as $t \to 0+$ for all compactly supported f. Hence $\Gamma_{(\varepsilon)}$ has the stated form. $\blacksquare$

Now set

$$\Delta = \lim_{\varepsilon \to 0+} \Delta^{(\varepsilon)} = \sum_i \frac{\partial^2}{\partial x_i^2} + \sum_{i,j} \beta_i(x) \frac{\partial}{\partial x_i} + \gamma(x)$$

where

(5.24) $\qquad \beta_i(x) = \lim_{\varepsilon \to 0+} b_i^{(\varepsilon)}(x) \quad , \quad \gamma(x) = \lim_{\varepsilon \to 0+} c^{(\varepsilon)}(x) .$

In specific examples it will be clear that these last limits exist, but it will be instructive to determine general conditions under which they exist and, more particularly, to determine the nature of these limits. Denote by $[\xi]_p$ the component in $\mathcal{A}_p$ of any ξ in $\mathcal{L}(\mathfrak{H})$; then the product

(5.25) $\qquad a \circ b = [ab]_{p+q} \qquad (a \in \mathcal{A}_p \ , \ b \in \mathcal{A}_q)$

extends linearly to give a second multiplicative structure on $\mathcal{L}(\mathfrak{H})$ with respect to which it becomes a graded algebra, denoted by $\mathcal{G}r(\mathcal{L}(\mathfrak{H}))$.

(5.26) Lemma.
Let ϕ be a function in $C^\infty(\mathbb{R}^n, \mathcal{A}_0 \oplus \cdots \oplus \mathcal{A}_{rq})$ such that

$$\phi(x) = \sum_{|\alpha|=q-p} x^\alpha \xi_\alpha + O(|x|^{q-p+1})$$

near the origin. Then

$$\lim_{\varepsilon \to 0+} \varepsilon^p \big(\lambda(\varepsilon) \phi(\varepsilon x) \lambda(\varepsilon)^{-1} \big) a = \left(\sum_{|\alpha|=q-p} x^\alpha [\xi_\alpha]_{rq} \right) \circ a$$

for all x in $\mathbb{R}^n$ and a in $\mathcal{L}(\mathfrak{H})$. Conversely, if $\phi(0) \neq 0$ and

$$\lim_{\varepsilon \to 0+} \varepsilon^p \big(\lambda(\varepsilon) \phi(\varepsilon x) \lambda(\varepsilon)^{-1} \big)$$

exists in $\mathcal{L}(\mathfrak{H})$, then ϕ has range in $\mathcal{A}_0 \otimes \ldots \otimes \mathcal{A}_{rp}$.

Proof. Fix a in $\mathcal{A}_k$. Then $\lambda(\varepsilon)^{-1} a = \varepsilon^{k/r} a$, and $\phi(\varepsilon x) a : \mathbb{R}^n \to \mathcal{A}_0 \oplus \cdots \oplus \mathcal{A}_{rq+k}$. Thus

$$\varepsilon^p \big(\lambda(\varepsilon) \phi(\varepsilon x) \lambda(x)^{-1} \big) a = \varepsilon^{k/r+q} \sum_{|\alpha|=q-p} x^\alpha \lambda(\varepsilon)(\xi_\alpha a) + O(\varepsilon) ;$$

consequently,

$$\lim_{\varepsilon \to 0+} \varepsilon^p \big(\lambda(\varepsilon) \phi(\varepsilon x) \lambda(x)^{-1} \big) a = \sum_{|\alpha|=q-p} x^\alpha [\xi_\alpha a]_{rq+k} = \left(\sum_{|\alpha|=q-p} x^\alpha [\xi_\alpha]_{rq} \right) \circ a ,$$

in view of the definition of multiplication on $\mathcal{G}r(\mathcal{L}(\mathfrak{H}))$. Since this holds for each k, the result is true in general. The converse is proved analogously. $\blacksquare$

In the specific examples related to the Dirac operator that we shall study, lemma (5.26) will ensure that

$$\Delta = \sum_i \frac{\partial^2}{\partial x_i^2} + \sum_{i,j} \xi_{ij} x_i \frac{\partial}{\partial x_j} + \sum_{i,j} \left(\sum_k \eta_{ik} \eta_{jk} \right) x_i x_j + \zeta$$

where ξ_{ij}, η_{ij} and ζ multiply as elements of $\mathcal{G}r(\mathcal{L}(\mathfrak{H}))$. Such an operator does not satisfy (5.6)(ii), so theorem (5.9) cannot be used to guarantee that Δ has a unique heat kernel, though with more care in the estimates in the proof of (5.9) we could in fact prove (5.9)(i) for operators such as Δ having unbounded lower order coefficients. But *one of the basic features of Getzler's idea is that Δ can be recognized as a classically occurring operator* even though the Δ from which it is obtained is quite arbitrary. The existence and uniqueness of the heat kernel $\boldsymbol{\Gamma} = \boldsymbol{\Gamma}(x, y; t)$ for Δ is then part of the 'well-known' theory for such classical operators, and consequently does not need a separate proof. The following result relating $\Gamma_{(\varepsilon)}$ and $\boldsymbol{\Gamma}$ is easier to appreciate than to prove, so we postpone its proof to the end of this section.

(5.27) Theorem.
If the limits in (5.24) exist, then
$$\lim_{\varepsilon \to 0+} \Gamma_{(\varepsilon)}(x, y; t) = \boldsymbol{\Gamma}(x, y; t)$$
for all sufficiently small t.

Together with lemma (5.26), this result enables us to link $\Gamma(x, y; t)$ with $\boldsymbol{\Gamma}(x, y; t)$. In fact, let

$$\boldsymbol{\Gamma}(0, 0; t) = \frac{1}{(4\pi t)^{n/2}} \left(I + a_1 t + \cdots + a_m t^m + O(t^{m+1}) \right)$$

be the asymptotic expansion of $\boldsymbol{\Gamma}$ at the origin. Then, by (5.23), the corresponding expansion for $\Gamma_{(\varepsilon)}$ is given by

$$\Gamma_{(\varepsilon)}(0, 0; t) = \varepsilon^n \lambda(\varepsilon) \boldsymbol{\Gamma}(0, 0; \varepsilon^2 t) \lambda(\varepsilon)^{-1}$$

$$= \frac{1}{(4\pi t)^{n/2}} \left(I + \varepsilon^2 \lambda(\varepsilon) a_1 \lambda(\varepsilon)^{-1} t + \cdots \right.$$
$$\left. + \varepsilon^{2m} \lambda(\varepsilon) a_m \lambda(\varepsilon)^{-1} t^m + O(\varepsilon t^{m+1}) \right)$$

provided $m > n/2$, since $\|\lambda(\varepsilon)\| \le \varepsilon^{-n}$. Consequently,

$$(5.28) \qquad \boldsymbol{\Gamma}(0, 0; t) = \lim_{\varepsilon \to 0+} \frac{1}{(4\pi t)^{n/2}} \left(\sum_{k=0}^{[\frac{1}{2}n]} \varepsilon^{2k} \lambda(\varepsilon) a_k \lambda(\varepsilon)^{-1} t^k \right).$$

In practice $\Gamma(0,0;t)$ is easily calculated. Let $g = [g_{ij}(x)]$ be a Riemannian metric on $\mathbb{R}^n$ which is asymptotically Euclidean, and let $(\mathfrak{H}, \tau)$ be a representation of $\mathrm{Spin}(n)$ on an $\mathfrak{A}_n$-module $\mathfrak{H}$ such that

$$\tau(a)e_j\tau(a)^{-1} = \sigma(a)e_j \qquad (a \in \mathrm{Spin}(n)) \ .$$

The heat kernel for $-D^2$ as an elliptic operator on $C^\infty(\mathbb{R}^n, \mathfrak{H})$ will be denoted by $\Gamma_\tau = \Gamma_\tau(x, y; t)$; the heat kernel for the corresponding operator

$$\lim_{\varepsilon \to 0+} \varepsilon^2 M_\varepsilon(-D^2) M_\varepsilon^{-1}$$

will be denoted analogously by $\Gamma_\tau = \Gamma_\tau(x, y; t)$. In the applications of the asymptotic estimate for $\Gamma_\tau(0,0;t)$ that will be made in the next section we are free to choose the coordinate system near the origin. Thus we shall assume that these *coordinates are normal*, so that the results of (1.39) can be used.

(5.29) (A) Spinor Laplacian. As $(\mathfrak{H}, \tau) = (\Lambda^*(\mathbb{C}^m), S)$ in this case (see (1.35)(A)), $\mathcal{L}(\mathfrak{H})$ is a realization of the complex Clifford algebra $\mathfrak{C}_n$, $n = 2m$. For (5.22)(i) we can thus use the decomposition

$$\mathfrak{C}_n = \mathfrak{C}_n^{(0)} \oplus \mathfrak{C}_n^{(1)} \oplus \cdots \oplus \mathfrak{C}_n^{(n)}$$

of $\mathfrak{C}_n$ into its subspaces $\mathfrak{C}_n^{(k)}$ of k-multivectors; so $r = 1$, and the associated graded algebra $\mathcal{G}r(\mathfrak{C}_n)$ is just $\mathfrak{C}_n$ with the wedge product (7.16) in chapter 1 for multiplication. Consequently, $\mathcal{G}r(\mathfrak{C}_n)$ is isomorphic as an algebra to the exterior algebra $\Lambda^*(\mathbb{C}^n)$, $n = 2m$.

Near the origin the spinor Laplacian takes the form

$$D^2 = -\sum_{i=1}^n \left(\frac{\partial}{\partial x_i} + \tfrac{1}{2} \sum_j x_j R_{ij}(0) \right)^2 + \mathcal{R}_S$$

with careful control on the coefficients of $\mathcal{R}_S$. But

$$\lambda(\varepsilon) : e_{\alpha_1} \cdots e_{\alpha_k} \longrightarrow (1/\varepsilon^k) e_{\alpha_1} \cdots e_{\alpha_k} \qquad (\varepsilon > 0) \ .$$

Consequently, by (5.26), $\varepsilon^2 M_\varepsilon \mathcal{R}_S M_\varepsilon^{-1} \to 0$ as $\varepsilon \to 0$, and so

$$(*) \qquad \lim_{\varepsilon \to 0+} \varepsilon^2 M_\varepsilon(-D^2) M_\varepsilon^{-1} = \sum_i \left(\left(\frac{\partial}{\partial x_i} \right)^2 + \tfrac{1}{4} \sum_{j,k} x_j x_k R_{ij}(0) R_{ik}(0) \right)$$

$$+ \tfrac{1}{2} \sum_{i,j} \left(x_i \frac{\partial}{\partial x_j} - x_j \frac{\partial}{\partial x_i} \right) R_{ij}(0)$$

with $R_{ij}(0) \in \mathfrak{C}_n^{(2)} \subseteq \mathcal{G}r(\mathfrak{C}_n)$. Now $\mathfrak{C}_n^{(2)}$ generates a nilpotent abelian subalgebra of $\mathcal{G}r(\mathfrak{C}_n)$, so

$$(**) \qquad \sum_i \left(\left(\frac{\partial}{\partial x_i} \right)^2 + \tfrac{1}{4} \sum_{j,k} x_j x_k R_{ij}(0) R_{ik}(0) \right)$$

is the harmonic oscillator (5.4) with the matrix $C = [c_{ij}]$ of scalars replaced by the matrix $\frac{1}{2}i\Omega(0) = [\frac{1}{2}iR_{jk}(0)]$, $i = \sqrt{-1}$, of elements from this nilpotent abelian subalgebra. Thus the heat kernel, say Γ, for (∗∗) is just the heat kernel Γ_H for (5.4) with C replaced by $\frac{1}{2}i\Omega(0)$; in particular, the only function of x in $\Gamma(x, 0; t)$ is

$$\exp\left[-\frac{1}{4t}\Big(L\big(i\Omega(0)t\big)x, x\Big)\right] .$$

Now $x \to (L(i\Omega(0)t)x, x)$ is fixed by any rotation of x, i.e., is an $O(n)$-invariant. Consequently,

$$x_j \frac{\partial}{\partial x_k} - x_k \frac{\partial}{\partial x_j} : \Big(L\big(i\Omega(0)t\big)x, x\Big) \longrightarrow 0 \qquad (t \geq 0) ,$$

in which case
$$\Gamma_S(x, 0; t)$$

$$= \frac{1}{(4\pi t)^{n/2}} \det\left(\frac{i\Omega(0)t}{\sinh(i\Omega(0)t)}\right)^{1/2} \exp\left[-\frac{1}{4t}\Big(L\big(i\Omega(0)t\big)x, x\Big)\right]$$

$$= \Gamma(x, 0; t) .$$

since Γ_S is the heat kernel of (∗). Hence the asymptotic expansion

$$\Gamma_S(0, 0; t) = \frac{1}{(4\pi t)^{n/2}}\big(I + a_1 t + \cdots + a_m t^m + O(t^{m+1})\big)$$

at the origin of the heat kernel Γ_S for the negative spinor Laplacian satisfies

$$\lim_{\varepsilon \to 0+}\left(\sum_{k=0}^{[\frac{1}{2}n]} \varepsilon^{2k}\lambda(\varepsilon)a_k t^k\right) = \det\left(\frac{i\Omega(0)t}{\sinh(i\Omega(0)t)}\right)^{1/2} .$$

(5.29) **(B) Hodge Laplacian.** In this case $(\mathfrak{H}, \tau) = (\Lambda^*(\mathbb{C}^n), \sigma)$ and the 2^{2n} reduced products

$$\tilde{e}_\alpha e_\beta = \tilde{e}_{\alpha_1} \cdots \tilde{e}_{\alpha_j} e_{\beta_1} \cdots e_{\beta_k}$$

form a basis for $\mathcal{L}(\mathfrak{H})$ (see chapter 4 (2.8)(ii)). But $\Lambda^*(\mathbb{C}^n)$ is a complex $\mathfrak{A}_{2n}$-module relative to

$$e_j = \mu_j - \mu_j^* \quad , \quad i\tilde{e}_j = i(\mu_j + \mu_j^*) \qquad (1 \leq j \leq n) .$$

Hence $\mathcal{L}(\mathfrak{H})$ is a realization of the complex Clifford algebra $\mathfrak{C}_{2n}$. For (5.22)(i) we shall thus use the decomposition

$$\mathcal{L}(\mathfrak{H}) = \mathcal{A}_0 \oplus \mathcal{A}_1 \oplus \cdots \oplus \mathcal{A}_{2n} ,$$

where $r = 2$ and

$$\mathcal{A}_\ell = \mathrm{span}_{\mathbb{C}}\{\tilde{e}_\alpha e_\beta : |\alpha| + |\beta| = \ell\} = \mathfrak{C}_{2n}^{(\ell)} .$$

It is easily seen that the associated graded algebra $\mathcal{G}r(\mathcal{L}(\mathfrak{S}))$ is isomorphic as an algebra to $\Lambda^*(\mathbb{C}^{2n})$ with exterior product as multiplication.

Near the origin the Hodge Laplacian takes the form

$$dd^* + d^*d = \sum_i \Big(\frac{\partial}{\partial x_i}\Big)^2 - \tfrac{1}{8}\sum_{i,j,k,\ell} R_{ijk\ell}(0)e_i e_j \tilde{e}_k \tilde{e}_\ell R_\sigma \ .$$

But $\lambda(\varepsilon) = (1/\varepsilon^k)^{1/2}$ on $\mathcal{A}_k$; consequently, by (5.26),

$$\lim_{\varepsilon\to 0+} \varepsilon^2 M_\varepsilon (dd^* + d^*d)M_\varepsilon^{-1} = \sum_i \Big(\frac{\partial}{\partial x_i}\Big)^2 - \tfrac{1}{2}\sum_{i<j} e_i e_j R_{ij}(0)$$

with e_i, e_j and $R_{ij}(0)$ multiplying as elements of $\mathcal{G}r(\mathbb{C}_{2n})$, i.e., as wedge products. To calculate the heat kernel of this last operator we use the same strategy as in the previous example, first finding the heat kernel for a scalar-valued operator and then passing to the $\mathcal{G}r(\mathbb{C}_{2n})$-valued case. But the scalar operator is just

$$\sum_i \Big(\frac{\partial}{\partial x_i}\Big)^2 - \lambda = \Delta_n - \lambda \qquad (\lambda \in \mathbb{C})\ ,$$

which has heat kernel

$$\frac{1}{(4\pi t)^{n/2}}\exp\Big[-\frac{1}{4t}|x-y|^2 - \lambda t\Big] = e^{-\lambda t} W_t(x-y)\ .$$

Now consider the operator

$$\Delta_n - \sum_{i<j} a_{ij} e_i e_j = \Delta_n - X\ ,$$

for some skew-symmetric matrix $[a_{ij}]$ of scalars, so that each $tX = \sum_{i<j} t a_{ij} e_i e_j$, $t > 0$, can be regarded as an element of spin(n). With respect to multiplication in $\mathcal{G}r(\mathbb{C}_n)$, however,

$$e^{-tX} = Pf(-tX) = I - tX + \frac{t^2}{2!}X \wedge X - \cdots$$

where $Pf(X)$ is Pfaffian of X defined earlier in section 8 of chapter 1 (see chapter 1, (8.24)). Thus the heat kernel for $\Delta_n - X$ is $Pf(tX)W_t(x-y)$. Hence as in example (**A**) we deduce finally that

$$\Gamma_\sigma(x, y; t) = Pf\big(-2t\Omega(0)\big)W_t(x-y)\ .$$

Consequently, the asymptotic estimate

$$\Gamma_\sigma(0, 0; t) = \frac{1}{(4\pi t)^{n/2}}\big(I + a_1 t + \cdots + a_m t^m + O(t^{m+1})\big)$$

at the origin for the heat hernel Γ_σ of the Hodge Laplacian satisfies

$$\lim_{\varepsilon\to 0+}\Big(\sum_{k=0}^{[\frac{1}{2}n]} \varepsilon^{2k}\lambda(\varepsilon)a_k\lambda(\varepsilon)^{-1}t^k\Big) = Pf\big(-2t\Omega(0)\big)\ .$$

Similar calculations can be done also for twisted versions of these operators.

The proof of (5.27) is still lacking, of course. Actually, Getzler supplies two proofs of this result. The first uses estimates obtained by probabilistic arguments. A direct analytic proof can be given, however, based on the ideas used in the proof of (5.9)(i), should one wish to avoid a probabilistic proof. The second proof exploits the fact that the functions $\phi_0, \phi_1, \ldots, \phi_m$ in lemma (5.18) have 'computable' expressions in terms of the coefficients of Δ (see Notes and remarks for references).

6 The index theorem for Dirac operators

Let E and F be Hilbert bundles over a compact oriented Riemannian manifold M (without boundary) and let ∂ be an elliptic differential operator from sections of E to sections of F; ∂ could also be an elliptic pseudo-differential operator. Then, as is well-known, the kernel and co-kernel of ∂ are finite-dimensional and both consist of smooth sections. Thus the index,

$$(6.1) \qquad \text{index } \partial = \dim(\ker \partial) - \dim(\text{co-ker } \partial) \, ,$$

of ∂ is well-defined. For classical geometric differential operators this index can be expressed in terms of topological invariants of M; more generally, the Atiyah-Singer index theorem describes the index of any elliptic operator ∂ in topological terms. Our purpose in this section is to state and prove the index theorem, but only for graded Dirac operators. The proof is almost entirely analytic, using only a minimal number of results from differential geometry. From these special cases, however, the general index theorem then follows by K-theoretic arguments.

Let $\partial : \mathcal{C}^\infty(M, E) \to \mathcal{C}^\infty(M, F)$ be for the moment an arbitrary elliptic operator of order 1, say. Our first task will be to obtain an analytic expression for the index of ∂. This is easily accomplished with standard results from the theory of elliptic operators. Let $\partial^* : \mathcal{C}^\infty(M, F) \to \mathcal{C}^\infty(M, E)$ be the adjoint of ∂. Then $\partial^*\partial$ and $\partial\partial^*$ are non-negative self-adjoint operators on $\mathcal{C}^\infty(M, E)$ and $\mathcal{C}^\infty(M, F)$ respectively, each having discrete spectrum; the respective eigenspaces

$$\mathcal{E}_\lambda = \left\{ \phi \in \mathcal{C}^\infty(M, E) : (\partial^*\partial)\phi = \lambda\phi \right\} \, ,$$

and

$$\mathcal{F}_\mu = \left\{ \psi \in \mathcal{C}^\infty(M, F) : (\partial\partial^*)\psi = \mu\psi \right\}$$

are finite-dimensional. But

$$(\partial\partial^*)\partial\phi = \lambda\partial\phi \quad (\phi \in \mathcal{E}_\lambda) \,, \qquad (\partial^*\partial)\partial^*\psi = \mu\partial^*\psi \quad (\psi \in \mathcal{F}_\mu) \,.$$

Consequently, $\partial\phi$ will be a non-trivial eigenfunction of $\partial\partial^*$ having eigenvalue λ, provided $\lambda \neq 0$; similarly, $\partial^*\psi$ will be a non-trivial eigenfunction of $\partial^*\partial$ with eigenvalue μ whenever $\mu \neq 0$. Hence $\partial^*\partial$ and $\partial\partial^*$ have the same non-zero eigenvalues, and each has the same multiplicity. On the other hand, since

$$\int_M |\partial\phi(\xi)|^2 d\,\mathrm{vol} = \int_M (\partial^*\partial\phi(\xi), \phi(\xi)) d\,\mathrm{vol} = 0$$

when $(\partial^*\partial)\phi = 0$, it is clear that the 0-eigenspace of $\partial^*\partial$ coincides with the kernel of ∂; similarly, the 0-eigenspace of $\partial\partial^*$ coincides with the kernel of ∂^*. In particular, therefore,

$$(6.2) \qquad \mathrm{index}\,\partial = \dim(\mathcal{E}_0) - \dim(\mathcal{F}_0) \,.$$

Now let $e^{t\partial^*\partial}$ and $e^{t\partial\partial^*}$ be the semigroups of operators on $L^2(M, E)$ and $L^2(M, F)$ having respective generators $\partial^*\partial$ and $\partial\partial^*$. By the functional calculus of elliptic operators,

$$(e^{t\partial^*\partial}f)(\xi) = \int_M K_1(\xi, \zeta; t)f(\zeta) \, d\,\mathrm{vol}(\zeta)$$

is a trace-class integral operator having smooth kernel K_1; furthermore

$$\mathrm{trace}\,(e^{t\partial^*\partial}) = \int_M \mathrm{tr}\big(K_1(\xi, \xi; t)\big) d\,\mathrm{vol}(\xi) = \sum_j e^{-t\lambda_j} \qquad (t > 0)$$

with the sum taken over all eigenvalues $\lambda_1, \lambda_2, \ldots$ of $\partial^*\partial$, repeating each eigenvalue according to its multiplicity. In fact,

$$K_1(\xi, \zeta; t) = \sum_j e^{-t\lambda_j} \phi_j(\xi)\overline{\phi_j(\zeta)}$$

where $\{\phi_j\}$ is an orthonormal basis of $L^2(M, E)$ of eigenfunctions of $\partial^*\partial$. Similarly,

$$\mathrm{tr}(e^{t\partial\partial^*}) = \int_M \mathrm{tr}\big(K_2(\xi, \xi; t)\big) d\,\mathrm{vol}(\xi) = \sum_j e^{-t\mu_j} \qquad (t > 0)$$

with

$$K_2(\xi, \zeta; t) = \sum_j e^{-t\mu_j} \psi_j(\xi)\overline{\psi_j(\zeta)} \,.$$

But the non-zero eigenvalues coincide and have the same multiplicity. Hence

$$(6.3)\ \mathrm{index}\,\partial = \mathrm{tr}(e^{t\partial^*\partial}) - \mathrm{tr}(e^{t\partial\partial^*})$$

$$= \int_M \Big(\mathrm{tr}\big(K_1(\xi, \xi; t)\big) - \mathrm{tr}\big(K_2(\xi, \xi; t)\big)\Big) d\,\mathrm{vol}(\xi) \qquad (t > 0) \,,$$

expressing the index of ∂ as the integral of the difference of the traces of the heat kernels of the second-order operators $\partial^*\partial$ and $\partial\partial^*$. Already this expression looks as though a link has been forged with ideas developed in the previous section (the difference between the manifold setting and the Euclidean is not significant). But the connection with the previous section is still illusory because the integrand in (6.3) is a difference of terms which individually may be very complicated although their difference need not be. To illustrate the problem and find its solution, let us consider the Gauss–Bonnet theorem.

In extending the classical Gauss–Bonnet formula, Chern showed that the Euler characteristic $\chi(M)$ of M, a topological invariant, could be expressed as

$$(6.4) \qquad \chi(M) = \int_M C(\xi)\, d\operatorname{vol}(\xi)$$

where $C = C(\xi)$ is a homogeneous polynomial of degree m, $\dim M = 2m$, in the components of the curvature tensor of M. Now let $\partial = d - d^*$ be the Hodge–deRham operator on the space $\Omega(M)$ of smooth forms on M. On $\Omega(M)$, ∂ is self-adjoint, and $(-\partial^2)$ is just the Hodge Laplacian $dd^* + d^*d$. But as an operator $\partial : \Omega_+(M) \to \Omega_-(M)$ mapping even forms to odd forms, its adjoint ∂^* maps $\Omega_-(M)$ to $\Omega_+(M)$; consequently, $(-\partial^*\partial)$ is the restriction of the Hodge Laplacian to $\Omega_+(M)$, while $(-\partial\partial^*)$ is its restriction to $\Omega_-(M)$. As McKean and Singer then proved,

$$\text{index } (d - d^*) = \chi(M) = \int_M \operatorname{tr}\big(K_+(\xi,\xi;t)\big) - \operatorname{tr}\big(K_-(\xi,\xi;t)\big)\, d\operatorname{vol}(\xi)$$

for all $t > 0$, where $K_+ = K_+(x,y;t)$ is the heat kernel for $-\partial^*\partial :$ $\Omega_+(M) \to \Omega_+(M)$ and $K_- = K_-(x,y;t)$ is the corresponding kernel for the adjoint operator. Presumably then some 'fantastic cancellation' must occur in this last integrand so that

$$\text{index } (d - d^*) = \int_M C(\xi)\, d\operatorname{vol}(\xi) \ .$$

Patodi showed that such cancellation did occur, using a direct but complicated argument; later, Gilkey considerably clarified the argument using invariance theory. However, in the earlier parts of this book, we have met all the ideas which have arisen in this analytic approach to the Gauss–Bonnet theorem. Indeed, $\Omega_+(M)$ and $\Omega_-(M)$ are just the ± 1-eigenspaces of the Clifford involution η on $\Omega(M)$ determined by the involution on $\Lambda^*(\mathbf{R}^n)$ which is $(-1)^k$ on $\Lambda^k(\mathbf{R}^n)$, while $d - d^*$ is the graded Dirac operator associated with this involution. But if $K = K(x,y;t)$ is

the heat kernel for the Hodge Laplacian as an operator on $\Omega(M)$, then
$$K_+ = \tfrac{1}{2}(I + \eta)K \quad , \quad K_- = \tfrac{1}{2}(I - \eta)K \; ;$$
as a result, the *supertrace* $\mathrm{tr}_s\, K$ of K satisfies
$$\mathrm{tr}_s\big(K(\xi,\xi;t)\big) = \mathrm{tr}\big(\eta K(\xi,\xi;t)\big) = \mathrm{tr}\big(K_+(\xi,\xi;t)\big) - \mathrm{tr}\big(K_-(\xi,\xi;t)\big) \, ,$$
(see chapter 4, (2.11)). Formally, therefore,
$$\mathrm{index}\,(d - d^*) = \lim_{t\to 0+} \int_M \mathrm{tr}_s\big(K(\xi,\xi;t)\big)\, d\mathrm{vol}(\xi)$$
$$= \int_M \lim_{t\to 0+} \mathrm{tr}_s\big(K(\xi,\xi;t)\big)\, d\,\mathrm{vol}(\xi)$$
which reduces to (6.4) provided
$$(6.5) \qquad \lim_{t\to 0+} \mathrm{tr}_s\big(K(\xi,\xi;t)\big) = C(\xi) \qquad (\xi \in M) \, .$$

Thus, by combining the difference of the traces into a single supertrace
of the heat kernel associated with a graded Dirac operator, we have established the true link with the asymptotic estimates developed in the
previous section because limits such as (6.5) are exactly what was calculated there (with $\mathbb{R}^n$ instead of M). But (6.5) is a *local* result depending
on the behavior of K in a neighborhood of x, and, as we shall see, *local* asymptotic estimates for heat kernels on M are essentially the same
as *local* asymptotic estimates on $\mathbb{R}^n$. Hence the heat kernel approach
to the index theorem has been revealed. From now on therefore, M
will be a compact oriented n-dimensional Riemannian manifold without
boundary, $n = 2m$, having a Spin structure if necessary.

For the moment, let $\mathfrak{H}$ be an arbitrary $\mathfrak{A}_n$-module on which there is a
unitary representation τ of $\mathrm{Spin}(n)$ and an involution η compatible with
its $\mathfrak{A}_n$-module structure such that
$$\tau(a)e_j\tau(a)^{-1} = \sigma(a)e_j \quad , \quad \eta\big(\tau(a)u\big) = \tau(a)\eta(u)$$
for all a in $\mathrm{Spin}(n)$ and in $\mathfrak{H}$. Then
$$\eta : [p, u] \longrightarrow [p, \eta(u)]$$
extends η to an involution on $E_{\mathfrak{H}}$ which will now be compatible with its
$\mathfrak{A}(M)$-module structure in the sense that
$$\eta\big([p, e_j][p, u]\big) = [p, e_j]\eta\big([p, u]\big) \, .$$
Let $E_{\mathfrak{H}}^+$ and $E_{\mathfrak{H}}^-$ be the respective ± 1-eigenspaces of η; in fact, $E_{\mathfrak{H}}^+ = E_{\mathfrak{H}_+}$, while $E_{\mathfrak{H}}^- = E_{\mathfrak{H}_-}$. Thus we obtain a graded Dirac operator
$$(6.6) \qquad\qquad D_+ : C^\infty(M, E_{\mathfrak{H}}^+) \longrightarrow C^\infty(M, E_{\mathfrak{H}}^-)$$
and its counterpart D_- by restricting the Dirac operator D_M on
$C^\infty(M, E_{\mathfrak{H}})$ to sections of these eigenspaces. This operator (6.6) is the

one whose index we shall attempt to compute. But, since D_- is the adjoint of D_+,

$$\text{index } D_+ = \dim(\ker D_+) - \dim(\ker D_-)$$

and the proof just outlined for the Gauss–Bonnet theorem now goes over intact to this more general case. Let $K(\xi,\zeta;t)$ be the heat kernel for $(-D_M^2)$ as an operator on $C^\infty(M,E_{\mathfrak{H}})$. Then the respective heat kernels $K_+ = K_+(\xi,\zeta;t)$ and $K_- = K_-(\xi,\zeta;t)$ for the restriction of $(-D_M^2)$ to $C^\infty(M,E_{\mathfrak{H}}^+)$ and to $C^\infty(M,E_{\mathfrak{H}}^-)$ are given by

$$K_+ = \tfrac{1}{2}(I+\eta)K \quad,\quad K_- = \tfrac{1}{2}(I-\eta)K .$$

Consequently,

$$\text{tr}_s\big(K(\xi,\xi;t)\big) = \text{tr}\big(\eta K(\xi,\xi;t)\big) = \text{tr}\big(K_+(\xi,\xi;t) - K_-(\xi;\xi;t)\big)$$

and

$$(6.7) \qquad \text{index } D_+ = \int_M \text{tr}_s\big(K(\xi,\xi;t)\big) d\,\text{vol}\,(\xi) \qquad (t>0) .$$

But if $K(\xi,\xi;t)$ is known to have an asymptotic expansion

$$(6.8) \quad K(\xi,\xi;t) = \frac{1}{(4\pi t)^{n/2}}\big(I + a_1(\xi)t + \cdots + a_m(\xi)t^m + O(t^{m+1})\big)$$

at each fixed ξ in M, then it follows immediately that

$$(6.9) \quad \text{index } D_+ = \frac{1}{(4\pi t)^{n/2}} \int_M \text{tr}_s\big(a_m(\xi)\big) d\,\text{vol}\,(\xi) \qquad (n=2m) .$$

In many important examples, however, the involution η on $\mathfrak{H}$ and the properties of the decomposition (5.22) are so intimately related that

$$(6.10) \qquad \text{tr}_s\big(a_m(\xi)\big) = \lim_{\varepsilon\to 0+} \varepsilon^n \lambda(\varepsilon)a_m(\xi)\lambda(\varepsilon)^{-1} .$$

Hence in this case,

$$(6.11) \quad \text{index } D_+ = \frac{1}{(4\pi t)^{n/2}} \int_M \big(\lim_{\varepsilon\to 0+} \varepsilon^n \lambda(\varepsilon)a_m(\xi)\lambda(\varepsilon)^{-1}\big) d\,\text{vol}\,(x) ,$$

and in particular examples this integrand can be evaluated using the form that (5.28) has in these examples. These will establish the Atiyah–Singer index theorem.

Let us now be more specific, establishing (6.8) first. Since we are interested in the asymptotic behavior of $K(\xi_0,\xi_0;t)$ as $t\to 0+$ for *fixed* ξ_0 in M, choose a normal coordinate system for a neighborhood U_0 of ξ_0 corresponding to an open set U in $\mathbb{R}^n$, with the fixed ξ_0 corresponding to the origin in $\mathbb{R}^n$; and then extend the Riemannian metric $g = [g_{ik}(x)]$ on U coming from that on U_0 to an asymptotically Euclidean Riemannian metric g on all of $\mathbb{R}^n$. The elliptic operator $-D_{\mathbb{R}^n}^2$ on $C^\infty(\mathbb{R}^n,\mathfrak{H})$ determined by this metric coincides on U with $-D_M^2$, and has heat kernel

$\Gamma(x, y; t)$. By (5.9)(ii),

$$(6.12) \qquad \Gamma(0, 0; t) = \frac{1}{(4\pi t)^{n/2}} \left(I + a_1 t + \cdots + a_m t^m + O(t^{m+1}) \right).$$

The asymptotic expansion (6.8) will follow from this.

(6.13) Theorem.
   The heat kernel K for $(-D_M^2)$ satisfies

$$K(\xi_0, \xi_0; t) = \frac{1}{(4\pi t)^{n/2}} \left(I + a_1 t + \cdots + a_m t^m + O(t^{m+1}) \right)$$

where $a_1, \ldots, a_m$ are defined by (6.12) for $(-D_{\mathbb{R}^n}^2)$.

Proof. Without loss of generality we can assume $U = \{x : |x| < 2R\}$ for some R. Then, by (5.9)(i),

$$\|\Gamma(x, 0; t)\| \leq \frac{\text{const.}}{(4\pi t)^{n/2}} \exp\left(-\frac{\delta R}{t} \right) \qquad (|x| = R)$$

for some $\delta > 0$. Now let ξ_x be the point in U_0 corresponding to x in U. There is a corresponding estimate

$$\|K(\xi_x, \xi_0; t)\| \leq \frac{\text{const.}}{(4\pi t)^{n/2}} \exp\left(-\frac{\beta R}{t} \right) \qquad (|x| = R)$$

for some possibly different $\beta > 0$. Hence there exists $\alpha > 0$ such that

$$\|K(\xi_x, \xi_0; t) - \Gamma(x, 0; t)\| \leq \frac{\text{const.}}{(4\pi t)^{n/2}} \exp\left(-\frac{\alpha R}{t} \right) \qquad (|x| = R).$$

Set

$$F(x, t) = K(\xi_x, \xi_0; t) - \Gamma(x, 0; t) \qquad \left(|x| < \tfrac{3}{4} R \right).$$

This function $F = F(x, t)$ is a solution of the heat equation $(\frac{\partial}{\partial t} + D_{\mathbb{R}^n}^2)F = 0$ for which $F(x, 0) \equiv 0$ and

$$(6.14) \qquad \|F(x, t)\| \leq \frac{\text{const.}}{(4\pi t)^{n/2}} \exp\left(-\frac{\alpha R}{t} \right) \qquad (|x| = R).$$

If the maximum principle were available to us, this inequality would persist for $|x| < R$; unfortunately, the maximum principle does not apply. Nonetheless, general results for Green's function on a compact manifold with boundary ensure that (6.14) still holds inside the sphere $|x| = R$. Hence

$$K(\xi_0, \xi_0; t) = \Gamma(0, 0; t) + O\left(t^{-m} \exp\left(-\frac{\alpha R}{t} \right) \right),$$

from which the theorem follows immediately. ∎

   After all the preliminaries stretching back over several chapters we come to one of the principal results of this book. Consider the Dirac operator on the space $\mathcal{C}^\infty(M, E_{\mathfrak{S}_n})$ of sections of the spinor bundle $E_{\mathfrak{S}_n}$, taking as involution

$$(6.15) \qquad \eta(u) = i^m(e_1 \ldots e_{2m})u = \left(\prod_{j=1}^m (\mu_j^* \mu_j - \mu_j \mu_j^*)\right)u$$

on $\mathfrak{S}_n \cong \Lambda^*(\mathbb{C}^m)$ (see chapter 4,(2.8)(iii)). Its ± 1-eigenspaces are $\Lambda_\pm^*(\mathbb{C}^m)$, the irreducible Spin($n$)-modules having signature $s_\pm$. Now let

$$(6.15)(i) \qquad \mathfrak{C}_n = \mathfrak{C}_n^{(0)} \oplus \mathfrak{C}_n^{(1)} \oplus \cdots \oplus \mathfrak{C}_n^{(n)}$$

be the decomposition of the complex Clifford algebra $\mathfrak{C}_n = \mathcal{L}(\Lambda^*(\mathbb{C}^m))$, $n = 2m$, used in (5.29)(**A**), and let $\{R_{ikpq}(\xi)\}$ be the Riemannian curvature tensor of M at each ξ. Then with obvious notation,

$$R_{jk}(\xi) = \tfrac{1}{2} \sum_{p<q} R_{jkpq}(\xi) e_p e_q$$

defines a spin(n)-valued section of the Clifford algebra bundle $\mathfrak{A}_n(M)$; consequently,

$$i\Omega(\xi) = \big[i R_{jk}(\xi)\big]_{j,k=1}^n \qquad (i = \sqrt{-1}\,)$$

defines a symmetric $(n \times n)$-matrix of sections of the complex Clifford algebra bundle $\mathfrak{A}_n(M)$. To exploit (5.29)(**A**) we define

$$(6.16) \qquad \xi \longrightarrow \det\left(\frac{i\Omega(\xi)t}{\sinh i\Omega(\xi)t}\right)^{1/2},$$

using the wedge product in $\mathfrak{C}_n(M)$ *as multiplication* in (5.5)(i). Hence (6.16) is simply a *form* on Ω which has arisen from the asymptotic expansion at ξ of the heat kernel associated with the 'principal part'

$$-\sum_j \left(\frac{\partial}{\partial x_j} - \tfrac{1}{2}\sum_k x_k R_{jk}(\xi)\right)^2$$

of the spinor Laplacian near ξ', so to speak; let

$$\left(\det\left(\frac{i\Omega(\xi)t}{\sinh i\Omega(\xi)t}\right)^{1/2}\right)_n$$

be the top degree of this form, a scalar.

(6.17) Theorem.

(Index theorem for spin manifolds) *Let M be a compact n-dimensional oriented Riemannian spin manifold without boundary, $n = 2m$, and let*

$$D_+ : \mathcal{C}^\infty(M, E_{\Lambda_+^*}) \longrightarrow \mathcal{C}^\infty(M, E_{\Lambda_-^*})$$

be the graded Dirac operator on sections of the spinor bundle. Then

$$\text{index } D_+ = \frac{1}{(2\pi i)^m} \int_M \left(\det \left(\frac{i\Omega(\xi)}{\sinh i\Omega(\xi)} \right)^{1/2} \right)_n d\text{vol}\,(\xi) \,.$$

Proof. Fix ξ in M temporarily, and take normal coordinates about ξ with ξ corresponding to the origin in $\mathbf{R}^n$. Then by (6.9) we have to evalute $\text{tr}_s(a_m(\xi))$ where $a_m(\xi)$ is the coefficient of t^m in the asymptotic expansion (6.12) and the supertrace is defined in terms of the involution (6.15). On the other hand, by (5.29)(**A**),

$$\det \left(\frac{i\Omega(\xi)t}{\sinh i\Omega(\xi)t} \right)^{1/2} = \lim_{\varepsilon \to 0+} \left(\sum_{k=0}^m \varepsilon^{2k} \lambda(\varepsilon) a_k(\xi) \lambda(\varepsilon)^{-1} t^k \right) \,.$$

Let us take for granted for the moment the following lemma.

(6.18) Lemma.
   The supertrace $\text{tr}_s(A) = \text{tr}(\eta A)$ with respect to the involution (6.15) of an element

$$A = A_0 + A_1 + \cdots + A_n \qquad (A_k \in \mathbf{C}_n^{(k)})$$

in $\mathbf{C}_n$ is given by

$$(-2i)^m A_n = \text{tr}_s(A) e_1 \ldots e_n \,;$$

intuitively, therefore, $\text{tr}_s(A) = (-2i)^m$ top degree (A).

By (6.18) together with (5.26) therefore,

$$\text{tr}_s \left(\lim_{\varepsilon \to 0} \sum_{k=0}^m \varepsilon^{2k} \lambda(\varepsilon) a_k(\xi) \lambda(\varepsilon)^{-1} t^k \right) = \text{tr}_s \big(a_m(\xi) \big) t^m \,.$$

On the other hand, by (6.18) again

$$\text{tr}_s \left(\det \left(\frac{i\Omega(\xi)t}{\sinh i\Omega(\xi)t} \right)^{1/2} \right)$$

$$= (-2it)^m \text{ top degree} \left(\det \left(\frac{i\Omega(\xi)}{\sinh i\Omega(\xi)t} \right)^{1/2} \right) \,.$$

Hence

$$\text{tr}_s \big(a_m(\xi) \big) = (-2it)^m \left(\det \left(\frac{i\Omega(\xi)}{\sinh i\Omega(\xi)} \right)^{1/2} \right)_n \,,$$

from which the index theorem follows immediately using (6.9) and freeing ξ. ∎

Proof of (6.18). If $A = A_n$, say $A = \alpha e_1 \cdots e_n$, $\alpha \in \mathbf{C}$, then

$$\eta A = \alpha i^n (e_1 \ldots e_n)^2 = \lambda i^{3m} = (-i)^m \alpha \,.$$

In this case,

$$\operatorname{tr}_s(A) = (-i)^m \alpha \dim \mathfrak{S}_n = (-2i)^m \alpha .$$

Now suppose $A = e_\alpha = e_{i_1} \cdots e_{i_k}$, $k < n$. Then

$$\eta A = i^m (e_1 \cdots e_n)(e_{i_1} \cdots e_{i_k}) = \text{const. } e_\beta$$

where $\alpha \cup \beta = \{1, \ldots, n\}$. Thus we have to show that $\operatorname{tr}(e_\beta) = 0$. But $e_\beta : \Lambda_\pm^* \to \Lambda_\mp^*$ when $|\beta|$ is odd, so certainly $\operatorname{tr}(e_\beta) = 0$ in this case. But, if $|\beta|$ is even and $e_\beta = e_{i_1} \cdots e_{i_r}$, then

$$e_\beta = e_{i_1} \cdots e_{i_r} = (-1)^{|\beta|-1} e_{i_2} \cdots e_{i_r} e_{i_1} = -e_{i_2} \cdots e_{i_r} e_{i_1} .$$

(see chapter 1,(3.4)). Since it is always true that

$$\operatorname{tr}\big(e_{i_1}(e_{i_2} \cdots e_{i_r})\big) = \operatorname{tr}\big((e_{i_2} \cdots e_{i_r})e_{i_1}\big) ,$$

we deduce that $\operatorname{tr}(e_\beta) = 0$ unless $e_\beta = 1$.   ∎

The Gauss–Bonnet theorem can be proved in exactly the same way replacing (5.29)(**A**) by (5.29)(**B**). The twisted case of the Dirac operator on spinor-valued sections has also been proved by Getzler using these same ideas, and Roe shows how to deduce the Hirzebruch signature theorem as well. These special cases together with K-theory then establish the index theorem in complete generality, but we spare the reader (and the authors) the details.

Notes and remarks for chapter 5

1. There are many treatments of differential geometry. The books by Kobayashi–Nomizu ([68]) and by Poor ([87]) are standard references, but Boothby's book ([14]) is very readable and closer to the spirit of this chapter. Roe ([90]) gives a concise introduction to all the ideas needed; he includes also a computation of the Bochner–Weitzenböck formula, as does Poor. For normal coordinates one can consult [68] or [35].

2. Again much of this consists of standard material from differential geometry for which [68] is a basic reference. Our treatment was heavily influenced by the book of Freed and Uhlenbeck ([37]) as well as by papers of Bourguignon ([15], [16]). A detailed explicit discussion of Spin structures is given in [27]. The work of Bourguignon emphasizes how strongly the algebraic properties of the Spin(n)-equivariant symbol mapping $A : \mathcal{A} \otimes \mathbb{C}^n \to \mathcal{B}$ determine the properties of $\eth_A$ on any manifold on which it can be defined: the so-called 'natural' property of differential operators (see also [97]). This idea completely

unifies the Euclidean and non-Euclidean, but lack of time and space prevented detailed discussion.

3. The idea of using Clifford algebras to extend to $\mathbb{R}^n$, $n > 2$, the classical fractional linear transformations on $\mathbb{C} \sim \mathbb{R}^2$ goes back to Vahlen and Maas, but it has been brought to the fore more recently by Ahlfors ([2], [3]), and independently by Takahashi ([98]).

4. The original work of Bargmann, Kunze and Stein is in [9] and in [73]. A detailed discussion of the general theory is contained in Knapp's book ([66]) and, more particularly, of intertwining operators, in further papers of Kunze–Stein and Knapp–Stein. The realization of discrete series representations in the kernel of a Dirac operator goes back to Parthasarathy ([85]) and of course to Bargmann in the case of $SL(2, \mathbb{R})$. Use of the Cauchy–Szegő transform as an equivariant mapping between representations is due to Knapp–Wallach ([67]), and Helgason has developed extensively the theory of Poisson transforms (see, for instance, [55]). But the idea that many more of the exceptional irreducible unitary representations of a semi-simple Lie group should be contained in the kernel of elliptic operators such as the $\eth_r$ is developed in the series of papers ([30], [31], [47], [48]), and is still being developed (see also the work of Slebarski [92], [93]). The results presented in this section do not even begin to scratch the surface of the body of known results.

5. Asymptotic estimates for heat kernels are established in Friedman's book ([38]) and in many other papers (see, for instance, [5], [51],...). Use of the rescaling operator $\lambda(\varepsilon)$ has been common for some time, but the treatment here is based on the work of Getzler ([43], [44], [45]); an unpublished paper of Parker was very helpful, too ([84]). A proof of the results needed to establish (5.16) and (5.18) can be found in [38] and [86] (see also Roe's book for a straightforward proof).

6. The Atiyah–Singer index theorem was announced in [8] and proofs were given in subsequent papers. The book of Booss and Bleecker gives a discursive development of index theorems and all related material ([13]); it is highly recommended for browsing. The analytic approach began with the papers of McKean–Singer and Patodi ([81], [86]). Invariant-theoretic ideas and much additional material are discussed in Gilkey's book ([49]; see also [6]). A probabilistic approach has been developed by Bismut ([12]), and a more equivariant approach via principal bundles has been given by Berline and Vergne ([11]). The use of super-symmetric ideas goes back to Witten ([104]), whose ideas were then taken up by Alvarez-Gaume ([4]), as well as by Get-

zler (*loc. cit.*). In all these more recent analytic proofs, one of the main goals has been to give some clear analytic reason for the appearance of the integrand in the statement of the index theorem. Finally, Seeley's paper ([91]) gives a discussion of the relation of the index theorem to classical elliptic boundary value problems; this paper should be very appealing to readers who have learned the ideas in chapter 2 before those in chapter 5. Estimates necessary for Green's functions on compact manifolds with boundary are established by Arima and Greiner ([5], [51]). The relation between the index theorem for the Dirac operator and discrete series representations has been extensively discussed by Atiyah and Schmid ([7]).

References

[1] J.F. ADAMS, P.D. LAX and R.S. PHILIPS, On matrices whose real linear combinations are non-singular, *Proc. Amer. Math. Soc.*, **16** (1965), 318–22.

[2] L.V. AHLFORS, Möbius transformations in ℝ expressed through 2 × 2 matrices of Clifford Numbers, *Complex Variables*, **5** (1986), 215–24.

[3] L.V. AHLFORS, Möbius transformations and Clifford Numbers, *Differential Geometry and Complex Analysis*, Springer-Verlag, Berlin–Heidelberg, 1985.

[4] L. ALVAREZ-GAUME, Supersymmetry and the Atiyah–Singer index theorem, *Comm. Math. Phys.*, **90** (1983), 161–73.

[5] R. ARIMA, On general boundary value problem for parabolic equations, *J. Math. Kyoto Univ.*, **4** (1964), 207–43.

[6] M. ATIYAH, R. BOTT and V. PATODI, On the heat equation and the index theorem, *Inventiones Math.*, **19** (1973), 279–330.

[7] M. ATIYAH and W. SCHMID, A geometric construction of the discrete series for semi-simple Lie groups, *Inventiones*, **42** (1977), 1–62.

[8] M. ATIYAH and I.M. SINGER, The index of elliptic operators on compact manifolds, *Bull. Amer. Math. Soc.*, **69** (1963), 422–33.

[9] V. BARGMANN, Irreducible unitary representations of the Lorentz group, *Ann. of Math.*, **48** (1947), 568–640.

[10] A.F. BEARDON, *The Geometry of Discrete Groups*, Springer-Verlag, New York–Heidelberg–Berlin, 1983.

[11] N. BERLINE and M. VERGNE, A computation of the equivariant index of the Dirac operator, *Bull. Soc. Math. France*, **113** (1985), 305–45.

[12] J.M. BISMUT, The Atiyah-Singer theorems for classical elliptic operators: a probabilistic approach, *J. Funct. Anal.*, **57** (1984), 56–99.

[13] B. BOOSS and D.D. BLEECKER, *Topology and Analysis, The Atiyah-Singer Index Formula and Gauge-Theoretic Physics*, Universitext, Springer-Verlag, New York–Berlin–Heidelberg–Tokyo, 1985.

[14] W.M. BOOTHBY, *An introduction to differential manifolds and Riemannian geometry*, Academic Press, 1975.

[15] J.-P. BOURGUIGNON, Opérateurs différentiels et théorèmes d'annulation, *Exposé VIII, Séminaire E.N.S. (1977-78)*, 135–46.

[16] J.-P. BOURGUIGNON, Formulas de Weitzenböck en Dimension 4, *Géom. Riemanniene de dimension 4*, CEDIC, Paris, 1981.

[17] F. BRACKX, R. DELANGHE and F. SOMMEN, *Clifford Analysis*, Pitman, London, 1983.

[18] R. BRAUER and H. WEYL, Spinors in n Dimensions, *Amer. J. of Math.*, **57** (1935), 425–49.

[19] T. BROCKER and T. TOM DIECK, *Representations of Compact Lie Groups*, Springer-Verlag, New York, 1985.

[20] A.P. CALDERÓN, Cauchy integrals on Lipschitz curves and related operators, *Proc. Nat. Acad. Sci. USA*, **74** (1977), 1324–7.

[21] A.P. CALDERÓN and A. ZYGMUND, On higher gradients of harmonic functions, *Studia Math.*, **24** (1964), 212–26.

[22] C. CHEVALLEY, *The Construction and Study of Certain Important Algebras*, Mathematical Society of Japan, 1955.

[23] W.K. CLIFFORD, On the classification of geometric algebras, in *Mathematical Papers*, (R. Tucker, ed.), Macmillan, London, (1882), pp. 397–401.

[24] R.R. COIFMAN, A. MCINTOSH and Y. MEYER, L'intégrale de Cauchy définit un opérateur borné sur L^2 pour les courbes Lipschitziennes, *Annals of Math.*, **116** (1982), 361–87.

[25] R.R. COIFMAN and G. WEISS, *Analyse harmonique noncommutative sur certains espaces homogènes*, Lecture Notes Math., No.242, Springer-Verlag, Berlin, 1971.

[26] R.R. COIFMAN and G. WEISS, Extensions of Hardy spaces and their use in analysis, *Bull. Amer. Math. Soc.*, **83** (1977), 569–645.

[27] L. DABROWSKI and A. TRAUTMAN, Spinor structures on spheres and projective spaces, *J. Math. Phys.*, **27** (8) (1986), 2022-28.

[28] B.E.J. DAHLBERG, Estimates of harmonic measure, *Arch. Rat. Mech. Anal.*, **65** (1977), 275–88.

[29] K.M. DAVIS, J.E. GILBERT and R.A. KUNZE, Elliptic differential operators in harmonic analysis, I. Generalized Cauchy–Riemann systems. Submitted to *Amer. J. Math.*

[30] K.M. DAVIS, J.E. GILBERT and R.A. KUNZE, Invariant differential operators in harmonic analysis on real hyperbolic space. *Proc. Int. Conf. Harmonic Analysis and Operator Algebras, Aust. Nat. Univ., Canberra, 1987*, 58–78.

[31] K.M. DAVIS, J.E. GILBERT and R.A. KUNZE, Harmonic analysis and exceptional representations of semi-simple Lie groups. *Proc. Int. Conf. Harmonic Analysis and Operator Algebras, Aust. Nat. Univ., Canberra, 1987*, 79–91.

[32] E. D'HOKER, Decompositions of representations into basis representations for the classical groups, *J. Math. Phys.*, **25** (1) (1984), 1–12.

[33] P.A.M. DIRAC, The quantum theory of the electron, I, II, *Proc. Roy. Soc. London*, A**117** (1928), 610–24; A**118** (1928), 351–61.

[34] PETER L. DUREN, *Theory of H^p Spaces*, Academic Press, New York, 1970

[35] D.B.A. EPSTEIN, Natural tensors on Riemannian manifolds, *J. Diff. Geom.*, **10** (1975), 631–45.

[36] C. FEFFERMAN and E.M. STEIN, H^p spaces of several variables, *Acta Math.*, **129** (1972), 137–93.

[37] D.S. FREED and K.K. UHLENBECK, *Instantons and Four-Manifolds*, MSRI Publications No.1, Springer-Verlag, New York, 1984.

[38] A. FRIEDMAN, *Partial Differential Equations of Parabolic Type*, Prentice Hall, Englewood Cliffs, N.J., 1964.

[39] R. FUETER, Die Funktionentheorie der Differentialgleichungen $\Delta u = 0$ and $\Delta\Delta u = 0$ mit vier reel en Variablen, *Comm. Math. Helv.*, **7** (1934), 307–30.

[40] R. FUETER, Zur Theorie der regulären Funktionen einer Quaternionenvariablen, *Monat. für Math. und Phys.*, **43** (1935), 69–74.

[41] J.B. GARNETT, *Bounded Analytic Functions*, Academic Press, New York, 1981.

[42] S. GELBART, A theory of Stiefel harmonics, *Trans. Amer. Math. Soc.*, **192** (1974), 29–50.

[43] E. GETZLER, Pseudo-differential operators on super-manifolds and the Atiyah–Singer index theorem, *Comm. Math. Phys.*, **92** (1983), 163–78.

[44] E. GETZLER, A short proof of the local Atiyah–Singer index theorem, *Topology*, **25** (1986), 111–17.

[45] E. GETZLER, The local Atiyah–Singer index theorem, *Seminar 11, Critical phenomena, random systems, gauge theories* (K. Otserwalder, R. Skora, eds.), Elsevier Science Publishers, B.V., 1986, 967–74.

[46] J.E. GILBERT, Polynomial invariant theory and Taylor series. Submitted for publication.

[47] J.E. GILBERT, R.A. KUNZE, R.J. STANTON and P.A. TOMAS, Higher Gradients and Representations of Lie Groups in *Conference on Harmonic Analysis in Honor of Antoni Zygmund*, Vol.II (W. Beckner, A.P. Calderon, R. Fefferman, P. Jones, eds.) Wadsworth International (1981), 416–26.

[48] J.E. GILBERT, R.A. KUNZE and P.A. TOMAS, Intertwining kernels and invariant differential operators in representation theory, Chap. 7 in *Probability Theory and Harmonic Analysis*, (J.-A. Chao, W. Woyczynski, eds), Marcel Dekker, (1985), 91–112.

[49] P.B. GILKEY, *Invariance Theory, The Heat Equation, and the Atiyah–Singer Index Theorem*, Math. Lect. Series, Vol.11, Publish or Perish Inc., Wilmington, Delaware, 1984.

[50] K.R. GOODEARL, *Real and Complex C*-algebras*, Shiva Math. Series No.5, Birkhäuser, 1982.

[51] P. GREINER, An asymptotic expansion for the Heat equation, *Arch. Rat. Mech. Anal.*, **41** (1971), 163–218.

[52] M. GROMOV and B. LAWSON, Positive scalar curvature and the Dirac operator on complete Riemannian manifolds, *Publ. Math. IHES*, **58** (1983), 295–408.

[53] K.I. GROSS and D. ST. P. RICHARDS, Special functions of matrix argument. I: algebraic induction, zonal polynomials, and hypergeometric functions, *Trans. Amer. Math. Soc.*, **301** (1987), 781–811.

[54] F. REESE HARVEY, *Spinors and Calibrations*, Academic Press, New York, 1989.

[55] S. HELGASON, A duality for symmetric spaces with applications to group representations, *Advances in Math.*, **5** (1970), 1–154.

[56] S. HELGASON, *Differential geometry, Lie groups, and symmetric spaces*, Academic Press, New York, 1978.

[57] S. HELGASON, *Groups and geometric analysis*, Academic Press, New York, 1984.

[58] C.S. HERZ, Bessel functions of matrix argument, *Ann. of Math.*, **61** (1955), 474–523.

[59] R. HOWE, Remarks on classical invariant theory, *Transactions of the AMS*, **313** (1989), 539–70.

[60] J.E. HUMPHREYS, *Introduction to Lie Algebras and Representation Theory*, Springer-Verlag, New York–Heidelberg–Berlin, 1972.

[61] DALE HUSEMOLLER, *Fibre Bundles*, Springer-Verlag, New York, 1966.

[62] A.T. JAMES, Special functions of matrix and single argument in Statistics, *Theory and Applications of Special Functions* (R.A. Askey, ed.), Academic Press, New York, 1975, 497–520.

[63] C.E. KENIG, Weighted H^p spaces on Lipschitz domains, *Amer. J. Math.*, **102** (1980), 129–63.

[64] C.E. KENIG, Elliptic boundary value problems on Lipschitz domains, in *Beijing Lectures in Harmonic Analysis*, Princeton Univ. Press (1986), 131–83.

[65] C.E. KENIG and J. PIPHER, Hardy spaces and the Dirichlet problem on Lipschitz domains, *Rev. Mat. Iberoamericana*, **3** (1987), 191–247.

[66] A.W. KNAPP, *Representation theory of Semi-simple groups*, Princeton University Press, Princeton, N.J., 1987.

[67] A.W. KNAPP and N.R. WALLACH, Szegő kernels associated with Discrete series, *Inventiones*, **34** (1976),163–200.

[68] S. KOBAYASHI and K. NOMIZU, *Foundations of Differential Geometry*, Vol. 1, Interscience Publ., New York, 1963.

[69] T.H. KOORNWINDER, *Two-variable analogues of the Classical Orthogonal Polynomials, Theory and Applications of Special Functions* (R.A. Askey, ed.), Academic Press, New York, 1975, 435–96.

[70] PAUL KOOSIS, *Introduction to H^p Spaces*, Cambridge University Press, New York, 1980.

[71] A. KORÁNYI and S. VÁGI, Group-theoretic remarks on Riesz systems on balls, *Proc. Amer. Math. Soc.*, **85** (1982), 200–5.

[72] B. KOSTANT, Lie group representations on polynomial rings, *Amer. J. Math.*, **85** (1963), 327–404.

[73] R.A. KUNZE and E.M. STEIN Uniformly bounded representations and harmonic analysis of the 2×2 unimodular group, *Amer. J. Math.*, **82** (1960), 1–62.

[74] Ü. KURAN, On Brelot–Choquet axial polynomials, *J. London Math. Soc.*, (2) **4** (1971), 15–26.

[75] R. LIPSCHITZ, Correspondence, from an ultramundane correspondent, *Annals of Math.*, **69** (1959), 247–51.

[76] D.E. LITTLEWOOD, *The theory of group characters and matrix representations of groups*, 2nd edition, Oxford Univ. Press, Oxford 1950.

[77] H. MAAS, Spherical functions and quadratic forms, *J. Indian Math. Soc.*, **20** (1956), 117–62.

[78] I.G. MACDONALD, Commuting differential operators and zonal spherical functions, Lecture Notes Math. No.1271, Springer-Verlag, Berlin, 1987, 189–200.

[79] M. MARCUS, *Finite Dimensional Multilinear Algebra, Part II*, Marcel Dekker, New York, 1975.

[80] ALAN MCINTOSH, Clifford algebras and the higher dimensional Cauchy integral, in *Approximation Theory and Function Spaces*, Banach Center Publications, Warsaw, 1986.

[81] H.P. MCKEAN JR. and I.M. SINGER, Curvature and eigenvalues of the Laplacian, *J. Diff. Geom.*, **1** (1967), 43–69.

[82] M.A.M. MURRAY, The Cauchy integral, Calderón commutators, and conjugations of singular integrals in $\mathbb{R}^n$, *Transactions of the AMS*, **289** (1985), 497–518.

[83] O.T. O'MEARA, *Introduction to Quadratic Forms*, Springer-Verlag, Berlin, 1963.

[84] T.H. PARKER, Super-symmetry and the Gauss–Bonnet theorem, (unpublished manuscript).

[85] R. PARTHASARATHY, Dirac operator and the discrete series, *Ann. of Math.*, **96** (1972), 1 30.

[86] V.K. PATODI, Curvature and the eigenforms of the Laplace operator, *J. Diff. Geom.*, **5** (1971), 233–49.

[87] W.A. POOR, *Differential Geometric Structures*, McGraw-Hill, New York, 1981.

[88] I.R. PORTEOUS, *Topological Geometry*, 2nd edition, Cambridge University Press, Cambridge, 1981.

[89] C. PROCESI, *A primer of invariant theory*, Brandeis Lecture Notes, **1** (1982).

[90] JOHN ROE, *Elliptic Operators, Topology, and Asymptotic Methods*, Longman, New York, 1988.

[91] R.T. SEELEY, Elliptic singular integral equations, *Proc. Symp. Pure Math., Vol X (Singular Integrals)*, Amer. Math. Soc. Publ., (1967), 308–15.

[92] S. SLEBARSKI, Dirac operators on a compact Lie group, *Bull. Lond. Math. Soc.*, **17** (1985), 579–83.

[93] S. SLEBARSKI, The Dirac operator on homogeneous spaces and representations of reductive Lie groups I, II, *Amer. J. Math.*, **109** (1987), 283–302, 499–520.

[94] E.M. STEIN, *Singular integrals and differentiability properties of functions*, Princeton University Press, Princeton, N.J., 1970.

[95] E.M. STEIN and G. WEISS, Generalization of the Cauchy–Riemann equations and representations of the rotation group, *Amer. J. Math.*, **90** (1968), 163–96.

[96] E.M. STEIN and G. WEISS, *Introduction to Fourier Analysis on Euclidean Spaces*, Princeton University Press, Princeton, N.J., 1971.

[97] P. STREDDER, Natural differential operators on Riemannian manifolds and representations of the orthogonal and special orthogonal groups, *J. Diff. Geom.*, **10** (1975), 647–60.

[98] R. TAKAHASHI, Série discrète pour les groupes de Lorentz $SO_0(2n, 1)$, *Colloque sur les Fonctions Spheriques et la Theorie des Groupes*, Nancy, 5–9 janvier 1971.

[99] T. TON-THAT, Lie group representations and harmonic polynomials of a matrix variable, *Trans. Amer. Math. Soc.*, **216** (1976), 1–46.

[100] H.W. TURNBULL, *The Theory of Determinants, Matrices and Invariants*, 3rd edition, Dover Publications, 1960.

[101] G.C. VERCHOTA, Layer potentials and boundary value problems for Laplace's equation in Lipschitz domains, *J. Funct. Anal.*, **59** (1984), 572–611.

[102] P. WEISS, An extension of Cauchy's integral formula by means of Maxwell's stress tensor, *J. London Math. Soc.*, **21** (1946), 210–218.

[103] H. WEYL, *Classical Groups*, 2nd edition, Princeton University Press, Princeton, N.J., 1953.

[104] E. WITTEN, Supersymmetry and Morse theory, *J. Diff. Geom.*, **17** (1982), 661–92.

Index

$\mathfrak{A}_n$, C^* algebra structure of, 55–65

$\mathfrak{A}_n$, Banach algebra structure of, 53–65

$\mathfrak{A}_n$-module, 60

$\mathfrak{A}_n$-module, structure of, 62–3

$\mathfrak{A}_n$: tangential and normal components of, on $\partial \mathbb{R}^n_+$, 125

$\mathfrak{A}_n$: universal Clifford algebra for $\mathbb{R}^n$, 49–65

A_q weight, 118

Adams, J.F., 206, 244

adjoint Dirac operator, $\overline{D}$, 207

Ahlfors, L., 318

algebraically over-determined operator, 206

Alvarez–Gaume, L., 318

analytic over-determinedness, 223

Arima, R., 319

associated vector bundle, 266–7

Atiyah, M., 244, 318–9

Atiyah–Singer index theorem, 244

Atiyah–Singer index theorem, for spin manifolds, 315–6

Atiyah–Singer index theorem, for graded Dirac operators, 309–17

axial polynomial, 164

Bargmann, V., 284, 318

Beardon, A., 85

Bergman space, 287–8

Berline, N., 318

Bianchi identities, 259

Bismut, J., 318

Bleecker, D., 318

BMO, 133

Bochner, S., 199

Bochner–Weitzenböck formula, 249, 258, 317

Booss, B., 318

Boothby, W., 317

Borel–Pompieu theorem, 101

Bourguignon, J.-P., 317

Brackx, F, 141

Brauer, R., 94, 141

Bruhat decomposition, 280

Bruhat decomposition, for $\mathrm{Spin}_0(n, 1)$, 278

Bröcker, T., 85, 201

Burkholder-Gundy-Silverstein theorem, classical, 114

Burkholder-Gundy-Silverstein theorem, for $H^p(\mathbb{R}^n_+, \mathfrak{H})$, 122

C^* algebra, 55

$\mathfrak{C}_n$: universal Clifford algebra for $\mathbb{C}^n$, 56

Calderón, A.P., 126, 135, 141, 239

Calderón–Zygmund type operators, 289–90
Capelli operator, 181
Capelli, 181
Cartan composition, 159, 222
Cartan, E., 143, 150, 163
Cartan-Schur-Weyl theorem, 163
Cauchy integral operator, for $\mathbb{R}^n_+$, 121
Cauchy integral operator, for a Lipschitz domain in $\mathbb{R}^n$, 242–3
Cauchy integral operator, on Tan $H^p(\mathbb{R}^{n-1})$, 125
Cauchy integral theorem, for Clifford analytic functions, 101
Cauchy–Binet identity, 180, 230
Cauchy–Riemann operators, 87–93
Cauchy–Szegö kernel, 293
Cauchy–Szegö transform, 293, 295
Cayley mapping, 70
Cayley operator, 180–1
Cayley, 69
Cayley, A., 181
Chern, S.S., 311
Chevalley, C., 85, 181
class 1 property, 164
class 1 property, for harmonic polynomials of matrix argument, 198
class 1 property, of $\mathcal{H}_m(\mathbb{R}^n)$, 171
Clifford algebra, 8
Clifford algebra, $\mathbb{Z}(2)$-grading of, 14
Clifford algebra, as a Lie algebra, 67–8
Clifford algebra, center of, 27–8
Clifford algebra, dimension of, 10, 11, 31
Clifford algebra, even and odd subspaces of, 13
Clifford algebra, ideal structure of, 28–30, 32
Clifford algebra, matrix realizations of, 18–22
Clifford algebra, universal, 12
Clifford algebra bundle, over a Riemannian manifold, 264, 267
Clifford algebra norm, 267–8

Clifford analytic function, definition of, 97
Clifford analytic functions, Cauchy integral theorem for, 101
Clifford analytic functions, Cauchy theorem for, 103
Clifford analytic functions, maximum modulus principle for, 104
Clifford analytic functions, Mean-value theorem for, 103
Clifford analytic functions, Morera's theorem for, 103
Clifford analytic functions, subharmonicity properties of, 105–8
Clifford analytic functions, Weierstrass theorem for, 104
Clifford group, 38, 39, 42–5
Clifford group, definition of, 42
Clifford module, 60
Clifford module, structure of, 62–3
Clifford norm, on $\mathfrak{A}_n$, 53, 267–8
Clifford norm, on $\mathfrak{C}_n$, 56
Clifford operator norm, in $\mathfrak{C}_n$, 56, 128
Clifford operator norm, on $\mathfrak{A}_n$, 53
Clifford semigroup, 39
Clifford semigroup, definition of, 41
Clifford, W.K., 8, 85
Coifman, R.R., 126, 134, 141, 201
complex Spinor space, 151, 206
conformal group, 36–38
conformal map, 36
conformal sphere, 37
conformal transformations, 277, 281–2
conjugate Poisson integral, 110
conjugate Poisson kernel, 110, 121
conjugation, 17, 128
connection, 250
connection, Euclidean, 250–1
connection, Riemannian, 251
covariant derivatives, 253
creation and annihilation operators, 15
critical index of subharmonicity, 232–44
critical index of subharmonicity, defined, 233

curvature operator, 254, 257
Dahlberg, B., 141
Delanghe, R., 141
derived representation, on a complex
  G-module, 146
determined elliptic operator, 100
dilation, of $\mathbb{R}^n$, 278
dimension, of a Clifford algebra, 10,
  11, 31
Dirac $\mathcal{D}$ operator, 207
Dirac operator, basis independence
  of, 95
Dirac operator, standard Euclidean,
  96
Dirac operators 93–7
Dirac operators, on hyperbolic and
  spherical space, 272–84
Dirac, P.A.M., 83, 93, 141
Dirichlet problem, for Dirac operator
  in a Lipschitz domain in $\mathbb{R}^n$, 243
Dirichlet problem, for Laplace's
  equation in a Lipschitz domain in
  $\mathbb{R}^n$, 135–7
div-curl system, 224
divergence, on a Riemannian
  manifold, 256
Duren, P., 141
eigenspace representation, of
  $\mathrm{Spin}(n) \circledS \mathbb{R}^n$, 219
elementary symmetric polynomials,
  69, 178
elliptic operator, 205
equivalence, of first order differential
  operators, 204–5
equivalent G-modules, 144
equivariant transformation, 144
Euclidean motion group, 36
Euler characteristic, 311
Euler operator, 181
Fatou theorem, 110
Fefferman, C., 141
fiber norm, 267
Fischer inner product, 164
Flett, T.M., 109
Fourier transform, for $\mathbb{R}$, 111
Fourier transform, for $\mathbb{R}^{n-1}$, 121
Freed, D., 317

Friedman, A., 318
Fueter operators, 89, 92
Fueter, R., 89, 140–1
fundamental representations, 147
fundamental representations, of
  $\mathrm{Spin}(n)$, 159–60, 231
G-invariant inner product, 145
Garnett, J., 141
Gauss–Bonnet theorem, 311, 317
Gauss–Weierstrass kernel, 297
Gegenbauer polynomials, 168, 173
Gegenbauer polynomials, of matrix
  argument, 191
Gegenbauer, L., 199
Gelbart, S., 199, 201
Generalized Cauchy–Riemann (GCR)
  operator, 206, 233
Generalized Cauchy–Riemann (GCR)
  operator, rotation-invariant, 236–7
Generalized Cauchy–Riemann (GCR)
  system, 233
generalized Cauchy–Riemann (GCR)
  systems, 140
Getzler, E., 248, 305, 309, 317–19
Gilkey, P., 311, 318
Goodearl, K.R., 85
grad, on a Riemannian manifold, 255
graded Dirac operator, 97, 203, 208,
  210
graded Dirac operator, determined by
  an involution, 209–13
graded Dirac operators,
  rotation-invariance of, 216–18
Grassmann, 8
Greiner, P., 319
Gross, K., 190
group of similarities, 37
group representation, 144
H^p functions, analytic
  over-determinedness of, 115, 118
H^p spaces of Clifford analytic
  functions, and critical index of
  subharmonicity, 242–3
H^p spaces of Clifford analytic
  functions, boundary regularity of,
  242–3

H^p spaces, of Clifford analytic functions in $\mathbb{R}^n_+$, 119–25

H^p theory, classical, 108–19

H^p theory, for Clifford analytic functions in a Lipschitz domain in $\mathbb{R}^n$, 135–41

H^p theory, for Lipschitz domains in $\mathbb{C}$, 115–119

Hadamard three-circles theorem, 109

Hamilton, 8

Hardy, G.H., 108

Hardy–Littlewood maximal function, 112, 122–3

Harish–Chandra, 181

harmonic oscillator, 297

harmonic polynomials of matrix argument, 193–200

Harvey, F. Reese, 85

heat kernel, existence and uniqueness of, 298–301

heat kernel, of a second order elliptic operator, 296–301

heat kernels, asymptotics for, 296–309

heighest weight, of a representation, 150

Helgason, S., 85, 201, 318

Herz, C., 190, 199, 201

higher gradients operators, 225, 228

higher gradients operators, critical index of subharmonicity of, 239–42

highest weight submodule, 159

highest weight vector, 150–1

Hilbert bundle, 267

Hilbert transform, for $\mathbb{R}$, 111

Hilbert transform for a Lipschitz domain in $\mathbb{R}^n$, L^p estimates for, 126–35

Hilbert transform, for a Lipschitz domain in $\mathbb{R}^n$, 126, 242–3

Hilbert–Schmidt inner product, 146, 204

Hirzebruch signature theorem, 317

Hodge Laplacian, 260, 263, 307–8, 311–2

Hodge–deRham (d, d^*)-system, 99–100, 217–18, 222, 224, 243–4, 260, 311

Hodge–deRham (d, d^*)-system, critical index of, 237–9

Hodge–deRham (d, d^*)-system, subharmonicity of, 237–9

horizontal lift, 269

horizontal vectors, 269

Howe, R., 201

Husemoller, D., 85

hyperbolic metric, 273

hypergeometric function, 168

hypergeometric function, of matrix argument, 190

imaginary unit, 50

induced representation, 286

injectively elliptic operator, 205

integration by parts theorem, 101

intertwining operator, 286–7, 290–4

inversion, 37, 279

irreducible unitary representation, 145–6

isometry, 35

James, A., 201

k-multivectors, 50, 153, 159, 262

Kenig, C.E., 115, 141, 244

Knapp, A., 318

Kobayashi, S., 317

Koornwinder, T., 201

Koosis, P., 141

Korányi, A., 141

Kostant, B., 201

Kunze, R., 284, 318

Laplacian, factorization of, 91–3

Laplacian, on a Riemannian manifold, 254–5

Laplacian, on a Riemannian manifold, defined, 256

Levi, E.E., 299

Lichnerowicz' formula, 259

limits of discrete series representations, 288

linear fractional transformations, 275–84

Lipschitz domain, in $\mathbb{C}$, 115

Lipschitz domain, in $\mathbb{R}^n$, 126

Lipschitz, R., 72, 85

$\mathcal{M}_{\mathbb{C}}(G)$, 145–6

$\mathcal{M}_{\mathbb{C}}(\mathrm{Spin}(n))$, 220

Maas, H., 201, 318
Macdonald, I., 201
Marcus, M., 85
Maxwell stress tensor, 244
Maxwell's reciprocal theorem, 245
McIntosh, A., 126, 141
McKean, H.P., 311, 318
Merryfield, K., 290
Meyer, Y., 126, 134, 141
Minkowski space, 6
Muckenhoupt, B., 118
Murray, M., 141
Möbius group, 37–8
Möbius transformation, 37
Neumann problem, for Laplace's
  equation in a Lipschitz domain in
  $\mathbb{R}^n$, 136
Nomizu, K., 317
non-tangential limit, 120
non-tangential limits, 110
non-tangential maximal function, 114,
  122
norm function on $\mathfrak{A}_n$ and $\mathfrak{C}_n$, 128
norm function, 41
normal coordinate system, 261
normalized basis, for a quadratic
  space, 7
$O(n)$-harmonic function, 194
$O(n)$-harmonic functions, 227
O'Meara, O.T., 85
operator of Dirac type, 203
operator of Dirac type, definition of,
  208–9
orthogonal group $O(V, Q)$, 33–8
orthogonal transformations, 32–8
orthonormal frame bundle, 265
orthonormal frame bundle, oriented,
  265
Parasarathy, R., 318
Parker, T., 318
Patodi, V., 301, 311, 318
Pauli matrices, 9, 25, 93
periodicity theorem, for $\mathfrak{A}_n$, 57
Pfaffian, 72, 308
Pfaffian, mapping properties of, 72–7
pin group, 47
Pin(V, Q), 47

Pipher, J., 245
Planck's constant, 296
Pochhammer symbol, 168, 190
Poincaré group, 36
Poisson integral, 110
Poisson kernel, 110, 121
Poisson transform, 293
polynomial representations of
  $GL(n, \mathbb{R})$, 163
polynomials of matrix argument,
  $SL(r)$-invariant, 180
polynomials of matrix argument,
  173–93
polynomials of matrix argument,
  determinantally homogeneous,
  174, 182, 184
Poor, W., 317
Porteous, I.R., 85
principal anti-automorphism, 17
principal automorphism, 17
principal fiber bundle, 264
Procesi, C., 201
pseudo-scalar (e_v), 24
quadratic form, 6
quadratic space, 6
quadratic space, degenerate, 7
quadratic space, non-degenerate, 7
quadratic space, normalized basis for,
  7
quadratic space, radical of, 7, 22
quaternion-valued regular function, 89
quaternionic structure, on a complex
  G-module, 146
quaternions ($\mathbb{H}$), 9
Quillen, D., 296
$\mathfrak{R}H^p(\mathbb{R})$, 113
radical, of a quadratic space, 7, 22
Radon–Hurwitz numbers, 206
real projective space, 266
real spinor space, 206 real structure,
  on a complex G-module, 146
reduced products, 11
reflection, 33
representation theory, for Spin$_0(n, 1)$,
  284–96
representation, finite dimensional, 144

representation, infinite dimensional, 144
representation, of a group, 144
representations of $O(n)$, 162–3
representations of $SO(n)$, 162–3
representations of $SO(n)$, double-valued, 162
representations of $SO(n)$, single-valued, 162
reproducing kernel property, 91
reversion, 18
Richards, D.St.P., 201
Riemannian curvature tensor, 257
Riemannian metric, for spherical space, 273
Riesz transforms, 121
rigid motion, 35
Roe, J., 141, 317–8
rotation, 35
rotation, of $\mathbb{R}^n$, 278
rotation-invariant first-order systems, 213–19
rotation-invariant operators of Dirac type, 215–19
rotation-invariant operators, and $\mathrm{Spin}(n)$-submodules, 221
Schmid, W., 319
Schur's Lemma, 145
Schur, I., 145, 163, 179
Seeley, R., 319
signature, of a representation, 147, 150–1
similarity transformation, 37
Singer, I., 244, 311, 318
single layer potential; double layer potential, 135–6
Slebarski, S., 318
Sommen, F., 141
special Lipschitz domain, in $\mathbb{R}^n$, 127
spherical harmonics, 165, 225, 228–30
spherical harmonics, representations of $SO(n)$ on, 167–73
$\mathrm{Spin}(n)$, complex, 163
$\mathrm{Spin}(n)$, maximal abelian subgroup T_n of, 148
$\mathrm{Spin}(n)$, quaternionic, 163
$\mathrm{Spin}(n)$, real, 163

$\mathrm{Spin}(n)$, representations of, 163
$\mathrm{Spin}(n)$, Weyl group of, 150
spin group, 46
Spin group, 46–9
Spin group, as a classical Lie group, 77–84
Spin group, as a Lie group, 65–84
spin representations of $\mathrm{Spin}(n)$, 215
Spin structure, on a Riemannian manifold, 268
$\mathrm{Spin}(V, Q)$, 46
$\mathrm{Spin}_0(V, Q)$: connected component of $\mathrm{Spin}(V, Q)$, 49
spinor Laplacian, 259, 262, 306
spinor space, complex, 60
spinor space, real, 60
$\mathrm{Spoin}(V, Q)$, 47
standard basis, for $\mathbb{R}^n$, 51
standard Dirac operator, 96
standard Dirac operator, ellipticity of, 254
standard Dirac operator, essential self-adjointness of, 254
standard Dirac operator, on a Riemannian manifold, 250, 253–4
standard Dirac operator, on hyperbolic space, 273
standard Dirac operator, on spherical space, 274
standard Dirac operator, realizations of, 99–100
Stein, E.M., 134, 140–1, 201, 209, 220, 221, 222, 244, 284, 318
Stein–Weiss $\overline{\partial}_\tau$ operators, 222
stereographic projection, 273–4
structure map, 146
subharmonic function, 91
subharmonicity property, of a first-order system, 233
submodule, irreducible, 144
submodule, reducible, 144
supersymmetry, 212
supertrace, 212, 312–3, 316
symbol mapping, of δ_A, 205
Takahashi, R., 85, 318
Tan $H^p(\mathbb{R}^{n-1})$, 125
Taylor series expansions, 164–6, 185–6

tent space theory, 134
δ_A, 204
δ_τ, 220–32
tom Dieck, T., 85, 201
Ton-That, T., 201
total derivative, on a Riemannian
  manifold, 270
trace-harmonic function, 195
transformer, 42
transformers, 42–45, 275
translation, 35
translation, of $\mathbb{R}^n$, 278
Turnbull, H., 201
twisted Dirac operator, 211
twisted spinor Laplacian,
  260–1
two-fold covering, 46
Uhlenbeck, K., 317
ultraspherical polynomials, 168
unitary representation, 145
unitary representation, complex type,
  147

unitary representation, real or
  quaternionic type, 146–7
universal Clifford algebra, 12
universal Clifford algebra,
  constructions of, 13–17
Vahlen, K., 318
Vandermonde, 190
Verchota, G., 137, 141
Vergne, M., 318
Vági, S., 141
Wallach, N., 318
weak-type (1,1), 112
weight, of a representation of Spin(n),
  149
weighted Hardy space, 288
Weiss, G., 140, 141, 201, 209, 220,
  221, 222, 244, 245
Weyl group element, 279
Weyl group of Spin(n), 150
Weyl, H., 94, 141, 143, 163, 175, 201
Witten, E., 318
Zygmund, A., 135, 239